Exosomes in Cancers Therapy

Exosomes in Cancers Therapy

Editors

Farrukh Aqil
Ramesh C. Gupta

Basel • Beijing • Wuhan • Barcelona • Belgrade • Novi Sad • Cluj • Manchester

Editors
Farrukh Aqil
University of Louisville
Louisville, KY
USA

Ramesh C. Gupta
University of Louisville
Louisville, KY
USA

Editorial Office
MDPI AG
Grosspeteranlage 5
4052 Basel, Switzerland

This is a reprint of articles from the Special Issue published online in the open access journal *Cancers* (ISSN 2072-6694) (available at: https://www.mdpi.com/journal/cancers/special_issues/Exosomes_Cancers_Therapy).

For citation purposes, cite each article independently as indicated on the article page online and as indicated below:

Lastname, A.A.; Lastname, B.B. Article Title. *Journal Name* **Year**, *Volume Number*, Page Range.

ISBN 978-3-7258-2347-5 (Hbk)
ISBN 978-3-7258-2348-2 (PDF)
doi.org/10.3390/books978-3-7258-2348-2

Contents

About the Editors

Farrukh Aqil

Farrukh Aqil, Ph.D., is an Associate Professor at the University of Louisville, specializing in cancer prevention and therapy. His research focuses on utilizing exosomes for drug delivery in cancer treatment, as well as developing combinatorial approaches using natural compounds and chemotherapeutic agents. With over 100 peer-reviewed publications and multiple editorial roles, Dr. Aqil is a recognized leader in his field, particularly in the innovative use of milk-derived exosomes for targeting cancer and neurological disorders.

Ramesh C. Gupta

Ramesh C. Gupta, Ph.D., is a Professor and Endowed Chair of Cancer Research at the University of Louisville. His work focuses on cancer prevention and therapeutics. Dr. Gupta has always worked with cutting-edge technology, pioneering sensitive methods to sequence tRNAs, followed by ultrasensitive ^{32}P-postlabeling to measure DNA damage by environmental carcinogens. His current research is centered on developing new prevention and treatment strategies, utilizing exosomes as nanocarriers for the oral delivery of both standard chemotherapeutic drugs and natural agents with therapeutic potential. He has published extensively in high-impact journals and has been recognized for his innovative approaches to cancer research, particularly in the development of novel therapeutic strategies.

Preface

The field of cancer therapy has undergone tremendous transformation over the past few decades and yet significant challenges remain, particularly in addressing drug resistance, targeting difficult-to-treat cancers, and improving therapeutic delivery. One of the most promising developments in recent years is the emergence of exosomes—small extracellular vesicles (EVs)—as potential drug delivery systems.

This Special Issue is dedicated to exploring the potential of exosomes in cancer diagnosis and therapy. The exosomes' potential as biomarkers, diagnostic tools, and therapeutic vehicles offers new hope in the fight against cancer. Their ability to cross biological barriers, such as the blood–brain barrier, and to carry both chemotherapeutic drugs and natural agents represents a breakthrough in cancer treatment. This Special Issue explores the mechanisms by which exosomes operate, their role in cancer progression and metastasis, and their application in therapeutic delivery. Moreover, it highlights the challenges of scaling exosome-based therapies for clinical use and explores alternative sources for large-scale production such as bovine milk and colostrum. These alternative approaches hold great promise for making exosome-based treatments more accessible and cost-effective.

The contributors to this Special Issue bring together a wealth of expertise from both academia and industry, and we are excited to present the latest findings in preclinical and clinical studies related to exosome-based cancer therapy. We hope that this Special Issue will serve as a valuable resource for scientists, clinicians, and students alike, promoting further exploration of exosomes and their potential to transform cancer therapy.

Farrukh Aqil and Ramesh C. Gupta
Editors

Editorial

Exosomes in Cancer Therapy

Farrukh Aqil [1,2,*] and Ramesh C. Gupta [1,3,*]

[1] UofL Health—Brown Cancer Center, University of Louisville, Louisville, KY 40202, USA
[2] Department of Medicine, University of Louisville, Louisville, KY 40202, USA
[3] Department of Pharmacology and Toxicology, University of Louisville, Louisville, KY 40202, USA
*   Correspondence: farrukh.aqil@louisville.edu (F.A.); rcgupta@louisville.edu (R.C.G.)

Citation: Aqil, F.; Gupta, R.C. Exosomes in Cancer Therapy. *Cancers* **2022**, *14*, 500. https://doi.org/10.3390/cancers14030500

Received: 13 January 2022
Accepted: 18 January 2022
Published: 20 January 2022

Publisher's Note: MDPI stays neutral with regard to jurisdictional claims in published maps and institutional affiliations.

Exosomes or small extracellular vesicles (EVs) are natural nanoparticles and known to play essential roles in intercellular communications, carrying a cargo of a broad variety of lipids, proteins, metabolites, RNAs (mRNA, miRNA, tRNA, long non-coding RNA), and DNAs (mtDNA, ssDNA, dsDNA). This is an emerging field. Recent studies have been conducted on exosomes demonstrating physical and biological stability and suitable tolerability, simplicity of preparation, possibility of commercial scale-up and functionalization for tumor-targeting. These features make exosomes ideal nanoparticles for drug delivery, with wide therapeutic applications. These findings have invigorated researchers to explore the exosomes and their roles under both physiological and pathological conditions in greater detail.

Exosomes participate in complex biological responses and have been proposed for drug-delivery purposes, as they can be loaded with both small molecules and macromolecules, which endorse their use as therapeutic tools to treat various diseases, including cancer. There have been inherent problems associated with other nanoparticle delivery systems. Thus, for exosomes to be accepted as a drug carrier in clinics, the development of biocompatible and economically viable exosomes, which are effective and well-tolerated in vivo, must be demonstrated. Exosomes have many of the desirable features, such as a long circulating half-life, the intrinsic ability to target tissues, biocompatibility, and minimal or no inherent toxicity issues, overcoming the limitations observed with the majority of other delivery systems.

In the last decade, there has been an exponential growth in the field of exosomes, with about 21,000 publications on exosomes listed in PubMed alone and over 4200 in the year 2020 (Figure 1). Due to their nano-size (30–150 nm) and biological functions, exosomes have been used as nano-carriers for small molecules and macromolecules (siRNA and pDNA) in cancer therapy in pre-clinical studies, as well as biomarkers for cancer diagnosis and prognosis. Progress in the use of exosomes in clinical studies has been slow. Kalluri's laboratory recently reported a scalable production of exosomes from mesenchymal stem cells (MSCs) using a bio-reactor [1] and listed a clinical trial in the National Institutes of health website (NCT03608631) with the exosome-mediated delivery of si$KRAS^{G12D}$ against pancreatic cancer [2]. The abundance of exosomes is orders of magnitude higher in bovine milk [3] and colostrum powder [4,5] compared to cell culture media. In this regard, the research article by Kandimalla and colleagues [5] presented in this Special Issue showed the utility of exosomes isolated from bovine colostrum powder for delivery of the therapeutic drug paclitaxel, which is of high clinical relevance. This article highlights the tumor targetability of exosomes and successfully showed that an oral functionalized exosomal formulation of paclitaxel significantly improved the efficacy and mitigated immunotoxicity, while providing a user-friendly, cost-effective alternative to an intravenous bolus dose standard-of-care paclitaxel and abraxane. Reviewing the progress from the discovery to the therapeutic development of exosomes, Jan et al. [6] summarized the valuable information on exosome donor cell types, exosome cargoes, cargo loading, routes of exosome administration, and the engineering of exosomal surfaces for specific peptides that increase target specificity and, as such, therapeutic efficacy.

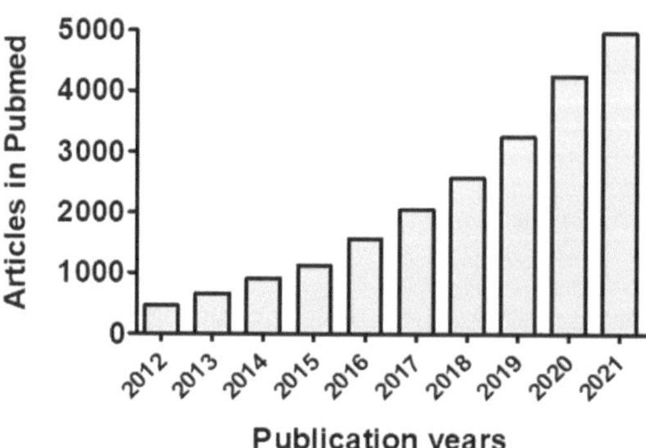

Figure 1. List of publications on exosomes in PubMed in last 10 years.

Exosomes derived from cancer cells carry the cargo reflective of genetic alterations in cancer cells and could likely serve as a biomarker in the early detection of cancer. To further highlight this area of exosome research, Guerrini and colleagues [7] discuss the role of exosomes as one of the most intriguing cancer biomarkers in modern oncology for early cancer diagnosis, prognosis and treatment monitoring. They discussed the application of plasmonic devices exploiting surface-enhanced Raman spectroscopy (SERS) as the optosensing technique for the structural interrogation and characterization of the heterogeneous nature of exosomes. Using a similar concept, Bondhopadhyay and colleagues [8] reviewed the role of exosomes in communication between tumor cells in the breast cancer tumor microenvironment. They highlighted the role of exosomes in breast carcinogenesis and how exosomes could be used or targeted by recent immunotherapeutics to achieve promising intervention strategies.

The role of Rab proteins and endocytosis process was discussed by Sinha et al. [9]. In this article, authors reviewed the potential of exosomes in many aspects of cancer biology including exosome biogenesis, cargo, Rab-dependent and Rab-independent secretion of endosomes and exosomal internalization. They further show that exosomes could migrate to distal parts and propagate oncogenic signaling and epigenetic regulation, modulate the tumor micro-environment and facilitate the immune escape, tumor progression and drug resistance responsible for cancer progression. The exosomes have been detected in all the bodily fluids, and play different roles based on their origin. In the research manuscript, Cacic and colleagues [10] showed that platelet-derived microparticles were internalized by THP-1 cells, and resulted in increased levels of different miRNAs such as miR-125a, miR-125b, and miR-199.

Recent advances have confirmed exosomes immunotherapy as a feasible, safe option leading to both innate and adaptive immune responses. Exosomes possess different properties according to their source and the cargo they carry. The review by Giacobino et al. [11], provides an up-to-date summary of exosome use in cancer immunotherapy involving the use of exosomes to transport molecules that are able to trigger an immune response and damage cancer cells. Furthermore, besides basic notions regarding cancer immunotherapy, this article focuses on the potential of exosome-based therapeutic vaccines in the treatment of cancer patients, overviewing the clinically relevant trials. This approach may represent a potential target for future anti-cancer therapy. On a similar line, Yao et al. [12] reviewed dendritic-cell-derived exosome vaccines exhibiting better antitumor efficacy in pre-clinical animal models. This review further highlights recent clinical trials with DC exosomes as cancer vaccines and discuss why they only showed limited clinical efficacy in advanced cancer patients. Since clinical studies failed to induce tumor-specific T-cell responses, these observations could be helpful in future clinical studies on the fight against cancer.

Tumor cell exosomes have also been shown to contain danger-associated molecular patterns (DAMPs), which are released in response to cellular stress to alert the immune system to the dangerous cell. Linder and Strandmann [13] shed the light on this aspect in their review. They discuss the role of exosomes in the defense mechanism of heat shock protein 70 (HSP70), as HSP70-positive T-EVs are known to trigger anti-tumor immune responses. The release of DAMPs, including HSP70, may also induce chronic inflammation or suppress immune cell activity, promoting tumor growth. They summarize the current knowledge on soluble, membrane-bound, and EV-associated HSP70 regarding their functions in regulating tumor-associated immune cells in the tumor microenvironment. Additionally, a valuable discussion on the immunotherapies that aimed to target HSP70 and its receptors for cancer treatment is presented.

As discussed above, while exosomes have shown utility in the diagnosis, and treatment of various cancers, recent evidence reveals that cancer-cell-derived exosomes can change the behavior of target cells. The review represented by Burgos-Ravanal [14] discussed that exosomes isolated from aggressive cancer cells can transfer their "traits" to less aggressive cancer cells and convert them into more malignant tumor cells. This review further highlights the role of exosomes in drug resistance, and provides a valuable discussion on why pharmacological therapies are often ineffective. Besides highlighting how inhibiting exosome production could interfere in reduced metastasis and drug resistance, this review highlights exosomes that can be used for therapeutic and prognostic purposes in cancer.

In summary, this Special Issue comprises informative research and authoritative review articles written by an international group of expert scientists and comprehensively discusses exosome biogenesis and protein-sorting, the isolation of exosomes, better loading efficiency, and targeted delivery of drugs and their roles in cancer diagnosis, progression, metastasis and treatment. The articles represent preclinical in vitro and in vivo data to demonstrate exosomes as an oral delivery vehicle for cancer therapeutics. Various reviews debate/discuss strategies where the use of exosomes led to 'cheerful' results when exploring diagnosis and treatment options for different human cancers.

Author Contributions: Conceptualization, F.A. and R.C.G.; validation, F.A. and R.C.G.; resources, R.C.G.; writing—original draft preparation, F.A.; writing—review and editing, F.A. and R.C.G.; funding acquisition, R.C.G. All authors have read and agreed to the published version of the manuscript.

Funding: This work was supported by the NCI grant R44-CA-221487 and, in part, the Agnes Brown Duggan Endowment (to R.C.G.).

Conflicts of Interest: The authors declare no conflict of interest.

References

1. Mendt, M.; Kamerkar, S.; Sugimoto, H.; McAndrews, K.M.; Wu, C.C.; Gagea, M.; Yang, S.; Blanko, E.V.R.; Peng, Q.; Ma, X.; et al. Generation and testing of clinical-grade exosomes for pancreatic cancer. *JCI Insight* **2018**, *3*, e99263. [CrossRef] [PubMed]
2. Tang, M.; Chen, Y.; Li, B.; Sugimoto, H.; Yang, S.; Yang, C.; LeBleu, V.S.; McAndrews, K.M.; Kalluri, R. Therapeutic targeting of STAT3 with small interference RNAs and antisense oligonucleotides embedded exosomes in liver fibrosis. *FASEB J.* **2021**, *35*, e21557. [CrossRef] [PubMed]
3. Munagala, R.; Aqil, F.; Jeyabalan, J.; Gupta, R.C. Bovine milk-derived exosomes for drug delivery. *Cancer Lett.* **2016**, *371*, 48–61. [CrossRef] [PubMed]
4. Munagala, R.; Aqil, F.; Jeyabalan, J.; Kandimalla, R.; Wallen, M.; Tyagi, N.; Wilcher, S.; Yan, J.; Schultz, D.J.; Spencer, W.; et al. Exosome-mediated delivery of RNA and DNA for gene therapy. *Cancer Lett.* **2021**, *505*, 58–72. [CrossRef] [PubMed]
5. Kandimalla, R.; Aqil, F.; Alhakeem, S.S.; Jeyabalan, J.; Tyagi, N.; Agrawal, A.; Yan, J.; Spencer, W.; Bondada, S.; Gupta, R.C. Targeted Oral Delivery of Paclitaxel Using Colostrum-Derived Exosomes. *Cancers* **2021**, *13*, 3700. [CrossRef] [PubMed]
6. Jan, A.T.; Rahman, S.; Badierah, R.; Lee, E.J.; Mattar, E.H.; Redwan, E.M.; Choi, I. Expedition into Exosome Biology: A Perspective of Progress from Discovery to Therapeutic Development. *Cancers* **2021**, *13*, 1157. [CrossRef] [PubMed]
7. Guerrini, L.; Garcia-Rico, E.; O'Loghlen, A.; Giannini, V.; Alvarez-Puebla, R.A. Surface-Enhanced Raman Scattering (SERS) Spectroscopy for Sensing and Characterization of Exosomes in Cancer Diagnosis. *Cancers* **2021**, *13*, 2179. [CrossRef]

8. Bondhopadhyay, B.; Sisodiya, S.; Alzahrani, F.A.; Bakhrebah, M.A.; Chikara, A.; Kasherwal, V.; Khan, A.; Rani, J.; Dar, S.A.; Akhter, N.; et al. Exosomes: A Forthcoming Era of Breast Cancer Therapeutics. *Cancers* **2021**, *13*, 4672. [CrossRef]
9. Sinha, D.; Roy, S.; Saha, P.; Chatterjee, N.; Bishayee, A. Trends in Research on Exosomes in Cancer Progression and Anticancer Therapy. *Cancers* **2021**, *13*, 326. [CrossRef]
10. Cacic, D.; Reikvam, H.; Nordgard, O.; Meyer, P.; Hervig, T. Platelet Microparticles Protect Acute Myelogenous Leukemia Cells against Daunorubicin-Induced Apoptosis. *Cancers* **2021**, *13*, 1870. [CrossRef]
11. Giacobino, C.; Canta, M.; Fornaguera, C.; Borros, S.; Cauda, V. Extracellular Vesicles and Their Current Role in Cancer Immunotherapy. *Cancers* **2021**, *13*, 2280. [CrossRef] [PubMed]
12. Yao, Y.; Fu, C.; Zhou, L.; Mi, Q.S.; Jiang, A. DC-Derived Exosomes for Cancer Immunotherapy. *Cancers* **2021**, *13*, 3667. [CrossRef] [PubMed]
13. Linder, M.; von Strandmann, E.P. The Role of Extracellular HSP70 in the Function of Tumor-Associated Immune Cells. *Cancers* **2021**, *13*, 4721. [CrossRef] [PubMed]
14. Burgos-Ravanal, R.; Campos, A.; Diaz-Vesga, M.C.; Gonzalez, M.F.; Leon, D.; Lobos-Gonzalez, L.; Leyton, L.; Kogan, M.J.; Quest, A.F.G. Extracellular Vesicles as Mediators of Cancer Disease and as Nanosystems in Theranostic Applications. *Cancers* **2021**, *13*, 3324. [CrossRef] [PubMed]

Review

Expedition into Exosome Biology: A Perspective of Progress from Discovery to Therapeutic Development

Arif Tasleem Jan [1,†], Safikur Rahman [2,†], Raied Badierah [3], Eun Ju Lee [4], Ehab H. Mattar [3], Elrashdy M. Redwan [3,*] and Inho Choi [4,*]

[1] School of Biosciences and Biotechnology, Baba Ghulam Shah Badshah University, Rajouri 185234, India; atasleem@bgsbu.ac.in

[2] Department of Botany, MS College, BR Ambedkar Bihar University, Muzaffarpur, Bihar 842001, India; shafique@ynu.ac.kr

[3] Biological Sciences Department, Faculty of Science, and Laboratory University Hospital, King Abdulaziz University, P.O. Box 80203, Jeddah 21589, Saudi Arabia; rbadierah@kau.edu.sa (R.B.); emattar@kau.edu.sa (E.H.M.)

[4] Department of Medical Biotechnology and Research Institute of Cell Culture, Yeungnam University, Gyeongsan 38541, Korea; gorapadoc0315@ynu.ac.kr

[*] Correspondence: lradwan@kau.edu.sa (E.M.R.); inhochoi@ynu.ac.kr (I.C.)

[†] Authors contributed equally to the study.

Citation: Jan, A.T.; Rahman, S.; Badierah, R.; Lee, E.J.; Mattar, E.H.; Redwan, E.M.; Choi, I. Expedition into Exosome Biology: A Perspective of Progress from Discovery to Therapeutic Development. *Cancers* **2021**, *13*, 1157. https://doi.org/10.3390/cancers13051157

Academic Editors: Farrukh Aqil and Ramesh Chandra Gupta

Received: 27 January 2021
Accepted: 3 March 2021
Published: 8 March 2021

Publisher's Note: MDPI stays neutral with regard to jurisdictional claims in published maps and institutional affiliations.

Simple Summary: Exosomes symbolize membrane-enclosed entities of endocytic origin. They play an important role in the intracellular communication by shuttling proteins, nucleic acids, etc., between cells of different tissues and organs. Recent studies have revealed an interplay between cell and exosomes; thereby highlighted their importance in disease diagnosis and possible implication for use in therapeutics. They are currently been explored for the strategic development of platforms towards their employment in achieving the target specific delivery of therapeutics. This review summarizes the composition, biogenesis and trafficking of exosomes in different cellular backgrounds and explores their multifarious role as drug delivery vehicles towards achieving correct functionality and efficacy of the therapeutic molecules. Additionally, it discusses genetic engineering platforms for employment in the designing of optimal delivery modules for their application in the delivery of drugs as part of anticancer therapy.

Abstract: Exosomes are membrane-enclosed distinct cellular entities of endocytic origin that shuttle proteins and RNA molecules intercellularly for communication purposes. Their surface is embossed by a huge variety of proteins, some of which are used as diagnostic markers. Exosomes are being explored for potential drug delivery, although their therapeutic utilities are impeded by gaps in knowledge regarding their formation and function under physiological condition and by lack of methods capable of shedding light on intraluminal vesicle release at the target site. Nonetheless, exosomes offer a promising means of developing systems that enable the specific delivery of therapeutics in diseases like cancer. This review summarizes information on donor cell types, cargoes, cargo loading, routes of administration, and the engineering of exosomal surfaces for specific peptides that increase target specificity and as such, therapeutic delivery.

Keywords: antigen; cancer; exosomes; immune response; therapeutics

1. Introduction

Extracellular vesicles (EVs) represent a heterogenous population of membranous structures of varying sizes and cellular origin [1]. Their secretion into the extracellular milieu provides a means of mediating intercellular communication. Exosomes are a subset of EVs that were introduced to the scientific world as vesicles released from mature blood reticulocytes expressing transferrin receptor [2]. Exosomes develop intracellularly as multivesicular bodies (MVBs) that undergo fusion with cell membrane for their release into the extracellular space [3,4]. Exosomes are homogenous in shape with size ranging in between 30 to 150 nm compared to microvesicles and apoptotic bodies that exhibit substantial variation in size (from 100 to 1000 nm and 50 to 500 nm, respectively) [5,6]. They were initially determined to be definite intracellular entities by electron microscopy (EM) [7].

Morphologically, exosomes are "saucer-like" or "deflated football shaped" in whole-mount EM images, though their collapsed appearance is probably caused by sample preparation procedures, as SEM (scanning electron microscopy) showed them to be perfectly spherical [8]. Irrespective of their cellular background, exosomes display specific components on their surface and sequester molecules such as nucleic acids, cytokines, and other bioactive compounds. Following their secretion by epithelial, endothelial, and cells of other sources [9], exosomes make their way into body fluids such as blood, bile, bronchoalveolar lavage, urine, and breast milk [10–12], and their transport to distant sites facilitate cell-to-cell communication that influence physiologies and pathologies [13]. In addition to their role in intercellular communication [8], their potential diagnostic and therapeutic applications are of great interest to researchers. The present review was undertaken to provide an overview of the composition, biogenesis, and trafficking of exosomes, and to provide insight into the marked changes they undergo in diseased state and a detailed summary of their therapeutic applications with respect to types of cells and therapeutic cargoes, methods of loading, and possible administration routes. In addition, we discuss methods used to engineer exosomes with enhanced specificities and their current therapeutic statuses in the context of different diseases.

2. Composition

Exosomes constitute a subcomponent of the secretome [14], and their composition is dictated by the functional status of the cell (rested, stimulated, transformed, or stressed) [13]. Although the composition of exosomes are highly dependent on their origin, they all contain specific sets of endocytic proteins and nucleic acids (DNA, RNA), and are enclosed by a membrane of plasma membrane origin (Figure 1).

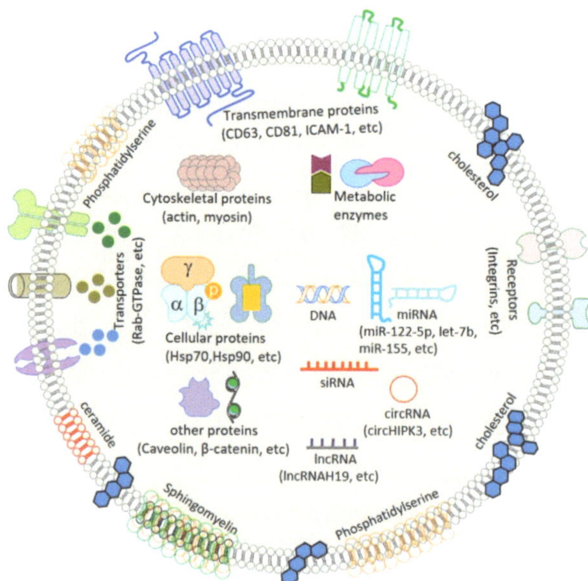

Figure 1. General representation of the exosome structure.

A wide range of methods are employed to separate exosomes from cell culture and body fluids (Table 1). Analyses of their composition by fluorescence-activated cell sorting (FACS), Western blot, and mass spectrometry have revealed them to have a series of tetraspanins (CD9, -26, -58 and others), RAB proteins, heat shock proteins (Hsp70, -90), endosome-associated proteins (Alix, TSG101), annexins, cytoskeletal elements (actin, tubulin), the lysosomal protein (Lamp2b), and the intercellular adhesion molecule (ICAM-1) and co-stimulatory molecules of T-cell origin such as CD86 [15–18]. Surface proteins such as heat shock protein, $\alpha 4\beta 1$ (surface localized protein) on reticulocytes, A33 on enterocytes, and P-selectin on platelets are signatures of cell-specific exosomes [19–21]. Protcomic analyses of exosomes have shown them to possess surface-anchored sheddases, such as ADAM (a disintegrin and metalloproteinase), matrix metalloproteinases (MMPs), and MHC II molecules [22–24].

In addition to their role in extracellular matrix (ECM) remodeling, MMPs have been associated with intra- and intercellular communication via the proteasomal processing of exosome contents [25]. Enzymatic proteins, such as pyruvate kinases and peroxidases, have also been reported in human dendritic cells (DCs) and enterocyte-derived exosomes. In addition to displaying an array of intracellular proteins, exosomes contain DNA, and a wide range of non-coding RNAs (miRNAs, lncRNAs, and circRNAs). lncRNAs have emerged as regulatory RNA molecules with functions often related to cell differentiation and cell cycle regulation, whereas circRNAs act as competitive inhibitors of miRNAs during regulation of protein function [12,26–28]. Furthermore, exosome membranes are rich in lipids such as phosphatidylserine and cholesterol [29]. At the time of writing, the exosome database (http://www.exocarta.org; accessed on 20 December 2020) contained 9769 entries for proteins, 3408 for mRNAs, 2838 for miRNAs, and 1116 lipid entries. The presence of such a wide range of proteins, mRNAs, and miRNAs suggest enormous heterogeneity in terms of exosomal contents, the local expression of proteins and lipids, and the uniqueness of exosomes.

Table 1. Exosome isolation methods and their advantages and disadvantages.

Extraction Method	Advantages	Disadvantages	Reference(s)
Ultracentrifugation (UC; Differential centrifugation)	High Purity	Low yield, time-consuming, requires costly instruments	[30–33]
Density gradient centrifugation	Satisfactory purity	Low yield, time-consuming	[30,34,35]
Size elusion chromatography (SEC)	Relatively gentle	Unable to differentiate exosomes from particles of similar size	[35–37]
Filtration (Non-porous membrane-based)	Simple, time saving	Low yield, high contamination	[38]
Polymeric precipitation	High yield	Low purity than SEC	[39]
Affinity capture (Vn-96 peptide-based)	Simple and time-saving, high yield, high purity	Costly, unsatisfactory recovery	[40–42]
Immunoaffinity capture (Antibody-based)	Simple and time-saving, high yield, high purity	Costly, non-specificity of Abs	[43–45]

3. Biogenesis

The most accepted model of exosome biogenesis involves membrane orientation and inward budding. According to this model, budding events during exosome formation occur in a reverse membrane orientation, similar to that observed during apoptosis [22,46,47] and the release of milk fat globules from the epithelial cells of mammary glands [48]. Budding events during exosome formation involves phosphatidylserine flipping from the inner to the outer plasma membrane leaflet. Furthermore, electron microscopic observations have revealed the fusion profiles of late endosomes with the plasma membrane of antigen-presenting cells (APCs [15]), cytotoxic T-lymphocytes (CTLs [49]), dendritic cells (DCs [50]), and platelets [51]. Exosome production occurs in an active or passive manner, that is, with or without protein involvement. Active production involves a heterooligomeric protein complex referred to as endosomal sorting complex required for transport (ESCRT) and fusion of multivesicular bodies (MVBs) with the plasma membrane to enable exosome release. Ubiquitination is one of the sorting mechanisms that results in the incorporation of endosomal proteins into MVBs. The loading of monoubiquitinated entities into MVB compartments is achieved by four different ESCRTs (ESCRT- 0, I, II, and III) that interact with accessory proteins such as Vps-4 (vacuolar protein sorting-4) and ALIX (programmed cell death 6 interacting protein, also called PDCD6IP) [52–54]. A complex comprising ESCRT-0, HRS (hepatocyte growth factor regulated tyrosine kinase substrate), and STAM1 (signal transducing adapter molecule 1) aids in the recognition of ubiquitinated transmembrane proteins for incorporation into endosomal membrane [54]. Reportedly, ESCRT-I and II recruitment drive membrane budding, whereas ESCRT-III is required for bud scission [54–56]. The recruitment of ESCRT-III by ESCRT-I and II occurs with the involvement of ALIX, a protein that causes simultaneous binding of ESCRT-III to TSG101 (tumor susceptibility gene 101 and a component of ESCRT-I) [57]. After exosome membrane formation, ESCRT dissociates from MVB membrane and contributes to the transport of new cargos. ATPase VPS-4 (adenosine triphosphatase vacuolar protein sorting-4) is required for the dissociation of ESCRT from MVB membrane, which represents the first step of the ESCRT recycling machinery [54,58].

The production of exosomes involves ten stages; (1) endosomal membrane invagination, (2) budding of intraluminal vesicles (ILVs), (3) loading of different entities (DNA, non-coding RNAs, proteins, etc.), (4) formation of multivesicular bodies (MVBs; ESCRT-0,I, II, & 4, Vps-4, ALIX), (5) docking and fusion of MVB that have escaped fusion with lysosomal components to the plasma membrane (PM; Rab and SNARE proteins), (6) release of exosomes into the extracellular milieu, (7) exosome-receptor interaction, (8) receptor-mediated exosome uptake by the recipient, (9) exosome internalization, and (10) release of exosome contents in cytoplasm (Figure 2).

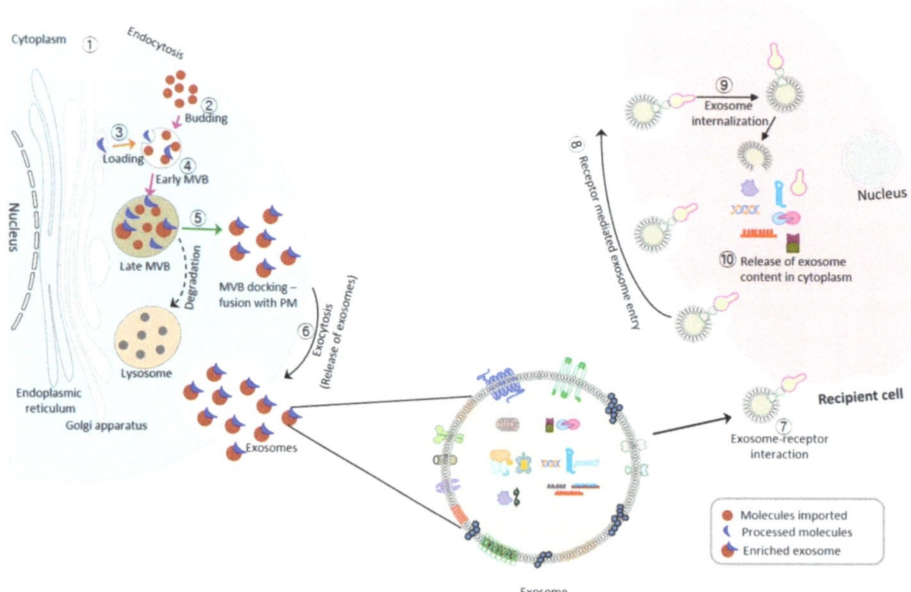

Figure 2. Exosome biogenesis and uptake at recipient surfaces.

Passive exosome formation involves the participation of lipids (ceramide), tetraspanins (CD63), and heat shock proteins independently of ESCRT [59–61]. Studies have shown localization of lipid metabolizing enzyme sphingomyelinase (SMase) and phospholipase D2 (PLD2) to MVB membrane induces the inward curvature required for exosome formation [62–64]. Concomitant inactivation of different ESCRT components using RNAi helped in establishing the independent nature of exosome biogenesis, as knockdown of different ESCRT components did not affect CD63 accumulation or suppress MVB formation [65,66]. Studies by Wehman et al. painted a mixed picture of this RNAi-based strategy as ESCRT-0 and I silencing were found partially suppressing the shedding, but have no effect on ESCRT- II or III [67]. The dependence or independence of exosome biogenesis on the ESCRT machinery has been extensively studied and discussed elsewhere [56,68,69].

4. Exosome Trafficking

Fusion of MVBs with the plasma membrane results in the release of exosomes into the extracellular milieu. Although the mechanism that drives this fusion is unknown, the secretion of acetylcholinesterase tagged exosomes from reticulocytes was found to depend on the function of VAMP-7 (vesicle associated molecular pattern-7) [70]. Recent studies on exosomes carrying WNT3A morphogen revealed that their release is dependent on R-SNARE (soluble N-ethylmaleimide sensitive fusion attachment protein receptor) protein (also called Ykt6) [71–73]. Furthermore, MVB–plasma membrane fusion was found to be mediated by a ternary SNARE (t-SNARE) complex formed by v-SNARE (vesicle SNARE) and t-SNARE [73–77]. After the two membranes make contact, the energy barrier required for their fusion is overcome by the SNARE complex due to its association with the V_0 subunit of V-type ATPase. The ability of V-type ATPase to overcome this energy barrier was found to be independent of its proton pump activity [78]. Other key regulatory components of the exosome secretion pathway include Rab proteins, e.g., Rab11 and Rab27b, which play key roles in the docking of MVBs to the plasma membrane [79].

Exosomes are rich in Rab GTPases, particularly Rab4 and Rab5, which are believed to be regulators of membrane trafficking [80]. Raposo et al. reported that plasma membrane fusion with MHC-II enriched MVBs in B-lymphocytes results in exosome release [15], and Zitvogel et al. reported stimulation of T-cell response by the components of exosomes from DCs [81]. Savina et al. deciphered the presence of Rab11 in exosome secretions [82], and in another study, though calcium transients were found to trigger exosome release, Rab 27 and Rab35 acted as regulatory GTPases for exosome secretion [83–88]. In addition, Alix and Vps4 (components of the ESCRT pathway) were reported to play an important role in exosome secretion [89], which was found to be regulated by P2X receptor activation by LPS-induced ATP on monocytes and neutrophils, and by TLR4 activation on dendritic cells [9,10,79,90].

5. Immunomodulatory Effect of Exosomes

Insights of the role of exosomes have revealed their importance as regulator of different biological processes under physiologic and pathologic conditions. Exosomes release into the extracellular milieu influences cellular morphology by interfering with cell signaling components and by modulating recipient gene expressions and functions and the cell differentiation program. Exosomes have been reported to influence infections [91–93], tumor development and metastasis [94–98], neurodegenerative diseases [99–102], inflammation, and autoimmune disorders [103–106]. In addition, they play crucial role in intracellular communication and in the pathogenesis of several diseases as they can transfer signals (cytokines, proteins, lipids, nucleic acids, and infectious agents) from cells to nearby or distant locations [91,107,108]. In one study, exosomes derived from immunocytes were found to contain a minimum of 98 immunogenic molecules [109]. The immunological functions of exosomes are highly dependent on their membrane proteins and cells of origin, and their stabilities in the extracellular space enable them to carry cargoes to distant cells [110]. Furthermore, the regulatory effects of exosomes involve cross-talk between different immune cells, for example, between B-lymphocyte-derived exosomes and CD8$^+$ cytotoxic cells [111] and between T-cell-derived exosomes and DCs [112–118]. Here, we summarize the involvements of exosomes derived from mesenchymal stem cells (MSCs) and immune cells in cell-to-cell communication and immune system stimulation and suppression (Figure 3).

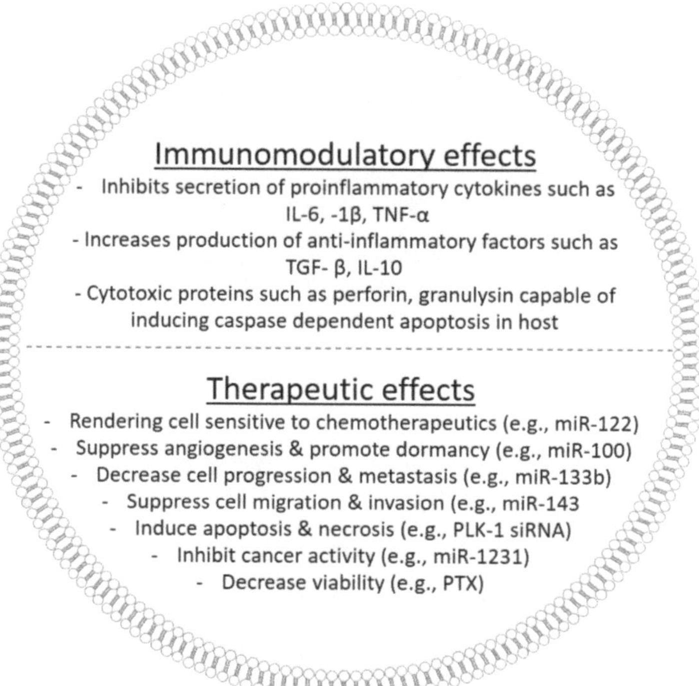

Figure 3. Immunomodulatory and tumor inhibitory effects of exosome loaded with different therapeutic cargoes.

5.1. MSC-Derived Exosomes

Mesenchymal stem cells (MSCs) are multipotent stromal cells sourced from bone marrow, adipose tissues, placenta, or umbilical cord (Table 2). Their regenerative capacities underlie their importance in immune modulation [106,119–122]. The immunomodulatory effects of MSC-derived exosomes on peripheral blood mononuclear cells (PBMCs) have been well established. Exosomes from healthy human bone marrow are essential for the interaction between MSCs and PBMCs. Furthermore, MSC-derived exosomes can modulate the activities of lymphocytes, macrophages, neutrophils, DCs, and natural killer (NK) cells [123]. The ability of MSC-derived exosomes to inhibit the secretion of pro-inflammatory cytokines such as interleukin-6 (IL-6), interleukin-1beta (IL-1β) [124], and tumor necrosis factor-alpha (TNF-α) and to increase the production of anti-inflammatory factors such as transforming growth factor-beta (TGF-β) and interleukin-10 (IL-10) have been well described [125]. In addition, MSC-derived exosomes induce conversion of T-helper-1 (Th1) to T-helper-2 (Th2) cells and reduces potential of T-cells to differentiate into effector T-cells (Th17, capable of producing IL-17). The exosomes induce the proliferation and differentiation of CD4$^+$ cells into Th2 cells, and thereby, suppress differentiation of Th1 to Th17 cells, which are known to participate in autoimmune response. Furthermore, an increase in the regulatory T-cells (Tregs) was also observed in the interaction between Th-cells and exosomes. Together, studies have revealed that MSC-derived exosomes have favorable immunomodulatory properties [106,120–126], and thus, they are considered as potential therapeutic candidates in many pathological contexts and as a convenient means of delivering therapeutics, enzymes, and genes to targeted cells [127]. Interestingly, recent evidence suggests that MSC-derived exosomes offer a potentially safe means of treating graft-versus-host disease (GvHD) [128].

Table 2. Characteristics of exosomes derived from human mesenchymal stem cells (MSCs) and other human immune cells.

Source of Exosomes	Markers	Characteristic miRNAs	Cargo/Pathway	Role	Reference(s)
MSC-derived exosomes (BM-MSCs-exo, AD-MSCs-exo, UC-MSCs-exo, and PL-MSCs-exo)	CD9,CD34,CD44,CD63,CD81,CD90, CD105, ALix, TSG101, OCN, OPN, BMP-7, NKG2D ULBPs	miR-155, miR-146	P13K/AKT/AKT mTOR, TGF-β/Smad/β-catenin, STAT3/Bcl-2/Beclin1, IL-6, ERK1/2, P38, MAPK	Immunosuppressive	[129–133]
1. Mature DEXs	CD63, CD81, CD82, αMβ2, MFG-E8	miR-155	Syntenin Gi2α, β-catenin	Immunostimulatory	[134]
2. Immature DEXs	Annexins, CD63, Alix, TSG101, Calnexin and CCR-7	miR-125b-5p, miR-146a, and miR-148	Syntenin Gi2α, β-catenin	Immunosuppressive	[135]
NKC-derived exosomes (NK-exo)	CD56	miR-186, miR-328, miR-21, miR-29a	Granulysin (GNLY), TGF-β, granzymes (Gzm-A & Gzm-B), perforin (PFN)	Immunostimulatory	[136,137]
Treg-derived exosomes (Treg-exo)	CD25 and CTLA-4	miRNA-155, Let-7b, Let-7d	IL-10, IL-35, and TGF-β	Immunostimulatory	[138,139]

BM, Bone Marrow; AD, Adipose Tissue; UC, Umbilical cord; PL, Placenta; DEX, Dendritic cell derived exosomes.

5.2. DC-Derived Exosomes

Exosomes secreted by immune cells such as mature DCs displaying MHC molecules on their surface can act as antigen-presenting vesicles, thereby activate lymphocytes and initiate innate or adaptive immune responses [118,134,140]. DC-derived exosomes can bind antigenic peptides either by direct capture or by indirect antigen processing through parent DCs [141]. DC-derived exosomes displaying MHC II molecules mediate CD4$^+$ helper cell activation by interacting with lymphocyte function-associated antigen 1 (LFA-1) expressed on the surface of T-cells [142]. In the context of antigen-presenting properties, DC-derived exosomes have greater immunostimulatory effect than intact DCs [143], and in the absence of antigen-presenting cells (APCs), exosomes can activate CD8$^+$ lymphocytes, which supports a report that exosomes contain high levels of class I MHC proteins and ICAM-1 [110]. On the other hand, immature DC-derived exosomes have opposite effects on the immune system, as their cargoes are enriched with self-antigens and anti-inflammatory factors that might promote or induce immune tolerance. The immature DC-derived exosomes were also found to contain low levels of MHC II and co-stimulatory CD86$^+$ molecules, and thus, were incapable of inducing immune response and instead had immunosuppressive effects [104,135]. In the background of allograft transplantation, immature DC-derived exosomes have been shown to promote allograft survival by secreting anti-inflammatory cytokine IL-10, and thus, suppressing T-cell proliferation [144]. It appears that DC-derived exosomes participate in the modulation of helper and cytotoxic T-cell immune responses, and thus, maintain immune tolerance.

5.3. NK-Derived Exosomes

NK cells are innate immune cells that play a central role in immune response. These cells exhibit natural cytotoxicity that enables them to lyse malignant and virus-infected cells without prior sensitization [145]. Also, activated NK cells can mediate immune response indirectly by secreting pro-inflammatory cytokines and chemokines that modulate adaptive cell-mediated immune response [146]. It has also been reported NK-derived exosomes have anti-tumor effects similar to those of NK cells [136]. In a recent study, activated NK cell-derived exosomes loaded with cytotoxic proteins, such as perforin (PFN), granulysin (GNLY), and granzymes (Gzm-A and Gzm-B) induced caspase-dependent apoptosis on entry into target cells [137]. A comparative study on the effect of resting and activated NK cells on tumor cells revealed that activated NK cell-derived exosomes contain high levels of FasL (Fas ligand) and perforin molecules with cytotoxic lysing activity against cancer cells, especially in hematologic malignancies, such as leukemia and lymphoma [147]. Furthermore, it has been suggested that understanding of the cytotoxic activities of NK-derived exosomes at the molecular level would undoubtedly aid in the development of immunotherapeutic strategies for the treatment of cancers and viral infections [148–150].

5.4. Treg-Derived Exosomes

Treg cells (suppressive T-cells) compose a subset of T-cells that play crucial immunomodulatory role by maintaining self-antigen tolerance and in preventing autoimmunity by inhibiting the proliferation of effector T-cells (i.e., CD4$^+$ and CD8$^+$ cells) [151]. Like other immune cells, Treg cells are capable of releasing exosomes, which markedly outnumber those released by other T-cell subpopulation [152–154]. The secretion of exosomes by Treg cells is highly dependent on hypoxia, calcium levels, and IL-2 [155–157]. Recent studies on the proteomic profile of Treg-derived exosomes have shown that these exosomes contain most components of the parent cell and transport several molecules such as miRNAs, CD73$^+$, CD25$^+$, and CD125$^+$ (also known as cytotoxic T-lymphocyte-associated protein 4 (CTLA-4)) with marked immunomodulatory effect [139,158].

Recently, Treg-derived exosomes were reported to be enriched with miRNAs (e.g., miRNA-155, Let-7b, and Let-7d) as compared with parental Tregs, and when transferred to conventional effector cells, these specific miRNAs suppressed IFN-γ production and the expression of effector genes, thereby, inhibited T-cell proliferation [139]. An analysis of the Treg-derived exosomes showed high expression of CD73$^+$, which perform an essential function in immune modulation by enhancing the production of adenosine (an anti-inflammatory modulator) that potently suppresses the proliferation and function of T-cells and block the production of IFN-γ and IL-2 [158].

6. Exploiting Exosomes for Therapeutics

The utilization of exosomes as drug delivery vehicles requires proper understanding of their production in different cellular backgrounds to achieve correct functionality and efficacy of the therapeutic cargoes. The following section summarizes the considerations that should be borne in mind to achieve targeted drug delivery.

6.1. Choice of Cells

In addition to stability in body fluids, reduced immune-stimulatory activity and minimal inflammatory response are prerequisites of therapeutic exosomes, and correct donor cell choice is a steppingstone toward achieving these developmental targets (Table 3).

Table 3. Exosomes in therapeutics. The table summarizes sources, cargoes, loading mechanisms, and effects observed for exosomes from different cell types.

Exosome Source	Cargo and Loading Mechanism	Effect Observed	Reference(s)
Mesechymal Stem Cell	miR-124 (Transfection)	Reduction of cell migration & self-renewal	[159]
	Anti-miR-9 (Transfection)	Reversal of chemoresistance	[160]
	miR-146b (Transfection)	Reduction of progression & metastasis	[161]
	miR-133b (Transfection)	Suppression of progression	[162]
	PLK-1 siRNA (Electroporation)	Induction of apoptosis & necrosis	[163]
	Paclitaxel (Incubation)	Growth inhibition of human pancreatic adenocarcinoma cell	[164]

Table 3. *Cont.*

Exosome Source	Cargo and Loading Mechanism	Effect Observed	Reference(s)
Dendritic Cell	BACE1 siRNA (Electroporation)	Knockdown of specific gene after specific siRNA delivery to the brain for AD	[165]
	VEGF siRNA (Electroporation)	Suppression of tumor growth in breast cancer	[166]
	GAPDH siRNA (Electroporation)	Knockdown of specific gene after specific siRNA delivery to the brain for AD	[165]
	Doxorubicin (Electroporation)	Specific drug delivery to the tumor site & inhibited tumor growth	[167]
HEK293	Let-7a mimic (Transfection)	Target EGPR-expressing cancerous tissues with nucleic acid drugs for breast cancer	[168]
HEK293T	BCR-ABL siRNA (Transfection)	Overcome pharmacological resistance in CML cells	[169]
Mouse lymphoma cell	Curcumin (Mixing)	Increase anti-inflammatory activity	[170]

Human cell lines such as HeLa and HEK293 and murine melanoma cell lines like B16-F1, B16-F10, and B16-BL6 are commonly used to produce exosomes [168,171–179]. In terms of immunogenic properties, immature DCs acts as a suitable donor cell alternative for exosome production [135]. Additionally, surface modification of locally expressed peptides enable exosomes to be used for targeted drug delivery [165,167]. DC-derived exosomes engineered to locally express rabies virus glycoproteins have been utilized to deliver siRNA across the blood-brain barrier in murine models [165]. However, despite their attractive characteristics, production at large-scale for clinical use is restricted due to technical difficulties [167]. To scale up production for clinical use, MSCs offer a possible alternative as they produce large number of exosomes [160,161,180–182]. The use of MSC-derived exosomes to deliver drugs to glioblastoma (GBM) xenograft tumors significantly reduced tumor size [161]. Although exosomes provide a platform for developing new therapeutic strategies, scale-up of MSC-derived exosome production is mostly hampered by technical difficulties [183,184], and manufacturing challenges remain to be properly addressed [7]. In this regard, a combination of tissue-specific targeting and scalability to large-scale production appears to be an appropriate developmental target.

6.2. Choice of Therapeutic Cargoes

Several therapeutic cargoes have been loaded into exosome-based delivery systems. Utilization of the abilities of exosomes to carry interfering RNAs [185,186] and deliver therapeutic cargoes offer a potential means of treating different cancers [187]. Several research groups have investigated the use of exosomes to carry siRNA for gene-based therapy [165,174,176,187–190]. Exosome-mediated delivery of siRNA not only reduces the risk of degradation, but substantially increases bioavailability and delivery efficiency. When MAPK1-siRNA was delivered using plasma or cell-based exosomes, a significant reduction in MAPK1 gene expression was observed in peripheral blood mononuclear cells [174]. In fibrosarcoma cells, gene knockdown by exosome-mediated delivery of RAD51 or RAD52-siRNA reduced viability and proliferation [176]. In a similar study, exosomes carrying the siRNAs of glyceraldehyde-3-phosphate dehydrogenase (GAPDH; the housekeeping gene) or β-site APP cleaving enzyme -1 (BACE1; an Alzheimer's disease-associated gene) downregulated targeted protein level in neurons [165]. Also, the risk of hepatitis C virus (HCV) infection was reduced in liver cells treated with exosomes containing short hairpin RNAs (shRNAs) against viral entry receptor and the replicative machinery of HCV [49,176].

Dysregulation of the expression profiles of miRNAs is a characteristic of a large number of cancers [191,192], and subsequent studies reported that the exosome-based targeted delivery of miRNAs suppressed symptoms in different disease models [185]. Encapsulation of miR-150 in exosomes suppressed T-cell populations and reduced endothelial cell migration, and treatment of T-cells with the conditioned media of miR-122 transduced HEK293T cells increased miR-122 gene expression several-fold and suppressed hepatic inflammation, necrosis, and fibrosis [172,193,194]. Exosome-based delivery of miR-214 to hepatic stellate cells suppressed fibrosis by downregulating CCN2 expression [195,196], and miRNAs had tumor-suppressive effects when miR-143 or let-7a were transported to prostate and breast cancers in vivo [168,173]. However, no effect was observed when normal prostate epithelial cells were treated with exosome-encapsulated miR-143 [173]. MSC exosome (MSC^{exos})-mediated delivery of miR-133b was found to be effective for treating brain ischemia in mice [182], and exosome-mediated miRNA transfer from activated immune cells effectively induced epigenetic changes that influence convalescent plasma response to virus in COVID-19 [197].

In a systematic review, Khalaj et al. [198] reported that exosomes extracted from mesenchymal stem cells derived from bone marrow or umbilical cord ameliorate lung injury in experimental models by (1) attenuating inflammation (reducing pro-inflammatory cytokine levels, neutrophil infiltration, and macrophage polarization); (2) regenerating alveolar epithelium (by reducing apoptosis and stimulating surfactant production); (3) reducing microvascular permeability (by upregulating endothelial cell junction protein levels); and (4) preventing fibrosis (reducing fibrin production). The authors attributed these differential effects to the release of EV cargoes and identified several of the factors responsible, which included miRs126, -30b, -3p, -145, -27a-3p, syndecan-1, hepatocyte growth factor, and angiopoietin-1 [198]. Exosomal delivery of miR-146b inhibited tumor growth in a xenograft model of GBM [161,199], and the delivery of anti-miRs against miR-9 (an oncogenic miRNA) to GBM cells increased their susceptibility to chemotherapeutics like temozolomide [160]. The desired output highlights the communicative role played by exosomes in interaction between MSCs and GBM cells irrespective of the presence of gap junctions. These observations show that the exosomal delivery of miRNAs offers a promising means of delivering anti-cancer and anti-COVID-19 agents. Nevertheless, knowledge of the mechanisms of miRNA loading into exosomes would undoubtedly improve results. In particular, we suggest investigations be conducted to identify and characterize the EXO-motifs that direct the targeted exosome-based deliveries of miRNAs.

Exosomes containing chemotherapeutics like doxorubicin have shown growth inhibitory effect on xenografted breast and colon adenocarcinoma tumors [167,200]. Enhancement in the efficiency of chemotherapeutic agents like doxorubicin achieved by direct delivery of immature DC-derived exosomes effectively reduced side effects on non-targeted organs, especially the heart [167,201]. An exosome preparation of JSI-124 (a STAT3 inhibitor) effectively reduced tumor volume in a murine model of GBM [170,202,203], and notably, exosomes containing 5-fluorocytidine (a prodrug) facilitated its conversion to 5-fluorouracil and 5-fluoro-deoxyuridine and resulted in tumor cell apoptosis in an orthotopic model of schwannoma [175,204]. Furthermore, an exosome-based co-treatment offer another means of treating malignancies, and exosomes loaded with super paramagnetic iron oxide nanoparticles (SPIONs) were shown to have potential use as an MRI cancer imaging agent [177,205].

6.3. Exosome Loading Procedures

The loading of therapeutic cargoes into exosomes involves the use of classical incubation and electroporation methods and transfection reagents and the modern techniques of donor cell transfection or activation [177,188]. However, simple incubation with a cargo is sometimes sufficient to load exosomes (Table 3). The best example of this is provided by curcumin, a natural compound with an anti-inflammatory effect, which can be loaded by simple incubation for 5 min at 22 °C, presumably because curcumin rearranges membrane lipids and alters membrane fluidity [206,207]. The encapsulation efficiency for the drug doxorubicin was higher for exosome-mimetic bioengineered nanovesicles generated from filtered monocytes or macrophages [194,200]. On the other hand, the loading of small-sized cargoes, such as miR-150, was efficiently achieved by simple incubation [200,208].

Efficient loading of therapeutic cargoes into exosomes can also be achieved by electroporation at 150–700 V [165,186], but the effectiveness of cargo loading depends on the donor cell type [167,174,176], exosome type, and cell concentrations [165,167,177,209]. Quantification of cell delivery using fluorescently labeled siRNA revealed higher uptake than by chemical reagent-based transfection [163,174]. An analysis of the cell viability after electroporation of exosomes with therapeutic cargoes was used to investigate the efficiency of the technique [176]. Although it seems to be a suitable clinical option, electroporation is known to have adverse effects on the integrities of exosomes and cargoes, for example, it has been reported to induce exosome and siRNA aggregation. In fact, after optimizing delivery parameters and using trehalose medium to minimize exosome aggregation [177], siRNA retention in exosomes was only <0.05% [210]. Nevertheless, the loading of drugs like doxorubicin by electroporation is still considered a better option than incubation or chemically based transfection methods, because it better maintains the functionality of the drug [167]. The use of chemical-based transfection methods to load therapeutic cargoes such as siRNA into exosomes has restricted usage because they are less efficient than that achieved using HiPerFect transfection reagent-based methods [174,176]. Although Lipofectamine 2000-based siRNA loading was reported to alter gene expression in recipients, leftover micelles generated during exosome preparation prevented quantification of the effects of siRNA cargoes at target sites [174,176].

Transfection of donor cells with appropriate cargoes to obtain cargo-loaded exosomes appears to offer an acceptable means of therapeutic exosome production [211]. Destined for secretion, transfection of donor cells with the overexpression construct facilitates entry of therapeutic cargo into the lumen or its labeling to the surface of exosomes [161,168]. In most studies, miRNAs are transfected as overexpression constructs in miRNA expression vectors and then loaded into exosomes [161,162,172,173,195]. Exosomes produced from MSCs transfected with a construct carrying miR-146b were found to restrict tumor growth effectively [161]. In a similar study, let-7a containing exosomes with a surface expressed target peptide efficiently delivered cargo to epidermal growth factor receptor (EGFR) expressing breast cancer cells [168]. Elevated miR-214 expression achieved by transfecting cells with anti-miRs seems to be a promising alternative to transfecting donor cells with pre-miR-214 [160,195]. Though transfection of donor cells seems appropriate for exosome loading for in vivo studies, engineering cells to express desired surface molecules and carry maximum therapeutic load is time-consuming. Thus, non-autologous exosome producing methods are required to generate non-immunogenic exosomes with specific targeting characteristics for clinical use.

Studies that used activated donor cells to generate exosomes have shown them to be less appropriate choice for exosome production as they are capable of transferring therapy resistance to drug sensitive cells via, proteins, that increases DNA repair and tumor cell survival along with disposal of the pro-apoptotic proteins. Using this methodology, stimulation of THP-1 cells using inflammatory stimulants caused an increase in miR-150 levels in vesicles [193], and in another study, co-culture of brain extracts from rats that had undergone middle cerebral artery occlusion show increased miR-133b levels [182]. Hypoxia is a characteristic of tumor microenvironment [212–214] and is believed to enhance release of exosomes. Studies that used hypoxic condition to generate exosomes have revealed them to be enriched with CD81, CD63 and HSP70 markers [215–217]. Although hypoxic microenvironment alter the miRNA cargoes of exosomes from different cells [215], exosomes generated under hypoxic conditions were found to be enriched in IL-8 and IGFBP3 mRNAs and proteins, which promote the proliferation and migration of angiogenic cells in vitro [218,219].

6.4. Exosome Administration Routes

Conventional routes are required to administer drug-loaded exosomes. In addition, to the efforts being made to increase stability during long-term storage, research is also being conducted to identify means of delivering drugs to tumors located in fragile tissues [220,221]. Administration of exosome-based therapeutics via intravenous injection has been commonly used to deliver drugs to brain, pancreas, and tumors in other tissues [165,167,168,172,198,222–226], and the endogenic origins of exosomes help them escape removal by immune cells [227]. Exosome-based delivery of therapeutics increases drug stability and enables high drug loadings in body fluids [227], and lack of lymphatic drainage and the presence of fewer blood vessels aid in the retention of exosomes in tumorigenic tissues [12,228,229], which enhances their therapeutic efficacies. Upon administration through an intravenous mode, the half-life of exosome-based therapeutic cargo in circulation was approximately two minutes [178]. The distribution of exosomes to lungs, liver, spleen, and bone marrow and their later accumulation in liver and then lungs, suggests a clearance mode similar to that of synthetic liposomes [178,230,231]. Accumulation in liver has also been reported in studies on the administration of EGFR-bearing exosomes with high affinity for hepatic tissues, and in tumor tissues in a xenograft model of breast cancer [168,232]. Despite their exhaustion in circulation within short span of time, the presence of therapeutic cargo in tumor vasculogenesis appears to program bone marrow-derived MSCs [233]. In addition, modifications, such as PEGylation, aimed at increasing their half-lives, are still warranted [234].

The intra-tumoral injection (another appropriate administration technique) of exosome-encapsulated therapeutics for the treatment of different cancers resulted in successful reduction in tumor volumes [161,173,175,235,236]. The combined use of intratumoral injection and tumor resection further reduces the risk of tumor recurrence [161,237]. The oral administration of exosomes potently induce intestinal stem cell proliferation after stable passage through the gut in a murine model of colitis [238]. Administration of exosomes loaded intraperitoneally with curcumin increased their bioavailability by improving their stability in the circulation [170]. Intranasal administration of exosomes encapsulating curcumin or Stat3 inhibitor for delivery to microglial cells reduced inflammation in brain [202], and the subcutaneous administration of MHC II over-expressing exosomes proved effective in murine melanoma [179,202,239]. The exosomes loaded with therapeutic cargo exerts their effects at the target with in a short span of time after its delivery to the target [165,178]. Adoption of exosomes in clinical settings requires characterization of exosome protein compositions in order to avoid adverse effects in patients.

7. Increased Specificity by Exosome Engineering

The expression of targeting peptides or proteins on exosome surface is a prerequisite for the specific delivery of therapeutic cargoes and in avoiding the adverse effects associated with chemotherapeutic agents on normal cells surrounding tumors. Although many studies have been performed on the exosome-based delivery of therapeutic cargoes, few have addressed the engineering of exosomes to achieve the target-specific delivery of therapeutic cargoes [240–244]. The exosome-engineering aimed at inserting a peptide correctly into exosomes, while avoiding cleavage of peptide regions, is accomplished by expressing the target peptide as a fusion product with the surface localized lysosomal associated membrane protein-2b (Lamp-2b) [245,246]. This bioengineering approach helps to enhance the uptake of exosomes and as such treatment specificities in tissues of interest. An excellent example of this phenomenon is provided by RVG and iRGD peptides, which when engineered on immature DC-derived exosomes helps to target therapeutics to the brain and tumor tissues [165,167]. The expressions of hemagglutinin, myc-tag, and peptide (epidermal growth factor; EGF or GE11) as a fusion protein with platelet-derived growth factor receptor (PDGFR) on the surface of exosomes effectively targeted drugs to tumors [168]. With ability to bind specifically to EGFR-upregulated cells in tumor tissues, GE11-mediated delivery of therapeutic cargoes proceeds without activating the

EGF-receptor [168], and thus, this method of delivery appears to be appropriate for treating different types of cancers [247].

U937 or Raw264.7 cell-derived exosomes or exosome mimetic nanoparticles expressing surface LFA-1 induced significant reduction in the tumor volume when used to deliver chemotherapeutics to tumor cells [200]. LFA-1 facilitates binding of exosomes to endothelial cell adhesion molecules and has been used to deliver therapeutics to rapidly growing tumors with extensive neovascularization [200]. The cell-specific characteristics of exosomes facilitate the delivery of therapeutics more specifically to tumor tissues. Transfection of the CIITA gene to induce the expression of MHC II in murine melanoma cells resulted in the production of exosomes expressing high surface levels of the MHC II protein [179]. The study indicated that MHC II has two functions, that is, as a targeting peptide to deliver cargoes to specific destinations and as a therapeutic [179]. Exosomes derived from choroid plexus epithelial cells expressing folate receptor-α (FRα) were reported to transport cargo to brain parenchyma cells after passaging through the choroid plexus [239]. The ability to cross the blood–brain barrier (BBB) or choroid plexus and the surface expression of targeting peptides on exosomes hold great promise for drug delivery to the brain [165,239,248]. The surface expression of tetraspanin proteins can be used as an alternative method to engineer exosomes that deliver therapeutics to tumor tissues [222]. Similarly, utilizing target specific antibodies to coat the surface of exosomes provides another means of avoiding the laborious procedure of modifying membrane proteins.

8. Advancement in the Therapeutic Uses of Exosomes

Many commercial enterprises have been established to exploit the exosome-based delivery of therapeutics. Codial BioSciences (Cambridge, MA, USA) has devised a specific platform called engEx™ for engineering exosomes to deliver different therapeutics entities [249]. exoSTING—a therapeutic entity developed on exosome backbone with minimal cytotoxicity is viewed as a promising therapeutic delivery candidate in the treatment of cancer [249]. Exosomes carrying therapeutic cargoes have also been subjected to clinical trials (Table 4). In a phase I study, DC-derived exosomes (DEX) loaded with MAGE3 antigenic peptides were administered to stage III/IV melanoma patients [250]. Studies performed on the intradermal and subcutaneous administration of DEX revealed an increased number of natural killer cells (NKCs) and reconstitution of NKG2D expression on NK and CD8$^+$ T-cells. Autologous exosome production from these non-toxic cells was achieved successfully using standard manufacturing protocols [250]. In a phase II study of DC-derived exosomes (DEX2) loaded with the chemotherapeutic metronomic cyclophosphamide, DEX2 encapsulation increased the immunostimulatory effect of the drug on T-cells (NCT01159228). In addition, the application of ascites-derived exosomes (AEX) together with GM-CSF was found to have greater cytotoxic T-cell response in colorectal cancer than AEX alone [251]. Furthermore, exosome-based treatment was subjected to clinical trials in malignant glioma. Implantation of glioma cells isolated from resected tumor tissue into the abdomen of glioma patients treated with drug-inhibiting insulin-like growth factor receptor-1 (IGF-1) induced apoptosis in implanted cells, and this was followed by exosome release from these cells that stimulated the immune system to induce a T-cell mediating antitumor response (NCT01550523).

A joint venture between PureTech Health and Roche aimed at developing novel exosome technologies, led to the development of milk exosome-based technology for the oral administration of antisense oligonucleotides [252], and this technology is considered to have the potential to enhance treatment efficacies and reduce toxicities as compared with conventional intravenous injection. In addition, plant-derived exosomes were assessed for potential use as cancer treatments at the James Graham Brown Cancer Center. Orally administered exosomes containing curcumin were tested for therapeutic effectiveness against colorectal cancer (NCT01294072) and evaluated for their effects on oral mucositis and pain after chemotherapy for head and neck cancers (NCT01668849). These trials, which are ongoing and completed, respectively, have demonstrated good safety profiles in clinical settings, and relevance of continuing the development of exosome-based drug delivery systems.

Table 4. Clinical trial data of exosomes used to treat various diseases.

Exosome Source	Condition	Payload	Phase, Patients	Clinical Trial Identifier
MSCs	Multiple organ failure	NA	NA ($n = 60$)	NCT04356300
	Severe COVID-19 Pneumonia	NA	Phase 1 ($n = 24$)	NCT04276987
	Periodontitis	NA	Phase 1 ($n = 10$)	NCT04270006
	Dry Eye	NA	Phase 1 ($n = 27$)	NCT04213248
	Type I Diabetes Mellitus	NA	Phase 1 ($n = 20$)	NCT02138331
	Metastatic Pancreatic cancer	KRAS G12D siRNA	Phase 1 ($n = 28$)	NCT03608631
	Macular Holes	NA	Phase 1 ($n = 44$)	NCT03437759
	Cerebrovascular disorders	NA	Phase 1/2 ($n = 5$)	NCT03384433
	Diabetic Nephropathy	Placebo	NA ($n = 38$)	NCT04562025
Dendritic Cell	Sepsis	Antibiotics	NA ($n = 50$)	NCT02957279
	Non-small cell lung cancer	Antigens	Phase 2 ($n = 41$)	NCT01159288
		MAGE tumor antigens		
	Metastatic melanoma	MAGE 3 peptides		
Plant	Colorectal cancer	Curcumin	Phase 1 ($n = 7$)	NCT01294072
	Obesity	NA	NA ($n = 160$)	NCT02706262
	Head & Neck cancer	Grape extract	Phase I ($n = 60$)	NCT01668849
	Polycystic ovary syndrome	Ginger & Aloe	NA ($n = 176$)	NCT03493984

Source: https://www.clinicaltrials.gov (accessed on 24 December 2020). NA = Not available.

9. Conclusions

Exosomes are considered as versatile carriers due to their immunogenic nature and abilities to traverse biological barriers (e.g., the blood–brain barrier) and migrate to tissues or areas with no blood supply (e.g., dense cartilage matrix). Exosomes encapsulate many cargo types (DNAs, RNAs, proteins, and lipids) and transport them via body fluids to nearby or distant cells. Their biocompatibilities and the genetic engineering possibilities that prevent unwanted exosome accumulation and enable selective targeting, have encouraged researchers to develop exosome-based drug delivery systems. Selection of the source and optimization of the isolation methods are currently being explored towards achieving enhancement in the production of exosomes with distinct characteristics and functionalities. Studies are currently being undertaken on the potential therapeutic use of exosome derived from human tissues as drug carriers. However, such investigations are hampered by lack of suitable isolation methods and drug uptake discrepancies. Currently, the use of hollow fiber-based bioreactors offer an attractive means of harvesting exosomes with reproducible characteristics. As effectiveness of therapeutic cargo depends on the source of generation of exosomes and its release at target site, efforts are required to understand exosome generation in different cellular backgrounds and their drug uptake at the target tissues. Exosomes exhibit a lipid bilayer structure with embedded characteristic surface protein signatures that promote uptake at target sites. Given the complexity of exosomes, internalization of exosomes loaded with therapeutic cargoes can be achieved by incorporating cell-penetrating peptides (CPPs), such as arginine-rich CPPs, which stimulate micropinocytosis at target sites, onto their surfaces. Investigations are required to determine the optimal dosage, administration methods, and kinetic characteristics, and to further investigate the effects of environmental conditions, such as pH, on the efficiency of cargo delivery. Moreover, comprehensive investigations of the properties of cells used for exosome production and the functionalities of exosomes are needed to ensure target-specific delivery of therapeutics in the context of personalized medicine. Furthermore, the standardization of large-scale production and purification procedures would undoubtedly improve exosome reproducibility and aid in the development of exosome-based cancer therapeutics. Finally, investigations aimed at elucidating the mechanisms that govern the specific delivery of exogenously administered exosomes, their biodistribution, and pharmacokinetics would help to achieve the developmental transition of exosomes to the clinical level.

Author Contributions: Conceptualization and writing, I.C.; F.M.R.; writing and figure preparation, A.T.J. and S.R.; editing and suggestion of improvements, R.B.; E.J.L. and E.H.M. All authors have read and agreed to the published version of the manuscript.

Funding: This research was supported by Basic Science Research Program through the National Research Foundation of Korea (NRF) funded by the Ministry of Education (2020R1A6A1A03044512) and by a National Research Foundation of Korea (NRF) grant funded by the Korean government (MSIP; Grant No. NRF-2018R1A2B6001020).

Acknowledgments: Author A.T.J. is grateful to DST-SERB for financial support (CRG/2019/004106) that helped in establishing the infrastructural facilities.

Conflicts of Interest: The authors declare no conflict of interest.

References

1. Sun, H.; Burrola, H.; Wu, J.; Ding, W.Q. Extracellular vesicles in the development of cancer therapeutics. *Int. J. Mol. Sci.* **2020**, *21*, 6097. [CrossRef]
2. Blanc, L.; De Gassart, A.; Géminard, C.; Bette-Bobillo, P.; Vidal, M. Exosome release by reticulocytes–an integral part of the red blood cell differentiation system. *Blood Cells Mol. Dis.* **2005**, *35*, 21–26. [CrossRef] [PubMed]
3. Pan, B.T.; Johnstone, R.M. Fate of the transferrin receptor during maturation of sheep reticulocytes in vitro: Selective externalization of the receptor. *Cell* **1983**, *33*, 967–978. [CrossRef]
4. Johnstone, R.M.; Adam, M.; Hammond, J.R.; Orr, L.; Turbide, C. Vesicle formation during reticulocyte maturation. Association of plasma membrane activities with released vesicles (exosomes). *J. Biol. Chem.* **1987**, *262*, 9412–9420. [CrossRef]
5. Lee, Y.; Morrison, B.M.; Li, Y.; Lengacher, S.; Farah, M.H.; Hoffman, P.N.; Liu, Y.; Tsingalia, A.; Jin, L.; Zhang, P.W.; et al. Oligodendroglia metabolically support axons and contribute to neurodegeneration. *Nature* **2012**, *487*, 443–448. [CrossRef]
6. Sun, W.; Luo, J.D.; Jiang, H.; Duan, D.D. Tumor exosomes: A double-edged sword in cancer therapy. *Acta Pharmacol. Sin.* **2018**, *39*, 534–541. [CrossRef] [PubMed]
7. Akers, J.C.; Gonda, D.; Kim, R.; Carter, B.S.; Chen, C.C. Biogenesis of extracellular vesicles (EV): Exosomes, microvesicles, retrovirus-like vesicles, and apoptotic bodies. *J. Neurooncol.* **2013**, *113*, 1–11. [CrossRef] [PubMed]
8. Raposo, G.; Stoorvogel, W. Extracellular vesicles: Exosomes, microvesicles, and friends. *J. Cell Biol.* **2013**, *200*, 373–383. [CrossRef]
9. Bobrie, A.; Colombo, M.; Raposo, G.; Thery, C. Exosome secretion: Molecular mechanisms and roles in immune responses. *Traffic* **2011**, *12*, 1659–1668. [CrossRef] [PubMed]
10. Thery, C.; Ostrowski, M.; Segura, E. Membrane vesicles as conveyors of immune responses. *Nat. Rev. Immunol.* **2009**, *9*, 581–593. [CrossRef]
11. Mathivanan, S.; Ji, H.; Simpson, R.J. Exosomes: Extracellular organelles important in intercellular communication. *J. Proteom.* **2010**, *73*, 1907–1920. [CrossRef]
12. Vlassov, A.V.; Magdaleno, S.; Setterquist, R.; Conrad, R. Exosomes: Current knowledge of their composition, biological functions, and diagnostic and therapeutic potentials. *Biochim. Biophys. Acta* **2012**, *1820*, 940–948. [CrossRef]
13. Bang, C.; Thum, T. Exosomes: New players in cell-cell communication. *Int. J. Biochem. Cell Biol.* **2012**, *44*, 2060–2064. [CrossRef]
14. Cosme, J.; Liu, P.P.; Gramolini, A.O. The cardiovascular exosome: Current perspectives and potential. *Proteomics* **2013**, *13*, 1654–1659. [CrossRef]
15. Raposo, G.; Nijman, H.W.; Stoorvogel, W.; Liejendekker, R.; Harding, C.V.; Melief, C.J.; Geuze, H.J. B lymphocytes secrete antigen-presenting vesicles. *J. Exp. Med.* **1996**, *183*, 1161–1172. [CrossRef] [PubMed]
16. Huang, Y.; Cheng, L.; Turchinovich, A.; Mahairaki, V.; Troncoso, J.C.; Pletnikova, O.; Haughey, N.J.; Vella, L.J.; Hill, A.F.; Zheng, L.; et al. Influence of species and processing parameters on recovery and content of brain tissue-derived extracellular vesicles. *J. Extracell. Vesicles* **2020**, *9*, 1785746. [CrossRef] [PubMed]
17. Hu, W.; Huang, F.; Ning, L.; Hao, J.; Wan, J.; Hao, S. Enhanced immunogenicity of leukemia-derived exosomes via transfection with lentiviral vectors encoding costimulatory molecules. *Cell. Oncol.* **2020**. [CrossRef]
18. Chanteloup, G.; Cordonnier, M.; Isambert, N.; Bertaut, A.; Hervieu, A.; Hennequin, A.; Luu, M.; Zanetta, S.; Coudert, B.; Bengrine, L.; et al. Monitoring HSP70 exosomes in cancer patients' follow up: A clinical prospective pilot study. *J. Extracell. Vesicles* **2020**, *9*, 1766192. [CrossRef]

19. Balbinotti, H.; Cadore, N.A.; Dutra, C.S.; Da Silva, D.A.S.; Ferreira, H.B.; Zaha, A.; Monteiro, K.M. Protein Profiling of Extracellular Vesicles Associated With Cisplatin Resistance in Lung Cancer. *Anticancer Res.* **2020**, *40*, 5509–5516. [CrossRef]

20. Sadovska, L.; Eglītis, J.; Linē, A. Extracellular Vesicles as Biomarkers and Therapeutic Targets in Breast Cancer. *Anticancer. Res.* **2015**, *35*, 6379–6390. [PubMed]

21. Urabe, F.; Kosaka, N.; Ito, K.; Kimura, T.; Egawa, S.; Ochiya, T. Extracellular vesicles as biomarkers and therapeutic targets for cancer. *Am. J. Physiol. Physiol.* **2020**, *318*, C29–C39. [CrossRef] [PubMed]

22. Thery, C.; Zitvogel, L.; Amigorena, S. Exosomes: Composition, biogenesis and function. *Nat. Rev. Immunol.* **2002**, *2*, 569–579. [CrossRef] [PubMed]

23. Erin, N.; Ogan, N.; Yerlikaya, A. Secretomes reveal several novel proteins as well as TGF-β1 as the top upstream regulator of metastatic process in breast cancer. *Breast Cancer Res. Treat.* **2018**, *170*, 235–250. [CrossRef] [PubMed]

24. Han, K.Y.; Chang, J.H.; Azar, D.T. Proteomics-based Characterization of the Effects of MMP14 on the Protein Content of Exosomes from Corneal Fibroblasts. *Protein Pept. Lett.* **2020**. [CrossRef]

25. Shimoda, M.; Khokha, R. Proteolytic factors in exosomes. *Proteomics* **2013**, *13*, 1624–1636. [CrossRef] [PubMed]

26. Wojtuszkiewicz, A.; Schuurhuis, G.J.; Kessler, F.L.; Piersma, S.R.; Knol, J.C.; Pham, T.V.; Jansen, G.; Musters, R.J.; van Meerloo, J.; Assaraf, Y.G. Exosomes secreted by apoptosis-resistant acute myeloid leukemia (AML) blasts harbor regulatory network proteins potentially involved in antagonism of apoptosis. *Mol. Cell Proteom.* **2016**, *15*, 1281–1298. [CrossRef]

27. Yang, M.; Song, D.; Cao, X.; Wu, R.; Liu, B.; Ye, W.; Wu, J.; Yue, X. Comparative proteomic analysis of milk-derived exosomes in human and bovine colostrum and mature milk samples by iTRAQ-coupled LC-MS/MS. *Food Res. Int.* **2017**, *92*, 17–25. [CrossRef]

28. Kogure, T.; Yan, I.K.; Lin, W.L.; Patel, T. Extracellular vesicle-mediated transfer of a novel long noncoding RNA TUC339: A mechanism of intercellular signaling in human hepatocellular cancer. *Genes Cancer* **2013**, *4*, 261–272. [CrossRef] [PubMed]

29. Schorey, J.S.; Bhatnagar, S. Exosome function: From tumor immunology to pathogen biology. *Traffic* **2008**, *9*, 871–881. [CrossRef]

30. Van Deun, J.; Mestdagh, P.; Sormunen, R.; Cocquyt, V.; Vermaelen, K.; Vandesompele, J.; Bracke, M.; De Wever, O.; Hendrix, A. The impact of disparate isolation methods for extracellular vesicles on downstream RNA profiling. *J. Extracell. Vesicles* **2014**, *3*, 24858. [CrossRef]

31. Fraser, K.; Jo, A.; Giedt, J.; Vinegoni, C.; Yang, K.S.; Peruzzi, P.; Chiocca, E.A.; Breakefield, X.O.; Lee, H.; Weissleder, R. Characterization of single microvesicles in plasma from glioblastoma patients. *Neuro-Oncology* **2019**, *21*, 606–615. [CrossRef]

32. Webber, J.; Clayton, A. How pure are your vesicles? *J. Extracell. Vesicles* **2013**, *2*, 19861. [CrossRef]

33. Tian, Y.; Gong, M.; Hu, Y.; Liu, H.; Zhang, W.; Zhang, M.; Hu, X.; Aubert, D.; Zhu, S.; Wu, L.; et al. Quality and efficiency assessment of six extracellular vesicle isolation methods by nano-flow cytometry. *J. Extracell. Vesicles* **2020**, *9*, 1697028. [CrossRef]

34. Van Deun, J.; Mestdagh, P.; Agostinis, P.; Akay, Ö.; Anand, S.; Anckaert, J.; Martinez, Z.A.; Baetens, T.; Beghein, E.; Bertier, L.; et al. EV-TRACK: Transparent reporting and centralizing knowledge in extracellular vesicle research. *Nat. Methods* **2017**, *14*, 228–232. [CrossRef] [PubMed]

35. Simonsen, J.B. What are we looking at? extracellular vesicles, lipoproteins, or both? *Circ. Res.* **2017**, *121*, 920–922. [CrossRef]

36. Flaherty, S.E., 3rd; Grijalva, A.; Xu, X.; Ables, E.; Nomani, A.; Ferrante, A.W., Jr. A lipase-independent pathway of lipid release and immune modulation by adipocytes. *Science* **2019**, *363*, 989–993. [CrossRef]

37. Takov, K.; Yellon, D.M.; Davidson, S.M. Comparison of small extracellular vesicles isolated from plasma by ultracentrifugation or size-exclusion chromatography: Yield, purity and functional potential. *J. Extracell. Vesicles* **2019**, *8*, 1560809. [CrossRef] [PubMed]

38. Coumans, F.A.W.; Brisson, A.R.; Buzas, E.I.; Dignat-George, F.; Drees, E.E.E.; El-Andaloussi, S.; Emanueli, C.; Gasecka, A.; Hendrix, A.; Hill, A.F.; et al. Methodological guidelines to study extracellular vesicles. *Circ. Res.* **2017**, *120*, 1632–1648. [CrossRef] [PubMed]

39. Lobb, R.J.; Becker, M.; Wen, S.W.; Wong, C.S.; Wiegmans, A.P.; Leimgruber, A.; Möller, A. Optimized exosome isolation protocol for cell culture supernatant and human plasma. *J. Extracell. Vesicles* **2019**, *4*, 27031. [CrossRef] [PubMed]

40. Ghosh, A.; Davey, M.; Chute, I.C.; Griffiths, S.G.; Lewis, S.; Chacko, S.; Barnett, D.; Crapoulet, N.; Fournier, S.; Joy, A.; et al. Rapid isolation of extracellular vesicles from cell culture and biological fluids using a synthetic peptide with specific affinity for heat shock proteins. *PLoS ONE* **2014**, *9*, e110443. [CrossRef]

41. Bijnsdorp, I.V.; Maxouri, O.; Kardar, A.; Schelfhorst, T.; Piersma, S.R.; Pham, T.V.; Vis, A.; van Moorselaar, R.J.; Jimenez, C.R. Feasibility of urinary extracellular vesicle proteome profiling using a robust and simple, clinically applicable isolation method. *J. Extracell. Vesicles* **2017**, *6*, 1313091. [CrossRef]

42. Stokman, M.F.; Bijnsdorp, I.V.; Schelfhorst, T.; Pham, T.V.; Piersma, S.R.; Knol, J.C.; Giles, R.H.; Bongers, E.M.H.F.; Knoers, N.V.A.M.; Lilien, M.R.; et al. Changes in the urinary extracellular vesicle proteome are associated with nephronophthisis-related ciliopathies. *J. Proteom.* **2019**, *192*, 27–36. [CrossRef]

43. Fang, S.; Tian, H.; Li, X.; Jin, D.; Li, X.; Kong, J.; Yang, C.; Yang, X.; Lu, Y.; Luo, Y.; et al. Clinical application of a microfluidic chip for immunocapture and quantification of circulating exosomes to assist breast cancer diagnosis and molecular classification. *PLoS ONE* **2017**, *12*, e0175050. [CrossRef]

44. Dorayappan, K.D.P.; Gardner, M.L.; Hisey, C.L.; Zingarelli, R.A.; Smith, B.Q.; Lightfoot, M.D.S.; Gogna, R.; Flannery, M.M.; Hays, J.; Hansford, D.J.; et al. A microfluidic chip enables isolation of exosomes and establishment of their protein profiles and associated signaling pathways in ovarian cancer. *Cancer Res.* **2019**, *79*, 3503–3513. [CrossRef]

45. Hisey, C.L.; Dorayappan, K.D.P.; Cohn, D.E.; Selvendiran, K.; Hansford, D.J. Microfluidic affinity separation chip for selective capture and release of labelfree ovarian cancer exosomes. *Lab Chip* **2018**, *18*, 3144–3153. [CrossRef] [PubMed]

46. Aupeix, K.; Hugel, B.; Martin, T.; Bischoff, P.; Lill, H.; Pasquali, J.L.; Freyssinet, J.M. The significance of shed membrane particles during programmed cell death in vitro, and in vivo, in HIV-1 infection. *J. Clin. Investig.* **1997**, *99*, 1546–1554. [CrossRef] [PubMed]
47. Zhang, Y.; Liu, Y.; Liu, H.; Tang, W.H. Exosomes: Biogenesis, biologic function and clinical potential. *Cell Biosci.* **2019**, *9*, 19. [CrossRef]
48. Mather, I.H. Proteins of the milk-fat-globule membrane as markers of mammary epithelial cells and apical plasma membrane. In *The Mammary Gland: Development, Regulation and Function*; Neville, M.C., Daniel, C.W., Eds.; Plenum Press: New York, NY, USA, 1987; pp. 217–267.
49. Peters, P.J.; Geuze, H.J.; Van der Donk, H.A.; Slot, J.W.; Griffith, J.M.; Stam, N.J.; Clevers, H.C.; Borst, J. Molecules relevant for T cell-target cell interaction are present in cytolytic granules of human T lymphocytes. *Eur. J. Immunol.* **1989**, *19*, 1469–1475. [CrossRef] [PubMed]
50. Thery, C.; Regnault, A.; Garin, J.; Wolfers, J.; Zitvogel, L.; Ricciardi-Castagnoli, P.; Raposo, G.; Amigorena, S. Molecular characterization of dendritic cell-derived exosomes. Selective accumulation of the heat shock protein hsc73. *J. Cell Biol.* **1999**, *147*, 599–610. [CrossRef]
51. Heijnen, H.F.; Schiel, A.E.; Fijnheer, R.; Geuze, H.J.; Sixma, J.J. Activated platelets release two types of membrane vesicles: Microvesicles by surface shedding and exosomes derived from exocytosis of multivesicular bodies and alpha-granules. *Blood* **1999**, *94*, 3791–3799. [CrossRef]
52. Babst, M.; Katzmann, D.J.; Estepa-Sabal, E.J.; Meerloo, T.; Emr, S.D. Escrt-III: An endosome-associated heterooligomeric protein complex required for mvb sorting. *Dev. Cell* **2002**, *3*, 271–282. [CrossRef]
53. Hosseini, H.M.; Fooladi, A.A.; Nourani, M.R.; Ghanezadeh, F. The role of exosomes in infectious diseases. *Inflamm. Allergy Drug Targets* **2013**, *12*, 29–37. [CrossRef]
54. Cashikar, A.G.; Shim, S.; Roth, R.; Maldazys, M.R.; Heuser, J.E.; Hanson, P.I. Structure of cellular ESCRT-III spirals and their relationship to HIV budding. *eLife* **2014**, *3*. [CrossRef]
55. Wideman, J.G.; Leung, K.F.; Field, M.C.; Dacks, J.B. The cell biology of the endocytic system from an evolutionary perspective. *Cold Spring Harb. Perspect Biol.* **2014**, *6*, a016998. [CrossRef]
56. Hurley, J.H. ESCRTs are everywhere. *EMBO J.* **2015**, *34*, 2398–2407. [CrossRef]
57. McCullough, J.; Fisher, R.D.; Whitby, F.G.; Sundquist, W.I.; Hill, C.P. ALIX-CHMP4 interactions in the human ESCRT pathway. *Proc. Natl. Acad. Sci. USA* **2008**, *105*, 7687–7691. [CrossRef] [PubMed]
58. Hanson, P.I.; Cashikar, A. Multivesicular body morphogenesis. *Annu. Rev. Cell Dev. Biol.* **2012**, *28*, 337–362. [CrossRef] [PubMed]
59. Wei, D.; Zhan, W.; Gao, Y.; Huang, L.; Gong, R.; Wang, W.; Zhang, R.; Wu, Y.; Gao, S.; Kang, T. RAB31 marks and controls an ESCRT-independent exosome pathway. *Cell Res.* **2020**. [CrossRef]
60. Addi, C.; Presle, A.; Frémont, S.; Cuvelier, F.; Rocancourt, M.; Milin, F.; Schmutz, S.; Chamot-Rooke, J.; Douché, T.; Duchateau, M.; et al. The Flemmingsome reveals an ESCRT-to-membrane coupling via ALIX/syntenin/syndecan-4 required for completion of cytokinesis. *Nat. Commun.* **2020**, *11*, 1941. [CrossRef]
61. Bánová Vulić, R.; Zdurienčíková, M.; Tyčiaková, S.; Benada, O.; Dubrovčáková, M.; Lakota, J.; Škultéty, Ľ. Silencing of carbonic anhydrase I enhances the malignant potential of exosomes secreted by prostatic tumour cells. *J. Cell. Mol. Med.* **2019**, *23*, 3641–3655. [CrossRef]
62. Trajkovic, K.; Hsu, C.; Chiantia, S.; Rajendran, L.; Wenzel, D.; Wieland, F.; Schwille, P.; Brugger, B.; Simons, M. Ceramide triggers budding of exosome vesicles into multivesicular endosomes. *Science* **2008**, *319*, 1244–1247. [CrossRef]
63. Ghossoub, R.; Lembo, F.; Rubio, A.; Gaillard, C.B.; Bouchet, J.; Vitale, N.; Slavik, J.; Machala, M.; Zimmermann, P. Syntenin-ALIX exosome biogenesis and budding into multivesicular bodies are controlled by ARF6 and PLD2. *Nat. Commun.* **2014**, *5*, 3477. [CrossRef]
64. Groot, M.; Lee, H. Sorting Mechanisms for MicroRNAs into Extracellular Vesicles and Their Associated Diseases. *Cells* **2020**, *9*. [CrossRef] [PubMed]
65. Stuffers, S.; Sem Wegner, C.; Stenmark, H.; Brech, A. Multivesicular endosome biogenesis in the absence of ESCRTs. *Traffic* **2009**, *10*, 925–937. [CrossRef]
66. Babst, M. MVB vesicle formation: ESCRT-dependent, ESCRT-independent and everything in between. *Curr. Opin. Cell Biol.* **2011**, *23*, 452–457. [CrossRef]
67. Wehman, A.M.; Poggioli, C.; Schweinsberg, P.; Grant, B.D.; Nance, J. The P4-ATPase TAT-5 inhibits the budding of extracellular vesicles in C. elegans embryos. *Curr. Biol.* **2011**, *21*, 1951–1959. [CrossRef] [PubMed]
68. Colombo, M.; Raposo, G.; Thery, C. Biogenesis, secretion, and intercellular interactions of exosomes and other extracellular vesicles. *Annu. Rev. Cell Dev. Biol.* **2014**, *30*, 255–289. [CrossRef] [PubMed]
69. Kowal, J.; Tkach, M.; Thery, C. Biogenesis and secretion of exosomes. *Curr. Opin. Cell Biol.* **2014**, *29*, 116–125. [CrossRef]
70. Fader, C.M.; Sanchez, D.G.; Mestre, M.B.; Colombo, M.I. TI-VAMP/VAMP7 and VAMP3/cellubrevin: Two v-SNARE proteins involved in specific steps of the autophagy/multivesicular body pathways. *Biochim. Biophys. Acta* **2009**, *1793*, 1901–1916. [CrossRef] [PubMed]
71. Gross, J.C.; Chaudhary, V.; Bartscherer, K.; Boutros, M. Active Wnt proteins are secreted on exosomes. *Nat. Cell Biol.* **2012**, *14*, 1036–1045. [CrossRef]
72. Ruiz-Martinez, M.; Navarro, A.; Marrades, R.M.; Viñolas, N.; Santasusagna, S.; Muñoz, C.; Ramírez, J.; Molins, L.; Monzo, M. YKT6 expression, exosome release, and survival in non-small cell lung cancer. *Oncotarget* **2016**, *7*, 51515–51524. [CrossRef]

73. Santos, M.F.; Rappa, G.; Karbanová, J.; Kurth, T.; Corbeil, D.; Lorico, A. VAMP-associated protein-A and oxysterol-binding protein-related protein 3 promote the entry of late endosomes into the nucleoplasmic reticulum. *J. Biol. Chem.* **2018**, *293*, 13834–13848. [CrossRef]

74. De Paoli, S.H.; Tegegn, T.Z.; Elhelu, O.K.; Strader, M.B.; Patel, M.; Diduch, L.L.; Tarandovskiy, I.D.; Wu, Y.; Zheng, J.; Ovanesov, M.V.; et al. Dissecting the biochemical architecture and morphological release pathways of the human platelet extracellular vesiculome. *Cell Mol. Life Sci.* **2018**, *75*, 3781–3801. [CrossRef]

75. Duan, M.J.; Yan, M.L.; Wang, Q.; Mao, M.; Su, D.; Sun, L.L.; Li, K.X.; Qu, Y.; Sun, Q.; Zhang, X.Y.; et al. Overexpression of miR-1 in the heart attenuates hippocampal synaptic vesicle exocytosis by the posttranscriptional regulation of SNAP-25 through the transportation of exosomes. *Cell Commun. Signal* **2018**, *16*, 91. [CrossRef]

76. Limanaqi, F.; Biagioni, F.; Gambardella, S.; Ryskalin, L.; Fornai, F. Interdependency Between Autophagy and Synaptic Vesicle Trafficking: Implications for Dopamine Release. *Front. Mol. Neurosci.* **2018**, *11*, 299. [CrossRef]

77. Xu, P.; Hankins, H.M.; MacDonald, C.; Erlinger, S.J.; Frazier, M.N.; Diab, N.S.; Piper, R.C.; Jackson, L.P.; MacGurn, J.A.; Graham, T.R. COPI mediates recycling of an exocytic SNARE by recognition of a ubiquitin sorting signal. *Elife* **2017**, *6*. [CrossRef]

78. Marshansky, V.; Futai, M. The V-type H+-ATPase in vesicular trafficking: Targeting, regulation and function. *Curr. Opin. Cell Biol.* **2008**, *20*, 415–426. [CrossRef] [PubMed]

79. Hu, G.; Gong, A.Y.; Roth, A.L.; Huang, B.Q.; Ward, H.D.; Zhu, G.; Larusso, N.F.; Hanson, N.D.; Chen, X.M. Release of luminal exosomes contributes to TLR4-mediated epithelial antimicrobial defense. *PLoS Pathog.* **2013**, *9*, e1003261. [CrossRef] [PubMed]

80. Vidal, M.J.; Stahl, P.D. The small GTP-binding proteins Rab4 and ARF are associated with released exosomes during reticulocyte maturation. *Eur. J. Cell Biol.* **1993**, *60*, 261–267. [PubMed]

81. Zitvogel, L.; Regnault, A.; Lozier, A.; Wolfers, J.; Flament, C.; Tenza, D.; Ricciardi-Castagnoli, P.; Raposo, G.; Amigorena, S. Eradication of established murine tumors using a novel cell-free vaccine: Dendritic cell-derived exosomes. *Nat. Med.* **1998**, *4*, 594–600. [CrossRef]

82. Savina, A.; Vidal, M.; Colombo, M.I. The exosome pathway in K562 cells is regulated by Rab11. *J. Cell Sci.* **2002**, *115*, 2505–2515.

83. Hsu, C.; Morohashi, Y.; Yoshimura, S.; Manrique-Hoyos, N.; Jung, S.; Lauterbach, M.A.; Bakhti, M.; Gronborg, M.; Mobius, W.; Rhee, J.; et al. Regulation of exosome secretion by Rab35 and its GTPase-activating proteins TBC1D10A-C. *J. Cell Biol.* **2010**, *189*, 223–232. [CrossRef] [PubMed]

84. Ostrowski, M.; Carmo, N.B.; Krumeich, S.; Fanget, I.; Raposo, G.; Savina, A.; Moita, C.F.; Schauer, K.; Hume, A.N.; Freitas, R.P.; et al. Rab27a and Rab27b control different steps of the exosome secretion pathway. *Nat. Cell Biol.* **2010**, *12*, 19–30. [CrossRef]

85. Augimeri, G.; La Camera, G.; Gelsomino, L.; Giordano, C.; Panza, S.; Sisci, D.; Morelli, C.; Győrffy, B.; Bonofiglio, D.; Andò, S.; et al. Evidence for Enhanced Exosome Production in Aromatase Inhibitor-Resistant Breast Cancer Cells. *Int. J. Mol. Sci.* **2020**, *21*. [CrossRef] [PubMed]

86. Matikainen, S.; Nyman, T.A.; Cypryk, W. Inflammasomes: Exosomal miRNAs loaded for action. *J. Cell Biol.* **2020**, *219*. [CrossRef]

87. Tsuruda, M.; Yoshino, H.; Okamura, S.; Kuroshima, K.; Osako, Y.; Sakaguchi, T.; Sugita, S.; Tatarano, S.; Nakagawa, M.; Enokida, H. Oncogenic effects of RAB27B through exosome independent function in renal cell carcinoma including sunitinib-resistant. *PLoS ONE* **2020**, *15*, e0232545. [CrossRef] [PubMed]

88. Underwood, R.; Wang, B.; Carico, C.; Whitaker, R.H.; Placzek, W.J.; Yacoubian, T.A. The GTPase Rab27b regulates the release, autophagic clearance, and toxicity of α-synuclein. *J. Biol. Chem.* **2020**, *295*, 8005–8016. [CrossRef]

89. Baietti, M.F.; Zhang, Z.; Mortier, E.; Melchior, A.; Degeest, G.; Geeraerts, A.; Ivarsson, Y.; Depoortere, F.; Coomans, C.; Vermeiren, E.; et al. Syndecan-syntenin-ALIX regulates the biogenesis of exosomes. *Nat. Cell Biol.* **2012**, *14*, 677–685. [CrossRef] [PubMed]

90. Bhatnagar, S.; Schorey, J.S. Exosomes released from infected macrophages contain Mycobacterium avium glycopeptidolipids and are proinflammatory. *J. Biol. Chem.* **2007**, *282*, 25779–25789. [CrossRef] [PubMed]

91. Wang, J.; Yao, Y.; Chen, X.; Wu, J.; Gu, T.; Tang, X. Host derived exosomes-pathogens interactions: Potential functions of exosomes in pathogen infection. *Biomed. Pharm.* **2018**, *108*, 1451–1459. [CrossRef]

92. Goodlet, K.J.; Bansal, S.; Arjuna, A.; Nailor, M.D.; Buddhdev, B.; Abdelrazek, H.; Mohamed, H.; Omar, A.; Walia, R.; Mohanakumar, T.; et al. COVID-19 in a Lung Transplant Recipient: Exploring the Diagnostic Role of Circulating Exosomes and the Clinical Impact of Advanced Immunosuppression. *Transpl. Infect. Dis.* **2020**, e13480. [CrossRef] [PubMed]

93. Zou, X.; Yuan, M.; Zhang, T.; Zheng, N.; Wu, Z. EVs Containing Host Restriction Factor IFITM3 Inhibited ZIKV Infection of Fetuses in Pregnant Mice through Trans-placenta Delivery. *Mol. Ther.* **2020**. [CrossRef] [PubMed]

94. Osaki, M.; Okada, F. Exosomes and Their Role in Cancer Progression. *Yonago Acta Med.* **2019**, *62*, 182–190. [CrossRef]

95. Tian, W.; Liu, S.; Li, B. Potential Role of Exosomes in Cancer Metastasis. *Biomed Res. Int.* **2019**, *2019*, 4649705. [CrossRef] [PubMed]

96. Guo, J.; Duan, Z.; Zhang, C.; Wang, W.; He, H.; Liu, Y.; Wu, P.; Wang, S.; Song, M.; Chen, H.; et al. Mouse 4T1 Breast Cancer Cell-Derived Exosomes Induce Proinflammatory Cytokine Production in Macrophages via miR-183. *J. Immunol.* **2020**. [CrossRef]

97. Wang, Y.; Zhang, M.; Zhou, F. Biological functions and clinical applications of exosomal long non-coding RNAs in cancer. *J. Cell Mol. Med.* **2020**. [CrossRef] [PubMed]

98. Wei, Q.; Wei, L.; Zhang, J.; Li, Z.; Feng, H.; Ren, L. EphA2-enriched exosomes promote cell migration and are a potential diagnostic serum marker in pancreatic cancer. *Mol. Med. Rep.* **2020**, *22*, 2941–2947. [CrossRef] [PubMed]

99. Jan, A.T.; Malik, M.A.; Rahman, S.; Yeo, H.R.; Lee, E.J.; Abdullah, T.S.; Choi, I. Perspective Insights of Exosomes in Neurodegenerative Diseases: A Critical Appraisal. *Front. Aging Neurosci.* **2017**, *9*, 317. [CrossRef]

100. Abdel-Haq, H. The Potential of Liquid Biopsy of the Brain Using Blood Extracellular Vesicles: The First Step Toward Effective Neuroprotection Against Neurodegenerative Diseases. *Mol. Diagn. Ther.* **2020**. [CrossRef] [PubMed]
101. Fuller, O.K.; Whitham, M.; Mathivanan, S.; Febbraio, M.A. The Protective Effect of Exercise in Neurodegenerative Diseases: The Potential Role of Extracellular Vesicles. *Cells* **2020**, *9*. [CrossRef]
102. Guedes, V.A.; Devoto, C.; Leete, J.; Sass, D.; Acott, J.D.; Mithani, S.; Gill, J.M. Extracellular Vesicle Proteins and MicroRNAs as Biomarkers for Traumatic Brain Injury. *Front. Neurol.* **2020**, *11*, 663. [CrossRef] [PubMed]
103. Anel, A.; Gallego-Lleyda, A.; de Miguel, D.; Naval, J.; Martinez-Lostao, L. Role of Exosomes in the Regulation of T-cell Mediated Immune Responses and in Autoimmune Disease. *Cells* **2019**, *8*. [CrossRef]
104. Li, X.B.; Zhang, Z.R.; Schluesener, H.J.; Xu, S.Q. Role of exosomes in immune regulation. *J. Cell Mol. Med.* **2006**, *10*, 364–375. [CrossRef] [PubMed]
105. Liu, H.; Li, R.; Liu, T.; Yang, L.; Yin, G.; Xie, Q. Immunomodulatory Effects of Mesenchymal Stem Cells and Mesenchymal Stem Cell-Derived Extracellular Vesicles in Rheumatoid Arthritis. *Front. Immunol.* **2020**, *11*, 1912. [CrossRef] [PubMed]
106. Tavasolian, F.; Hosseini, A.Z.; Soudi, S.; Naderi, M. miRNA-146a Improves Immunomodulatory Effects of MSC-derived Exosomes in Rheumatoid Arthritis. *Curr. Gene Ther.* **2020**. [CrossRef]
107. Corrado, C.; Raimondo, S.; Chiesi, A.; Ciccia, F.; De Leo, G.; Alessandro, R. Exosomes as intercellular signaling organelles involved in health and disease: Basic science and clinical applications. *Int. J. Mol. Sci.* **2013**, *14*, 5338–5366. [CrossRef]
108. Boriachek, K.; Islam, M.N.; Moller, A.; Salomon, C.; Nguyen, N.T.; Hossain, M.S.A.; Yamauchi, Y.; Shiddiky, M.J.A. Biological Functions and Current Advances in Isolation and Detection Strategies for Exosome Nanovesicles. *Small* **2018**, *14*. [CrossRef]
109. Li, Q.; Wang, H.; Peng, H.; Huyan, T.; Cacalano, N.A. Exosomes: Versatile Nano Mediators of Immune Regulation. *Cancers* **2019**, *11*, 1557. [CrossRef] [PubMed]
110. Sprent, J. Direct stimulation of naive T cells by antigen-presenting cell vesicles. *Blood Cells Mol. Dis.* **2005**, *35*, 17–20. [CrossRef]
111. Saunderson, S.C.; McLellan, A.D. Role of Lymphocyte Subsets in the Immune Response to Primary B Cell-Derived Exosomes. *J. Immunol.* **2017**, *199*, 2225–2235. [CrossRef]
112. Mittelbrunn, M.; Gutierrez-Vazquez, C.; Villarroya-Beltri, C.; Gonzalez, S.; Sanchez-Cabo, F.; Gonzalez, M.A.; Bernad, A.; Sanchez-Madrid, F. Unidirectional transfer of microRNA-loaded exosomes from T cells to antigen-presenting cells. *Nat. Commun.* **2011**, *2*, 282. [CrossRef]
113. Kugeratski, F.G.; Kalluri, R. Exosomes as mediators of immune regulation and immunotherapy in cancer. *FEBS J.* **2020**. [CrossRef] [PubMed]
114. Qian, X.; An, N.; Ren, Y.; Yang, C.; Zhang, X.; Li, L. Immunosuppressive Effects of Mesenchymal Stem Cells-derived Exosomes. *Stem Cell Rev. Rep.* **2020**. [CrossRef]
115. Cui, X.; Wang, S.; Zhao, N.; Wang, Z.; Huang, M.; Liu, Y.; Qin, J.; Shan, Z.; Teng, W.; Li, Y. Thyrocyte-derived exosome-targeted dendritic cells stimulate strong CD4(+) T lymphocyte responses. *Mol. Cell Endocrinol.* **2020**, *506*, 110756. [CrossRef] [PubMed]
116. Fu, C.; Peng, P.; Loschko, J.; Feng, L.; Pham, P.; Cui, W.; Lee, K.P.; Krug, A.B.; Jiang, A. Plasmacytoid dendritic cells cross-prime naive CD8 T cells by transferring antigen to conventional dendritic cells through exosomes. *Proc. Natl. Acad. Sci. USA* **2020**, *117*, 23730–23741. [CrossRef]
117. Than, U.T.T.; Le, H.T.; Hoang, D.H.; Nguyen, X.H.; Pham, C.T.; Bui, K.T.V.; Bui, H.T.H.; Nguyen, P.V.; Nguyen, T.D.; Do, T.T.H.; et al. Induction of Antitumor Immunity by Exosomes Isolated from Cryopreserved Cord Blood Monocyte-Derived Dendritic Cells. *Int. J. Mol. Sci.* **2020**, *21*. [CrossRef]
118. Wu, H.; Xu, Z.; Wang, Z.; Ren, Z.; Li, L.; Ruan, Y. Exosomes from dendritic cells with Mettl3 gene knockdown prevent immune rejection in a mouse cardiac allograft model. *Immunogenetics* **2020**. [CrossRef]
119. Salem, H.K.; Thiemermann, C. Mesenchymal stromal cells: Current understanding and clinical status. *Stem Cells* **2010**, *28*, 585–596. [CrossRef] [PubMed]
120. Chen, Y.; Li, J.; Ma, B.; Li, N.; Wang, S.; Sun, Z.; Xue, C.; Han, Q.; Wei, J.; Zhao, R.C. MSC-derived exosomes promote recovery from traumatic brain injury via microglia/macrophages in rat. *Aging* **2020**, *12*. [CrossRef] [PubMed]
121. Golchin, A. Cell-Based Therapy for Severe COVID-19 Patients: Clinical Trials and Cost-Utility. *Stem Cell Rev. Rep.* **2020**. [CrossRef]
122. Guy, R.; Offen, D. Promising Opportunities for Treating Neurodegenerative Diseases with Mesenchymal Stem Cell-Derived Exosomes. *Biomolecules* **2020**, *10*. [CrossRef] [PubMed]
123. Zhang, B.; Yin, Y.; Lai, R.C.; Tan, S.S.; Choo, A.B.; Lim, S.K. Mesenchymal stem cells secrete immunologically active exosomes. *Stem Cells Dev.* **2014**, *23*, 1233–1244. [CrossRef] [PubMed]
124. Chen, W.; Huang, Y.; Han, J.; Yu, L.; Li, Y.; Lu, Z.; Li, H.; Liu, Z.; Shi, C.; Duan, F.; et al. Immunomodulatory effects of mesenchymal stromal cells-derived exosome. *Immunol. Res.* **2016**, *64*, 831–840. [CrossRef]
125. Del Fattore, A.; Luciano, R.; Pascucci, L.; Goffredo, B.M.; Giorda, E.; Scapaticci, M.; Fierabracci, A.; Muraca, M. Immunoregulatory Effects of Mesenchymal Stem Cell-Derived Extracellular Vesicles on T Lymphocytes. *Cell Transplant.* **2015**, *24*, 2615–2627. [CrossRef] [PubMed]
126. Duffy, M.M.; Ritter, T.; Ceredig, R.; Griffin, M.D. Mesenchymal stem cell effects on T-cell effector pathways. *Stem Cell Res. Ther.* **2011**, *2*, 34. [CrossRef]
127. Mendt, M.; Rezvani, K.; Shpall, E. Mesenchymal stem cell-derived exosomes for clinical use. *Bone Marrow Transplant.* **2019**, *54*, 789–792. [CrossRef] [PubMed]

128. Kordelas, L.; Rebmann, V.; Ludwig, A.K.; Radtke, S.; Ruesing, J.; Doeppner, T.R.; Epple, M.; Horn, P.A.; Beelen, D.W.; Giebel, B. MSC-derived exosomes: A novel tool to treat therapy-refractory graft-versus-host disease. *Leukemia* **2014**, *28*, 970–973. [CrossRef] [PubMed]

129. Yin, K.; Wang, S.; Zhao, R.C. Exosomes from mesenchymal stem/stromal cells: A new therapeutic paradigm. *Biomark. Res.* **2019**, *7*, 8. [CrossRef] [PubMed]

130. Tiechao, J.; Zhongyu, W.; Ji, S. Human bone marrow mesenchymal stem cell-derived exosomes stimulate cutaneous wound healing mediates through TGF-β/signaling pathway. *Stem Cell Res. Ther.* **2020**, *11*, 1723–1726.

131. Dimitrios, T.; Lorraine, O.; Driss, C. Mesenchymal stem cells derived extracellular vesicles for tissue engineering and regenerative medicine application. *Cells* **2020**, *9*, 3390–4099.

132. Saadeldin, I.M.; Oh, H.J.; Lee, B.C. Embryonic-maternal cross-talk via exosomes: Potential implications. *Stem Cells Clon. Adv. Appl.* **2015**, *8*, 103–107.

133. Siqi, Y.; Cheng, J.; Peipei, W. Human umbilical cord mesenchymal stem cells and exosomes: Bioactive ways of tissue injury repair. *Am. J. Transl. Res.* **2019**, *11*, 1230–1240.

134. Segura, E.; Amigorena, S.; Théry, C. Mature dendritic cells secrete exosomes with strong ability to induce antigen-specific effector immune responses. *Blood Cells Mol. Dis.* **2005**, *35*, 89–93. [CrossRef]

135. Yin, W.; Ouyang, S.; Li, Y.; Xiao, B.; Yang, H. Immature dendritic cell-derived exosomes: A promise subcellular vaccine for autoimmunity. *Inflammation* **2013**, *36*, 232–240. [CrossRef]

136. Fais, S. NK cell-released exosomes: Natural nanobullets against tumors. *Oncoimmunology* **2013**, *2*, e22337. [CrossRef]

137. Jong, A.Y.; Wu, C.H.; Li, J.; Sun, J.; Fabbri, M.; Wayne, A.S.; Seeger, R.C. Large-scale isolation and cytotoxicity of extracellular vesicles derived from activated human natural killer cells. *J. Extracell. Vesicles* **2017**, *6*, 1294368. [CrossRef] [PubMed]

138. Whiteside, T.L. Human regulatory T cells (Treg) and their response to cancer. *Exp. Rev. Precis. Med. Drug Dev.* **2019**, *4*, 215–228. [CrossRef] [PubMed]

139. Okoye, I.S.; Coomes, S.M.; Pelly, V.S.; Czieso, S.; Papayannopoulos, V.; Tolmachova, T.; Seabra, M.C.; Wilson, M.S. MicroRNA-containing T-regulatory-cell-derived exosomes suppress pathogenic T helper 1 cells. *Immunity* **2014**, *41*, 89–103. [CrossRef]

140. Gao, W.; Liu, H.; Yuan, J.; Wu, C.; Huang, D.; Ma, Y.; Zhu, J.; Ma, L.; Guo, J.; Shi, H.; et al. Exosomes derived from mature dendritic cells increase endothelial inflammation and atherosclerosis via membrane TNF-α mediated NF-κB pathway. *J. Cell Mol. Med.* **2016**, *20*, 2318–2327. [CrossRef] [PubMed]

141. Morelli, A.E.; Larregina, A.T.; Shufesky, W.J.; Sullivan, M.L.; Stolz, D.B.; Papworth, G.D.; Zahorchak, A.F.; Logar, A.J.; Wang, Z.; Watkins, S.C.; et al. Endocytosis, intracellular sorting, and processing of exosomes by dendritic cells. *Blood* **2004**, *104*, 3257–3266. [CrossRef]

142. Nolte-'t Hoen, E.N.; Buschow, S.I.; Anderton, S.M.; Stoorvogel, W.; Wauben, M.H. Activated T cells recruit exosomes secreted by dendritic cells via LFA-1. *Blood* **2009**, *113*, 1977–1981. [CrossRef]

143. Quah, B.J.; O'Neill, H.C. The immunogenicity of dendritic cell-derived exosomes. *Blood Cells Mol. Dis.* **2005**, *35*, 94–110. [CrossRef] [PubMed]

144. Yang, X.; Meng, S.; Jiang, H.; Zhu, C.; Wu, W. Exosomes derived from immature bone marrow dendritic cells induce tolerogenicity of intestinal transplantation in rats. *J. Surg. Res.* **2011**, *171*, 826–832. [CrossRef]

145. Federici, C.; Shahaj, E.; Cecchetti, S.; Camerini, S.; Casella, M.; Iessi, E.; Camisaschi, C.; Paolino, G.; Calvieri, S.; Ferro, S.; et al. Natural-Killer-Derived Extracellular Vesicles: Immune Sensors and Interactors. *Front. Immunol.* **2020**, *11*, 262. [CrossRef] [PubMed]

146. Paul, S.; Lal, G. The Molecular Mechanism of Natural Killer Cells Function and Its Importance in Cancer Immunotherapy. *Front. Immunol.* **2017**, *8*, 1124. [CrossRef]

147. Lugini, L.; Cecchetti, S.; Huber, V.; Luciani, F.; Macchia, G.; Spadaro, F.; Paris, L.; Abalsamo, L.; Colone, M.; Molinari, A.; et al. Immune surveillance properties of human NK cell-derived exosomes. *J. Immunol.* **2012**, *189*, 2833–2842. [CrossRef] [PubMed]

148. Zhu, L.; Kalimuthu, S.; Gangadaran, P.; Oh, J.M.; Lee, H.W.; Baek, S.H.; Jeong, S.Y.; Lee, S.W.; Lee, J.; Ahn, B.C. Exosomes Derived from Natural Killer Cells Exert Therapeutic Effect in Melanoma. *Theranostics* **2017**, *7*, 2732–2745. [CrossRef]

149. Li, S.; Li, S.; Wu, S.; Chen, L. Exosomes Modulate the Viral Replication and Host Immune Responses in HBV Infection. *Biomed Res. Int.* **2019**, *2019*, 2103943. [CrossRef]

150. Wang, G.; Hu, W.; Chen, H.; Shou, X.; Ye, T.; Xu, Y. Cocktail Strategy Based on NK Cell-Derived Exosomes and Their Biomimetic Nanoparticles for Dual Tumor Therapy. *Cancers* **2019**, *11*. [CrossRef] [PubMed]

151. Romano, M.; Fanelli, G.; Albany, C.J.; Giganti, G.; Lombardi, G. Past, Present, and Future of Regulatory T Cell Therapy in Transplantation and Autoimmunity. *Front. Immunol.* **2019**, *10*, 43. [CrossRef]

152. Li, P.; Liu, C.; Yu, Z.; Wu, M. New Insights into Regulatory T Cells: Exosome- and Non-Coding RNA-Mediated Regulation of Homeostasis and Resident Treg Cells. *Front. Immunol.* **2016**, *7*, 574. [CrossRef] [PubMed]

153. Azimi, M.; Ghabaee, M.; Moghadasi, A.N.; Noorbakhsh, F.; Izad, M. Immunomodulatory function of Treg-derived exosomes is impaired in patients with relapsing-remitting multiple sclerosis. *Immunol. Res.* **2018**, *66*, 513–520. [CrossRef] [PubMed]

154. Tung, S.L.; Boardman, D.A.; Sen, M.; Letizia, M.; Peng, Q.; Cianci, N.; Dioni, L.; Carlin, L.M.; Lechler, R.; Bollati, V.; et al. Regulatory T cell-derived extracellular vesicles modify dendritic cell function. *Sci. Rep.* **2018**, *8*, 6065. [CrossRef] [PubMed]

155. Savina, A.; Furlan, M.; Vidal, M.; Colombo, M.I. Exosome release is regulated by a calcium-dependent mechanism in K562 cells. *J. Biol. Chem.* **2003**, *278*, 20083–20090. [CrossRef] [PubMed]

156. King, H.W.; Michael, M.Z.; Gleadle, J.M. Hypoxic enhancement of exosome release by breast cancer cells. *BMC Cancer* **2012**, *12*, 421. [CrossRef]
157. Fontenot, J.D.; Rasmussen, J.P.; Gavin, M.A.; Rudensky, A.Y. A function for interleukin 2 in Foxp3-expressing regulatory T cells. *Nat. Immunol.* **2005**, *6*, 1142–1151. [CrossRef] [PubMed]
158. Smyth, L.A.; Ratnasothy, K.; Tsang, J.Y.; Boardman, D.; Warley, A.; Lechler, R.; Lombardi, G. CD73 expression on extracellular vesicles derived from CD4+ CD25+ Foxp3+ T cells contributes to their regulatory function. *Eur. J. Immunol.* **2013**, *43*, 2430–2440. [CrossRef]
159. Lee, H.K.; Finniss, S.; Cazacu, S.; Bucris, E.; Ziv-Av, A.; Xiang, C.; Bobbitt, K.; Rempel, S.A.; Hasselbach, L.; Mikkelsen, T.; et al. Mesenchymal stem cells deliver synthetic microRNA mimics to glioma cells and glioma stem cells and inhibit their cell migration and self-renewal. *Oncotarget* **2013**, *4*, 346–361. [CrossRef]
160. Munoz, J.L.; Bliss, S.A.; Greco, S.J.; Ramkissoon, S.H.; Ligon, K.L.; Rameshwar, P. Delivery of functional anti-miR-9 by mesenchymal stem cell-derived exosomes to glioblastoma multiforme cells conferred chemosensitivity. *Mol. Ther. Ther. - Nucleic Acids* **2013**, *2*, e126. [CrossRef]
161. Katakowski, M.; Buller, B.; Zheng, X.; Lu, Y.; Rogers, T.; Osobamiro, O.; Shu, W.; Jiang, F.; Chopp, M. Exosomes from marrow stromal cells expressing miR-146b inhibit glioma growth. *Cancer Lett.* **2013**, *1*, 201–204. [CrossRef]
162. Xu, H.; Zhao, G.; Zhang, Y.; Jiang, H.; Wang, W.; Zhao, D.; Hong, J.; Yu, H.; Qi, L. Mesenchymal stem cellderived exosomal microRNA-133b suppresses glioma progression via Wnt/β-catenin signaling pathway by targeting EZH2. *Stem Cell Res. Ther.* **2019**, *1*, 381. [CrossRef]
163. Greco, K.A.; Franzen, C.A.; Foreman, K.E.; Flanigan, R.C.; Kuo, P.C.; Gupta, G.N. PLK-1 silencing in bladder cancer by siRNA delivered with exosomes. *Urology* **2016**, *91*, 241.e1–241.e7. [CrossRef]
164. Pascucci, L.; Cocce, V.; Bonomi, A.; Ami, D.; Ceccarelli, P.; Ciusani, E.; Viganò, L.; Locatelli, A.; Sisto, F.; Doglia, S.M.; et al. Paclitaxel is incorporated by mesenchymal stromal cells and released in exosomes that inhibit in vitro tumor growth: A new approach for drug delivery. *J. Control Release* **2014**, *192*, 262–270. [CrossRef] [PubMed]
165. Alvarez-Erviti, L.; Seow, Y.; Yin, H.; Betts, C.; Lakhal, S.; Wood, M.J. Delivery of siRNA to the mouse brain by systemic injection of targeted exosomes. *Nat. Biotechnol.* **2011**, *29*, 341–345. [CrossRef] [PubMed]
166. Wang, Y.; Chen, X.; Tian, B.; Liu, J.; Yang, L.; Zeng, L.; Chen, T.; Hong, A.; Wang, X. Nucleolin targeted extracellular vesicles as a versatile platform for biologics delivery to breast cancer. *Theranostics* **2017**, *7*, 1360–1372. [CrossRef] [PubMed]
167. Tian, Y.; Li, S.; Song, J.; Ji, T.; Zhu, M.; Anderson, G.J.; Wei, J.; Nie, G. A doxorubicin delivery platform using engineered natural membrane vesicle exosomes for targeted tumor therapy. *Biomaterials* **2014**, *35*, 2383–2390. [CrossRef]
168. Ohno, S.; Takanashi, M.; Sudo, K.; Ueda, S.; Ishikawa, A.; Matsuyama, N.; Fujita, K.; Mizutani, T.; Ohgi, T.; Ochiya, T.; et al. Systemically injected exosomes targeted to EGFR deliver antitumor microRNA to breast cancer cells. *Mol. Ther.* **2013**, *21*, 185–191. [CrossRef]
169. Bellavia, D.; Raimondo, S.; Calabrese, G.; Forte, S.; Cristaldi, M.; Patinella, A.; Memeo, L.; Manno, M.; Raccosta, S.; Diana, P.; et al. Interleukin 3- receptor targeted exosomes inhibit in vitro and in vivo Chronic Myelogenous Leukemia cell growth. *Theranostics* **2017**, *7*, 1333–1345. [CrossRef] [PubMed]
170. Sun, D.; Zhuang, X.; Xiang, X.; Liu, Y.; Zhang, S.; Liu, C.; Barnes, S.; Grizzle, W.; Miller, D.; Zhang, H.G. A novel nanoparticle drug delivery system: The anti-inflammatory activity of curcumin is enhanced when encapsulated in exosomes. *Mol. Ther.* **2010**, *18*, 1606–1614. [CrossRef]
171. Maguire, C.A.; Balaj, L.; Sivaraman, S.; Crommentuijn, M.H.; Ericsson, M.; Mincheva-Nilsson, L.; Baranov, V.; Gianni, D.; Tannous, B.A.; Sena-Esteves, M.; et al. Microvesicle-associated AAV vector as a novel gene delivery system. *Mol. Ther.* **2012**, *20*, 960–971. [CrossRef] [PubMed]
172. Pan, Q.; Ramakrishnaiah, V.; Henry, S.; Fouraschen, S.; de Ruiter, P.E.; Kwekkeboom, J.; Tilanus, H.W.; Janssen, H.L.; van der Laan, L.J. Hepatic cell-to-cell transmission of small silencing RNA can extend the therapeutic reach of RNA interference (RNAi). *Gut* **2012**, *61*, 1330–1339. [CrossRef]
173. Kosaka, N.; Iguchi, H.; Yoshioka, Y.; Hagiwara, K.; Takeshita, F.; Ochiya, T. Competitive interactions of cancer cells and normal cells via secretory microRNAs. *J. Biol. Chem.* **2012**, *287*, 1397–1405. [CrossRef]
174. Wahlgren, J.; De, L.K.T.; Brisslert, M.; Vaziri Sani, F.; Telemo, E.; Sunnerhagen, P.; Valadi, H. Plasma exosomes can deliver exogenous short interfering RNA to monocytes and lymphocytes. *Nucleic Acids Res.* **2012**, *40*, e130. [CrossRef] [PubMed]
175. Mizrak, A.; Bolukbasi, M.F.; Ozdener, G.B.; Brenner, G.J.; Madlener, S.; Erkan, E.P.; Strobel, T.; Breakefield, X.O.; Saydam, O. Genetically engineered microvesicles carrying suicide mRNA/protein inhibit schwannoma tumor growth. *Mol. Ther.* **2013**, *21*, 101–108. [CrossRef]
176. Shtam, T.A.; Kovalev, R.A.; Varfolomeeva, E.Y.; Makarov, E.M.; Kil, Y.V.; Filatov, M.V. Exosomes are natural carriers of exogenous siRNA to human cells in vitro. *Cell Commun. Signal.* **2013**, *11*, 88. [CrossRef] [PubMed]
177. Hood, J.L.; Scott, M.J.; Wickline, S.A. Maximizing exosome colloidal stability following electroporation. *Anal. Biochem.* **2014**, *448*, 41–49. [CrossRef]
178. Takahashi, Y.; Nishikawa, M.; Shinotsuka, H.; Matsui, Y.; Ohara, S.; Imai, T.; Takakura, Y. Visualization and in vivo tracking of the exosomes of murine melanoma B16-BL6 cells in mice after intravenous injection. *J. Biotechnol.* **2013**, *165*, 77–84. [CrossRef] [PubMed]

179. Lee, Y.S.; Kim, S.H.; Cho, J.A.; Kim, C.W. Introduction of the CIITA gene into tumor cells produces exosomes with enhanced anti-tumor effects. *Exp. Mol. Med.* **2011**, *43*, 281–290. [CrossRef]
180. Yeo, R.W.; Lai, R.C.; Zhang, B.; Tan, S.S.; Yin, Y.; Teh, B.J.; Lim, S.K. Mesenchymal stem cell: An efficient mass producer of exosomes for drug delivery. *Adv. Drug Deliv. Rev.* **2013**, *65*, 336–341. [CrossRef]
181. Chen, T.S.; Arslan, F.; Yin, Y.; Tan, S.S.; Lai, R.C.; Choo, A.B.; Padmanabhan, J.; Lee, C.N.; de Kleijn, D.P.; Lim, S.K. Enabling a robust scalable manufacturing process for therapeutic exosomes through oncogenic immortalization of human ESC-derived MSCs. *J. Transl. Med.* **2011**, *9*, 47. [CrossRef]
182. Xin, H.; Li, Y.; Buller, B.; Katakowski, M.; Zhang, Y.; Wang, X.; Shang, X.; Zhang, Z.G.; Chopp, M. Exosome-mediated transfer of miR-133b from multipotent mesenchymal stromal cells to neural cells contributes to neurite outgrowth. *Stem Cells* **2012**, *30*, 1556–1564. [CrossRef] [PubMed]
183. Zhu, W.; Huang, L.; Li, Y.; Zhang, X.; Gu, J.; Yan, Y.; Xu, X.; Wang, M.; Qian, H.; Xu, W. Exosomes derived from human bone marrow mesenchymal stem cells promote tumor growth in vivo. *Cancer Lett.* **2012**, *315*, 28–37. [CrossRef] [PubMed]
184. Bruno, S.; Collino, F.; Deregibus, M.C.; Grange, C.; Tetta, C.; Camussi, G. Microvesicles derived from human bone marrow mesenchymal stem cells inhibit tumor growth. *Stem Cells Dev.* **2013**, *22*, 758–771. [CrossRef]
185. Valadi, H.; Ekstrom, K.; Bossios, A.; Sjostrand, M.; Lee, J.J.; Lotvall, J.O. Exosome-mediated transfer of mRNAs and microRNAs is a novel mechanism of genetic exchange between cells. *Nat. Cell Biol.* **2007**, *9*, 654–659. [CrossRef]
186. Neumann, E.; Schaefer-Ridder, M.; Wang, Y.; Hofschneider, P.H. Gene transfer into mouse lyoma cells by electroporation in high electric fields. *EMBO J.* **1982**, *1*, 841–845. [CrossRef]
187. Lin, Q.; Chen, J.; Zhang, Z.; Zheng, G. Lipid-based nanoparticles in the systemic delivery of siRNA. *Nanomedicine* **2014**, *9*, 105–120. [CrossRef]
188. Rao, D.D.; Vorhies, J.S.; Senzer, N.; Nemunaitis, J. siRNA vs. shRNA: Similarities and differences. *Adv. Drug Deliv. Rev.* **2009**, *61*, 746–759. [CrossRef]
189. Ozpolat, B.; Sood, A.K.; Lopez-Berestein, G. Liposomal siRNA nanocarriers for cancer therapy. *Adv. Drug Deliv. Rev.* **2014**, *66*, 110–116. [CrossRef] [PubMed]
190. Tabernero, J.; Shapiro, G.I.; LoRusso, P.M.; Cervantes, A.; Schwartz, G.K.; Weiss, G.J.; Paz-Ares, L.; Cho, D.C.; Infante, J.R.; Alsina, M.; et al. First-in-humans trial of an RNA interference therapeutic targeting VEGF and KSP in cancer patients with liver involvement. *Cancer Discov.* **2013**, *3*, 406–417. [CrossRef] [PubMed]
191. Moller, H.G.; Rasmussen, A.P.; Andersen, H.H.; Johnsen, K.B.; Henriksen, M.; Duroux, M. A systematic review of microRNA in glioblastoma multiforme: Micro-modulators in the mesenchymal mode of migration and invasion. *Mol. Neurobiol.* **2013**, *47*, 131–144. [CrossRef]
192. Henriksen, M.; Johnsen, K.B.; Andersen, H.H.; Pilgaard, L.; Duroux, M. MicroRNA expression signatures determine prognosis and survival in glioblastoma multiforme–a systematic overview. *Mol. Neurobiol.* **2014**, *50*, 896–913. [CrossRef] [PubMed]
193. Zhang, Y.; Liu, D.; Chen, X.; Li, J.; Li, L.; Bian, Z.; Sun, F.; Lu, J.; Yin, Y.; Cai, X.; et al. Secreted monocytic miR-150 enhances targeted endothelial cell migration. *Mol. Cell* **2010**, *39*, 133–144. [CrossRef]
194. Bryniarski, K.; Ptak, W.; Jayakumar, A.; Pullmann, K.; Caplan, M.J.; Chairoungdua, A.; Lu, J.; Adams, B.D.; Sikora, E.; Nazimek, K.; et al. Antigen-specific, antibody-coated, exosome-like nanovesicles deliver suppressor T-cell microRNA-150 to effector T cells to inhibit contact sensitivity. *J. Allergy Clin Immunol.* **2013**, *132*, 170–181. [CrossRef]
195. Chen, L.; Charrier, A.; Zhou, Y.; Chen, R.; Yu, B.; Agarwal, K.; Tsukamoto, H.; Lee, L.J.; Paulaitis, M.E.; Brigstock, D.R. Epigenetic regulation of connective tissue growth factor by MicroRNA-214 delivery in exosomes from mouse or human hepatic stellate cells. *Hepatology* **2014**, *59*, 1118–1129. [CrossRef] [PubMed]
196. Gressner, O.A.; Gressner, A.M. Connective tissue growth factor: A fibrogenic master switch in fibrotic liver diseases. *Liver Int.* **2008**, *28*, 1065–1079. [CrossRef]
197. Askenase, P.W. COVID-19 therapy with mesenchymal stromal cells (MSC) and convalescent plasma must consider exosome involvement: Do the exosomes in convalescent plasma antagonize the weak immune antibodies? *J. Extracell. Vesicles* **2020**, *10*, e12004. [CrossRef] [PubMed]
198. Khalaj, K.; Figueira, R.L.; Antounians, L.; Lauriti, G.; Zani, A. Systematic review of extracellular vesicle-based treatments for lung injury: Are EVs a potential therapy for COVID-19? *J. Extracell. Vesicles* **2020**, *9*, 1795365. [CrossRef]
199. Katakowski, M.; Zheng, X.; Jiang, F.; Rogers, T.; Szalad, A.; Chopp, M. MiR-146b-5p suppresses EGFR expression and reduces in vitro migration and invasion of glioma. *Cancer Investig.* **2010**, *28*, 1024–1030. [CrossRef]
200. Jang, S.C.; Kim, O.Y.; Yoon, C.M.; Choi, D.S.; Roh, T.Y.; Park, J.; Nilsson, J.; Lotvall, J.; Kim, Y.K.; Gho, Y.S. Bioinspired exosome-mimetic nanovesicles for targeted delivery of chemotherapeutics to malignant tumors. *ACS Nano* **2013**, *7*, 7698–7710. [CrossRef]
201. Tacar, O.; Sriamornsak, P.; Dass, C.R. Doxorubicin: An update on anticancer molecular action, toxicity and novel drug delivery systems. *J. Pharm. Pharmacol.* **2013**, *65*, 157–170. [CrossRef]
202. Zhuang, X.; Xiang, X.; Grizzle, W.; Sun, D.; Zhang, S.; Axtell, R.C.; Ju, S.; Mu, J.; Zhang, L.; Steinman, L.; et al. Treatment of brain inflammatory diseases by delivering exosome encapsulated anti-inflammatory drugs from the nasal region to the brain. *Mol. Ther.* **2011**, *19*, 1769–1779. [CrossRef]

203. Blaskovich, M.A.; Sun, J.; Cantor, A.; Turkson, J.; Jove, R.; Sebti, S.M. Discovery of JSI-124 (cucurbitacin I), a selective Janus kinase/signal transducer and activator of transcription 3 signaling pathway inhibitor with potent antitumor activity against human and murine cancer cells in mice. *Cancer Res.* **2003**, *63*, 1270–1279.
204. Hamstra, D.A.; Lee, K.C.; Tychewicz, J.M.; Schepkin, V.D.; Moffat, B.A.; Chen, M.; Dornfeld, K.J.; Lawrence, T.S.; Chenevert, T.L.; Ross, B.D.; et al. The use of 19F spectroscopy and diffusion-weighted MRI to evaluate differences in gene-dependent enzyme prodrug therapies. *Mol. Ther.* **2004**, *10*, 916–928. [CrossRef]
205. Rachakatla, R.S.; Balivada, S.; Seo, G.M.; Myers, C.B.; Wang, H.; Samarakoon, T.N.; Dani, R.; Pyle, M.; Kroh, F.O.; Walker, B.; et al. Attenuation of mouse melanoma by A/C magnetic field after delivery of bi-magnetic nanoparticles by neural progenitor cells. *ACS Nano* **2010**, *4*, 7093–7104. [CrossRef]
206. Jaruga, F.; Sokal, A.; Chrul, S.; Bartosz, G. Apoptosis-independent alterations in membrane dynamics induced by curcumin. *Exp. Cell Res.* **1998**, *245*, 303–312. [CrossRef]
207. Barry, J.; Fritz, M.; Brender, J.R.; Smith, P.E.; Lee, D.K.; Ramamoorthy, A. Determining the effects of lipophilic drugs on membrane structure by solid-state NMR spectroscopy: The case of the antioxidant curcumin. *J. Am. Chem. Soc.* **2009**, *131*, 4490–4498. [CrossRef]
208. Jo, W.; Jeong, D.; Kim, J.; Cho, S.; Jang, S.C.; Han, C.; Kang, J.Y.; Gho, Y.S.; Park, J. Microfluidic fabrication of cell-derived nanovesicles as endogenous RNA carriers. *Lab Chip* **2014**, *14*, 1261–1269. [CrossRef] [PubMed]
209. El Andaloussi, S.; Lakhal, S.; Mager, I.; Wood, M.J. Exosomes for targeted siRNA delivery across biological barriers. *Adv. Drug Deliv. Rev.* **2013**, *65*, 391–397. [CrossRef] [PubMed]
210. Kooijmans, S.A.; Stremersch, S.; Braeckmans, K.; de Smedt, S.C.; Hendrix, A.; Wood, M.J.; Schiffelers, R.M.; Raemdonck, K.; Vader, P. Electroporation-induced siRNA precipitation obscures the efficiency of siRNA loading into extracellular vesicles. *J. Control Release* **2013**, *172*, 229–238. [CrossRef]
211. Villarroya-Beltri, C.; Gutierrez-Vazquez, C.; Sanchez-Cabo, F.; Perez-Hernandez, D.; Vazquez, J.; Martin-Cofreces, N.; Martinez-Herrera, D.J.; Pascual-Montano, A.; Mittelbrunn, M.; Sanchez-Madrid, F. Sumoylated hnRNPA2B1 controls the sorting of miRNAs into exosomes through binding to specific motifs. *Nat. Commun.* **2013**, *4*, 2980. [CrossRef] [PubMed]
212. Xue, M.; Chen, W.; Xiang, A.; Wang, R.; Chen, H.; Pan, J.; Pang, H.; An, H.; Wang, X.; Hou, H.; et al. Hypoxic exosomes facilitate bladder tumor growth and development through transferring long noncoding RNA-UCA1. *Mol. Cancer* **2017**, *16*, 143. [CrossRef]
213. Chen, J.T.; Liu, C.C.; Yu, J.S.; Li, H.H.; Lai, M.C. Integrated omics profiling identifies hypoxia-regulated genes in HCT116 colon cancer cells. *J. Proteom.* **2018**, *188*, 139–151. [CrossRef]
214. Huang, Z.; Yang, M.; Li, Y.; Yang, F.; Feng, Y. Exosomes derived from hypoxic colorectal cancer cells transfer Wnt4 to normoxic cells to elicit a prometastatic phenotype. *Int. J. Biol. Sci.* **2018**, *14*, 2094–2102. [CrossRef] [PubMed]
215. Hsu, Y.; Hung, J.; Chang, W.; Lin, Y.; Pan, Y.; Tsai, P.; Wu, C.; Kuo, P. Hypoxic lung cancer-secreted exosomal miR-23a increased angiogenesis and vascular permeability by targeting prolyl hydroxylase and tight junction protein ZO-1. *Oncogene* **2017**, *36*, 4929–4942. [CrossRef]
216. Tadokoro, H.; Umezu, T.; Ohyashiki, K.; Hirano, T.; Ohyashiki, J.H. Exosomes derived from hypoxic leukemia cells enhance tube formation in endothelial cells. *J. Biol. Chem.* **2013**, *288*, 34343–34351. [CrossRef] [PubMed]
217. Rong, L.; Li, R.; Li, S.; Luo, R. Immunosuppression of breast cancer cells mediated by transforming growth factor-β in exosomes from cancer cells. *Oncol. Lett.* **2016**, *11*, 500–504. [CrossRef]
218. Kucharzewska, P.; Christianson, H.C.; Welch, J.E.; Svensson, K.J.; Fredlund, E.; Ringnér, M.; Mörgelin, M.; Bourseau-Guilmain, E.; Bengzon, J.; Belting, M. Exosomes reflect the hypoxic status of glioma cells and mediate hypoxia-dependent activation of vascular cells during tumor development. *Proc. Natl. Acad. Sci. USA* **2013**, *110*, 7312–7317. [CrossRef] [PubMed]
219. Witwer, K.W.; Buzas, E.I.; Bemis, L.T.; Bora, A.; Lasser, C.; Lotvall, J.; Nolte-'t Hoen, E.N.; Piper, M.G.; Sivaraman, S.; Skog, J.; et al. Standardization of sample collection, isolation and analysis methods in extracellular vesicle research. *J. Extracell. Vesicles* **2013**, *2*. [CrossRef]
220. Sokolova, V.; Ludwig, A.K.; Hornung, S.; Rotan, O.; Horn, P.A.; Epple, M.; Giebel, B. Characterisation of exosomes derived from human cells by nanoparticle tracking analysis and scanning electron microscopy. *Colloids Surf. B Biointerfaces* **2011**, *87*, 146–150. [CrossRef]
221. Kalra, H.; Adda, C.G.; Liem, M.; Ang, C.S.; Mechler, A.; Simpson, R.J.; Hulett, M.D.; Mathivanan, S. Comparative proteomics evaluation of plasma exosome isolation techniques and assessment of the stability of exosomes in normal human blood plasma. *Proteomics* **2013**, *13*, 3354–3364. [CrossRef] [PubMed]
222. Rana, S.; Yue, S.; Stadel, D.; Zoller, M. Toward tailored exosomes: The exosomal tetraspanin web contributes to target cell selection. *Int. J. Biochem. Cell Biol.* **2012**, *44*, 1574–1584. [CrossRef]
223. Bagheri, E.; Abnous, K.; Farzad, S.A.; Taghdisi, S.M.; Ramezani, M.; Alibolandi, M. Targeted doxorubicin-loaded mesenchymal stem cells-derived exosomes as a versatile platform for fighting against colorectal cancer. *Life Sci.* **2020**, *261*, 118369. [CrossRef] [PubMed]
224. Dasgupta, D.; Nakao, Y.; Mauer, A.S.; Thompson, J.M.; Sehrawat, T.S.; Liao, C.Y.; Krishnan, A.; Lucien, F.; Guo, Q.; Liu, M.; et al. IRE1A Stimulates Hepatocyte-Derived Extracellular Vesicles That Promote Inflammation in Mice With Steatohepatitis. *Gastroenterology* **2020**. [CrossRef] [PubMed]
225. Fan, C.; Joshi, J.; Li, F.; Xu, B.; Khan, M.; Yang, J.; Zhu, W. Nanoparticle-Mediated Drug Delivery for Treatment of Ischemic Heart Disease. *Front. Bioeng. Biotechnol.* **2020**, *8*, 687. [CrossRef]

226. Zhang, K.; Shao, C.X.; Zhu, J.D.; Lv, X.L.; Tu, C.Y.; Jiang, C.; Shang, M.J. Exosomes function as nanoparticles to transfer miR-199a-3p to reverse chemoresistance to cisplatin in hepatocellular carcinoma. *Biosci. Rep.* **2020**, *40*. [CrossRef] [PubMed]
227. El-Andaloussi, S.; Lee, Y.; Lakhal-Littleton, S.; Li, J.; Seow, Y.; Gardiner, C.; Alvarez-Erviti, L.; Sargent, I.L.; Wood, M.J. Exosome-mediated delivery of siRNA in vitro and in vivo. *Nat. Protoc.* **2012**, *7*, 2112–2126. [CrossRef] [PubMed]
228. Maeda, H.; Wu, J.; Sawa, T.; Matsumura, Y.; Hori, K. Tumor vascular permeability and the EPR effect in macromolecular therapeutics: A review. *J. Control Release* **2000**, *65*, 271–284. [CrossRef]
229. Kaasgaard, T.; Andresen, T.L. Liposomal cancer therapy: Exploiting tumor characteristics. *Expert Opin. Drug Deliv.* **2010**, *7*, 225–243. [CrossRef]
230. Chonn, A.; Semple, S.C.; Cullis, P.R. Association of blood proteins with large unilamellar liposomes in vivo. Relation to circulation lifetimes. *J. Biol. Chem.* **1992**, *267*, 18759–18765. [CrossRef]
231. Takino, T.; Nakajima, C.; Takakura, Y.; Sezaki, H.; Hashida, M. Controlled biodistribution of highly lipophilic drugs with various parenteral formulations. *J. Drug Target.* **1993**, *1*, 117–124. [CrossRef]
232. Natarajan, A.; Wagner, B.; Sibilia, M. The EGF receptor is required for efficient liver regeneration. *Proc. Natl. Acad. Sci. USA* **2007**, *104*, 17081–17086. [CrossRef] [PubMed]
233. Peinado, H.; Aleckovic, M.; Lavotshkin, S.; Matei, I.; Costa-Silva, B.; Moreno-Bueno, G.; Hergueta-Redondo, M.; Williams, C.; Garcia-Santos, G.; Ghajar, C.; et al. Melanoma exosomes educate bone marrow progenitor cells toward a pro-metastatic phenotype through MET. *Nat. Med.* **2012**, *18*, 883–891. [CrossRef] [PubMed]
234. Mortensen, J.H.; Jeppesen, M.; Pilgaard, L.; Agger, R.; Duroux, M.; Zachar, V.; Moos, T. Targeted antiepidermal growth factor receptor (cetuximab) immunoliposomes enhance cellular uptake in vitro and exhibit increased accumulation in an intracranial model of glioblastoma multiforme. *J. Drug Deliv.* **2013**, *2013*, 209205. [CrossRef]
235. Jella, K.K.; Nasti, T.H.; Li, Z.; Lawson, D.H.; Switchenko, J.M.; Ahmed, R.; Dynan, W.S.; Khan, M.K. Exosome-Containing Preparations From Postirradiated Mouse Melanoma Cells Delay Melanoma Growth In Vivo by a Natural Killer Cell-Dependent Mechanism. *Int. J. Radiat. Oncol. Biol. Phys.* **2020**, *108*, 104–114. [CrossRef] [PubMed]
236. Wang, Y.; Goliwas, K.F.; Severino, P.E.; Hough, K.P.; Van Vessem, D.; Wang, H.; Tousif, S.; Koomullil, R.P.; Frost, A.R.; Ponnazhagan, S.; et al. Mechanical strain induces phenotypic changes in breast cancer cells and promotes immunosuppression in the tumor microenvironment. *Lab. Investig.* **2020**. [CrossRef]
237. Hong, B.; Wiese, B.; Bremer, M.; Heissler, H.E.; Heidenreich, F.; Krauss, J.K.; Nakamura, M. Multiple microsurgical resections for repeated recurrence of glioblastoma multiforme. *Am. J. Clin. Oncol.* **2013**, *36*, 261–268. [CrossRef]
238. Ju, S.; Mu, J.; Dokland, T.; Zhuang, X.; Wang, Q.; Jiang, H.; Xiang, X.; Deng, Z.B.; Wang, B.; Zhang, L.; et al. Grape exosome-like nanoparticles induce intestinal stem cells and protect mice from DSS-induced colitis. *Mol. Ther.* **2013**, *21*, 1345–1357. [CrossRef]
239. Grapp, M.; Wrede, A.; Schweizer, M.; Huwel, S.; Galla, H.J.; Snaidero, N.; Simons, M.; Buckers, J.; Low, P.S.; Urlaub, H.; et al. Choroid plexus transcytosis and exosome shuttling deliver folate into brain parenchyma. *Nat. Commun.* **2013**, *4*, 2123. [CrossRef]
240. Chinnappan, M.; Srivastava, A.; Amreddy, N.; Razaq, M.; Pareek, V.; Ahmed, R.; Mehta, M.; Peterson, J.E.; Munshi, A.; Ramesh, R. Exosomes as drug delivery vehicle and contributor of resistance to anticancer drugs. *Cancer Lett.* **2020**, *486*, 18–28. [CrossRef]
241. Giassafaki, L.N.; Siqueira, S.; Panteris, E.; Psatha, K.; Chatzopoulou, F.; Aivaliotis, M.; Tzimagiorgis, G.; Müllertz, A.; Fatouros, D.G.; Vizirianakis, I.S. Towards analyzing the potential of exosomes to deliver microRNA therapeutics. *J. Cell Physiol.* **2020**. [CrossRef] [PubMed]
242. Hassanpour, M.; Rezaie, J.; Nouri, M.; Panahi, Y. The role of extracellular vesicles in COVID-19 virus infection. *Infect. Genet. Evol.* **2020**, *85*, 104422. [CrossRef] [PubMed]
243. Wani, T.U.; Mohi-Ud-Din, R.; Mir, R.H.; Itoo, A.M.; Mir, K.B.; Fazli, A.A.; Pottoo, F.H. Exosomes Harnessed as Nanocarriers for Cancer Therapy-Current Status and Potential for Future Clinical Applications. *Curr. Mol. Med.* **2020**. [CrossRef] [PubMed]
244. Yan, F.; Zhong, Z.; Wang, Y.; Feng, Y.; Mei, Z.; Li, H.; Chen, X.; Cai, L.; Li, C. Exosome-based biomimetic nanoparticles targeted to inflamed joints for enhanced treatment of rheumatoid arthritis. *J. Nanobiotechnol.* **2020**, *18*, 115. [CrossRef] [PubMed]
245. Salimi, F.; Forouzandeh Moghadam, M.; Rajabibazl, M. Development of a novel anti-HER2 scFv by ribosome display and in silico evaluation of its 3D structure and interaction with HER2, alone and after fusion to LAMP2B. *Mol. Biol. Rep.* **2018**, *45*, 2247–2256. [CrossRef]
246. Salunkhe, S.; Dheeraj; Basak, M.; Chitkara, D.; Mittal, A. Surface functionalization of exosomes for target-specific delivery and in vivo imaging & tracking: Strategies and significance. *J. Control Release* **2020**, *326*, 599–614. [CrossRef]
247. Nedergaard, M.K.; Hedegaard, C.J.; Poulsen, H.S. Targeting the epidermal growth factor receptor in solid tumor malignancies. *BioDrugs* **2012**, *26*, 83–99. [CrossRef]
248. Chen, Y.; Liu, L. Modern methods for delivery of drugs across the blood-brain barrier. *Adv. Drug Deliv. Rev.* **2012**, *64*, 640–665. [CrossRef] [PubMed]
249. Zhou, B.; Xu, K.; Zheng, X.; Chen, T.; Wang, J.; Song, Y.; Shao, Y.; Zheng, S. Application of exosomes as liquid biopsy in clinical diagnosis. *Signal Transduct. Target. Ther.* **2020**, *5*, 144. [CrossRef]
250. Escudier, B.; Dorval, T.; Chaput, N.; André, F.; Caby, M.-P.; Novault, S.; Flament, C.; Leboulaire, C.; Borg, C.; Amigorena, S.; et al. Vaccination of metastatic melanoma patients with autologous dendritic cell (DC) derived-exosomes: Results of the first phase I clinical trial. *J. Transl. Med.* **2005**, *3*, 10. [CrossRef] [PubMed]

251. Dai, S.; Wan, T.; Wang, B.; Zhou, X.; Xiu, F.; Chen, T.; Wu, Y.; Cao, X. More efficient induction of HLA-A*0201-restricted and carcinoembryonic antigen (CEA)-specific CTL response by immunization with exosomes prepared from heat-stressed CEA-positive tumor cells. *Clin. Cancer Res.* **2005**, *11*, 7554–7563. [CrossRef]
252. Agrawal, A.K.; Aqil, F.; Jeyabalan, J.; Spencer, W.A.; Beck, J.; Gachuki, B.W.; Alhakeem, S.S.; Oben, K.; Munagala, R.; Bondada, S.; et al. Milk-derived exosomes for oral delivery of paclitaxel. *Nanomed. Nanotechnol. Biol. Med.* **2017**, *13*, 1627–1636. [CrossRef] [PubMed]

Review

Extracellular Vesicles as Mediators of Cancer Disease and as Nanosystems in Theranostic Applications

Renato Burgos-Ravanal [1,2], América Campos [1,2,3], Magda C. Díaz-Vesga [1,2,4], María Fernanda González [1,2], Daniela León [2,5], Lorena Lobos-González [6], Lisette Leyton [1,2], Marcelo J. Kogan [2,5,*] and Andrew F. G. Quest [1,2,*]

1 Laboratorio de Comunicaciones Celulares, Centro de Estudios en Ejercicio, Metabolismo y Cáncer (CEMC), Programa de Biología Celular y Molecular, Facultad de Medicina, Universidad de Chile, Santiago 8380453, Chile; raburgos@uc.cl (R.B.-R.); ame.campos@ug.uchile.cl (A.C.); magdiaz@ug.uchile.cl (M.C.D.-V.); mariagonzalez1@ug.uchile.cl (M.F.G.); lleyton@med.uchile.cl (L.L.)
2 Centro Avanzado para Estudios en Enfermedades Crónicas (ACCDIS), Santiago 8380453, Chile; daniela.leon92@ucuenca.ec
3 Exosome Biology Laboratory, Centre for Clinical Diagnostics, UQ Centre for Clinical Research, Royal Brisbane and Women's Hospital, The University of Queensland, Brisbane 4029, Australia
4 Grupo de Investigación en Ciencias Básicas y Clínicas de la Salud, Pontificia Universidad Javeriana de Cali, Cali 760008, Colombia
5 Departamento de Química Farmacológica y Toxicológica, Facultad de Ciencias Químicas y Farmacéuticas, Universidad de Chile, Santos Dumont 964, Independencia, Santiago 8380494, Chile
6 Centro de Medicina Regenerativa, Facultad de Medicina, Universidad del Desarrollo-Clínica Alemana, Santiago 7590943, Chile; llobos@udd.cl
* Correspondence: mkogan@ciq.uchile.cl (M.J.K.); aquest@med.uchile.cl (A.F.G.Q.)

Citation: Burgos-Ravanal, R.; Campos, A.; Díaz-Vesga, M.C.; González, M.F.; León, D.; Lobos-González, L.; Leyton, L.; Kogan, M.J.; Quest, A.F.G. Extracellular Vesicles as Mediators of Cancer Disease and as Nanosystems in Theranostic Applications. *Cancers* **2021**, *13*, 3324. https://doi.org/10.3390/cancers13133324

Academic Editors: Farrukh Aqil and Ramesh Chandra Gupta

Received: 11 May 2021
Accepted: 20 June 2021
Published: 2 July 2021

Publisher's Note: MDPI stays neutral with regard to jurisdictional claims in published maps and institutional affiliations.

Simple Summary: Cancer is the second leading cause of death in humans, and in 2020, 9.8 million cancer-related deaths were reported worldwide. In the last 20 years, it has become apparent that small vesicles released by cancer cells, referred to as extracellular vesicles (EVs), are key players in cell–cell communication in the tumor environment, and as a consequence, research in this area has increased dramatically. This review summarizes the recent advances in our understanding of how EVs serve as mediators of communication between cancer cells and with their surroundings in order to promote the acquisition of specific characteristics that permit their aberrant behavior. In addition, we dwell on how EVs aid in the development of drug resistance, which is a frequent cause of treatment failure in chemotherapy. Finally, we discuss an exciting new area of research that envisions harnessing the unique characteristics of EVs for therapeutic and diagnostic purposes (theranostics). Taken together, the available literature suggests that advances in our understanding of EV biology in the next decades will likely be critical to achieving more effective treatments in cancer patients.

Abstract: Cancer remains a leading cause of death worldwide despite decades of intense efforts to understand the molecular underpinnings of the disease. To date, much of the focus in research has been on the cancer cells themselves and how they acquire specific traits during disease development and progression. However, these cells are known to secrete large numbers of extracellular vesicles (EVs), which are now becoming recognized as key players in cancer. EVs contain a large number of different molecules, including but not limited to proteins, mRNAs, and miRNAs, and they are actively secreted by many different cell types. In the last two decades, a considerable body of evidence has become available indicating that EVs play a very active role in cell communication. Cancer cells are heterogeneous, and recent evidence reveals that cancer cell-derived EV cargos can change the behavior of target cells. For instance, more aggressive cancer cells can transfer their "traits" to less aggressive cancer cells and convert them into more malignant tumor cells or, alternatively, eliminate those cells in a process referred to as "cell competition". This review discusses how EVs participate in the multistep acquisition of specific traits developed by tumor cells, which are referred to as "the hallmarks of cancer" defined by Hanahan and Weinberg. Moreover, as will be discussed, EVs play an important role in drug resistance, and these more recent advances may explain, at least in part, why pharmacological therapies are often ineffective. Finally, we discuss literature proposing the use of EVs for therapeutic and prognostic purposes in cancer.

Keywords: extracellular vesicles; hallmarks of cancer; drug resistance; theranostics

1. Introduction

Extracellular vesicles (EVs) were initially identified in the 1950s as a type of particle derived from platelets present in plasma [1]. Approximately 20 years later, these particles were still merely considered as "platelet dust" or an insignificant platelet by-product [2]. It took several years before the role of EVs was revealed to be very much the opposite of meaningless cell debris, as their fundamental role in regulating homeostasis at the local and systemic level became apparent [3,4]. EVs are generally described as a heterogeneous population of membrane-enclosed, non-replicating, and sub-micron sized structures, which are actively secreted by a wide variety of eukaryotic and prokaryotic organisms [5,6]. Moreover, EVs can be found in biological fluids, such as serum, plasma, urine, saliva, and breast milk, amongst others [7–10]. In general terms, EVs can be separated into three subtypes according to their biogenesis and biophysical properties [11], namely exosomes, microvesicles, and other small membrane-limited fragments, such as apoptotic bodies, which are generally thought to be less relevant to cell-to-cell communication [12,13].

Indeed, EVs can also induce important changes in recipient cells [4,14,15]. Specifically in cancer, EVs secreted by tumor cells promote the development of tumor-related features in recipient cells and the acquisition of the cancer hallmarks described in the literature [16]. Furthermore, several studies have documented that cancer cells secrete increased levels of EVs when compared to normal cells [17,18]. Considering the aforementioned data and the fact that EVs play an important role in cancer progression, EVs can also be envisioned as appealing targets for developing non-invasive liquid biopsy strategies in patients with cancer. These micron-sized particles can be readily isolated from biofluids as mentioned, and they can be used to facilitate cancer diagnosis and surveillance. Moreover, they can serve to evaluate treatment efficacy, as well as identify patients prone to cancer relapse and/or resistance to therapy [19,20]. Interestingly, EVs have ultimately been described to display considerable potential as novel transport vehicles, which may be employed to deliver molecules or chemotherapeutic drugs in a targeted manner to tumors. In doing so, toxicity or adverse effects can be reduced in comparison to conventional treatment approaches [21,22]. Thus, this review will discuss literature relating to the role of EVs in promoting acquisition of the hallmarks of cancer and also the use of these vesicles in cancer therapy.

It should be mentioned that one of the many difficulties associated with the EV research field in recent years has been the considerable confusion that exists with respect to their nomenclature. This can be attributed largely to the lack of a consensus between the type of isolation used to purify EVs and the techniques used to distinguish between EV subtypes according to their biogenesis or release. To tackle this problem, several EV researchers decided to combine their knowledge to unify the currently used nomenclature [15,23]. This effort gave rise to the development of guidelines, which permit distinguishing between EVs according to their size, density, molecular cargo, or information regarding the cell of origin. In addition, these guidelines also determined that the terms "exosomes" or "microvesicles" should only be used, for example, when imaging techniques were used to confirm a specific biogenesis pathway [15,23]. Thus, in this review, we will refer to the terms "exosomes", "microvesicles", or "apoptotic bodies" only when data regarding their biogenesis is presented and confirmed. Alternatively, when such data are unclear or lacking, the term "EVs" will be used instead. In doing so, this review focuses the discussion predominantly, but not exclusively, on the effects of exosomes.

2. Extracellular Vesicles

Extracellular vesicles (EVs) are mainly featured as a heterogeneous population of membrane-enclosed, non-replicating, and sub-micron sized structures, which are actively secreted by a wide variety of eukaryotic and prokaryotic organisms [5,6]. In addition, EVs are mediators of communication between cells in physiological and pathological settings, and they transport a diverse array of biomolecules, including lipids, nucleic acids, carbohydrates and proteins [6,24]. Finally, EVs can be sorted into three different subtypes according to their biogenesis and biophysical properties (Figure 1) [11].

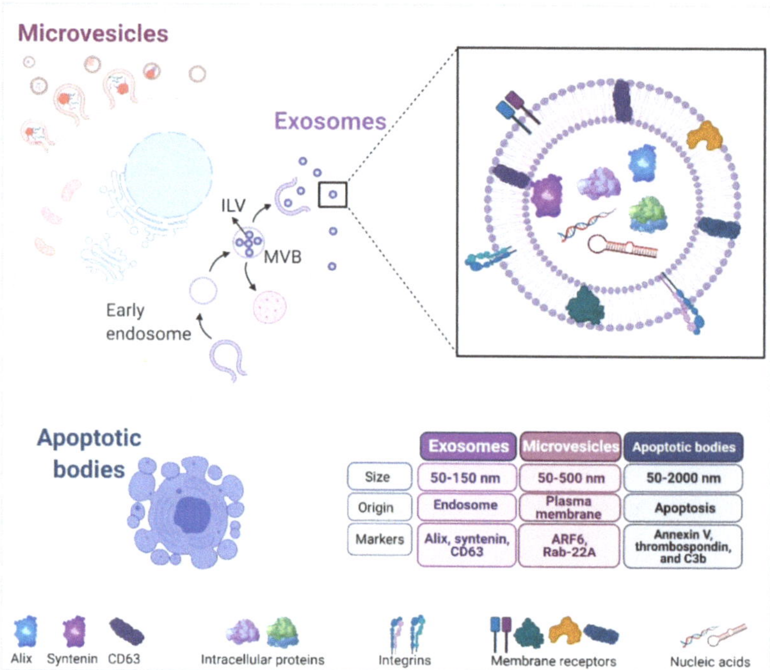

Figure 1. Extracellular vesicles are a heterogeneous population of cell-derived membrane vesicles. Extracellular vesicles (EVs) have classically been divided into three types according to their biogenesis and biophysical properties: exosomes, microvesicles, and apoptotic bodies. Recently, a new group of non-membranous nanoparticles of less than 50 nm, called exomeres, was identified. However, still, little is known about their biogenesis, and proteins that have been connected to exomeres must be characterized further in order to validate them as markers. For this reason, they are not included here. EVs are carriers of a variety of molecules, including proteins, nucleic acids, and lipids. The insert with a close-up view of exosomes shows some molecules commonly transported by them.

2.1. Exosome Biogenesis

Exosomes are currently considered the most studied subtype of nano-sized vesicles smaller than 150 nm, which originate by the inward budding of endosomes or multivesicular bodies (MVBs) toward the luminal space, which results in the formation of intraluminal vesicles (ILVs) that are also known as exosome precursors [11,25]. As a next step, these ILV-containing MVBs can either be redirected to degradation in the lysosome or fuse with the plasma membrane (PM), thus leading to the release of exosomes into the extracellular space [6,11]. Interestingly, exosome biogenesis and cargo sorting are closely related processes. In this regard, there are two well-known mechanisms of ILV formation that may depend or not on the presence of a particular set of Endosomal Sorting Complexes Required for Transport (ESCRT), namely complexes 0, I, II, and III [6,11]. The first mechanism involving ILV formation requires the presence of ESCRT-0, which has been described to select ubiquitinated proteins and segregate them into microdomains found on the endosomal membrane. In addition, ESCRT-I and II are held responsible for the binding of specific cargoes to the aforementioned microdomains. Subsequently, these complexes recruit the Alix protein, which aids in recruiting the ESCRT-III complex containing proteins involved in the last stages of ILV formation or vesicle budding and complex detachment from the endosomal membrane [11]. The second mechanism, also considered as being independent of ESCRT, requires the presence of Alix and transmembrane proteins, such as syntenin and syndecan, which are responsible for recruiting specific molecular cargoes (adhesion molecules, growth factors, integrins, etc.) along with the tetraspanin CD63, which eventually leads to ILV formation [11]. Recent evidence points towards the existence of a third mechanism of ILV biogenesis, which does not depend on components of the ESCRT complexes but rather involves the participation of membrane lipid microdomains or lipid rafts. The specific characteristics of the lipids involved favor inward bending of the MVB membrane and thereby promote ILV formation [11]. One of the main proteins involved in this lipid-dependent mechanism is the neutral sphingomyelinase, which is responsible for generating ceramide, a conical lipid with a small head group that favors bending toward the lumen of the MVB membrane [26].

2.2. EV Release from the Cell Surface

MVBs can either fuse with the plasma membrane for release of their content or with lysosomes for their subsequent destruction [6,27]. Several reports are available indicating that the final destination depends on factors such as the interaction with microtubules or the actin cytoskeleton, as well as the engagement of specific members of the Rab GTPase family of proteins [27]. Examples in the latter case include Rab27b, Rab11, and Rab35, which promote MVB motility and fusion with the plasma membrane in HeLa, K562 (bone marrow chronic myelogenous leukaemia cells), and Oli-neu (oligodendroglial) cells, respectively [28–30].

The second subtype of vesicles ranging in size 50–500 nm (up to 1000 nm), also known as microvesicles (MVs), ectosomes, oncosomes, or microparticles, are described to be released from the cell surface by blebbing from the plasma membrane and subsequent membrane fission [6]. Interestingly, MVs are formed by phospholipid redistribution, positioning phosphatidylserines to the outer leaflet followed by actin–myosin contraction [31,32]. In addition, MV biogenesis requires the participation of small GTPases, such as ADP-ribosylation factor 6 (ARF6) [33,34] and Ras-related proteins, e.g., Rab-22A [34,35]. Importantly, ESCRT complexes also participate in MV formation [32], increasing the level of complexity in EV subtype studies when evaluating vesicle biogenesis. In addition, MV release has been shown to involve Rho family members, such as RhoA, which promotes MV release via ROCK and ERK activation [27,36]. Moreover, RhoA, together with ARF6 and ARF1, increases myosin contractility, thereby favoring MV fission and the subsequent pinching-off from the plasma membrane [6,27,33,37].

Apoptotic bodies, referred to as the third subtype of EVs in the literature, vary widely in size ranging from 50 to 2000 nm in diameter and are ultimately produced by an essential physiological process, which is known as programmed cell death or apoptosis. One of the main features of apoptotic bodies is that mechanisms for specific sorting of organelles, RNA and DNA fragments can be detected, which are absent in other EV subtypes [32,38].

2.3. Exomeres

The discovery of exomeres was made possible by the development of new technologies to isolate and visualize EVs. In this regard, two studies report on the efficient isolation of exomeres by optimizing asymmetric-flow field-flow fractionation and ultracentrifugation protocols [39,40]. Exomeres are approximately 50 nm and smaller in size than EVs. In addition, they were shown to be highly enriched in calreticulin, argonaute proteins, amyloid precursor proteins, proteins associated with coagulation (for instance, factors

VIII and X), and enzymes involved in metabolism (e.g., glycolysis), especially glycolysis, and mammalian target of rapamycin complex 1 (mTORC1) metabolic pathways [40,41]. Moreover, several recent reports have shown that exomeres can carry nucleic acids, such as DNA, RNA, and miRNAs along with lipids, such as ceramide, esterified cholesterol triglycerides, and phosphatidylcholine [42]. Interestingly, exomeres are not limited by a lipid bilayer, but instead are enriched in certain types of lipids, which differ from those found in exosomes [41]. Although limited information is available concerning their biogenesis, the absence of a lipid bilayer suggests that exomeres cannot be classified as EVs but rather should be viewed as a new type of extracellular particle (EP). In addition, the absence of ESCRT components in these EPs suggests they are different from EVs derived from the plasma membrane or generated via the endocytic pathway [42]. Despite such differences, a novel role for exomeres has been proposed in cancer, since they were shown to promote tumor organoid growth in recipient cells [40].

Interestingly, a novel role for exomeres in the COVID-19 pandemic was suggested, as full-length angiotensin-converting enzyme 2 (ACE2) was reported to be contained in EVs from colorectal cancer cells. Specifically, these cells were able to shed ectodomain fragments of ACE2 that were enriched in exomeres [43]. Given that soluble human recombinant ACE2 can bind to SARS-CoV-2 [44], the binding of SARS-CoV-2 S protein to ACE2 fragments in EVs and exomeres may play an important role in controlling the infection [43]. A relevant question at this point is whether the ability to shed ACE2 fragments is limited to cancer cells and if so, thinking of treatments for SARS-CoV-2 infection, why this might be the case.

2.4. EVs in Cell Communication

EVs have emerged as essential players in cell-to-cell communication, because they represent a complex type of "biological package" capable of transporting a wide variety of molecules from one cell to another.

EVs can elicit cellular responses without the need to be internalized into a cell by two mechanisms referred to as soluble and juxtacrine signaling. Soluble signaling involves the proteolysis of an EV surface ligand and its subsequent binding to a cell membrane receptor, whereas juxtacrine signaling requires the juxtapositioning of ligands and receptors on opposing surfaces of the EVs and the target cell [45].

On the other hand, EV internalization by recipient cells involves at least four mechanisms: membrane fusion, phagocytosis, micropinocytosis, and endocytosis. For membrane fusion, the EV membrane directly merges with the cell plasma membrane and transfers cargo molecules to recipient cells. Protein members of the Rab family, Sec1/Munc-18-related proteins (SM proteins), Lamp-1, and SNAREs contribute to this process. Uptake by phagocytosis inevitably results in the fusion of the phagosome with the lysosomes and the degradation of EV content. Phagocytosis likely represents a process important for EV clearance by the immune system, given that the presence of phosphatidylserine (PS) on the outer EV surface promotes their uptake. Indeed, PS appears to represent an essential component of EVs for triggering their clearance by phagocytosis. Macropinocytosis is characterized by plasma membrane ruffling induced by growth factors or other signals. The resulting vesicles contain extracellular fluid and small particles. Macropinocytosis is induced by signaling cascades involving Rho family GTPases, which facilitate actin-driven membrane protrusion formation. The mechanism of EV macropinocytosis is dependent on Na+/H+ exchanger function, actin, Rac1 GTPase activity, cholesterol, dynamin, and low pH. Endocytosis is divided into two types of receptor-mediated processes: clathrin-mediated endocytosis (CME) and caveolin-dependent endocytosis (CDE). CME is produced by the interaction between ligands on the EV surface and specific receptors present on the plasma membrane that utilize clathrin and adaptor protein 2 (AP2) complexes for the subsequent formation of clathrin-coated vesicles (intracellular) to internalize EVs. The clathrin coat alters the structure of the plasma membrane to promote invagination and vesicle fission. Once inside the cell, the clathrin coat of the vesicles is removed to permit fusion with the endosome and transfer of cargo molecules. CDE requires the presence of caveolins, which associate with cholesterol-rich microdomains in the plasma membrane and form small flask-shaped invaginations together with cavins. Hence, CDE is sensitive to cholesterol depletion. In addition, dynamin 2 is a common regulator of endocytosis that has been implicated in CME and CDE [46].

In summary, soluble signaling, juxtacrine signaling, and membrane fusion are more likely to culminate in a cellular response, since EV components do not enter the endosomal-lysosomal degradation pathway directly.

EVs can modify the behavior of recipient cells depending on the biological message or cargo that is being transferred from the donor cell or tissue [47]. Specifically in cancer, EVs have been shown to play a critical role in cell-to-cell communication in the tumor microenvironment that permits the acquisition and maintenance of cell traits, which are referred to as the hallmarks of cancer.

2.5. Regulation of EV Release in Cancer

The number of EVs circulating in the blood of patients with different diseases is elevated compared to healthy subjects. For example, patients with breast, ovarian, gastric, prostate, liver, colon, and pancreas cancers have higher levels of exosomes in plasma than healthy donors [48]. Moreover, in gastric cancer, elevated levels of EVs were associated with more advanced stages of disease development [49,50]. In addition, patients with hematological malignancies have higher EV levels compared to healthy controls. Interestingly, among the latter patients, those with Hodgkin lymphomas, multiple myeloma, and primary myelofibrosis had a higher proportion of smaller EVs in blood samples [51], suggesting that vesicle size relates to function. Using scanning electron microscopy, normal human ovarian cells were found to release EVs from a few select areas of the plasma membrane, while ovarian serous adenocarcinoma cells release EVs from the entire cell surface [52]. Elevated EV release in cancer cells has been proposed to occur via a Ca2+-Munc13-4-Rab11-dependent pathway. Specifically, the expression of Munc13-4, a Ca2+-dependent Rab-binding protein, is elevated in cancer cells, which combined with the increased Ca2+ levels enhances exosome release from cancer cells [53]. Comparison of the breast cancer cells MCF-7 and MCF-7 LTED (Long-Term Estrogen Deprived, a cell line model for the resistance to aromatase inhibitors) revealed a significant increase in exosome secretion from the MCF-7 LTED cells. This was accompanied by an increase in Rab GTPase expression, which could represent another mechanism that permits increased exosome release from more malignant cells [54]. Finally, EV release from cancer cells can be increased by microenvironmental factors, such as hypoxia, increased glycolysis, an acidic microenvironment, calcium signaling, and irradiation [55].

3. EV-Mediated Function in Cancer

Extracellular vesicles have many physiological and pathophysiological functions. In cancer, EVs play an important role in many, if not all, stages of cancer development, including tumorigenesis, epithelial–mesenchymal transition, metastasis, and drug resistance. The available evidence also indicates that EVs play a role in many types of cancer, including gastric cancer, breast cancer, melanoma, and lung cancer, among many others. Moreover, EVs are involved in the acquisition of all the "hallmarks of cancer" (see Figure 2). Initially described by Hanahan and Weinberg (2000) [56] and updated in 2011 [16], these traits refer to several biological characteristics that are acquired by cancer cells during the multistep process leading to tumor development. In the following section, we will summarize evidence available from in vitro and in vivo studies indicating how EVs participate in these events (see Figure 2).

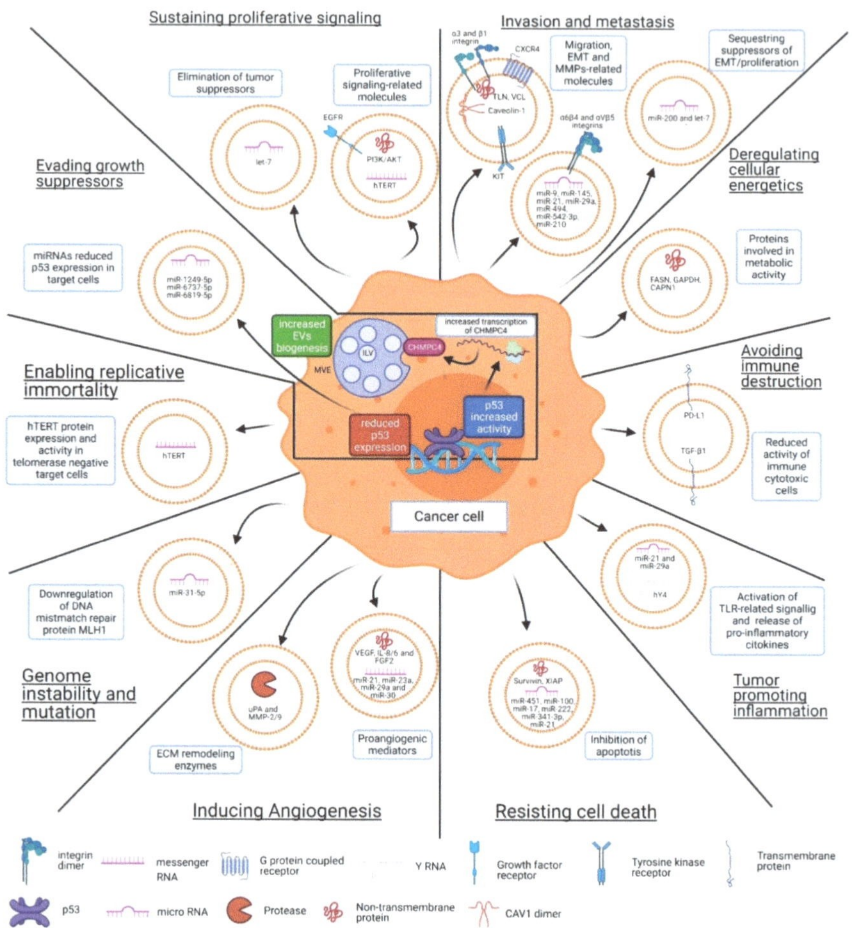

Figure 2. EV-mediated function in cancer. The term "hallmarks of cancer" described by Hanahan and Weinberg [56] refers to ten biological characteristics that are acquired by cancer cells during the multistep process leading to tumor development. The EV-mediated roles reported to date are shown here for each "hallmark of cancer": Sustaining proliferative signaling [57–63], Evading growth suppressors [64], Resisting cell death [65,66], Enabling replicative immortality [57,67], Inducing angiogenesis [58,68–72], Invasion and metastasis [14,73–79], Genome instability and mutation [80], Tumor promoting inflammation [81,82], Deregulating cellular energetics [83,84] and Avoiding immune destruction [85–87].

3.1. Sustaining Proliferative Signaling

Cancer cells acquire the ability to proliferate continuously and do so by generating their own signals, thus rendering themselves independent of external input. They may achieve this through a variety of strategies that do not necessarily involve EVs and have been reviewed elsewhere. EVs can promote cell proliferation in an autocrine manner in many types of cancer, including glioblastoma, breast adenocarcinoma, colorectal, and triple negative breast cancer [58,61,88,89]. Specifically, in some cancers, such as bladder, gastric, and non-small cell lung cancer, it has been shown that this increase in proliferation is through the activation of signaling pathways involving phosphatidylinositol 3-kinase (PI3K)/protein kinase B (AKT) or AMP-activated protein kinase (AMPK)/extracellular signal-regulated kinases (ERK) [62,90–92]. EVs transfer growth factor receptors, such as the epidermal growth factor receptor (EGFR), which promote receptor-dependent cell signaling [69]. In fact, the EGFR is widely present in EVs from various cancer cell lines [63]. Furthermore, a highly oncogenic isoform of the EGFR (EGFRvIII) is transferred through EVs in glioblastoma, which is a very aggressive cancer disease [58]. Intermediate signaling molecules, such as AKT, PI3K, and cyclins are also found within EVs [59,61,62], and they are likely transferred from cancer cells to target cells to activate proliferative signaling pathways. Nucleophosmin (NPM) is another oncoprotein that is highly enriched in EVs of several cancer cells and participates in many pathways involved in proliferation and growth suppression [63]. Furthermore, EVs may participate in the elimination of tumor suppressors, such as let-7, which is a microRNA precursor highly expressed in EVs from cancer cell lines [60].

The tumor microenvironment, which includes cells such as fibroblasts, myofibroblasts, endothelial, pericytes, and immune cells, is also important for tumor growth and development. EVs play an essential role in communication between tumor cells and the tumor microenvironment. For instance, HeLa cancer cell EVs increase Human Umbilical Vein Endothelial Cell (HUVEC) proliferation [93]. In addition, microvesicles from the cerebrospinal fluid of glioblastoma patients enhanced endothelial cell viability in vitro [94]. This is relevant, since angiogenesis promoted by endothelial cell proliferation increases tumorigenicity. Moreover, EVs isolated from non-small cell lung cancer cells (A549) increase the proliferation of the normal fibroblast cell line HLF1 [95]. Cancer cell EVs containing the mRNA for hTERT, the telomerase transcript, induce phenotypic changes, including increased proliferation and extension of the life span in fibroblasts. In addition, EVs isolated from the sera of patients with pancreatic and lung cancer also reportedly contain hTERT mRNA [57]. As previously stated, the inverse scenario has also been observed, namely that EVs from cancer-associated fibroblasts (CAFs) increase the proliferation of pancreatic and oral cancer cells [96,97]. In summary, the evidence presented highlights how EVs from cancer cells may act by several mechanisms in a paracrine manner to change the behavior of neighboring cancer cells or cells of the tumor microenvironment to enhance tumor growth.

3.2. Evading Growth Suppressors

Tumor suppressor genes act in many different ways to limit cell growth and proliferation. Thus, because the acquisition of these traits is key to the development and progression of cancer, tumor suppressor function is frequently reduced or eliminated in tumor cells. Some of the best-studied tumor suppressor proteins include the retinoblastoma (RB) protein and p53; both act as central control nodes within two key complementary regulatory circuits that determine whether cells proliferate or, alternatively, induce senescence and apoptosis [16].

Due to their relevance, many mechanisms have been identified that control the expression of these tumor suppressors; yet, to date no reports are available involving either EVs or exosomes in regulation of the RB protein or vice versa, RB in the regulation of EV composition. Alternatively, however, p53 has been shown to regulate the secretion, size, as well as the RNA and protein cargoes of tumor-derived EVs [98]. Proteomics analysis was used to identify proteins secreted in the culture media that are regulated by p53 in response to DNA damage in human non-small lung cancer cells. A more comprehensive analysis showed that exosomes isolated from the culture medium after p53 activation using ionizing radiation (IR) contained transcriptional targets of p53 (Maspin, PGK1, Eno1, and EF-1α), and unexpectedly, proteins encoded by genes that are not transcriptional targets of p53 (Hsp90β and CyPA). A p53-regulated gene product, tumor suppressor activated pathway-6 (TSAP6), was shown to increase exosome production in cells when p53 was activated in response to IR [99]. However, mechanisms that explain how TSAP6 increases exosome secretion have not yet been identified, although p53 is known to control the intracellular vesicle trafficking system by regulating components of the endosomal compartment (see details about EVs biogenesis in Section 1). The activation of p53 directly increases the transcription of the ESCRT-III subunit CHMP4C [100]. The ESCRT-III complex contains oligomers of small α-helical CHMP proteins, of which CHMP4 family members are the most abundant components [101]. ESCRT-III is required for the scission of the intraluminal vesicles (ILVs) into the MVB lumen during exosome biogenesis [6]. Human colorectal cancer cells expressing a dominant-negative mutation of p53 (R248W) were found to secrete exosomes enriched in several microRNAs (miRNAs), including miR-1246. The miR-1246-enriched exosomes are taken up by adjacent macrophages leading to their reprogramming into the anti-inflammatory, tumor-supportive M2-like phenotype characteristic of tumor-associated macrophages (TAMs) [102].

Beyond the ability of p53 to determine EV content, EVs are also known to regulate p53 activity. Bioinformatics analysis of proteome changes in astrocytes treated with glioblastoma (GBM)-derived EVs predicted the inhibition of p53. At the same time, significantly decreased Δ133p53 and increased p53β (truncated p53 isoforms) transcripts in astrocytes exposed to GBM-derived EVs were reported [103]. Changes in both truncated p53 isoforms suggest that astrocytes acquire a Senescence-Associated Secretory Phenotype that modifies the tissue microenvironment by secreting pro-inflammatory molecules, extracellular proteases, and extracellular matrix (ECM) components. In doing so, such "senescent" astrocytes promote tumor progression [103].

Exosomes from colon cancer cells transfected with a shRNA against p53 downregulated p53 expression in fibroblasts and promoted their proliferation. Among the miRNAs in exosomes from p53-deficient colon cancer cells, the upregulation of miR-1249-5p, miR-6737-5p, and miR-6819-5p was observed. Moreover, each of these miRNAs was shown individually to suppress p53 expression in fibroblasts [64]. These results reinforce the notion that p53 plays an active role in the control of exosomal RNA cargos.

3.3. Resisting Cell Death

Tumor cells develop strategies that limit or prevent apoptosis to survive and grow. One of these strategies involves EVs, since several studies have shown that EVs play a role in promoting resistance to cell death. Specifically, EVs are known to transport a defined set of miRNAs that transfer the resistance phenotype to sensitive cancer cells by altering cell cycle control and blocking apoptosis [104–108]. One of the anti-apoptotic pathways that has been linked to EV function is the inhibition of the c-Jun N-terminal kinase (JNK) pathway. Bone marrow-derived mesenchymal stem/stromal cell (BMSC)-derived exosomes have been shown to inhibit the JNK pathway and downregulate the expression and phosphorylation of Bcl-2-like protein 11 (Bim) [109]. Moreover, EVs can help prevent apoptosis under cell stress conditions. In EVs obtained from HeLa cervical carcinoma cells exposed to irradiation induced-stress, elevated levels of the inhibitor of apoptosis protein survivin were detected [110]. Finally, it has been reported that EVs derived from both bladder and gastric cancer cells inhibit cancer cell apoptosis by upregulating the expression of Bcl-2 and cyclin-D1 and downregulating Bax and caspase-3 [91,111].

3.4. Enabling Replicative Immortality

Cells in most normal cell lineages in the body can only divide a limited number of times, as defined by the "Hayflick" limit [112]. In cells in culture, repeated cycles of cell division induce initially senescence and then the crisis phase, which generally leads to cell death. However, cells that survive this crisis acquire an unlimited replicative potential. This transition is referred to as immortalization and is typical of cell lines that proliferate without developing senescence. The immortalization of cells, as occurs in tumors, is linked to their ability to maintain telomere regions, thereby avoiding senescence or apoptosis, and it is achieved by increasing telomerase expression. The telomeres are multiple tandem hexanucleotide repeats, which shorten progressively in non-immortalized cells after each cell division. Eventually, these regions lose the ability to protect the chromosome ends and generate unstable chromosome patterns that affect cell viability. The length of the telomer regions determines how many successive divisions a cell can undergo before telomeres are eroded and consequently lose their protective functions. Telomerase, the enzyme that adds telomere repeat segments to the ends of telomeric DNA, is almost absent in non-immortalized cells, but it is expressed at significant levels in human cancer cells, where it favors telomere maintenance [16].

As an example, breast epithelial cancer cells were treated with EVs purified from conditioned media of X-ray exposed cells. Compared to control cells, telomerase activity decreased in EV-treated cells. Moreover, exosome treatment with RNase prevented the effect on telomerase activity. These observations suggest that EVs transfer RNA-mediated information relating to telomerase activity between cells; however, unfortunately, this study did not provide any characterization of the exosomes/EVs [113].

The mRNA of the catalytic subunit of the telomerase reverse transcriptase (hTERT) is shuttled in exosomes from cancer cells to fibroblasts that do not express telomerase, where expression of the protein and activity are subsequently detected. Importantly, exosomes from the sera of patients with pancreatic or lung cancer contained hTERT mRNA as well. Telomerase activity induced phenotypic changes in target fibroblasts, including increased proliferation and delayed onset of senescence. In addition, telomerase activity protected the fibroblasts from DNA damage induced by phleomycin [57]. Later studies showed that hTERT is also present in amniotic fluid stem cell-derived EVs [67].

3.5. Inducing Angiogenesis

EVs participate in the regulation of pathological angiogenesis, as well as tumor angiogenesis. Hypoxia, a common feature of most solid malignant cancers, is generated by an imbalance between the altered oxygen supply capacity of the abnormal tumor vasculature and increased oxygen consumption of the tumor cells [114]. Therefore, hypoxia is a key driver of tumor angiogenesis [115,116]. Here, it should be noted that exosomes derived from hypoxic colorectal cancer cells promote angiogenesis in vitro and in vivo via Wnt/β-catenin signaling in endothelial cells [117]. In addition, another in vivo study showed that exosomes isolated from hypoxic lung cancer cells contained miR-23a, which increased angiogenesis [118]. In addition, exosomes derived from hypoxic leukemia cells were shown to enhance tube formation by human umbilical vein endothelial cells (HU-VECs) via a miR-210-dependent mechanism [119]. Nevertheless, it is important to consider that although hypoxia is important in the development of angiogenesis, it is not the only relevant factor, given that several different pro-angiogenic molecules are present in EVs from tumor cells that are independent of hypoxia.

EVs secreted by cancer cells contain pro-angiogenic mediators, including vascular endothelial growth factor (VEGFA), interleukin-8 (IL-8), interleukin 6 (IL-6) and fibroblast growth factor 2 (FGF2). Moreover, EVs can contain pro-angiogenic miRNAs, such as miR-21, miR-23a, miR-29a, and miR-30 [58,68–72]. Exosomes derived from gastric cancer cells deliver miR-130a to vascular cells to promote angiogenesis and tumor growth by targeting c-MYB both in vitro and in vivo [120]. In addition, exosomes that contain miR-205 from ovarian cancer cells significantly promoted angiogenesis in an in vivo model [121]. Additionally, using an in vivo nude mouse model, pancreatic cancer cell-derived exosomes carrying miR-27a were shown to promote angiogenesis [122]. In addition, recent research showed that the deleted in malignant brain tumors 1 protein (DMBT1) is enriched in EVs compared to the cancer cell of origin [63]. DMBT1 binds to pro-angiogenic factors and promotes adhesion, migration, proliferation, as well as angiogenesis [123]. Cancer cell-derived EVs also contain proangiogenic ECM remodeling enzymes, such as urokinase plasminogen activator (uPA), as well as the MMP2 and MMP9 [68]. Glioblastomas are among the most studied types of tumors known to release EVs carrying potent inducers of angiogenesis in vitro, ex vivo, and in vivo [124] that modify the phenotype of endothelial cells [58,94,125,126]. To date, two mechanisms have been proposed to understand how tumor-derived EVs may promote angiogenesis. First, the uptake by endothelial cells of exosomes derived from cancer cells is known to be increased. In addition, the expression of certain tetraspannins in cancer cell-derived EVs promotes the internalization of EVs by endothelial cells [127–131]. Therefore, EVs stimulate the transcription of genes related to angiogenesis and promote the migration and proliferation of endothelial cells. For example, EVs secreted by cancer cells were reported to transfer mutant EGFR to tumor endothelial cells, promoting mitogenic MAPK and AKT signaling [69]. Second, it has been reported that EVs mediate intercellular communication in the tumor microenvironment through mechanisms other than the transfer of their luminal cargos to recipient cells, in a manner independent on uptake [132,133]. Indeed, a recent study shows that cancer cell-derived EVs stimulate endothelial cell migration via the heparin-bound 189 amino acid isoform of VEGF, which, unlike other common VEGF isoforms, is enriched on the surface of EVs [134].

3.6. Invasion and Metastasis

The acquired ability to migrate and invade allows cancer cells to escape from the primary tumor and establish themselves at a new secondary site in a process commonly referred to as metastasis, which is responsible for 70–90% of all cancer-related deaths [16,135]. Numerous in vitro and in vivo studies show that EVs, in particular exosomes, play an important role in cell migration and metastasis in many types of cancer, including breast, glioblastoma, fibrosarcoma, nasopharyngeal, brain, melanoma, and colorectal, among others [89,136–139]. Some examples of how cancer cell-derived EVs modulate their environment are provided. For instance, the incubation of poorly metastatic B16F1 cells with EVs from the highly metastatic melanoma cell line B16F10 increased B16F1 metastasis to the lung after intravenous injection in mice [136]. Furthermore, exosomes obtained from the sera of prostate cancer patients increased significantly the invasiveness of DU145 prostate cancer cells in vitro compared to cells incubated with exosomes isolated from healthy individuals of the same age [140]. The loss of Rab27a in melanoma cell lines changes the size and protein composition of released exosomes [141]. Rab27a, a protein known to participate in exosome biogenesis [30], is overexpressed in melanomas. In addition, the loss of Rab27a in melanoma cell lines inhibited spontaneous metastasis in vivo, suggesting that Rab27a is important for the pro-invasive effects of exosomes produced by the wild-type cells [141].

EVs can modulate cell migration and metastasis through a variety of different mechanisms. These include the transfer of molecules that enhance migration, EMT-related molecules, MMPs, and miRNAs [14,73,76,78,142]. For instance, exosomes from a colorectal cancer cell line (HT-29), with high potential to induce liver metastasis, significantly increased in vitro migration and metastasis to the mouse liver of human colorectal Caco-2 cancer cells, which is a cell line with very low metastatic potential to the liver. This effect was proposed to be mediated by elevated levels of the C-X-C Motif Chemokine Receptor 4 in the exosomes [78]. In breast cancer, EVs from the highly metastatic cell line MDA-MB-231 containing caveolin-1 enhanced the migration and invasion in vitro of the less metastatic breast cancer cell line T47-D lacking caveolin-1 [14], providing evidence for the importance of caveolin-1 in the genesis of exosomes with elevated malignant potential. In prostate cancer, Integrin subunits $\alpha 3$ and $\beta 1$, Talin 1, and Vinculin, proteins all relevant to migration and invasion, were more abundant in EVs of the more aggressive PC3 cell line, compared to the exosomes from less aggressive LNPaC cells. Furthermore, EVs derived from each of these cell lines increased the invasion of non-cancerous cells, which was prevented when integrin subunit $\alpha 3$ was blocked. Additionally, integrin subunits $\alpha 3$ and $\beta 1$ are increased in EVs isolated from the urine of metastatic prostate cancer patients [142]. Interestingly, stromal cells from gastrointestinal tumors release exosomes containing the receptor tyrosine kinase proto-oncogene KIT (also called CD117), which increases MMP1 expression in smooth muscle cells, creating a positive feedback loop between stromal and tumor cells that favors tumor cell invasion [74]. Moreover, fibrosarcoma exosomes containing fibronectin, an important ECM protein, promoted cell adhesion and migration [143]. Finally, prostate cancer cells produce large oncosomes containing bioactive MMPs, in addition to other molecules that are important for cancer progression, such as caveolin-1 and ADP ribosylation factor 6 [144].

On the other hand, RNA-bearing exosomes are also important for cell migration and metastasis. In particular, some miRNAs are relevant in this context, such as miR-9, -145, -21, -29a, -494, and -542-3p. These miRNAs affect the expression of many different targets, such as cell–cell adhesion molecules, chemokine ligands, cell cycle regulators and angiogenesis-promoting proteins, which are all factors that contribute to metastasis [76,79]. Exosomes derived from primary lung tumors, carrying small nuclear RNAs, were shown to activate Toll-like receptor 3 (TLR3) in lung epithelial cells, inducing chemokine secretion and neutrophil recruitment to the lung that favors the formation of pre-metastatic niches in vivo [145]. In addition, exosomes derived from breast cancer cells were shown to transfer miR-105 to HUVEC cells, where miR-105 reduces the expression of the tight junction protein ZO-1 to promote vascular permeability that favors the spread of cancer cells [146]. Furthermore, exosomes from B16-F10 cells can also induce vascular leakiness, as evidenced by increased pulmonary endothelial permeability [147]. This was corroborated by Hoshino et al. using exosomes with tropism to the lung from the MDA-MB-231-derived human breast cancer cell lines 4175 and 1833, in a mouse model [77]. Taken together, this evidence suggests that exosomes initially increase vessel permeability in order to prepare the pre-metastatic niche.

Moreover, exosomes can also act as a scaffold for the attachment of metastatic cells [143]. In this respect, an interesting study shows that exosome release is important for autocrine cell migration. Specifically, using the chick embryo chorioallantoic membrane assay, as well as in vitro assays, the authors found that exosomes from H10T80 human fibrosarcoma cells enhanced directional migration and promoted adhesion assembly in an autocrine manner. Moreover, in these in vitro assays, exosomes promoted the migration of H10T80 cells by enhancing adhesion. Somewhat surprisingly, miR-210-containing exosomes from HCT-8 colon cancer cells with a more adhesive phenotype inhibited the MET and cell-surface adhesion of a subpopulation of HCT-8 cells with elevated metastatic potential in vitro [148]. These results suggest that exosomes may also reduce the adhesion of tumor cells and thereby favor their dissemination.

Exosomes can also function as vectors that sequester molecules to reduce their intracellular bioavailability, thereby altering the phenotype of the parent cell [75]. For example, the Let-7 and miR-200 miRNA levels observed in exosomes from the ovarian cancer cell lines SKOV-3 and OVCAR-3 were elevated compared to the intracellular levels. This is relevant, given that the let-7 miRNA family suppresses cell proliferation, while the miR-200 family suppresses EMT [75]. Thus, the elimination of these miRNAs through exosomes reduces their intracellular levels.

Regarding the role of exosomes in preparing the metastatic niche and colonization, Hood et al. provided evidence for the importance of melanoma-derived exosomes in promoting metastasis to lymph nodes in vivo. To this end, C57BL/6 mice were preconditioned by injecting into the left footpad exosomes isolated from B16F10 cell culture supernatants. The subsequent injection of B16F10 cells into the left footpad revealed that preconditioning increased melanoma cell recruitment to lymph nodes of the mice, which is a preferential site for melanoma metastasis [149]. Another study showed that the intravenous injection of B16F10-derived exosomes following orthotopic injection of B16F10 cells into C57BL/6 mice increased metastasis to the lung. Furthermore, the transplantation of bone marrow-derived cells (BMDC) treated with exosomes derived from B16F10 cells, after subcutaneous implantation with B16F10 cells, resulted in higher metastatic burden in vivo in the lung and ipsilateral lymph nodes, which was attributed to transfer of the MET oncoprotein [147]. In addition, during colonization, exosomal integrins are important for specific organ tropism. In particular, using a knock-down strategy, exosomes containing the $\alpha6\beta4$ integrin were shown to promote lung metastasis, while $\alpha v\beta5$ integrin presence was linked to liver metastasis. Furthermore, exosomal integrins were associated with the increased expression of genes related to metastasis, such as S100A8 and S100P, as well as elevated levels of the src protein and phosphorylation on Tyr-416 [77].

Finally, another mechanism by which exosomes can promote metastasis is by reducing the permeability of the vascular endothelial barrier, which is a topic that will be addressed further on in an independent section.

3.7. Genome Instability and Mutation

In physiological conditions, the genome maintenance systems find and repair defects in the DNA, maintaining very low rates of spontaneous mutation. Cancer cells often increase the rates of mutation through increased sensitivity to mutagenic agents or deregulation of components of the genome maintenance and repair machinery, or both [16]. Since the late 1990s, several types of defects that affect components of the DNA maintenance machinery, have been described [150]. For instance, DNA repair genes and mitotic checkpoint genes, such as the MutL homolog 1 (MLH1), the breast cancer susceptibility gene 1 (BRCA1), MYH (also known as MUTYH), and the xeroderma pigmentosum group A (XPA) all encode proteins that help to maintain genomic stability [151].

The MLH1 protein is one of seven DNA mismatch repair proteins (MLH1, MLH3, MSH2, MSH3, MSH6, PMS1, and PMS2) in humans. A heterodimer between MSH2 and MSH6/MSH3 first recognizes the DNA mismatch. The MSH2–MSH6 heterodimer allows the binding of a second heterodimer of MLH1 and PMS2/PMS3/MLH3. This protein complex formed between the two sets of heterodimers enables the initiation of repair of the mismatch defect in DNA [152]. EVs isolated from sorafenib-resistant renal cell carcinoma (RCC) cells contain high levels of the microRNA miR-31-5p. Treatment with miR-31-5p-containing EVs suffices to downregulate MLH1 expression in target cells [80]. This mechanism would presumably reduce the activity of the DNA mismatch repair system and lead to long-term accumulation of mutations, but this hypothesis has not yet been corroborated.

BRCA1 promotes the repair of DNA double-strand breaks (DSB) by homologous recombination. BRCA1 associates with BRCA1-associated RING domain protein 1 (BARD1) and other tumor suppressor proteins to initiate the nucleolytic resection of DNA lesions and the recruitment and regulation of the recombinase RAD51 [153], which catalyzes the insertion of single-stranded DNA (ssDNA) into sister chromatids. Using sister chromatid as the template, ssDNA is elongated, and junctions are formed between the two sister chromatids [154]. Recent studies show that BRCA1-deficient fibroblasts treated with uveal melanoma-derived and colorectal cancer-derived EVs transfer malignant traits to target cells, and the authors suggest that BRCA1 activity is necessary to prevent the detrimental effects of cancer-derived EVs in non-cancer cells [155,156].

To date, a literature search for evidence linking EVs to the control of the other two caretaker genes, MYH/XPA, did not yield any results.

3.8. Tumor-Promoting Inflammation

Cancer cell-derived EVs promote the generation and persistence of the inflammatory environment, which contributes to disease progression. Fabri et al. demonstrated that the miR-21 and miR-29a contained in exosomes derived from lung cancer cells bind to members of the Toll-like receptor (TLR) family on immune cells. TLR engagement triggers the activation of nuclear factor kappa-light chain-enhancer of activated B cells (NF-κβ), secretion of pro-metastatic inflammatory cytokines, and the transcription of genes that favor tumor proliferation and metastasis [81]. Another study showed that when monocytes are stimulated with EVs derived from oral squamous cell carcinoma (OSCC), the uptake of these EVs by monocytes leads to NF-κB activation and the generation of a pro-inflammatory environment, which was characterized by elevated levels of IL-6, monocyte chemoattractant protein 1 (MCP1), prostaglandin E2 (PEG2) and MMP9 [157]. Using RNA sequencing and proteomics analysis, Haderk et al., observed that expression of the Y RNA (small non-coding RNA) hY4 is increased in exosomes isolated from chronic lymphocytic leukemia (CLL) cells and from the culture supernatant of a CLL cell line. Additionally, when monocytes were treated with these exosomes, PD-L1 expression and cytokine release were induced, facilitating cancer-related inflammation [82].

In addition, the pro-inflammatory effects of tumor-derived exosomes that affect macrophage performance have been described. Wu et al., found that exosomes derived from gastric cancer cells induced macrophages to express higher levels of pro-inflammatory factors, such as IL-6 and TNF-α. These exosomes markedly increased the phosphorylation of NF-κB in macrophages and, additionally, activated macrophages in human peripheral blood monocytes via NF-κB [158]. Moreover, lung cancer cell-derived exosomes transform naïve mesenchymal stem cells (MSCs) into a new kind of pro inflammatory MSCs (P-MSCs) by activating TLR2/NF-κB signaling [159]. Recently, Pritchard et al. reported that lung tumor cells secrete exosomes that are taken up by macrophages and differentiate into tumor-associated M2 macrophages, which can promote inflammation in the tumor environment and immune suppression [160]. Together, these studies highlight how EVs play an important role as messengers in the communication between tumor cells and cells of the immune system. Such cell–cell communication promotes the genesis of a pro-inflammatory environment that permits the escape of tumor cells from destruction by the immune system.

3.9. Deregulating Cellular Energetics

Cancer cells exhibit remarkable metabolic plasticity that is necessary to generate energy and at the same time satisfy the biosynthetic requirements, which permit maintaining proliferation and/or metastatic spread [161]. In addition, particularly for cancer cells in a hypoxic environment, the enzymes of the glycolytic pathway are upregulated, and elevated release of lactate and pyruvate is observed, which leads to an acidification of the tumor environment [162]. In turn, the decrease in pH is associated with an increase in the secretion and uptake of EVs [163] that contain proteins involved in metabolism and miRNAs that target proteins related to metabolic activities of the cell [83,164]. Fatty acid synthase (FASN), a key enzyme involved in the de novo synthesis of FAs, is one of the most frequently identified proteins in EVs [83]. Additionally, not only the protein but also the mRNA of FASN has been identified in prostate cancer (PCa) cell-derived EVs [165], which suggests a possible role for these EVs in the lipogenesis of cancer cells. On the other hand, a study compared by proteomics analysis exosomes from non aggressive hepatocellular carcinoma cells with those released by aggressive cell lines and found that in the latter case, exosomes are enriched in enzymes involved in glycolysis, gluconeogenesis, and the pentose phosphate pathway [84]. Potentially, these exosomes may be more easily absorbed by the recipient cells, which translates into an increased uptake of these metabolic drivers that affect the metabolic profile of the recipient cells, as is the case for hepatocellular carcinoma cells [84]. However, the presence of glycolytic enzymes in EVs does not necessarily correlate with functional transfer, as shown in a proteomics analysis of adipocyte EVs, which suggested that both glucose oxidation and lactic acid release remained essentially unchanged in recipient cells after treatment with these EVs [166]. Therefore, it will be necessary to increase the number of studies both in vitro and in vivo to establish more conclusively whether EVs enriched in glycolytic enzymes are able to reprogram the metabolism of recipient cells and to what extent this capacity depends on the tumor cell origin.

3.10. Avoiding Immune Destruction

Exosomes can induce immune responses by regulating signals controlling both the adaptive and innate immune responses [167]. Tumors avoid being recognized by cytotoxic T cells as a strategy to escape destruction by the immune system. To do so, they can directly impair the functioning of antigen-presenting cells (APC) or cytotoxic T cells, or alternatively induce suppressor T cells. In all cases, efficient immune responses against cancer cells are blocked [168]. Several mechanisms have been described by which EVs participate in the evasion of the immune destruction of tumor cells. For instance, tumor-derived EVs induce immunosuppression by promoting the expansion of regulatory T cells (Treg) and depletion of anti-tumor CD8+ effector T cells, which in conjunction permit tumor escape [169]. Interestingly, metastatic melanomas release EVs, mainly in the form of exosomes, which transport programmed death-ligand 1 (PD-L1) on their surface and suppress CD8 T cell function [86,87]. Recently, a study showed that exosomes from Lewis lung carcinoma or 4T1 breast cancer cells impaired dendritic cell (DC) differentiation and promoted apoptosis [170]. Moreover, several studies have shown that exosomes from cancer cells can inhibit natural killer (NK) cell proliferation and cytotoxic functions, mainly through the downregulation of NK group 2 member D (NKG2D), which is a central mediator of NK cytotoxicity [171–177]. Xia et al., have shed light on a potentially new mechanism by which cancer-derived EVs may inhibit NK cell activity. Their study shows that exosomes isolated from the supernatants of primary cell cultures of tissue samples from patients with clear cell renal cell carcinoma obtained after nephrectomy display TGF-β1 on their surface, which may impair NK function by activating the Small Mothers Against Decapentaplegic (SMAD) pathway in these cells [85]. Despite their relevance, these results were obtained using in vitro approaches and need to be confirmed in in vivo settings. Elucidating the role of EVs in the evasion of cancer cell destruction by the immune system should aid in the development of new therapies that block evasion of the immune response by tumor cells, consequently enhancing anticancer treatment efficacy.

3.11. EVs and Thrombosis

Although the pro-thrombotic role of EVs is not considered a hallmark of cancer, presumably because it does not appear to contribute to cancer development, it does play an important role in determining cancer patient survival and for that reason is considered here.

A variety of studies have identified a role for EVs in modulating processes related to coagulation and hemostasis, as well as in pathologies associated with thromboembolic events, such as sepsis, atherosclerosis and cancer [178–180]. Thrombosis is one of the most common complications in cancer patients and represents the second leading cause of death in cancer patients in the United States [181–185]. The procoagulant activity of EVs is associated predominantly with the surface exposure of phosphatidylserine (PS), which facilitates the assembly of complexes, including the coagulation factors VIIIa, IXa, and X, as well as the prothrombinase factors Va, Xa, and II on the EV surface [179]. Moreover, tumor cells release EVs with tissue factor (TF) on their surface, which activates the extrinsic branch of the coagulation cascade [186,187]. Several in vitro and in vivo studies have linked the expression of TF on EVs to their pro-coagulant potential [188–192]. Indeed, circulating TF-positive EVs (TF+EVs) have been observed in leukemia [193], multiple myeloma [194], breast, pancreatic [195], ovarian [180] and lung [189] cancer.

3.12. EVs and Cell Competition

In any given tumor, several different cancer cell subpopulations coexist and, consequently, tumor subclones compete for available resources in a process denominated cell competition (CC). This process determines the relative fitness in neighboring cells and permits eliminating defective or damaged cells in communities to favor the proliferation and growth of the most competent cells [196]. Given that this will ultimately determine the nature of a tumor, some evidence relating to factors involved in CC mechanisms and the role of exosomes/EVs in that context will be discussed below.

One of the best characterized factors that regulates CC is the transmembrane protein Flower (hFWE). In humans, there are four splice variants of hFWE (1–4), and co-culture studies revealed that cells expressing hFWE2 or hFWE4 proliferate while triggering caspase-dependent apoptosis in cells expressing hFWE1 or hFWE3 [196]. Although this is perhaps one of the clearest examples illustrating how specific molecules participate in CC, there is unfortunately no published information available indicating that hFWE (1–4) are present in exosomes/EVs.

Bone morphogenetic proteins (BMPs) have also been previously associated with CC. In mammals, pluripotent cells with decreased BMP signaling are eliminated in the presence of WT cells [197]. Calcium-dependent activator protein for secretion 1 (CAPS1) protein promotes metastasis in colorectal cancer cells (CRCs), and exosomes derived from CAPS1-overexpressing CRCs increase the migration of normal colonic epithelial cells. Interestingly, proteomics analysis showed that the overexpression of CAPS1 downregulated the BMP4 cargo in exosomes [198]. Thus, these results suggest that CAPS-1 expressing cells restrict BMP4 export using exosomes to upregulate their own signaling or to prevent the functional transfer of this protein to neighboring cells.

Latent membrane protein 1 (LMP1), an oncogenic protein, plays an important role in malignant transformation. In AGS gastric cancer cell populations, LMP1-positive cells decreased gradually with each cell passage when the cells were co-cultured with LMP1-negative cells. The experiments performed to study this phenomenon suggest that LMP1-positive cells stimulate the proliferation of surrounding LMP1-negative, but not LMP1-positive cells, via EV-mediated EGFR activation [199].

YAP is a transcriptional co-activator that does not bind directly to DNA. The phosphorylation of YAP by LATS kinases can either prime the protein for binding to 14-3-3 proteins leading to cytoplasmic sequestration or ubiquitin-mediated protein degradation. Alternatively, however, active (non-phosphorylated) YAP translocates to the nucleus and binds mainly to transcription factors of the TEA domain family (TEAD). In the nucleus, the YAP–TEAD protein complex transcribes genes that control cell proliferation and apoptosis [200]. In co-culture conditions, cells expressing higher levels of YAP have enhanced growth and cause the elimination by apoptosis of cells expressing lower levels of this protein [201]. Wnt5a-enriched exosomes isolated from lymph node metastasis-derived gastric cancer (LNM-GC) cells induced YAP dephosphorylation in bone marrow-derived mesenchymal stem cells (BM-MSCs) [202]. Experiments performed in *Xenopus laevis* embryos have identified human frizzled-5 (hFz5) as the receptor for Wnt5a [203]. Thus, the autocrine stimulation of gastric cancer cells with Wnt5a-containing exosomes could function as an auto-stimulatory mechanism that increases the proliferation of specific subpopulations of cancer cells in metastatic tumors, which are mediated by the activation of hFz5 receptor and YAP-mediated intracellular signaling.

The non-canonical Wnt-planar cell polarity (PCP) pathway does not involve β-catenin but rather controls cell movement through the activation of RHOA, c-Jun N-terminal kinase (JNK), and nemo-like kinase (NLK)-dependent signaling cascades [204]. Exosomes, secreted from human fibroblasts, stimulate breast cancer cell (BCC) protrusive activity, motility, and metastasis via Wnt/PCP signaling in vitro. In orthotopic mouse models of breast cancer, the co-injection of BCCs with fibroblasts dramatically enhances metastasis in a manner dependent on Wnt/PCP signaling in BCCs. Surprisingly, exosome activity in BCCs was shown to be dependent on Wnt11 produced in BCCs. Proteomics analysis revealed that the fibroblast-derived exosomes do not contain Wnt11. The experiments carried out to elucidate the causes of this unexpected observation showed that fibroblast-derived exosomes are internalized by BCCs and then loaded with Wnt11 [205]. These results show how the interactions between different populations of cancer and stroma cells in complex biological systems can lead to modifications in the composition of exosomes/EVs. Finally, the incorporation of Wnt11 into exosomes/EVs may represent a key factor in determining fitness during CC.

Importantly, it should be noted that numerous other proteins found in exosomes or EVs involved in CC mechanisms were already mentioned in the sections on EV-mediated functions in cancer: see EGFR, MAPK (Section 3.5), p53 (Section 3.2), src (Section 3.6), and JNK (Section 3.3).

4. EVs in Cancer Drug Resistance

Chemotherapy is widely used to treat cancer, but the effectivity of such therapies is reduced in several types of cancer due to the development of drug resistance, which can be attributed to the activation of intrinsic or acquired mechanisms. Intrinsic resistance refers to the presence of resistance factors in tumor cells prior to chemotherapy that render the treatment ineffective. Acquired resistance, on the other hand, is developed during the treatment of tumors that were initially sensitive and can be caused either by mutations arising during treatment or through adaptive responses [206]. Moreover, tumors are extremely heterogeneous, so drug resistance can arise through the therapy-induced selection of a minor resistant subpopulation of cells that was present in the original tumor [207]. Alternatively, drug resistance can also be acquired by drug-sensitive cells via communication with drug-resistant cells (cancer or stromal) through EV-mediated transfer of resistance factors. Some of these EV-mediated mechanisms of drug resistance will be explored in the next sections.

4.1. EV-Mediated Drug Transport

Regardless of the route of anticancer drug administration, these drugs generally need to be taken up by cancer cells because they target an intracellular process. Note that membrane receptor antagonists are exceptions in this respect. The uptake of such drugs by cancer cells may involve active transport mechanisms or rely on simple diffusion because of high membrane permeability. Independent of the mechanism, these drugs must reach a threshold concentration to be effective. However, cancer cells are known to express multidrug resistance (MDR)–ATP binding-cassette (ABC) proteins that export drugs to the extracellular space. These transporters are membrane-bound proteins that consume ATP to eliminate a wide variety of molecules, even against steep concentration gradients [208]. This phenomenon results in decreased intracellular anticancer drug accumulation, which decreases or even abolishes drug effects. In this context, it is important to mention that an alternative drug export mechanism has been described involving EVs to eliminate the drugs in an ABC transporter-independent manner.

Shedden et al., (2003) were the first to report that anticancer drug resistance and the release of EVs could be mechanistically linked. In cancer cell lines, the expression of vesicle shedding-related genes is associated with chemosensitivity profiles. Furthermore, in the breast cancer cell line MCF7, the fluorescent chemotherapeutic agent doxorubicin was incorporated into EVs and released to the culture media [209]. Similarly, in vitro B-cell lymphoma cell lines efficiently extrude doxorubicin in exosomes [210].

Early studies suggested that cisplatin, once inside tumor cells, may be sequestered into acidic vesicles belonging to a secretory pathway. The treatment of human ovarian carcinoma cells with cisplatin showed that the exosomes released from cisplatin-resistant cells contained more than 2-fold higher platinum levels than those released from cisplatin-sensitive cells [211]. Moreover, exosomes released by drug-resistant melanoma cells that were previously treated with a fixed dose of cisplatin in culture contained varying amounts of the drug depending on the pH of the medium, and the level of cisplatin in the exosomes was higher in acidic culture medium [212]. Additionally, it was reported that mouse leukemia cell-derived exosomes can include paclitaxel and, interestingly, that the paclitaxel-containing exosomes reduced the proliferation of a human pancreatic cell line. These observations suggest that exosomes or EVs can be used to package and deliver active drugs [213].

4.2. EVs Transport Drug Efflux Pumps

ABC transporters can confer multidrug resistance to tumor cells. In addition, cancer cells can transmit resistance through horizontal transfer using EVs carrying drug efflux pumps. The first evidence for the transfer of ABC transporters between cancer cells was obtained studying human acute lymphoblastic leukemia cells. P-glycoprotein (P-gp)-containing "microparticles" were isolated from drug-resistant cells and then used to treat drug-sensitive cells. The results revealed that P-gp protein transfer coincided with reduced drug accumulation in recipient cells, confirming that the transfer of functional P-gp was mediated by EVs [214]. Later studies showed that exosomes from docetaxel-resistant human prostate cancer cell lines conferred resistance to previously sensitive target cells. In addition, this study revealed that P-gp was only present in exosomes derived from resistant but not docetaxel-sensitive cells [140]. Exosomes derived from doxorubicin-resistant (DXR) osteosarcoma cells are taken up by recipient cells, where they convey a doxorubicin-resistant phenotype. The treatment of doxorubicin-sensitive (DXS) osteosarcoma cells with exosomes derived from DXR cells reduced the sensitivity of the recipient cells to doxorubicin. Moreover, exosomes from DXR cells contain higher mRNA and protein levels of P-gp. In addition, both P-gp mRNA and protein levels increased in cells after treatment with DXR-derived exosomes [215].

P-gp is the best studied drug efflux pump; however, other members of the ABC transporter family have been identified in cancer cell-derived EVs/exosomes, too. GAIP interacting protein C terminus (GIPC) is a protein regulator of autophagy and the exocytotic pathways in cancer. The knockdown of GIPC in pancreatic cancer cells induces the overexpression and incorporation into exosomes of the ATP-binding cassette sub-family G member 2 (ABCG2). This finding opens up the possibility of horizontal transfer of ABCG2 via exosomes mediates drug resistance in pancreatic cancer [216].

In addition, exposure to the chemotherapeutic drug vincristine increases the secretion of ATP-binding cassette sub-family B member 1 (ABCB1)-enriched EVs by inducing dysregulation of the Ras-related proteins Rab8B and Rab5. The transfer of ABCB1 via exosomes helps sensitive cancer cells develop a drug-resistant phenotype [183].

4.3. EVs Transfer Pro-Survival Cargos

EV cargoes also include pro-survival factors, which decrease apoptosis sensitivity and increase cell viability, thus leading to resistance to anticancer drugs. Components of the PI3K/AKT pathway, an oncogenic signaling axis involved in cancer cell proliferation and survival, have been reported in EVs. Exosomes derived from HCC cells induced sorafenib resistance in vitro and in vivo by activating the HGF/c-Met/AKT signaling pathway and inhibiting sorafenib-induced apoptosis [217]. Triple negative breast cancer cell lines, resistant to docetaxel and doxorubicin, release EVs that induced resistance to the same drugs in recipient non-tumorigenic breast cells. The treatment with EVs from the resistant cells increased the expression of eight genes associated with the PI3K/AKT pathway [218].

BRAF is a component of the MAPK pathway involved in cell differentiation and survival. BRAF kinase inhibitors, such as vemurafenib and dabrafenib, are used in advanced melanoma treatment. Platelet-derived growth factor receptor β (PDGFRβ) is a receptor tyrosine kinase that induces activation of the PI3K/AKT pathway. Vella et al. showed that PDGFRβ can be transferred to recipient melanoma cells in EVs, resulting in a dose-dependent activation of PI3K/AKT signaling and escape from MAPK pathway inhibition by BRAF [219].

In addition, resistance to apoptosis is an escape mechanism by which cancer cells acquire drug resistance and thus contribute to cancer progression. Cancer-associated fibroblast (CAF)-EVs induced the drug resistance of gastric cancer cells by decreasing cisplatin-induced apoptosis. The proteomics analysis of CAF-derived EVs identified that annexin A6 plays a pivotal role in the drug resistance of gastric cancer cells via the activation of β1 integrin and the downstream intracellular signalling pathways, involving focal adhesion kinase (FAK) and the yes-associated protein (YAP). Consistently, the inhibition of FAK or YAP efficiently attenuated gastric cancer drug resistance in vitro and in vivo [220].

Survivin is a pro-survival protein member of the inhibitor of apoptosis (IAP) family that is present in EVs derived from different tumor types [221]. Paclitaxel treatment of triple negative breast cancer cells induces the secretion of EVs enriched in survivin, which increased the survival of serum-starved, as well as paclitaxel-treated fibroblasts and breast cancer cells [222].

4.4. EVs Mediate Drug Resistance via the Transfer of microRNAs

MicroRNAs (miRs) are well-established components of EVs, and their horizontal transfer favors the development of drug resistance. Sorafenib is a kinase inhibitor drug approved for the treatment of primary kidney cancer, advanced primary liver cancer, and advanced thyroid carcinoma. EVs derived from sorafenib-resistant (SR) cells were taken up by sorafenib-sensitive (SS) RCC cells and promoted drug resistance. Elevated miR-31-5p in EVs derived from SR cells downregulated the expression of MLH1, which is a gene commonly associated with hereditary nonpolyposis colorectal cancer in SS cells and thus promoted sorafenib resistance in vitro. In addition, low expression of MLH1 was observed in SR RCC cells and upregulation of MLH1 expression restored the sensitivity of resistant cell lines to sorafenib. Experiments in mice also confirmed that miR-31-5p could regulate drug sensitivity in vivo. Finally, miR-31-5p levels in circulating EVs from the plasma of RCC patients with progressive disease during sorafenib therapy were higher when compared with the levels observed prior to therapy [80].

Exosomes isolated from gemcitabine (GEM)-resistant human pancreatic cancer stem cells (R-CSCs) inhibited GEM-induced cell cycle arrest and apoptosis as well as promoted tube formation and cell migration in drug-sensitive human pancreatic cancer stem cells (S-CSCs). Elevated miR-210 levels were detected in R-CSC exosomes compared to S-CSCs exosomes, and MiR-210 levels in exosomes were dependent on the GEM doses used to treat cells. Moreover, treatment with R-CSC-derived exosomes increased miR-210 levels in recipient cells [223].

The aforementioned studies are only a few recently published examples of the increasing evidence linking cancer drug resistance to the presence of specific miRNA cargos in EVs. A more comprehensive summary of related information can be found in a recent article by Maacha et al. [221].

4.5. EV Interference in Immunotherapies

Specific EV surface antigens can be targeted by immunotherapy where they act as a "hunter" in monoclonal antibody-based therapies by diminishing antibody bioavailability. For instance, rituximab (anti-CD20 antibody) binds to CD20 on the surface of EVs and protects targeted lymphoma cells from rituximab-induced toxicity [224]. EVs secreted either by HER2-overexpressing breast carcinoma cells or present in the serum of breast cancer patients bind to trastuzumab. In vitro studies showed that HER2-containing EVs, but not EVs lacking HER2, prevent the reduction in cell proliferation induced by trastuzumab treatment, although no change in HER2 activation status was detected in EV-treated cells by Western blotting [225].

EVs are involved in additional ways in downregulating the immune response. Melanoma patients display different responses to the immune checkpoint inhibitor pembrolizumab (anti-PD-1). The detection of immune checkpoint ligand (PD-L1) on EVs early after therapy is indicative of whether the patients will respond or not to anti-PD-1 therapy. PD-L1 binds to PD-1 receptors on the surfaces of effector T cells, preventing their ability to target tumor cells for destruction. PD-L1 containing exosomes derived from melanoma cells inhibit the proliferation, cytokine production, and cytotoxicity of T cells. Pre-treatment of the exosomes with the anti-PD-L1 antibodies nearly abolished these effects. In vivo studies suggest that exosomal PD-L1 suppresses anti-tumor immunity systemically [86]. In addition, EVs from glioblastoma stem cells were found to contain PD-L1 and inhibit T cell proliferation and antigen-specific T cell responses [226]. These results suggest that by capturing the anti-PD-1 antibodies on their surface, EVs prevent this antibody from accessing the tumor, thereby permitting PD-L1 to bind to PD-1 on T cells and attenuate anti-tumor immune responses.

These findings further extend our understanding of the implications of EVs in the development of the disease. The composition of cancer-derived EVs can regulate patient responses to chemotherapy using one or more of the aforementioned mechanisms. With this in mind, one may predict that EVs will serve to predict or evaluate therapy efficacy, and as such will likely become powerful tools to improve cancer treatment. However, the clinical application of new techniques for rapid EV detection and characterization remains a pending issue.

5. EVs in Organ Tropism, Drug Delivery, Imaging and Theranosis

The intrinsic organ tropism of EVs and their potential physiological benefits, combined with drug loading and targeting strategies, provide multiple therapeutic benefits for drug delivery, such as greater cellular uptake and focalization, prolonged circulation time, immunomodulation, biocompatibility, and stability. Furthermore, EVs can be used as biological nanocarriers with the inclusion of active principles, nanoparticles, or imaging agents. As such, they can significantly improve the therapeutic efficacy and selectivity, as well as facilitate the early detection of multiple diseases, including cancer [227]. However, to consider the use of EVs in potential clinical applications, the effects discussed previously relating to the role of EVs in cancer and other pathologies need to be kept in mind.

5.1. EV Organ Tropism

EVs have emerged in recent years as potential tools for the delivery of different bioactive agents to target tissues and specific organs [228,229]. In this context, the cellular origin of EVs is key to determining the tropism toward specific organs. For instance, EVs from melanoma cells predominantly accumulate in the lungs, while EVs from dendritic cells tend to accumulate in the spleen [230]. Interestingly, EVs derived from tumor cells reportedly also show selective tropism toward the tumor tissue from which they originated. EVs from brain endothelial cells can cross the blood–brain barrier and accumulate in the brain and brain tumor tissue, while EVs from melanoma cells preferentially target metastatic melanoma tumors [229,231]. However, it is not clear whether the tumor cells from which EVs originate determine alone their tissue tropism. Garofalo et al. [232] observed the in vitro and in vivo targeting and accumulation of lung cancer cell-derived EVs in colon carcinoma cells and vice versa. This may be taken to suggest the existence of a generalized tropism for tumor-derived EVs toward any neoplastic tissue, regardless of the tumor type. Although the molecular basis for EV tropism is not fully understood, there have been some significant advances in discovering molecules involved in this process [233,234]. For example, integrins are cell surface adhesion molecules with a substantial role in determining EV organ tropism, particularly toward the lung and liver. In particular, the expression of $\alpha6\beta4$ and $\alpha6\beta1$ is important in the EV tropism toward the lungs, while $\alpha v\beta5$ promotes EV accumulation in the liver [77,235]. Exosomes from rat pancreatic carcinoma cells expressing the Tspan8-$\alpha4$ complex preferentially accumulate in the pancreas and lungs of rats [236]. There is also evidence showing that the cell migration-inducing and hyaluronan-binding protein (CEMIP), which is enriched in exosomes of brain-tropic metastasis-derived MDA-MB-231 breast cancer cells, promotes exosome accumulation in the brain by generating a pro-metastatic environment [237]. Additionally, expression of the programmed death-ligand 1 (PD-L1) in tumor-derived EVs is important for the suppression of T-cell activation and thereby avoiding the immunological anti-tumor responses [87]. These findings further extend our understanding of EV tropism, which opens up novel possibilities for the selective targeting of diagnostic/therapeutic agents to tumors.

5.2. EVs as Drug Delivery Vehicles

EVs have become novel biological delivery vehicles for several cargoes, due to the variety of natural properties that they possess. These vesicles have the intrinsic capacity to cross biological barriers and to transport various cargoes, protecting their content from degradation until reaching the target. Depending on their cellular origin, EVs are highly heterogeneous in content, and such variations contribute significantly to their uptake, organ tropism and immunomodulation [238]. EV tropism is determined by the presence on their surface of different adhesion and immunoregulatory molecules, as well as specific cell receptors, which contribute to enhancing their accumulation in specific tissues [239,240]. This characteristic combined with their small size favors EV accumulation in highly vascularized tissues with deficient lymphatic drainage, such as tumors. This phenomenon, referred to as the enhanced permeability and retention (EPR) effect, can be used as a strategy to increase targeting toward tumors [230]. EVs have been widely studied as drug delivery nanocarriers in cancer research, so recent and representative studies for each application of these vesicles in the delivery of proteins, genetic material, and chemotherapeutics drugs will be described (see Figure 3). In this field, Kim et al. developed a formulation of paclitaxel-loaded exosomes by the sonication and conjugation of an aminoethilanisamide–polyethylene glycol (AA-PEG) vector moiety to target the sigma receptor, which is overexpressed by lung cancer cells. The nanosystem (AA-PEG-exoPTX) possesses a remarkable ability to accumulate in cancer cells and demonstrates high anticancer efficacy in a mouse model of pulmonary metastasis [241].

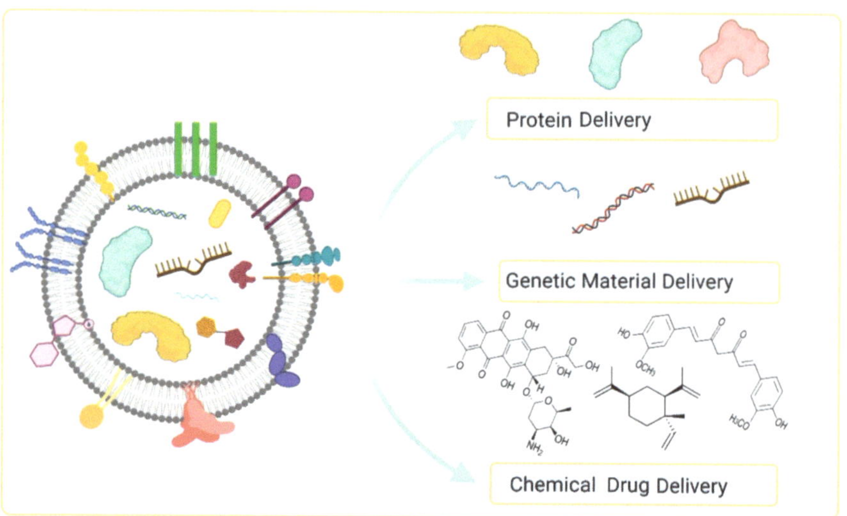

Figure 3. EVs are nanoscale structures with excellent biocompatibility and the ability to transport/deliver many different types of proteins, genetic material, and chemical drugs that can be used in cancer therapy.

With respect to protein delivery, Aspe et al. [242] engineered EVs from melanoma cells to overexpress survivin-T34A, which is a dominant-negative mutant variant of the inhibitor of apoptosis protein survivin that blocks the protein's function. Survivin overexpression plays an important role in the development of resistance to both chemo- and radiotherapy in pancreatic cancer. The authors observed that EVs containing either survivin-T34A alone or in combination with gemcitabine increased apoptosis in multiple pancreatic cancer cell lines, as well as enhanced the sensitivity of these cells to gemcitabine.

Beyond such applications, the use of EVs in site-specific drug delivery can be improved by protein engineering and modifying the vesicle surface by attaching additional ligands to improve EV targeting properties and their interaction with tumor cells [243]. For instance, glycosylphosphatidylinositol (GPI) anchored EV proteins such as decay-accelerating factor (known as CD55) were used by Kooijmans et al. [244] to attach anti-epidermal growth factor receptor (EGFR) nanobodies to EVs and thereby improve targeting to EGFR overexpressing epidermoid carcinoma A431 cells. They showed that the GPI-linked nanobodies were successfully displayed on EV surfaces and greatly improved EV binding to tumor cells in a manner dependent on EGFR density.

On the other hand, EVs readily transfer nucleic acids, such as DNA or RNA, to cells where they can cause specific genetic changes. Regarding genetic drug delivery,

Kamerkar et al. [245] engineered EVs known as iExosomes derived from fibroblast-like mesenchymal cells loaded by electroporation with siRNA or shRNA specific for the oncogenic GTPase KrasG12D, which is a common mutation in pancreatic cancer. The iExosomes showed enhanced targeting to oncogenic Kras-expressing cells, which was dependent on CD47 and the uptake facilitated by micropinocytosis. Subsequently, the treatment with iExosomes was shown to inhibit tumor growth and significantly increase the overall survival in multiple mouse models of pancreatic cancer.

5.3. EV Imaging for Cancer Diagnosis

Regarding the imaging of tumors, one of the major problems is the tremendous spatial heterogeneity combined with temporal variation, which leads to errors in the diagnosis and surgical treatment of tumors and thus represents major causes of therapy failure [246]. Since EVs permit detecting as little as a few hundred cancer cells, their application in cancer imaging represents a promising new approach. By attaching an optical reporter in the nanoscale dimension to the EVs and combining with optical imaging, robust diagnostic and prognostic modalities can be developed [243]. Using such approaches, tumor-targeted EVs can be monitored in real time to check their distribution and identify the precise location of tumors. Fluorescence is generally used for exosome tracking and imaging because of its great versatility and simple application by incubation of EVs with a variety of lipophilic fluorescent markers. In this field, generally small lipophilic fluorescent dyes, such as DiR, DiD and PKH67, have been used to label the membranes of EVs. Although these dyes are useful for distribution studies, clinical applications for diagnosis have yet to be developed [247,248]. Additionally, EV membranes have been labeled with fluorescent proteins, such as green fluorescent protein (GFP) or tandem dimer tomato (td Tomato) [249]. This type of labeling is considered more stable and suitable for evaluation in clinical applications. EVs can also be labeled using luciferase reporters in the cells of origin to produce bioluminescent proteins that are then included in the EVs and permit stable real-time monitoring [250,251].

Another alternative is the use of semiconductor quantum dots as optical reporters. They are more stable and have tunable optical properties that can be used for a wide range of applications, including in vivo imaging and diagnosis. For instance, Zong et al. [252] and Jiang et al. [253] obtained high-resolution images of breast tumor cells or their metastatic activity by loading either silicon or gold-carbon quantum dots, respectively, onto the outer membrane of the exosomes secreted by SKBR3 cells.

Superparamagnetic iron oxide nanoparticles (SPIONs) represent another interesting system for imaging. They have been effectively incorporated into EVs and then tracked in vivo by magnetic particle imaging and MRI, as has been shown for breast cancer [254] and melanomas [255].

Another type of nanomaterial that can be used for EV imaging is gold nanoparticles (AuNPs), which are highly versatile due to their tunability, biocompatibility, and unique optical properties [256]. The AuNP optical properties are due to the interaction of light with the electrons on the surface of the nanoparticles, which produces the collective oscillation of electrons, a phenomenon called surface plasmon resonance (SPR). This phenomenon leads to higher light absorption and scattering efficiency, thus making AuNPs excellent photoacoustic and Raman imaging agents [257,258]. On the other hand, gold exhibits a high absorption coefficient of X-rays, which make AuNPs useful as contrast agents for computerized tomography. AuNPs can be efficiently incorporated into EVs and then used for imaging, as well as tumor ablation in cancer therapy. In this field, Lara et al. [229] developed a double-labeling method to incorporate AuNPs indirectly into EVs by incubating them with cells and isolating them in EVs, which were then labeled with fluorescent dyes. This combination permitted analyzing the vesicle biodistribution and detecting the presence of small metastatic foci in the animal lungs by neutron activation analysis, NIR fluorescence, CT imaging and gold-enhanced microscopy imaging.

The use of EVs in imaging applications has been made possible by exploiting some of their natural properties. In particular, EVs have a mean size of 50–200 nm and can evade clearance by the mononuclear phagocytes, as well as favor passive extravasation in inflamed tissues [259]. The presence of immunomodulatory molecules, such as CD47 and PD-L1 ligand, on the EV surface aids significantly in avoiding phagocytosis and suppressing T cell activation, respectively [87,245]. EVs also possess a "tunable" surface that can be modified by adding targeting molecules, such as antibodies, aptamers, and ligands, all of which can favor specific EV accumulation in tumors, thereby avoiding undesirable off-target effects [260,261]. For these reasons, EVs are nowadays considered very appealing nanoscale tools for use as diagnostic sensors, as well as therapeutic vehicles in oncology.

5.4. EVs for Theranostic Applications

With the advances in nanotechnology and thanks to their unique properties, nanoparticles have become a promising tool in many areas in recent years, including theranosis. This novel concept, which combines the use of an agent for diagnosis and therapy in a single formulation, represents a great advance in personalized medicine. Existing evidence points towards the great potential of EVs both as diagnostic biomarkers and therapeutic tools. Such tumor-derived EVs have characteristic proteomic and genomic signatures, indicating that they represent suitable vectors for cancer diagnosis and prognosis [18,262]. In addition, because EVs can transfer various therapeutic compounds, as well as imaging agents, some researchers have proposed to exploit these vesicles as a tool for simultaneous therapy and active diagnosis (see Figure 4). EVs have been proposed as an ideal solution to overcome limitations of inorganic particles, including toxicity, off-target effects and immunogenicity [263].

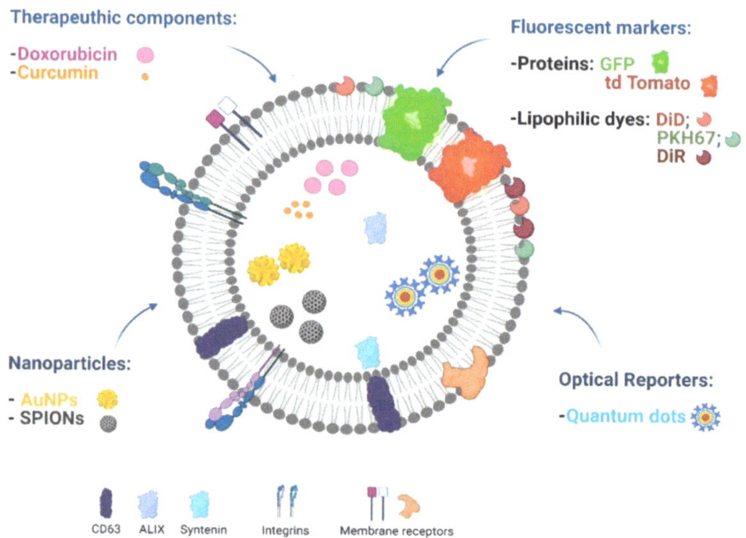

Figure 4. EVs as theranostic nanoplatforms. By combining the natural properties of exosomes with the use of drugs, imaging agents, or NPs, unique platforms can be generated. In combination with different targeting strategies, multiple therapeutic benefits can be achieved, such as improved targeting/uptake, immunomodulation, prolonged circulation time, easy tracking and better therapeutic effects. In doing so, side effects can be reduced, and the need to apply multiple drugs can be eliminated.

In this regard, Jia et al. [264] obtained glioma-targeting EVs with diagnostic and therapeutic potential by conjugating RGE, which is a peptide that binds to neuropilin-1 overexpressed on glioma cells, with the EV membrane by applying click chemistry. In addition, superparamagnetic iron nanoparticles (SPION) and curcumin were synchronously loaded into EVs by electroporation. The engineered system efficiently crossed the blood–brain barrier and provided good results for MRI-targeted imaging when applied to glioma cells and in orthotopic xenograft models. Additionally, SPION-mediated magnetic flow hyperthermia and curcumin-mediated effects combined lead to synergistic antitumor activity. Likewise, Wang et al. [234] designed a new platform for tumor-targeted chemo-photothermal therapy and imaging that was based on combining gold nanorods (AuNRs) with exosomes through a donor cell-assisted membrane modification strategy. First, the membrane of the donor cells was modified with RGD peptides and sulfhydryl groups. Then, the isolated exosomes (RGD-Exos-SH) were functionalized with AuNRs by the formation of Au-S bonds and coupling folic acid (FA) to improve the uptake efficiency by tumor cells. Further, doxorubicin (DOX) was loaded into exosomes by electroporation. Such designer exosomes showed effective accumulation at target tumor sites via dual ligand-mediated endocytosis, which were monitored in nude mice bearing tumor cell xenografts, using non-invasive near-infrared optical imaging. Moreover, the localized hyperthermia induced by the conjugated AuNRs during near-infrared irradiation increases the permeability of exosome membranes to enhance drug release, thereby preventing tumor relapse in a programable manner. Hence, the compatibility of EVs with different therapeutic agents and nanomaterials provides a unique opportunity to develop novel approaches in diagnosis and personalized treatment modalities.

6. EVs and Cancer Patient Survival

Cancer-derived EVs/exosomes are promising markers for diagnosis and prognosis in cancer and have been shown to predict the survival of patients. For example, in colorectal cancer, high levels of exosomes in the plasma of patients correlated with elevated presence of the carcino-embryonic antigen (CEA), and such patients tended to have shorter overall survival periods than patients with low exosome levels [265]. In addition, in lung cancer, the presence of the EGFR protein in exosomes from patient plasma has been suggested to represent a biomarker for lung cancer diagnosis [266]. Furthermore, high urinary exosomal levels of the long non-coding RNAs (lncRNAs) MALAT-1 and PCAT-1 correlated with decreased recurrence-free survival in non-invasive muscle bladder cancer (NIMBC) patients [267]. In pancreatic ductal carcinoma (PDAC), the lncRNA Sox2ot was identified in exosomes from plasma samples, and its presence there was closely associated with higher Classification of Malignant Tumors (TNM) stage and reduced overall survival rates of PDAC patients [268]. Combined analysis of exosomal miR-1290 and miR-375 reportedly predicts the overall survival of castration-resistant prostate cancer patients. Over the same follow-up period of 20 months, patients with high levels of both miRNAs had a general mortality rate of 80%, while patients with normal concentrations for both only had a mortality rate of 10% [269]. These are just a few examples from a rapidly growing research field illustrating how exosomes/EVs affect the survival of cancer patients.

On the other hand, tumor-derived exosomes may also be used to evaluate the response to surgery. For instance, their persistence after PDAC tumor resection is related to the presence of hidden metastases. Patients with more than 20% heparan sulfate proteoglycan glypican-1 (GPC1) positive exosomes in peripheral blood have been reported to have lower progression-free and overall survival [270]. Similarly, the detection of high levels of exosome-encapsulated miR-415a was also associated with reduced progression-free and overall survival [271]. In addition, the detection of exosomes containing miR-4525, miR-415a, and miR-21 in the portal vein identified more effectively patients at high risk for recurrence after surgery than did the detection in peripheral blood [272].

In summary, these examples illustrate how the targeted identification of specific proteins or miRNAs in exosomes may serve both diagnostic, as well as prognostic purposes. Indeed, it is important to mention that currently, there are 89 registered clinical trials [273] underway, studying exosomes in cancer patients and looking for markers that could be useful for diagnosis or prognosis. Of these trials, many focus on the prevalent cancer types in the lungs and prostate (14), breast (9), pancreas (8), and colon (6). As the results of these studies become available in the next few years, we may anticipate that a clearer picture should emerge connecting the presence of exosomes and their content to the survival of cancer patients.

7. Concluding Remarks

In summary, EVs are a heterogeneous population of membrane-enclosed, non-replicating, and sub-micron sized structures. EVs are actively released by virtually all cell types and by a wide variety of eukaryotic and prokaryotic organisms. EVs can be sorted into three different subtypes according to their biogenesis and biophysical properties: exosomes, microvesicles, and apoptotic bodies (Figure 1). A large number of studies show that EVs are active participants in cell communication. In the context of cancer biology, cancer cell-derived EV cargoes can change the behavior of target cells. The evidence provided shows that EVs are involved in the acquisition of all the "hallmarks of cancer", that is, biological characteristics acquired by cancer cells during tumor development. Consequently, more aggressive cancer cells can transfer their "traits" to less aggressive cancer cells and convert them into more malignant tumor cells (Figure 2). In addition, EVs play a role in the mechanisms of drug resistance, which can be acquired by drug-sensitive cells through EV-mediated transfer of resistance factors and other mechanisms, which aid in understanding why chemotherapy often fails. However, on the upside, these very characteristics of EVs combined with drug loading and targeting strategies provide unique opportunities for the delivery of different cargoes (Figure 3). Finally, EVs can be used as biological nanocarriers for both therapeutic and diagnostic purposes (theranosis) by including active principles, nanoparticles, as well as imaging agents (Figure 4). With this in mind, it would appear that such "designer" EVs will have a bright future in cancer medicine.

Author Contributions: Writing—original draft preparation, R.B.-R., A.C., M.C.D.-V., M.F.G. and D.L.; writing—review and editing, R.B.-R., A.C., M.C.D.-V., M.F.G., D.L., L.L.-G., L.L., M.J.K. and A.F.G.Q.; supervision, M.J.K. and A.F.G.Q.; funding acquisition, A.F.G.Q., L.L., L.L.-G., M.J.K., A.C., R.B.-R. and M.F.G. All authors have read and agreed to the published version of the manuscript.

Funding: Research in the laboratories was supported by Fondecyt grants 1210644 (A.F.G.Q.), 1200836 (L.L.), 1211223 (L.L.-G.), FONDAP grant 15130011 (A.F.G.Q., M.J.K., L.L., L.L.-G.), ANID postdoctoral fellowship award Becas Chile 2019 (AC), ANID PhD fellowship awards 21200147 (R.B.-R.), 21191341 (M.C.D.-V.), 21170292 (M.F.G.).

Acknowledgments: The authors also thank Alejandra Sandoval for the use of her paid account to make the figures created with BioRender.com accessed on 28 April 2021.

Conflicts of Interest: The authors declare no conflict of interest.

References

1. Chargaff, E.; West, R. The biological significance of the thromboplastic protein of blood. *J. Biol. Chem.* **1946**, *166*, 189–197. [CrossRef]
2. Wolf, P. The Nature and Significance of Platelet Products in Human Plasma. *Br. J. Haematol.* **1967**, *13*, 269–288. [CrossRef] [PubMed]
3. Yáñez-Mó, M.; Siljander, P.R.-M.; Andreu, Z.; Zavec, A.B.; Borràs, F.E.; Buzas, E.I.; Buzas, K.; Casal, E.; Cappello, F.; Carvalho, J.; et al. Biological Properties of Extracellular Vesicles and their Physiological Functions. *J. Extracell. Vesicles* **2015**, *4*, 27066. [CrossRef] [PubMed]
4. Aheget, H.; Mazini, L.; Martin, F.; Belqat, B.; Marchal, J.A.; Benabdellah, K. Exosomes: Their Role in Pathogenesis, Diagnosis and Treatment of Diseases. *Cancers* **2020**, *13*, 84. [CrossRef]
5. Shao, H.; Im, H.; Castro, C.M.; Breakefield, X.; Weissleder, R.; Lee, H. New Technologies for Analysis of Extracellular Vesicles. *Chem. Rev.* **2018**, *118*, 1917–1950. [CrossRef]
6. Van Niel, G.; D'Angelo, G.; Raposo, G. Shedding light on the cell biology of extracellular vesicles. *Nat. Rev. Mol. Cell Biol.* **2018**, *19*, 213–228. [CrossRef] [PubMed]
7. Caby, M.-P.; Lankar, D.; Vincendeau-Scherrer, C.; Raposo, G.; Bonnerot, C. Exosomal-like vesicles are present in human blood plasma. *Int. Immunol.* **2005**, *17*, 879–887. [CrossRef]

8. Gonzales, P.A.; Zhou, H.; Pisitkun, T.; Wang, N.S.; Star, R.A.; Knepper, M.A.; Yuen, P.S.T. Isolation and Purification of Exosomes in Urine. *Methods Mol. Biol.* **2010**, *641*, 89–99. [CrossRef]

9. Gomez, C.D.L.T.; Goreham, R.V.; Serra, J.J.B.; Nann, T.; Kussmann, M. "Exosomics"—A Review of Biophysics, Biology and Biochemistry of Exosomes with a Focus on Human Breast Milk. *Front. Genet.* **2018**, *9*, 92. [CrossRef]

10. Ding, X.-Q.; Wang, Z.-Y.; Xia, D.; Wang, R.-X.; Pan, X.-R.; Tong, J.-H. Proteomic Profiling of Serum Exosomes From Patients With Metastatic Gastric Cancer. *Front. Oncol.* **2020**, *10*, 1113. [CrossRef]

11. Mir, B.; Goettsch, C. Extracellular Vesicles as Delivery Vehicles of Specific Cellular Cargo. *Cells* **2020**, *9*, 1601. [CrossRef] [PubMed]

12. Roy, S.; Lin, H.-Y.; Chou, C.-Y.; Huang, C.-H.; Small, J.; Sadik, N.; Ayinon, C.M.; Lansbury, E.; Cruz, L.; Yekula, A.; et al. Navigating the Landscape of Tumor Extracellular Vesicle Heterogeneity. *Int. J. Mol. Sci.* **2019**, *20*, 1349. [CrossRef] [PubMed]

13. Xavier, C.P.R.; Caires, H.R.; Barbosa, M.A.G.; Bergantim, R.; Guimarães, J.E.; Vasconcelos, M.H. The Role of Extracellular Vesicles in the Hallmarks of Cancer and Drug Resistance. *Cells* **2020**, *9*, 1141. [CrossRef] [PubMed]

14. Campos, A.; Salomon, C.; Bustos, R.; Díaz, J.; Martínez, S.; Silva, V.; Reyes, C.; Diaz-Valdivia, N.; Varas-Godoy, M.; Lobos-Gonzalez, L.; et al. Caveolin-1-containing extracellular vesicles transport adhesion proteins and promote malignancy in breast cancer cell lines. *Nanomedicine* **2018**, *13*, 2597–2609. [CrossRef]

15. Théry, C.; Witwer, K.W.; Aikawa, E.; Alcaraz, M.J.; Anderson, J.D.; Andriantsitohaina, R.; Antoniou, A.; Arab, T.; Archer, F.; Atkin-Smith, G.K.; et al. Minimal information for studies of extracellular vesicles 2018 (MISEV2018): A position statement of the International Society for Extracellular Vesicles and update of the MISEV2014 guidelines. *J. Extracell. Vesicles* **2018**, *7*, 1535750. [CrossRef]

16. Hanahan, D.; Weinberg, R.A. Hallmarks of Cancer: The Next Generation. *Cell* **2011**, *144*, 646–674. [CrossRef]

17. Meng, X.; Pan, J.; Sun, S.; Gong, Z. Circulating exosomes and their cargos in blood as novel biomarkers for cancer. *Transl. Cancer Res.* **2018**, *7*, S226–S242. [CrossRef]

18. Huang, T.; Deng, C.-X. Current Progresses of Exosomes as Cancer Diagnostic and Prognostic Biomarkers. *Int. J. Biol. Sci.* **2019**, *15*, 1–11. [CrossRef]

19. Armstrong, D.; Wildman, D.E. Extracellular Vesicles and the Promise of Continuous Liquid Biopsies. *J. Pathol. Transl. Med.* **2018**, *52*, 1–8. [CrossRef]

20. Zhou, B.; Xu, K.; Zheng, X.; Chen, T.; Wang, J.; Song, Y.; Shao, Y.; Zheng, S. Application of exosomes as liquid biopsy in clinical diagnosis. *Signal. Transduct. Target. Ther.* **2020**, *5*, 1–14. [CrossRef]

21. Surman, M.; Drożdż, A.; Stępień, E.; Przybyło, M. Extracellular Vesicles as Drug Delivery Systems-Methods of Production and Potential Therapeutic Applications. *Curr. Pharm. Des.* **2019**, *25*, 132–154. [CrossRef] [PubMed]

22. Hernandez-Oller, L.; Seras-Franzoso, J.; Andrade, F.; Rafael, D.; Abasolo, I.; Gener, P.; Schwartz, S., Jr. Extracellular Vesicles as Drug Delivery Systems in Cancer. *Pharmaceutics* **2020**, *12*, 1146. [CrossRef]

23. Lötvall, J.; Hill, A.F.; Hochberg, F.; Buzás, E.I.; Di Vizio, D.; Gardiner, C.; Gho, Y.S.; Kurochkin, I.V.; Mathivanan, S.; Quesenberry, P.; et al. Minimal Experimental Requirements for Definition of Extracellular Vesicles and their Functions: A Position Statement from the International Society for Extracellular Vesicles. *J. Extracell. Vesicles* **2014**, *3*, 26913. [CrossRef]

24. Abels, E.R.; Breakefield, X.O. Introduction to Extracellular Vesicles: Biogenesis, RNA Cargo Selection, Content, Release, and Uptake. *Cell. Mol. Neurobiol.* **2016**, *36*, 301–312. [CrossRef] [PubMed]

25. Skryabin, G.O.; Komelkov, A.V.; Savelyeva, E.E.; Tchevkina, E.M. Lipid Rafts in Exosome Biogenesis. *Biochemistry* **2020**, *85*, 177–191. [CrossRef] [PubMed]

26. Trajkovic, K.; Hsu, C.; Chiantia, S.; Rajendran, L.; Wenzel, D.; Wieland, F.; Schwille, P.; Brugger, B.; Simons, M. Ceramide Triggers Budding of Exosome Vesicles into Multivesicular Endosomes. *Science* **2008**, *319*, 1244–1247. [CrossRef] [PubMed]

27. Catalano, M.; O'Driscoll, L. Inhibiting extracellular vesicles formation and release: A review of EV inhibitors. *J. Extracell. Vesicles* **2020**, *9*, 1703244. [CrossRef] [PubMed]

28. Savina, A.; Vidal, M.; Colombo, M.I. The exosome pathway in K562 cells is regulated by Rab11. *J. Cell Sci.* **2002**, *115*, 2505–2515. [CrossRef] [PubMed]

29. Hsu, C.; Morohashi, Y.; Yoshimura, S.-I.; Hoyos, N.M.; Jung, S.; Lauterbach, M.A.; Bakhti, M.; Grønborg, M.; Möbius, W.; Rhee, J.; et al. Regulation of exosome secretion by Rab35 and its GTPase-activating proteins TBC1D10A–C. *J. Cell Biol.* **2010**, *189*, 223–232. [CrossRef]

30. Ostrowski, M.; Carmo, N.; Krumeich, S.; Fanget, I.; Raposo, G.; Savina, A.; Moita, C.F.; Schauer, K.; Hume, A.; Freitas, R.P.; et al. Rab27a and Rab27b control different steps of the exosome secretion pathway. *Nat. Cell Biol.* **2009**, *12*, 19–30. [CrossRef] [PubMed]

31. Akers, J.C.; Gonda, D.; Kim, R.; Carter, B.S.; Chen, C.C. Biogenesis of extracellular vesicles (EV): Exosomes, microvesicles, retrovirus-like vesicles, and apoptotic bodies. *J. Neuro Oncol.* **2013**, *113*, 1–11. [CrossRef]

32. Petrovčíková, E.; Vičíková, K.; Leksa, V. Extracellular vesicles–biogenesis, composition, function, uptake and therapeutic applications. *Biologia* **2018**, *73*, 437–448. [CrossRef]

33. Muralidharan-Chari, V.; Clancy, J.; Plou, C.; Romao, M.; Chavrier, P.; Raposo, G.; D'Souza-Schorey, C. ARF6-Regulated Shedding of Tumor Cell-Derived Plasma Membrane Microvesicles. *Curr. Biol.* **2009**, *19*, 1875–1885. [CrossRef]

34. Tan, C.F.; Teo, H.S.; Park, J.E.; Dutta, B.; Tse, S.W.; Leow, M.K.-S.; Wahli, W.; Sze, S.K. Exploring Extracellular Vesicles Biogenesis in Hypothalamic Cells through a Heavy Isotope Pulse/Trace Proteomic Approach. *Cells* **2020**, *9*, 1320. [CrossRef]

35. Wang, T.; Gilkes, D.M.; Takano, N.; Xiang, L.; Luo, W.; Bishop, C.J.; Chaturvedi, P.; Green, J.J.; Semenza, G.L. Hypoxia-inducible factors and RAB22A mediate formation of microvesicles that stimulate breast cancer invasion and metastasis. *Proc. Natl. Acad. Sci. USA* **2014**, *111*, E3234–E3242. [CrossRef]
36. Li, B.; Antonyak, M.A.; Zhang, J.; Cerione, R.A. RhoA triggers a specific signaling pathway that generates transforming microvesicles in cancer cells. *Oncogene* **2012**, *31*, 4740–4749. [CrossRef] [PubMed]
37. Schlienger, S.; Campbell, S.; Claing, A. ARF1 regulates the Rho/MLC pathway to control EGF-dependent breast cancer cell invasion. *Mol. Biol. Cell* **2014**, *25*, 17–29. [CrossRef] [PubMed]
38. Willms, E.; Cabañas, C.; Mäger, I.; Wood, M.J.A.; Vader, P. Extracellular Vesicle Heterogeneity: Subpopulations, Isolation Techniques, and Diverse Functions in Cancer Progression. *Front. Immunol.* **2018**, *9*, 738. [CrossRef] [PubMed]
39. Zhang, H.; Freitas, D.; Kim, H.S.; Fabijanic, K.; Li, Z.; Chen, H.; Mark, M.T.; Molina, H.; Benito-Martin, A.; Bojmar, L.; et al. Identification of distinct nanoparticles and subsets of extracellular vesicles by asymmetric flow field-flow fractionation. *Nat. Cell Biol.* **2018**, *20*, 332–343. [CrossRef] [PubMed]
40. Zhang, Q.; Higginbotham, J.N.; Jeppesen, D.; Yang, Y.-P.; Li, W.; McKinley, E.T.; Graves-Deal, R.; Ping, J.; Britain, C.M.; Dorsett, K.A.; et al. Transfer of Functional Cargo in Exomeres. *Cell Rep.* **2019**, *27*, 940–954.e6. [CrossRef] [PubMed]
41. Zijlstra, A.; Di Vizio, D. Size matters in nanoscale communication. *Nat. Cell Biol.* **2018**, *20*, 228–230. [CrossRef] [PubMed]
42. Anand, S.; Samuel, M.; Mathivanan, S. Exomeres: A New Member of Extracellular Vesicles Family. In *Prokaryotic Cytoskeletons*; Springer Science and Business Media LLC: Berlin, Germany, 2021; Volume 97, pp. 89–97.
43. Zhang, Q.; Jeppesen, D.K.; Higginbotham, J.N.; Franklin, J.L.; Crowe, J.E.; Coffey, R.J. Angiotensin-converting Enzyme 2–containing Small Extracellular Vesicles and Exomeres Bind the Severe Acute Respiratory Syndrome Coronavirus 2 Spike Protein. *Gastroenterology* **2021**, *160*, 958–961.e3. [CrossRef]
44. Monteil, V.; Kwon, H.; Prado, P.; Hagelkrüys, A.; Wimmer, R.A.; Stahl, M.; Leopoldi, A.; Garreta, E.; Hurtado Del Pozo, C.; Prosper, F.; et al. Inhibition of SARS-CoV-2 Infections in Engineered Human Tissues Using Clinical-Grade Soluble Human ACE2. *Cell* **2020**, *181*, 905–913.e7. [CrossRef]
45. McKelvey, K.J.; Powell, K.L.; Ashton, A.W.; Morris, J.M.; McCracken, S.A. Exosomes: Mechanisms of Uptake. *J. Circ. Biomark.* **2015**, *4*, 7. [CrossRef]
46. Jadli, A.S.; Ballasy, N.; Edalat, P.; Patel, V.B. Inside(sight) of tiny communicator: Exosome biogenesis, secretion, and uptake. *Mol. Cell. Biochem.* **2020**, *467*, 77–94. [CrossRef] [PubMed]
47. Simons, M.; Raposo, G. Exosomes–vesicular carriers for intercellular communication. *Curr. Opin. Cell Biol.* **2009**, *21*, 575–581. [CrossRef]
48. Huang, M.-B.; Xia, M.; Gao, Z.; Zhou, H.; Liu, M.; Huang, S.; Zhen, R.; Wu, J.Y.; Roth, W.W.; Bond, V.C.; et al. Characterization of Exosomes in Plasma of Patients with Breast, Ovarian, Prostate, Hepatic, Gastric, Colon, and Pancreatic Cancers. *J. Cancer Ther.* **2019**, *10*, 382–399. [CrossRef]
49. Kim, H.K.; Song, K.S.; Park, Y.S.; Kang, Y.H.; Lee, Y.J.; Lee, K.R.; Ryu, K.W.; Bae, J.M.; Kim, S. Elevated levels of circulating platelet microparticles, VEGF, IL-6 and RANTES in patients with gastric cancer: Possible role of a metastasis predictor. *Eur. J. Cancer* **2003**, *39*, 184–191. [CrossRef]
50. Baran, J.; Baj-Krzyworzeka, M.; Weglarczyk, K.; Szatanek, R.; Zembala, M.; Barbasz, J.; Czupryna, A.; Szczepanik, A.; Zembala, M. Circulating tumour-derived microvesicles in plasma of gastric cancer patients. *Cancer Immunol. Immunother.* **2010**, *59*, 841–850. [CrossRef]
51. Caivano, A.; Laurenzana, I.; De Luca, L.; La Rocca, F.; Simeon, V.; Trino, S.; D'Auria, F.; Traficante, A.; Maietti, M.; Izzo, T.; et al. High serum levels of extracellular vesicles expressing malignancy-related markers are released in patients with various types of hematological neoplastic disorders. *Tumor Biol.* **2015**, *36*, 9739–9752. [CrossRef] [PubMed]
52. Giusti, I.; D'Ascenzo, S.; Dolo, V. Microvesicles as Potential Ovarian Cancer Biomarkers. *BioMed Res. Int.* **2013**, *2013*, 1–12. [CrossRef] [PubMed]
53. Messenger, S.W.; Woo, S.S.; Sun, Z.; Martin, T.F.J. A Ca2+-stimulated exosome release pathway in cancer cells is regulated by Munc13-4. *J. Cell Biol.* **2018**, *217*, 2877–2890. [CrossRef]
54. Augimeri, G.; La Camera, G.; Gelsomino, L.; Giordano, C.; Panza, S.; Sisci, D.; Morelli, C.; Győrffy, B.; Bonofiglio, D.; Andò, S.; et al. Evidence for Enhanced Exosome Production in Aromatase Inhibitor-Resistant Breast Cancer Cells. *Int. J. Mol. Sci.* **2020**, *21*, 5841. [CrossRef]
55. McAndrews, K.M.; Kalluri, R. Mechanisms associated with biogenesis of exosomes in cancer. *Mol. Cancer* **2019**, *18*, 1–11. [CrossRef] [PubMed]
56. Hanahan, D.; Weinberg, R.A. The Hallmarks of Cancer. *Cell* **2000**, *100*, 57–70. [CrossRef]
57. Gutkin, A.; Uziel, O.; Beery, E.; Nordenberg, J.; Pinchasi, M.; Goldvaser, H.; Henick, S.; Goldberg, M.; Lahav, M. Tumor cells derived exosomes contain hTERT mRNA and transform nonmalignant fibroblasts into telomerase positive cells. *Oncotarget* **2016**, *7*, 59173–59188. [CrossRef] [PubMed]
58. Skog, J.; Würdinger, T.; Van Rijn, S.; Meijer, D.H.; Gainche, L.; Curry, W.T., Jr.; Carter, B.S.; Krichevsky, A.M.; Breakefield, X.O. Glioblastoma microvesicles transport RNA and proteins that promote tumour growth and provide diagnostic biomarkers. *Nat. Cell Biol.* **2008**, *10*, 1470–1476. [CrossRef] [PubMed]
59. Meckes, D.G.; Shair, K.H.Y.; Marquitz, A.R.; Kung, C.-P.; Edwards, R.H.; Raab-Traub, N. Human tumor virus utilizes exosomes for intercellular communication. *Proc. Natl. Acad. Sci. USA* **2010**, *107*, 20370–20375. [CrossRef]

60. Ohshima, K.; Inoue, K.; Fujiwara, A.; Hatakeyama, K.; Kanto, K.; Watanabe, Y.; Muramatsu, K.; Fukuda, Y.; Ogura, S.-I.; Yamaguchi, K.; et al. Let-7 MicroRNA Family Is Selectively Secreted into the Extracellular Environment via Exosomes in a Metastatic Gastric Cancer Cell Line. *PLoS ONE* **2010**, *5*, e13247. [CrossRef]

61. Beckler, M.D.; Higginbotham, J.N.; Franklin, J.L.; Ham, A.-J.; Halvey, P.J.; Imasuen, I.E.; Whitwell, C.; Li, M.; Liebler, D.; Coffey, R.J. Proteomic Analysis of Exosomes from Mutant KRAS Colon Cancer Cells Identifies Intercellular Transfer of Mutant KRAS. *Mol. Cell. Proteom.* **2013**, *12*, 343–355. [CrossRef] [PubMed]

62. Choi, D.-Y.; You, S.; Jung, J.H.; Lee, J.C.; Rho, J.K.; Lee, K.Y.; Freeman, M.R.; Kim, K.P.; Kim, J. Extracellular vesicles shed from gefitinib-resistant nonsmall cell lung cancer regulate the tumor microenvironment. *Proteomics* **2014**, *14*, 1845–1856. [CrossRef] [PubMed]

63. Carvalho, A.S.; Baeta, H.; Silva, B.C.; Moraes, M.C.S.; Bodo, C.; Beck, H.C.; Rodriguez, M.S.; Saraswat, M.; Pandey, A.; Matthiesen, R. Extra-cellular vesicles carry proteome of cancer hallmarks. *Front. Biosci. Landmark Ed.* **2020**, *25*, 398–436.

64. Yoshii, S.; Hayashi, Y.; Iijima, H.; Inoue, T.; Kimura, K.; Sakatani, A.; Nagai, K.; Fujinaga, T.; Hiyama, S.; Kodama, T.; et al. Exosomal microRNAs derived from colon cancer cells promote tumor progression by suppressing fibroblast TP53 expression. *Cancer Sci.* **2019**, *110*, 2396–2407. [CrossRef]

65. Khoo, X.-H.; Paterson, I.C.; Goh, B.-H.; Lee, W.-L. Cisplatin-Resistance in Oral Squamous Cell Carcinoma: Regulation by Tumor Cell-Derived Extracellular Vesicles. *Cancers* **2019**, *11*, 1166. [CrossRef]

66. Wheatley, S.P.; Altieri, D.C. Survivin at a glance. *J. Cell Sci.* **2019**, *132*, 1–8. [CrossRef]

67. Radeghieri, A.; Savio, G.; Zendrini, A.; Di Noto, G.; Salvi, A.; Bergese, P.; Piovani, G. Cultured human amniocytes express hTERT, which is distributed between nucleus and cytoplasm and is secreted in extracellular vesicles. *Biochem. Biophys. Res. Commun.* **2017**, *483*, 706–711. [CrossRef] [PubMed]

68. Taraboletti, G.; D'Ascenzoy, S.; Giusti, I.; Marchetti, D.; Borsotti, P.; Millimaggi, D.; Giavazzi, R.; Pavan, A.; Dolo, V. Bioavailability of VEGF in Tumor-Shed Vesicles Depends on Vesicle Burst Induced by Acidic pH. *Neoplasia* **2006**, *8*, 96–103. [CrossRef] [PubMed]

69. Al-Nedawi, K.; Meehan, B.; Kerbel, R.S.; Allison, A.C.; Rak, J. Endothelial expression of autocrine VEGF upon the uptake of tumor-derived microvesicles containing oncogenic EGFR. *Proc. Natl. Acad. Sci. USA* **2009**, *106*, 3794–3799. [CrossRef] [PubMed]

70. Mineo, M.; Garfield, S.H.; Taverna, S.; Flugy, A.; De Leo, G.; Alessandro, R.; Kohn, E.C. Exosomes released by K562 chronic myeloid leukemia cells promote angiogenesis in a src-dependent fashion. *Angiogenesis* **2011**, *15*, 33–45. [CrossRef]

71. Li, C.C.Y.; Eaton, S.A.; Young, P.E.; Lee, M.; Shuttleworth, R.; Humphreys, D.; Grau, G.; Combes, V.; Bebawy, M.; Gong, J.; et al. Glioma microvesicles carry selectively packaged coding and non-coding RNAs which alter gene expression in recipient cells. *RNA Biol.* **2013**, *10*, 1333–1344. [CrossRef]

72. Ludwig, N.; Whiteside, T.L. Potential roles of tumor-derived exosomes in angiogenesis. *Expert Opin. Ther. Targets* **2018**, *22*, 409–417. [CrossRef]

73. Hakulinen, J.; Sankkila, L.; Sugiyama, N.; Lehti, K.; Keski-Oja, J. Secretion of active membrane type 1 matrix metalloproteinase (MMP-14) into extracellular space in microvesicular exosomes. *J. Cell. Biochem.* **2008**, *105*, 1211–1218. [CrossRef] [PubMed]

74. Atay, S.; Banskota, S.; Crow, J.; Sethi, G.; Rink, L.; Godwin, A.K. Oncogenic KIT-containing exosomes increase gastrointestinal stromal tumor cell invasion. *Proc. Natl. Acad. Sci. USA* **2014**, *111*, 711–716. [CrossRef]

75. Kobayashi, M.; Salomon, C.; Tapia, J.; Illanes, E.S.; Mitchell, M.D.; Rice, E.G. Ovarian cancer cell invasiveness is associated with discordant exosomal sequestration of Let-7 miRNA and miR-200. *J. Transl. Med.* **2014**, *12*, 4. [CrossRef]

76. Alečković, M.; Kang, Y. Regulation of cancer metastasis by cell-free miRNAs. *Biochim. Biophys. Acta BBA Bioenerg.* **2015**, *1855*, 24–42. [CrossRef] [PubMed]

77. Hoshino, A.; Costa-Silva, B.; Shen, T.-L.; Rodrigues, G.; Hashimoto, A.; Mark, M.T.; Molina, H.; Kohsaka, S.; Di Giannatale, A.; Ceder, S.; et al. Tumour exosome integrins determine organotropic metastasis. *Nature* **2015**, *527*, 329–335. [CrossRef] [PubMed]

78. Wang, X.; Ding, X.; Nan, L.; Wang, Y.; Wang, J.; Yan, Z.; Zhang, W.; Sun, J.; Zhu, W.; Ni, B.; et al. Investigation of the roles of exosomes in colorectal cancer liver metastasis. *Oncol. Rep.* **2015**, *33*, 2445–2453. [CrossRef] [PubMed]

79. Ma, Q.; Wu, H.; Xiao, Y.; Liang, Z.; Liu, T. Upregulation of exosomal microRNA-21 in pancreatic stellate cells promotes pancreatic cancer cell migration and enhances Ras/ERK pathway activity. *Int. J. Oncol.* **2020**, *56*, 1025–1033. [CrossRef] [PubMed]

80. He, J.; He, J.; Min, L.; He, Y.; Guan, H.; Wang, J.; Peng, X. Extracellular vesicles transmitted miR-31-5p promotes sorafenib resistance by targeting MLH1 in renal cell carcinoma. *Int. J. Cancer* **2020**, *146*, 1052–1063. [CrossRef] [PubMed]

81. Fabbri, M.; Paone, A.; Calore, F.; Galli, R.; Gaudio, E.; Santhanam, R.; Lovat, F.; Fadda, P.; Mao, C.; Nuovo, G.J.; et al. MicroRNAs bind to Toll-like receptors to induce prometastatic inflammatory response. *Proc. Natl. Acad. Sci. USA* **2012**, *109*, E2110–E2116. [CrossRef]

82. Haderk, F.; Schulz, R.; Iskar, M.; Cid, L.L.; Worst, T.; Willmund, K.V.; Schulz, A.; Warnken, U.; Seiler, J.; Benner, A.; et al. Tumor-derived exosomes modulate PD-L1 expression in monocytes. *Sci. Immunol.* **2017**, *2*, eaah5509. [CrossRef]

83. Kim, D.-K.; Lee, J.; Kim, S.R.; Choi, D.-S.; Yoon, Y.J.; Kim, J.H.; Go, G.; Nhung, D.; Hong, K.; Jang, S.C.; et al. EVpedia: A community web portal for extracellular vesicles research. *Bioinformatics* **2014**, *31*, 933–939. [CrossRef]

84. Zhang, J.; Lu, S.; Zhou, Y.; Meng, K.; Chen, Z.; Cui, Y.; Shi, Y.; Wang, T.; He, Q.Y. Motile hepatocellular carcinoma cells preferentially secret sugar metabolism regulatory proteins via exosomes. *Proteomics* **2017**, *17*, 1–36.

85. Xia, Y.; Zhang, Q.; Zhen, Q.; Zhao, Y.; Liu, N.; Li, T.; Hao, Y.; Zhang, J.; Luo, C.; Wu, X. Negative regulation of tumor-infiltrating NK cell in clear cell renal cell carcinoma patients through the exosomal pathway. *Oncotarget* **2017**, *8*, 37783–37795. [CrossRef] [PubMed]

86. Chen, G.; Huang, A.C.; Zhang, W.; Zhang, G.; Wu, M.; Xu, W.; Yu, Z.; Yang, J.; Wang, B.; Sun, H.; et al. Exosomal PD-L1 contributes to immunosuppression and is associated with anti-PD-1 response. *Nature* **2018**, *560*, 382–386. [CrossRef] [PubMed]

87. Xie, F.; Xu, M.; Lu, J.; Mao, L.; Wang, S. The role of exosomal PD-L1 in tumor progression and immunotherapy. *Mol. Cancer* **2019**, *18*, 146. [CrossRef]

88. Koga, K.; Matsumoto, K.; Akiyoshi, T.; Kubo, M.; Yamanaka, N.; Tasaki, A.; Nakashima, H.; Nakamura, M.; Kuroki, S.; Tanaka, M.; et al. Purification, characterization and biological significance of tumor-derived exosomes. *Anticancer Res.* **2005**, *25*, 3703–3707.

89. O'Brien, K.; Rani, S.; Corcoran, C.; Wallace, R.; Hughes, L.; Friel, A.; McDonnell, S.; Crown, J.; Radomski, M.W.; O'Driscoll, L. Exosomes from triple-negative breast cancer cells can transfer phenotypic traits representing their cells of origin to secondary cells. *Eur. J. Cancer* **2013**, *49*, 1845–1859. [CrossRef]

90. Qu, J.-L.; Qu, X.-J.; Zhao, M.-F.; Teng, Y.-E.; Zhang, Y.; Hou, K.-Z.; Jiang, Y.-H.; Yang, X.-H.; Liu, Y.-P.; Qu, J.-L.; et al. Gastric cancer exosomes promote tumour cell proliferation through PI3K/Akt and MAPK/ERK activation. *Dig. Liver Dis.* **2009**, *41*, 875–880. [CrossRef]

91. Yang, L.; Wu, X.-H.; Wang, D.; Luo, C.-L.; Chen, L.-X. Bladder cancer cell-derived exosomes inhibit tumor cell apoptosis and induce cell proliferation in vitro. *Mol. Med. Rep.* **2013**, *8*, 1272–1278. [CrossRef]

92. Dai, G.; Yao, X.; Zhang, Y.; Gu, J.; Geng, Y.; Xue, F.; Zhang, J. Colorectal cancer cell–derived exosomes containing miR-10b regulate fibroblast cells via the PI3K/Akt pathway. *Bull. Cancer* **2018**, *105*, 336–349. [CrossRef]

93. Lei, L.; Mou, Q. Exosomal taurine up-regulated 1 promotes angiogenesis and endothelial cell proliferation in cervical cancer. *Cancer Biol. Ther.* **2020**, *21*, 717–725. [CrossRef]

94. Liu, S.; Sun, J.; Lan, Q. Glioblastoma microvesicles promote endothelial cell proliferation through Akt/beta-catenin pathway. *Int. J. Clin. Exp. Pathol.* **2014**, *7*, 4857–4866.

95. Huang, J.; Ding, Z.; Luo, Q.; Xu, W. Cancer cell-derived exosomes promote cell proliferation and inhibit cell apoptosis of both normal lung fibroblasts and non-small cell lung cancer cell through delivering alpha-smooth muscle actin. *Am. J. Transl. Res.* **2019**, *11*, 1711–1723.

96. Richards, K.E.; Zeleniak, A.E.; Fishel, M.; Wu, J.; Littlepage, L.E.; Hill, R. Cancer-associated fibroblast exosomes regulate survival and proliferation of pancreatic cancer cells. *Oncogene* **2017**, *36*, 1770–1778. [CrossRef] [PubMed]

97. Li, Y.-Y.; Tao, Y.-W.; Gao, S.; Li, P.; Zheng, J.-M.; Zhang, S.-E.; Liang, J.; Zhang, Y. Cancer-associated fibroblasts contribute to oral cancer cells proliferation and metastasis via exosome-mediated paracrine miR-34a-5p. *EBioMedicine* **2018**, *36*, 209–220. [CrossRef] [PubMed]

98. Pavlakis, E.; Neumann, M.; Stiewe, T. Extracellular Vesicles: Messengers of p53 in Tumor–Stroma Communication and Cancer Metastasis. *Int. J. Mol. Sci.* **2020**, *21*, 9648. [CrossRef] [PubMed]

99. Yu, X.; Harris, S.L.; Levine, A.J. The Regulation of Exosome Secretion: A Novel Function of the p53 Protein. *Cancer Res.* **2006**, *66*, 4795–4801. [CrossRef] [PubMed]

100. Yu, X.; Riley, T.; Levine, A.J. The regulation of the endosomal compartment by p53 the tumor suppressor gene. *Febs. J.* **2009**, *276*, 2201–2212. [CrossRef] [PubMed]

101. Vietri, M.; Radulovic, M.; Stenmark, H. The many functions of ESCRTs. *Nat. Rev. Mol. Cell Biol.* **2020**, *21*, 25–42. [CrossRef] [PubMed]

102. Cooks, T.; Pateras, I.S.; Jenkins, L.M.; Patel, K.M.; Robles, A.I.; Morris, J.; Forshew, T.; Appella, E.; Gorgoulis, V.; Harris, C.C. Mutant p53 cancers reprogram macrophages to tumor supporting macrophages via exosomal miR-1246. *Nat. Commun.* **2018**, *9*, 1–15. [CrossRef]

103. Hallal, S.; Mallawaaratchy, D.M.; Wei, H.; Ebrahimkhani, S.; Stringer, B.W.; Day, B.W.; Boyd, A.W.; Guillemin, G.; Buckland, M.E.; Kaufman, K.L. Extracellular Vesicles Released by Glioblastoma Cells Stimulate Normal Astrocytes to Acquire a Tumor-Supportive Phenotype Via p53 and MYC Signaling Pathways. *Mol. Neurobiol.* **2019**, *56*, 4566–4581. [CrossRef]

104. Pazzaglia, E.U.; Pringle, A.J. The role of macrophages and giant cells in loosening of joint replacement. *Arch. Orthop. Trauma. Surgery* **1988**, *107*, 20–26. [CrossRef]

105. Cappellesso, R.; Tinazzi, A.; Giurici, T.; Simonato, F.; Guzzardo, V.; Ventura, L.; Crescenzi, M.; Chiarelli, S.; Fassina, A. Programmed cell death 4 and microRNA 21 inverse expression is maintained in cells and exosomes from ovarian serous carcinoma effusions. *Cancer Cytopathol.* **2014**, *122*, 685–693. [CrossRef]

106. Chen, W.-X.; Liu, X.-M.; Lv, M.-M.; Chen, L.; Zhao, J.-H.; Zhong, S.-L.; Ji, M.-H.; Hu, Q.; Luo, Z.; Wu, J.-Z.; et al. Exosomes from drug-resistant breast cancer cells transmit chemoresistance by a horizontal transfer of MicroRNAs. *PLoS ONE* **2014**, *9*, e95240. [CrossRef]

107. Liu, T.; Chen, G.; Sun, D.; Lei, M.; Li, Y.; Zhou, C.; Li, X.; Xue, W.; Wang, H.; Liu, C.; et al. Exosomes containing miR-21 transfer the characteristic of cisplatin resistance by targeting PTEN and PDCD4 in oral squamous cell carcinoma. *Acta Biochim. Biophys. Sin.* **2017**, *49*, 808–816. [CrossRef] [PubMed]

108. Allegra, A.; Ettari, R.; Innao, V.; Bitto, A. Potential Role of microRNAs in inducing Drug Resistance in Patients with Multiple Myeloma. *Cells* **2021**, *10*, 448. [CrossRef]

109. Wang, J.; Hendrix, A.; Hernot, S.; Lemaire, M.; De Bruyne, E.; Van Valckenborgh, E.; Lahoutte, T.; De Wever, O.; Vanderkerken, K.; Menu, E. Bone marrow stromal cell–derived exosomes as communicators in drug resistance in multiple myeloma cells. *Blood* **2014**, *124*, 555–566. [CrossRef]

110. Khan, S.; Jutzy, J.M.S.; Aspe, J.R.; McGregor, D.W.; Neidigh, J.W.; Wall, N.R. Survivin is released from cancer cells via exosomes. *Apoptosis* **2010**, *16*, 1–12. [CrossRef]
111. Qu, J.L.; Qu, X.J.; Qu, J.L.; Qu, X.J.; Zhao, M.F.; Teng, Y.E.; Zhang, Y.; Hou, K.Z.; Jiang, Y.H.; Yang, X.H.; et al. The role of cbl family of ubiquitin ligases in gastric cancer exosome-induced apoptosis of Jurkat T cells. *Acta Oncol.* **2009**, *48*, 1173–1180. [CrossRef]
112. Hayflick, L. The limited in vitro lifetime of human diploid cell strains. *Exp. Cell Res.* **1965**, *37*, 614–636. [CrossRef]
113. Al-Mayah, A.H.J.; Bright, S.J.; Bowler, D.A.; Slijepcevic, P.; Goodwin, E.; Kadhim, M.A. Exosome-Mediated Telomere Instability in Human Breast Epithelial Cancer Cells after X Irradiation. *Radiat. Res.* **2017**, *187*, 98–106. [CrossRef] [PubMed]
114. Bhandari, V.; Hoey, C.; Liu, L.Y.; Lalonde, E.; Ray, J.; Livingstone, J.; Lesurf, R.; Shiah, Y.J.; Vujcic, T.; Huang, X.; et al. Molecular landmarks of tumor hypoxia across cancer types. *Nat. Genet.* **2019**, *51*, 308–318. [CrossRef]
115. Ohyashiki, J.H.; Umezu, T.; Ohyashiki, K. Exosomes promote bone marrow angiogenesis in hematologic neoplasia: The role of hypoxia. *Curr. Opin. Hematol.* **2016**, *23*. [CrossRef]
116. Shao, C.; Yang, F.; Miao, S.; Liu, W.; Wang, C.; Shu, Y.; Shen, H. Role of hypoxia-induced exosomes in tumor biology. *Mol. Cancer* **2018**, *17*, 1–8. [CrossRef]
117. Huang, Z.; Feng, Y. Exosomes Derived From Hypoxic Colorectal Cancer Cells Promote Angiogenesis Through Wnt4-Induced β-Catenin Signaling in Endothelial Cells. *Oncol. Res. Featur. Preclin. Clin. Cancer Ther.* **2017**, *25*, 651–661. [CrossRef]
118. Hsu, Y.-L.; Hung, J.-Y.; Chang, W.-A.; Lin, Y.-S.; Pan, Y.-C.; Tsai, P.-H.; Wu, C.-Y.; Kuo, P.-L. Hypoxic lung cancer-secreted exosomal miR-23a increased angiogenesis and vascular permeability by targeting prolyl hydroxylase and tight junction protein ZO-1. *Oncogene* **2017**, *36*, 4929–4942. [CrossRef]
119. Tadokoro, H.; Umezu, T.; Ohyashiki, K.; Hirano, T.; Ohyashiki, J.H. Exosomes derived from hypoxic leukemia cells enhance tube formation in endothelial cells. *J. Biol. Chem.* **2013**, *288*, 34343–34351. [CrossRef]
120. Yang, H.; Zhang, H.; Ge, S.; Ning, T.; Bai, M.; Li, J.; Li, S.; Sun, W.; Deng, T.; Zhang, L.; et al. Exosome-Derived miR-130a Activates Angiogenesis in Gastric Cancer by Targeting C-MYB in Vascular Endothelial Cells. *Mol. Ther.* **2018**, *26*, 2466–2475. [CrossRef]
121. He, L.; Zhu, W.; Chen, Q.; Yuan, Y.; Wang, Y.; Wang, J.; Wu, X. Ovarian cancer cell-secreted exosomal miR-205 promotes metastasis by inducing angiogenesis. *Theranostics* **2019**, *9*, 8206–8220. [CrossRef] [PubMed]
122. Shang, D.; Xie, C.; Hu, J.; Tan, J.; Yuan, Y.; Liu, Z.; Yang, Z. Pancreatic cancer cell–derived exosomal microRNA-27a promotes angiogenesis of human microvascular endothelial cells in pancreatic cancer via BTG2. *J. Cell. Mol. Med.* **2020**, *24*, 588–604. [CrossRef] [PubMed]
123. Müller, H.; Hu, J.; Popp, R.; Schmidt, M.H.; Müller-Decker, K.; Mollenhauer, J.; Fisslthaler, B.; Eble, J.A.; Fleming, I. Deleted in Malignant Brain Tumors 1 is Present in the Vascular Extracellular Matrix and Promotes Angiogenesis. *Arter. Thromb. Vasc. Biol.* **2012**, *32*, 442–448. [CrossRef]
124. Lucero, R.; Zappulli, V.; Sammarco, A.; Murillo, O.D.; Cheah, P.S.; Srinivasan, S.; Tai, E.; Ting, D.T.; Wei, Z.; Roth, M.E.; et al. Glioma-Derived miRNA-Containing Extracellular Vesicles Induce Angiogenesis by Reprogramming Brain Endothelial Cells. *Cell Rep.* **2020**, *30*, 2065–2074.e4. [CrossRef] [PubMed]
125. Kucharzewska, P.; Christianson, H.C.; Welch, J.E.; Svensson, K.J.; Fredlund, E.; Ringnér, M.; Mörgelin, M.; Bourseau-Guilmain, E.; Bengzon, J.; Belting, M. Exosomes reflect the hypoxic status of glioma cells and mediate hypoxia-dependent activation of vascular cells during tumor development. *Proc. Natl. Acad. Sci. USA* **2013**, *110*, 7312–7317. [CrossRef] [PubMed]
126. Giusti, I.; Delle Monache, S.; Di Francesco, M.; Sanità, P.; D'Ascenzo, S.; Gravina, G.L.; Festuccia, C.; Dolo, V. From glioblastoma to endothelial cells through extracellular vesicles: Messages for angiogenesis. *Tumor Biol.* **2016**, *37*, 12743–12753. [CrossRef] [PubMed]
127. Gesierich, S.; Berezovskiy, I.; Ryschich, E.; Zöller, M. Systemic Induction of the Angiogenesis Switch by the Tetraspanin D6.1A/CO-029. *Cancer Res.* **2006**, *66*, 7083–7094. [CrossRef] [PubMed]
128. Nazarenko, I.; Rana, S.; Baumann, A.; McAlear, J.; Hellwig, A.; Trendelenburg, M.; Lochnit, G.; Preissner, K.T.; Zöller, M. Cell Surface Tetraspanin Tspan8 Contributes to Molecular Pathways of Exosome-Induced Endothelial Cell Activation. *Cancer Res.* **2010**, *70*, 1668–1678. [CrossRef]
129. Verweij, F.J.; van Eijndhoven, M.A.; Hopmans, E.S.; Vendrig, T.; Wurdinger, T.; Cahir-McFarland, E.; Kieff, E.; Geerts, D.; van der Kant, R.; Neefjes, J.; et al. LMP1 association with CD63 in endosomes and secretion via exosomes limits constitutive NF-κB activation. *Embo J.* **2011**, *30*, 2115–2129. [CrossRef]
130. Tkach, M.; Théry, C. Communication by Extracellular Vesicles: Where We Are and Where We Need to Go. *Cell* **2016**, *164*, 1226–1232. [CrossRef]
131. El Kharbili, M.; Robert, C.; Witkowski, T.; Danty-Berger, E.; Barbollat-Boutrand, L.; Masse, I.; Gadot, N.; de la Fouchardiere, A.; McDonald, P.C.; Dedhar, S.; et al. Tetraspanin 8 is a novel regulator of ILK-driven β1 integrin adhesion and signaling in invasive melanoma cells. *Oncotarget* **2017**, *8*, 17140–17155. [CrossRef]
132. Grosshans, C.A.; Cech, T.R. Metal ion requirements for sequence-specific endoribonuclease activity of the Tetrahymena ribozyme. *Biochemistry* **1989**, *28*, 6888–6894. [CrossRef]
133. Segura, E.; Guérin, C.; Hogg, N.; Amigorena, S.; Théry, C. CD8+Dendritic Cells Use LFA-1 to Capture MHC-Peptide Complexes from Exosomes In Vivo. *J. Immunol.* **2007**, *179*, 1489–1496. [CrossRef]
134. Ko, S.Y.; Lee, W.; Kenny, H.A.; Dang, L.H.; Ellis, L.M.; Jonasch, E.; Lengyel, E.; Naora, H. Cancer-derived small extracellular vesicles promote angiogenesis by heparin-bound, bevacizumab-insensitive VEGF, independent of vesicle uptake. *Commun. Biol.* **2019**, *2*, 1–17. [CrossRef]

135. Dillekås, H.; Rogers, M.; Straume, O. Are 90% of deaths from cancer caused by metastases? *Cancer Med.* **2019**, *8*, 5574–5576. [CrossRef]

136. Hao, S.; Ye, Z.; Li, F.; Meng, Q.; Qureshi, M.; Yang, J.; Xiang, J. Epigenetic transfer of metastatic activity by uptake of highly metastatic B16 melanoma cell-released exosomes. *Exp. Oncol.* **2006**, *28*, 126–131.

137. McCready, J.; Sims, J.D.; Chan, D.; Jay, D.G. Secretion of extracellular hsp90α via exosomes increases cancer cell motility: A role for plasminogen activation. *BMC Cancer* **2010**, *10*, 294. [CrossRef]

138. Higginbotham, J.N.; Beckler, M.D.; Gephart, J.D.; Franklin, J.L.; Bogatcheva, G.; Kremers, G.-J.; Piston, D.W.; Ayers, G.D.; McConnell, R.E.; Tyska, M.J.; et al. Amphiregulin Exosomes Increase Cancer Cell Invasion. *Curr. Biol.* **2011**, *21*, 779–786. [CrossRef]

139. Aga, M.; Bentz, G.L.; Raffa, S.; Torrisi, M.R.; Kondo, S.; Wakisaka, N.; Yoshizaki, T.; Pagano, J.S.; Shackelford, J. Exosomal HIF1α supports invasive potential of nasopharyngeal carcinoma-associated LMP1-positive exosomes. *Oncogene* **2014**, *33*, 4613–4622. [CrossRef] [PubMed]

140. Corcoran, C.; Rani, S.; O'Driscoll, L.; O'Brien, K.; O'Neill, A.; Prencipe, M.; Sheikh, R.; Webb, G.; McDermott, R.; Watson, W.; et al. Docetaxel-resistance in prostate cancer: Evaluating associated phenotypic changes and potential for resistance transfer via exosomes. *PLoS ONE* **2012**, *7*, e50999. [CrossRef] [PubMed]

141. Guo, D.; Lui, G.Y.L.; Lai, S.L.; Wilmott, J.S.; Tikoo, S.; Jackett, L.A.; Quek, C.; Brown, D.L.; Sharp, D.M.; Kwan, R.Y.Q.; et al. RAB27A promotes melanoma cell invasion and metastasis via regulation of pro-invasive exosomes. *Int. J. Cancer* **2019**, *144*, 3070–3085. [CrossRef] [PubMed]

142. Bijnsdorp, I.V.; Geldof, A.A.; Lavaei, M.; Piersma, S.R.; Van Moorselaar, R.J.A.; Jimenez, C.R. Exosomal ITGA3 interferes with non-cancerous prostate cell functions and is increased in urine exosomes of metastatic prostate cancer patients. *J. Extracell. Vesicles* **2013**, *2*, 1–10. [CrossRef]

143. Sung, B.H.; Ketova, T.; Hoshino, D.; Zijlstra, A.; Weaver, A.M. Directional cell movement through tissues is controlled by exosome secretion. *Nat. Commun.* **2015**, *6*, 7164. [CrossRef]

144. Di Vizio, D.; Morello, M.; Dudley, A.C.; Schow, P.W.; Adam, R.M.; Morley, S.; Mulholland, D.; Rotinen, M.; Hager, M.H.; Insabato, L.; et al. Large Oncosomes in Human Prostate Cancer Tissues and in the Circulation of Mice with Metastatic Disease. *Am. J. Pathol.* **2012**, *181*, 1573–1584. [CrossRef] [PubMed]

145. Liu, Y.; Gu, Y.; Han, Y.; Zhang, Q.; Jiang, Z.; Zhang, X.; Huang, B.; Xu, X.; Zheng, J.; Cao, X. Tumor Exosomal RNAs Promote Lung Pre-metastatic Niche Formation by Activating Alveolar Epithelial TLR3 to Recruit Neutrophils. *Cancer Cell* **2016**, *30*, 243–256. [CrossRef] [PubMed]

146. Zhou, W.; Fong, M.Y.; Min, Y.; Somlo, G.; Liu, L.; Palomares, M.R.; Yu, Y.; Chow, A.; O'Connor, S.T.F.; Chin, A.R.; et al. Cancer-secreted miR-105 destroys vascular endothelial barriers to promote metastasis. *Cancer Cell* **2014**, *25*, 501–515. [CrossRef]

147. Peinado, H.; Aleckovic, M.; Lavotshkin, S.; Matei, I.; Costa-Silva, B.; Moreno-Bueno, G.; Hergueta-Redondo, M.; Williams, C.; Garcia-Santos, G.; Ghajar, C.; et al. Melanoma exosomes educate bone marrow progenitor cells toward a pro-metastatic phenotype through MET. *Nat. Med.* **2012**, *18*, 883–891. [CrossRef] [PubMed]

148. Bigagli, E.; Luceri, C.; Guasti, D.; Cinci, L. Exosomes secreted from human colon cancer cells influence the adhesion of neighboring metastatic cells: Role of microRNA-210. *Cancer Biol. Ther.* **2016**, *17*, 1062–1069. [CrossRef]

149. Hood, J.L.; San, R.S.; Wickline, S.A. Exosomes released by melanoma cells prepare sentinel lymph nodes for tumor metastasis. *Cancer Res.* **2011**, *71*, 3792–3801. [CrossRef] [PubMed]

150. Kinzler, K.W.; Vogelstein, B. Cancer-susceptibility genes. Gatekeepers and caretakers. *Nature* **1997**, *386*, 761–763. [CrossRef] [PubMed]

151. Negrini, S.; Gorgoulis, V.G.; Halazonetis, T.D. Genomic instability–an evolving hallmark of cancer. *Nat. Rev. Mol. Cell Biol.* **2010**, *11*, 220–228. [CrossRef]

152. Pal, T.; Permuth-Wey, J.; Sellers, T.A. A review of the clinical relevance of mismatch-repair deficiency in ovarian cancer. *Cancer* **2008**, *113*, 733–742. [CrossRef]

153. Tarsounas, M.; Sung, P. The antitumorigenic roles of BRCA1–BARD1 in DNA repair and replication. *Nat. Rev. Mol. Cell Biol.* **2020**, *21*, 284–299. [CrossRef]

154. Wu, J.; Lu, L.-Y.; Yu, X. The role of BRCA1 in DNA damage response. *Protein Cell* **2010**, *1*, 117–123. [CrossRef] [PubMed]

155. Abdouh, M.; Floris, M.; Gao, Z.-H.; Arena, V.; Arena, M.; Arena, G.O. Colorectal cancer-derived extracellular vesicles induce transformation of fibroblasts into colon carcinoma cells. *J. Exp. Clin. Cancer Res.* **2019**, *38*, 1–22. [CrossRef] [PubMed]

156. Tsering, T.; Laskaris, A.; Abdouh, M.; Bustamante, P.; Parent, S.; Jin, E.; Ferrier, S.T.; Arena, G.; Burnier, J.V. Uveal Melanoma-Derived Extracellular Vesicles Display Transforming Potential and Carry Protein Cargo Involved in Metastatic Niche Preparation. *Cancers* **2020**, *12*, 2923. [CrossRef] [PubMed]

157. Momen-Heravi, F.; Bala, S. Extracellular vesicles in oral squamous carcinoma carry oncogenic miRNA profile and reprogram monocytes via NF-κB pathway. *Oncotarget* **2018**, *9*, 34838–34854. [CrossRef] [PubMed]

158. Wu, L.; Zhang, X.; Zhang, B.; Shi, H.; Yuan, X.; Sun, Y.; Pan, Z.; Qian, H.; Xu, W. Exosomes derived from gastric cancer cells activate NF-κB pathway in macrophages to promote cancer progression. *Tumour Biol.* **2016**, *37*, 12169–12180. [CrossRef] [PubMed]

159. Li, X.; Wang, S.; Zhu, R.; Li, H.; Han, Q.; Zhao, R.C. Lung tumor exosomes induce a pro-inflammatory phenotype in mesenchymal stem cells via NFκB-TLR signaling pathway. *J. Hematol. Oncol.* **2016**, *9*, 42. [CrossRef]

160. Pritchard, A.; Tousif, S.; Wang, Y.; Hough, K.; Khan, S.; Strenkowski, J.; Chacko, B.K.; Darley-Usmar, V.M.; Deshane, J.S. Lung Tumor Cell-Derived Exosomes Promote M2 Macrophage Polarization. *Cells* **2020**, *9*, 1303. [CrossRef] [PubMed]
161. Lehuédé, C.; Dupuy, F.; Rabinovitch, R.; Jones, R.G.; Siegel, P.M. Metabolic Plasticity as a Determinant of Tumor Growth and Metastasis. *Cancer Res.* **2016**, *76*, 5201–5208. [CrossRef]
162. Cardone, R.A.; Casavola, V.; Reshkin, S.J. The role of disturbed pH dynamics and the Na+/H+ exchanger in metastasis. *Nat. Rev. Cancer* **2005**, *5*, 786–795. [CrossRef] [PubMed]
163. Parolini, I.; Federici, C.; Raggi, C.; Lugini, L.; Palleschi, S.; De Milito, A.; Coscia, C.; Iessi, E.; Logozzi, M.; Molinari, A.; et al. Microenvironmental pH Is a Key Factor for Exosome Traffic in Tumor Cells. *J. Biol. Chem.* **2009**, *284*, 34211–34222. [CrossRef]
164. Fong, M.Y.; Zhou, W.; Liu, L.; Alontaga, A.Y.; Chandra, M.; Ashby, J.; Chow, A.; O'Connor, S.T.F.; Li, S.; Chin, A.R.; et al. Breast-cancer-secreted miR-122 reprograms glucose metabolism in premetastatic niche to promote metastasis. *Nat. Cell Biol.* **2015**, *17*, 183–194. [CrossRef]
165. Lázaro-Ibáñez, E.; Lunavat, T.R.; Jang, S.C.; Escobedo-Lucea, C.; La Cruz, J.O.-D.; Siljander, P.; Lötvall, J.; Yliperttula, M. Distinct prostate cancer-related mRNA cargo in extracellular vesicle subsets from prostate cell lines. *BMC Cancer* **2017**, *17*, 1–11. [CrossRef] [PubMed]
166. Lazar, I.; Clement, E.; Dauvillier, S.; Milhas, D.; Ducoux-Petit, M.; Legonidec, S.; Moro, C.; Soldan, V.; Dalle, S.; Balor, S.; et al. Adipocyte Exosomes Promote Melanoma Aggressiveness through Fatty Acid Oxidation: A Novel Mechanism Linking Obesity and Cancer. *Cancer Res.* **2016**, *76*, 4051–4057. [CrossRef] [PubMed]
167. Zhang, B.; Yin, Y.; Lai, R.C.; Lim, S.K. Immunotherapeutic potential of extracellular vesicles. *Front. Immunol.* **2014**, *5*, 518. [CrossRef]
168. Whiteside, T.L. Exosomes and tumor-mediated immune suppression. *J. Clin. Investig.* **2016**, *126*, 1216–1223. [CrossRef]
169. Wieckowski, E.U.; Visus, C.; Szajnik, M.; Szczepanski, M.J.; Storkus, W.; Whiteside, T.L. Tumor-Derived Microvesicles Promote Regulatory T Cell Expansion and Induce Apoptosis in Tumor-Reactive Activated CD8+ T Lymphocytes. *J. Immunol.* **2009**, *183*, 3720–3730. [CrossRef]
170. Ning, Y.; Shen, K.; Wu, Q.; Sun, X.; Bai, Y.; Xie, Y.; Pan, J.; Qi, C. Tumor exosomes block dendritic cells maturation to decrease the T cell immune response. *Immunol. Lett.* **2018**, *199*, 36–43. [CrossRef]
171. Hedlund, M.; Nagaeva, O.; Kargl, D.; Baranov, V.; Mincheva-Nilsson, L. Thermal- and Oxidative Stress Causes Enhanced Release of NKG2D Ligand-Bearing Immunosuppressive Exosomes in Leukemia/Lymphoma T and B Cells. *PLoS ONE* **2011**, *6*, e16899. [CrossRef]
172. Lundholm, M.; Schröder, M.; Nagaeva, O.; Baranov, V.; Widmark, A.; Mincheva-Nilsson, L.; Wikström, P. Prostate Tumor-Derived Exosomes Down-Regulate NKG2D Expression on Natural Killer Cells and CD8+ T Cells: Mechanism of Immune Evasion. *PLoS ONE* **2014**, *9*, e108925. [CrossRef]
173. Liu, Y.; Gu, Y.; Cao, X. The exosomes in tumor immunity. *OncoImmunology* **2015**, *4*, e1027472. [CrossRef]
174. Hong, C.-S.; Sharma, P.; Yerneni, S.S.; Simms, P.; Jackson, E.K.; Whiteside, T.L.; Boyiadzis, M. Circulating exosomes carrying an immunosuppressive cargo interfere with cellular immunotherapy in acute myeloid leukemia. *Sci. Rep.* **2017**, *7*, 1–10. [CrossRef]
175. Zhao, J.; Schlößer, H.; Wang, Z.; Qin, J.; Li, J.; Popp, F.C.; Popp, M.C.; Alakus, H.; Chon, S.-H.; Hansen, H.P.; et al. Tumor-derived extracellular vesicles inhibit natural killer cell function in pancreatic cancer. *Cancers* **2019**, *11*, 874. [CrossRef] [PubMed]
176. Sharma, P.; Diergaarde, B.; Ferrone, S.; Kirkwood, J.M.; Whiteside, T.L. Melanoma cell-derived exosomes in plasma of melanoma patients suppress functions of immune effector cells. *Sci. Rep.* **2020**, *10*, 1–11. [CrossRef]
177. Batista, I.; Quintas, S.; Melo, S. The Interplay of Exosomes and NK Cells in Cancer Biology. *Cancers* **2021**, *13*, 473. [CrossRef]
178. Anderson, H.C.; Mulhall, D.; Garimella, R. Role of extracellular membrane vesicles in the pathogenesis of various diseases, including cancer, renal diseases, atherosclerosis, and arthritis. *Lab. Investig.* **2010**, *90*, 1549–1557. [CrossRef] [PubMed]
179. Tripisciano, C.; Weiss, R.; Eichhorn, T.; Spittler, A.; Heuser, T.; Fischer, M.B.; Weber, V. Different Potential of Extracellular Vesicles to Support Thrombin Generation: Contributions of Phosphatidylserine, Tissue Factor, and Cellular Origin. *Sci. Rep.* **2017**, *7*, 1–11. [CrossRef] [PubMed]
180. Steidel, C.; Ender, F.; Rody, A.; von Bubnoff, N.; Gieseler, F. Biologically Active Tissue Factor-Bearing Larger Ectosome-Like Extracellular Vesicles in Malignant Effusions from Ovarian Cancer Patients: Correlation with Incidence of Thrombosis. *Int. J. Mol. Sci.* **2021**, *22*, 790. [CrossRef] [PubMed]
181. Prandoni, P.; Falanga, A.; Piccioli, A. Cancer and venous thromboembolism. *Lancet Oncol.* **2005**, *6*, 401–410. [CrossRef]
182. Grilz, E.; Königsbrügge, O.; Posch, F.; Schmidinger, M.; Pirker, R.; Lang, I.M.; Pabinger, I.; Ay, C. Frequency, risk factors, and impact on mortality of arterial thromboembolism in patients with cancer. *Haematologica* **2018**, *103*, 1549–1556. [CrossRef] [PubMed]
183. Wang, X.; Qiao, D.; Chen, L.; Xu, M.; Chen, S.; Huang, L.; Wang, F.; Chen, Z.; Cai, J.; Fu, L. Chemotherapeutic drugs stimulate the release and recycling of extracellular vesicles to assist cancer cells in developing an urgent chemoresistance. *Mol. Cancer* **2019**, *18*, 1–18. [CrossRef] [PubMed]
184. Osataphan, S.; Patell, R.; Chiasakul, T.; Khorana, A.A.; Zwicker, J.I. Extended thromboprophylaxis for medically ill patients with cancer: A systemic review and meta-analysis. *Blood Adv.* **2021**, *5*, 2055–2062. [CrossRef] [PubMed]
185. Shah, S.; Karathanasi, A.; Revythis, A.; Ioannidou, E.; Boussios, S. Cancer-Associated Thrombosis: A New Light on an Old Story. *Diseases* **2021**, *9*, 34. [CrossRef] [PubMed]

186. Tesselaar, M.E.T.; Romijn, F.P.H.T.M.; Van Der Linden, I.K.; Prins, F.A.; Bertina, R.M.; Osanto, S. Microparticle-associated tissue factor activity: A link between cancer and thrombosis? *J. Thromb. Haemost.* **2007**, *5*, 520–527. [CrossRef]
187. Gardiner, C.; Harrison, P.; Belting, M.; Böing, A.; Campello, E.; Carter, B.S.; Collier, M.E.; Coumans, F.; Ettelaie, C.; Van Es, N.; et al. Extracellular vesicles, tissue factor, cancer and thrombosis–discussion themes of the ISEV 2014 Educational Day. *J. Extracell. Vesicles* **2015**, *4*, 26901. [CrossRef]
188. Dávila, M.; Amirkhosravi, A.; Coll, E.; Desai, H.; Robles, L.; Colon, J.; Baker, C.H.; Francis, J.L. Tissue factor-bearing microparticles derived from tumor cells: Impact on coagulation activation. *J. Thromb. Haemost.* **2008**, *6*, 1517–1524. [CrossRef]
189. Thomas, G.; Panicot-Dubois, L.; Lacroix, R.; Dignat-George, F.; Lombardo, D.; Dubois, C. Cancer cell–derived microparticles bearing P-selectin glycoprotein ligand 1 accelerate thrombus formation in vivo. *J. Exp. Med.* **2009**, *206*, 1913–1927. [CrossRef]
190. Wang, J.-G.; Geddings, J.E.; Aleman, M.M.; Cardenas, J.C.; Chantrathammachart, P.; Williams, J.C.; Kirchhofer, D.; Bogdanov, V.; Bach, R.R.; Rak, J.; et al. Tumor-derived tissue factor activates coagulation and enhances thrombosis in a mouse xenograft model of human pancreatic cancer. *Blood* **2012**, *119*, 5543–5552. [CrossRef] [PubMed]
191. Thomas, G.M.; Brill, A.; Mezouar, S.; Crescence, L.; Gallant, M.; Dubois, C.; Wagner, D.D. Tissue factor expressed by circulating cancer cell-derived microparticles drastically increases the incidence of deep vein thrombosis in mice. *J. Thromb. Haemost.* **2015**, *13*, 1310–1319. [CrossRef]
192. Hisada, Y.; Ay, C.; Auriemma, A.C.; Cooley, B.C.; Mackman, N. Human pancreatic tumors grown in mice release tissue factor-positive microvesicles that increase venous clot size. *J. Thromb. Haemost.* **2017**, *15*, 2208–2217. [CrossRef] [PubMed]
193. Gheldof, D.; Mullier, F.; Bailly, N.; Devalet, B.; Dogné, J.-M.; Chatelain, B.; Chatelain, C. Microparticle bearing tissue factor: A link between promyelocytic cells and hypercoagulable state. *Thromb. Res.* **2014**, *133*, 433–439. [CrossRef]
194. Nielsen, T.; Kristensen, S.R.; Gregersen, H.; Teodorescu, E.M.; Christiansen, G.; Pedersen, S. Extracellular vesicle-associated procoagulant phospholipid and tissue factor activity in multiple myeloma. *PLoS ONE* **2019**, *14*, e0210835. [CrossRef]
195. Rousseau, A.; Van Dreden, P.; Khaterchi, A.; Larsen, A.K.; Elalamy, I.; Gerotziafas, G.T. Procoagulant microparticles derived from cancer cells have determinant role in the hypercoagulable state associated with cancer. *Int. J. Oncol.* **2017**, *51*, 1793–1800. [CrossRef]
196. Madan, E.; Pelham, C.J.; Nagane, M.; Parker, T.M.; Marques, R.C.; Fazio, K.; Shaik, K.; Yuan, Y.; Henriques, V.; Galzerano, A.; et al. Flower isoforms promote competitive growth in cancer. *Nat. Cell Biol.* **2019**, *572*, 260–264. [CrossRef] [PubMed]
197. Bowling, S.; Di Gregorio, A.; Sancho, M.; Pozzi, S.; Aarts, M.; Signore, M.; Schneider, M.D.; Martinez-Barbera, J.P.; Gil, J.; Rodríguez, T.A. P53 and mTOR signalling determine fitness selection through cell competition during early mouse embryonic development. *Nat. Commun.* **2018**, *9*, 1763. [CrossRef] [PubMed]
198. Wu, B.; Sun, D.; Ma, L.; Deng, Y.; Zhang, S.; Dong, L.; Chen, S. Exosomes isolated from CAPS1-overexpressing colorectal cancer cells promote cell migration. *Oncol. Rep.* **2019**, *42*, 2528–2536. [CrossRef] [PubMed]
199. Sato, Y.; Ochiai, S.; Murata, T.; Kanda, T.; Goshima, F.; Kimura, H. Elimination of LMP1-expressing cells from a monolayer of gastric cancer AGS cells. *Oncotarget* **2017**, *8*, 39345–39355. [CrossRef] [PubMed]
200. Boopathy, G.T.K.; Hong, W. Role of Hippo Pathway-YAP/TAZ Signaling in Angiogenesis. *Front. Cell Dev. Biol.* **2019**, *7*, 49. [CrossRef] [PubMed]
201. Liu, Z.; Yee, P.P.; Wei, Y.; Liu, Z.; Kawasawa, Y.I.; Li, W. Differential YAP expression in glioma cells induces cell competition and promotes tumorigenesis. *J. Cell Sci.* **2019**, *132*, jcs225714. [CrossRef] [PubMed]
202. Wang, M.; Zhao, X.; Qiu, R.; Gong, Z.; Huang, F.; Yu, W.; Shen, B.; Sha, X.; Dong, H.; Huang, J.; et al. Lymph node metastasis-derived gastric cancer cells educate bone marrow-derived mesenchymal stem cells via YAP signaling activation by exosomal Wnt5a. *Oncogene* **2021**, *40*, 2296–2308. [CrossRef]
203. He, X.; Saint-Jeannet, J.-P.; Wang, Y.; Nathans, J.; Dawid, I.; Varmus, H. A Member of the Frizzled Protein Family Mediating Axis Induction by Wnt-5A. *Science* **1997**, *275*, 1652–1654. [CrossRef] [PubMed]
204. Katoh, M. WNT/PCP signaling pathway and human cancer (Review). *Oncol. Rep.* **2005**, *14*, 1583–1588. [CrossRef]
205. Luga, V.; Zhang, L.; Viloria-Petit, A.M.; Ogunjimi, A.A.; Inanlou, M.R.; Chiu, E.; Buchanan, M.; Hosein, A.N.; Basik, M.; Wrana, J.L. Exosomes Mediate Stromal Mobilization of Autocrine Wnt-PCP Signaling in Breast Cancer Cell Migration. *Cell* **2012**, *151*, 1542–1556. [CrossRef] [PubMed]
206. Longley, D.B.; Johnston, P.G. Molecular mechanisms of drug resistance. *J. Pathol.* **2005**, *205*, 275–292. [CrossRef]
207. Holohan, C.; Van Schaeybroeck, S.; Longley, D.B.; Johnston, P.G. Cancer drug resistance: An evolving paradigm. *Nat. Rev. Cancer* **2013**, *13*, 714–726. [CrossRef] [PubMed]
208. Rijpma, S.R.; Heuvel, J.J.M.W.V.D.; Van Der Velden, M.; Sauerwein, R.W.; Russel, F.G.M.; Koenderink, J.B. Atovaquone and quinine anti-malarials inhibit ATP binding cassette transporter activity. *Malar. J.* **2014**, *13*, 359. [CrossRef]
209. Shedden, K.; Xie, X.T.; Chandaroy, P.; Chang, Y.T.; Rosania, G.R. Expulsion of small molecules in vesicles shed by cancer cells: Association with gene expression and chemosensitivity profiles. *Cancer Res.* **2003**, *63*, 4331–4337. [PubMed]
210. Koch, R.; Aung, T.; Vogel, D.; Chapuy, B.; Wenzel, D.; Becker, S.; Sinzig, U.; Venkataramani, V.; Von Mach, T.; Jacob, R.; et al. Nuclear Trapping through Inhibition of Exosomal Export by Indomethacin Increases Cytostatic Efficacy of Doxorubicin and Pixantrone. *Clin. Cancer Res.* **2016**, *22*, 395–404. [CrossRef]
211. Safaei, R.; Larson, B.J.; Cheng, T.C.; Gibson, M.A.; Otani, S.; Naerdemann, W.; Howell, S.B. Abnormal lysosomal trafficking and enhanced exosomal export of cisplatin in drug-resistant human ovarian carcinoma cells. *Mol. Cancer Ther.* **2005**, *4*, 1595–1604. [CrossRef]

212. Federici, C.; Petrucci, F.; Caimi, S.; Cesolini, A.; Logozzi, M.; Borghi, M.; D'Ilio, S.; Lugini, L.; Violante, N.; Azzarito, T.; et al. Exosome release and low pH belong to a framework of resistance of human melanoma cells to cisplatin. *PLoS ONE* **2014**, *9*, e88193. [CrossRef] [PubMed]

213. Pascucci, L.; Coccè, V.; Bonomi, A.; Ami, D.; Ceccarelli, P.; Ciusani, E.; Viganò, L.; Locatelli, A.; Sisto, F.; Doglia, S.M.; et al. Paclitaxel is incorporated by mesenchymal stromal cells and released in exosomes that inhibit in vitro tumor growth: A new approach for drug delivery. *J. Control. Release* **2014**, *192*, 262–270. [CrossRef]

214. Bebawy, M.; Combes, V.; Lee, E.; Jaiswal, R.; Gong, J.; Bonhoure, A.; Grau, G.E.R. Membrane microparticles mediate transfer of P-glycoprotein to drug sensitive cancer cells. *Leukemia* **2009**, *23*, 1643–1649. [CrossRef] [PubMed]

215. Torreggiani, E.; Roncuzzi, L.; Perut, F.; Zini, N.; Baldini, N. Multimodal transfer of MDR by exosomes in human osteosarcoma. *Int. J. Oncol.* **2016**, *49*, 189–196. [CrossRef]

216. Bhattacharya, S.; Pal, K.; Sharma, A.; Dutta, S.K.; Lau, J.S.; Yan, I.K.; Wang, E.; Elkhanany, A.; Alkharfy, K.M.; Sanyal, A.; et al. GAIP Interacting Protein C-Terminus Regulates Autophagy and Exosome Biogenesis of Pancreatic Cancer through Metabolic Pathways. *PLoS ONE* **2014**, *9*, e114409. [CrossRef] [PubMed]

217. Qu, Z.; Wu, J.; Wu, J.; Luo, D.; Jiang, C.; Ding, Y. Exosomes derived from HCC cells induce sorafenib resistance in hepatocellular carcinoma both in vivo and in vitro. *J. Exp. Clin. Cancer Res.* **2016**, *35*, 1–12. [CrossRef]

218. Ozawa, P.M.M.; Alkhilaiwi, F.; Cavalli, I.J.; Malheiros, D.; Ribeiro, E.M.D.S.F.; Cavalli, L.R. Extracellular vesicles from triple-negative breast cancer cells promote proliferation and drug resistance in non-tumorigenic breast cells. *Breast Cancer Res. Treat.* **2018**, *172*, 713–723. [CrossRef] [PubMed]

219. Vella, L.J.; Behren, A.; Coleman, B.; Greening, D.W.; Hill, A.; Cebon, J. Intercellular Resistance to BRAF Inhibition Can Be Mediated by Extracellular Vesicle–Associated PDGFRβ. *Neoplasia* **2017**, *19*, 932–940. [CrossRef]

220. Uchihara, T.; Miyake, K.; Yonemura, A.; Komohara, Y.; Itoyama, R.; Koiwa, M.; Yasuda, T.; Arima, K.; Harada, K.; Eto, K.; et al. Extracellular Vesicles from Cancer-Associated Fibroblasts Containing Annexin A6 Induces FAK-YAP Activation by Stabilizing β1 Integrin, Enhancing Drug Resistance. *Cancer Res.* **2020**, *80*, 3222–3235. [CrossRef]

221. Maacha, S.; Bhat, A.A.; Jimenez, L.; Raza, A.; Haris, M.; Uddin, S.; Grivel, J.-C. Extracellular vesicles-mediated intercellular communication: Roles in the tumor microenvironment and anti-cancer drug resistance. *Mol. Cancer* **2019**, *18*, 1–16. [CrossRef]

222. Kreger, B.T.; Johansen, E.R.; Cerione, R.A.; Antonyak, M.A. The Enrichment of Survivin in Exosomes from Breast Cancer Cells Treated with Paclitaxel Promotes Cell Survival and Chemoresistance. *Cancers* **2016**, *8*, 111. [CrossRef]

223. Yang, Z.; Zhao, N.; Cui, J.; Wu, H.; Xiong, J.; Peng, T. Exosomes derived from cancer stem cells of gemcitabine-resistant pancreatic cancer cells enhance drug resistance by delivering miR-210. *Cell. Oncol.* **2020**, *43*, 123–136. [CrossRef] [PubMed]

224. Aung, T.; Chapuy, B.; Vogel, D.; Wenzel, D.; Oppermann, M.; Lahmann, M.; Weinhage, T.; Menck, K.; Hupfeld, T.; Koch, R.; et al. Exosomal evasion of humoral immunotherapy in aggressive B-cell lymphoma modulated by ATP-binding cassette transporter A3. *Proc. Natl. Acad. Sci. USA* **2011**, *108*, 15336–15341. [CrossRef]

225. Ciravolo, V.; Huber, V.; Ghedini, G.C.; Venturelli, E.; Bianchi, F.; Campiglio, M.; Morelli, D.; Villa, A.; Della Mina, P.; Menard, S.; et al. Potential role of HER2-overexpressing exosomes in countering trastuzumab-based therapy. *J. Cell. Physiol.* **2012**, *227*, 658–667. [CrossRef]

226. Lubin, A.J.; Zhang, R.R.; Kuo, J.S. Extracellular Vesicles Containing PD-L1 Contribute to Immune Evasion in Glioblastoma. *Neurosurgery* **2018**, *83*, E98–E100. [CrossRef]

227. Lara, P.; Chan, A.; Cruz, L.; Quest, A.; Kogan, M. Exploiting the Natural Properties of Extracellular Vesicles in Targeted Delivery towards Specific Cells and Tissues. *Pharmaceutics* **2020**, *12*, 1022. [CrossRef] [PubMed]

228. Katakowski, M.; Buller, B.; Zheng, X.; Lu, Y.; Rogers, T.; Osobamiro, O.; Shu, W.; Jiang, F.; Chopp, M. Exosomes from marrow stromal cells expressing miR-146b inhibit glioma growth. *Cancer Lett.* **2013**, *335*, 201–204. [CrossRef] [PubMed]

229. Lara, P.; Palma-Florez, S.; Salas-Huenuleo, E.; Polakovicova, I.; Guerrero, S.; Lobos-Gonzalez, L.; Campos, A.; Muñoz, L.; Jorquera-Cordero, C.; Varas-Godoy, M.; et al. Gold nanoparticle based double-labeling of melanoma extracellular vesicles to determine the specificity of uptake by cells and preferential accumulation in small metastatic lung tumors. *J. Nanobiotechnol.* **2020**, *18*, 1–17. [CrossRef]

230. Wiklander, O.P.B.; Nordin, J.Z.; O'Loughlin, A.; Gustafsson, Y.; Corso, G.; Mäger, I.; Vader, P.; Lee, Y.; Sork, H.; Seow, Y.; et al. Extracellular vesicle in vivo biodistribution is determined by cell source, route of administration and targeting. *J. Extracell. Vesicles* **2015**, *4*, 26316. [CrossRef]

231. Yang, T.; Martin, P.; Fogarty, B.; Brown, A.; Schurman, K.; Phipps, R.; Yin, V.P.; Lockman, P.; Dai, S. Exosome Delivered Anticancer Drugs Across the Blood-Brain Barrier for Brain Cancer Therapy in Danio Rerio. *Pharm. Res.* **2015**, *32*, 2003–2014. [CrossRef]

232. Garofalo, M.; Villa, A.; Crescenti, D.; Marzagalli, M.; Kuryk, L.; Limonta, P.; Mazzaferro, V.; Ciana, P. Heterologous and cross-species tropism of cancer-derived extracellular vesicles. *Theranostics* **2019**, *9*, 5681–5693. [CrossRef] [PubMed]

233. Liu, Y.; Cao, X. Organotropic metastasis: Role of tumor exosomes. *Cell Res.* **2015**, *26*, 149–150. [CrossRef] [PubMed]

234. Wang, J.; Dong, Y.; Li, Y.; Li, W.; Cheng, K.; Qian, Y.; Xu, G.; Zhang, X.; Hu, L.; Chen, P.; et al. Designer Exosomes for Active Targeted Chemo-Photothermal Synergistic Tumor Therapy. *Adv. Funct. Mater.* **2018**, *28*, 1707360. [CrossRef]

235. Huang, R.; Rofstad, E.K. Integrins as therapeutic targets in the organ-specific metastasis of human malignant melanoma. *J. Exp. Clin. Cancer Res.* **2018**, *37*, 1–14. [CrossRef]

236. Rana, S.; Yue, S.; Stadel, D.; Zöller, M. Toward tailored exosomes: The exosomal tetraspanin web contributes to target cell selection. *Int. J. Biochem. Cell Biol.* **2012**, *44*, 1574–1584. [CrossRef] [PubMed]

237. Rodrigues, G.; Hoshino, A.; Kenific, C.M.; Matei, I.R.; Steiner, L.; Freitas, D.; Kim, H.S.; Oxley, P.R.; Scandariato, I.; Casanova-Salas, I.; et al. Tumour exosomal CEMIP protein promotes cancer cell colonization in brain metastasis. *Nat. Cell Biol.* **2019**, *21*, 1403–1412. [CrossRef]

238. Franzen, C.A.; Simms, P.E.; Van Huis, A.F.; Foreman, K.E.; Kuo, P.C.; Gupta, G.N. Characterization of Uptake and Internalization of Exosomes by Bladder Cancer Cells. *BioMed Res. Int.* **2014**, *2014*, 1–11. [CrossRef]

239. Escrevente, C.; Keller, S.; Altevogt, P.; Costa, J. Interaction and uptake of exosomes by ovarian cancer cells. *BMC Cancer* **2011**, *11*, 108. [CrossRef]

240. Matsumoto, A.; Takahashi, Y.; Nishikawa, M.; Sano, K.; Morishita, M.; Charoenviriyakul, C.; Saji, H.; Takakura, Y. Accelerated growth of B16 BL 6 tumor in mice through efficient uptake of their own exosomes by B16 BL 6 cells. *Cancer Sci.* **2017**, *108*, 1803–1810. [CrossRef] [PubMed]

241. Kim, M.S.; Haney, M.J.; Zhao, Y.; Yuan, D.; Deygen, I.; Klyachko, N.L.; Kabanov, A.V.; Batrakova, E.V. Engineering macrophage-derived exosomes for targeted paclitaxel delivery to pulmonary metastases: In vitro and in vivo evaluations. *Nanomedicine* **2018**, *14*, 195–204. [CrossRef] [PubMed]

242. Aspe, J.R.; Osterman, C.D.; Jutzy, J.M.; Deshields, S.; Whang, S.; Wall, N.R. Enhancement of Gemcitabine sensitivity in pancreatic adenocarcinoma by novel exosome-mediated delivery of the Survivin-T34A mutant. *J. Extracell. Vesicles* **2014**, *3*, 1–9. [CrossRef]

243. Hood, J.L. Post isolation modification of exosomes for nanomedicine applications. *Nanomedicine* **2016**, *11*, 1745–1756. [CrossRef]

244. Kooijmans, S.A.A.; Aleza, C.G.; Roffler, S.R.; Van Solinge, W.W.; Vader, P.; Schiffelers, R.M. Display of GPI-anchored anti-EGFR nanobodies on extracellular vesicles promotes tumour cell targeting. *J. Extracell. Vesicles* **2016**, *5*, 31053. [CrossRef] [PubMed]

245. Kamerkar, S.; LeBleu, V.S.; Sugimoto, H.; Yang, S.; Ruivo, C.; Melo, S.; Lee, J.J.; Kalluri, R. Exosomes facilitate therapeutic targeting of oncogenic KRAS in pancreatic cancer. *Nat. Cell Biol.* **2017**, *546*, 498–503. [CrossRef]

246. Limkin, E.J.; Sun, R.; Dercle, L.; Zacharaki, E.I.; Robert, C.; Reuzé, S.; Schernberg, A.; Paragios, N.; Deutsch, E.; Ferté, C. Promises and challenges for the implementation of computational medical imaging (radiomics) in oncology. *Ann. Oncol.* **2017**, *28*, 1191–1206. [CrossRef] [PubMed]

247. Pegtel, D.M.; Cosmopoulos, K.; Thorley-Lawson, D.A.; van Eijndhoven, M.A.J.; Hopmans, E.S.; Lindenberg, J.L.; de Gruijl, T.D.; Würdinger, T.; Middeldorp, J. Functional delivery of viral miRNAs via exosomes. *Proc. Natl. Acad. Sci. USA* **2010**, *107*, 6328–6333. [CrossRef] [PubMed]

248. Tian, T.; Zhu, Y.-L.; Zhou, Y.-Y.; Liang, G.-F.; Wang, Y.-Y.; Hu, F.-H.; Xiao, Z.-D. Exosome Uptake through Clathrin-mediated Endocytosis and Macropinocytosis and Mediating miR-21 Delivery. *J. Biol. Chem.* **2014**, *289*, 22258–22267. [CrossRef]

249. Lai, C.P.-K.; Kim, E.Y.; Badr, C.E.; Weissleder, R.; Mempel, T.R.; Tannous, B.A.; Breakefield, X.O. Visualization and tracking of tumour extracellular vesicle delivery and RNA translation using multiplexed reporters. *Nat. Commun.* **2015**, *6*, 7029. [CrossRef]

250. Lai, C.P.-K.; Tannous, B.A.; Breakefield, X.O. Noninvasive in vivo monitoring of extracellular vesicles. In *Breast Cancer*; Springer Science and Business Media LLC: Berlin, Germany, 2013; Volume 1098, pp. 249–258.

251. Gangadaran, P.; Li, X.J.; Kalimuthu, S.K.; Min, O.J.; Hong, C.M.; Rajendran, R.L.; Lee, H.W.; Zhu, L.; Baek, S.H.; Jeong, S.Y.; et al. New Optical Imaging Reporter-labeled Anaplastic Thyroid Cancer-Derived Extracellular Vesicles as a Platform for In Vivo Tumor Targeting in a Mouse Model. *Sci. Rep.* **2018**, *8*, 1–11. [CrossRef]

252. Zong, S.; Zong, J.; Chen, C.; Jiang, X.; Zhang, Y.; Wang, Z.; Cui, Y. Single molecule localization imaging of exosomes using blinking silicon quantum dots. *Nanotechnology* **2017**, *29*, 065705. [CrossRef] [PubMed]

253. Jiang, X.; Zong, S.; Chen, C.; Zhang, Y.; Wang, Z.; Cui, Y. Gold–carbon dots for the intracellular imaging of cancer-derived exosomes. *Nanotechnology* **2018**, *29*, 175701. [CrossRef] [PubMed]

254. Jung, K.O.; Jo, H.; Yu, J.H.; Gambhir, S.S.; Pratx, G. Development and MPI tracking of novel hypoxia-targeted theranostic exosomes. *Biomaterials* **2018**, *177*, 139–148. [CrossRef]

255. Hu, L.; Wickline, S.A.; Hood, J.L. Magnetic resonance imaging of melanoma exosomes in lymph nodes. *Magn. Reson. Med.* **2015**, *74*, 266–271. [CrossRef]

256. Singh, P.; Pandit, S.; Mokkapati, V.R.S.S.; Garg, A.; Ravikumar, V.; Mijakovic, I. Gold Nanoparticles in Diagnostics and Therapeutics for Human Cancer. *Int. J. Mol. Sci.* **2018**, *19*, 1979. [CrossRef] [PubMed]

257. Menon, J.; Jadeja, P.; Tambe, P.; Vu, K.; Yuan, B.; Nguyen, K.T. Nanomaterials for Photo-Based Diagnostic and Therapeutic Applications. *Theranostics* **2013**, *3*, 152–166. [CrossRef] [PubMed]

258. Zhao, L.; Kim, T.-H.; Kim, H.-W.; Ahn, J.-C.; Kim, S.Y. Surface-enhanced Raman scattering (SERS)-active gold nanochains for multiplex detection and photodynamic therapy of cancer. *Acta Biomater.* **2015**, *20*, 155–164. [CrossRef] [PubMed]

259. Li, Y.; Wang, J.; Wientjes, M.G.; Au, J.L.-S. Delivery of nanomedicines to extracellular and intracellular compartments of a solid tumor. *Adv. Drug Deliv. Rev.* **2012**, *64*, 29–39. [CrossRef] [PubMed]

260. Kooijmans, S.A.A.; Schiffelers, R.M.; Zarovni, N.; Vago, R. Modulation of tissue tropism and biological activity of exosomes and other extracellular vesicles: New nanotools for cancer treatment. *Pharmacol. Res.* **2016**, *111*, 487–500. [CrossRef]

261. Richardson, J.J.; Ejima, H. Surface Engineering of Extracellular Vesicles through Chemical and Biological Strategies. *Chem. Mater.* **2019**, *31*, 2191–2201. [CrossRef]

262. Choi, D.-S.; Kim, D.-K.; Kim, Y.-K.; Gho, Y.S. Proteomics of extracellular vesicles: Exosomes and ectosomes. *Mass Spectrom. Rev.* **2015**, *34*, 474–490. [CrossRef]

263. Meng, W.; He, C.; Hao, Y.; Wang, L.; Li, L.; Zhu, G. Prospects and challenges of extracellular vesicle-based drug delivery system: Considering cell source. *Drug Deliv.* **2020**, *27*, 585–598. [CrossRef] [PubMed]

264. Jia, G.; Han, Y.; An, Y.; Ding, Y.; He, C.; Wang, X.; Tang, Q. NRP-1 targeted and cargo-loaded exosomes facilitate simultaneous imaging and therapy of glioma in vitro and in vivo. *Biomaterials* **2018**, *178*, 302–316. [CrossRef]
265. Silva, J.; Garcia, V.; Rodriguez, M.; Compte, M.; Cisneros, E.; Veguillas, P.; Garcia, J.M.; Dominguez, G.; Campos-Martin, Y.; Cuevas, J.; et al. Analysis of exosome release and its prognostic value in human colorectal cancer. *Genes Chromosom. Cancer* **2012**, *51*, 409–418. [CrossRef] [PubMed]
266. Yamashita, T.; Kamada, H.; Kanasaki, S.; Maeda, Y.; Nagano, K.; Abe, Y.; Yoshioka, Y.; Tsutsumi, Y.; Katayama, S.; Inoue, M.; et al. Epidermal growth factor receptor localized to exosome membranes as a possible biomarker for lung cancer diagnosis. *Die Pharm.* **2013**, *68*, 969–973.
267. Georgantzoglou, N.; Pergaris, A.; Masaoutis, C.; Theocharis, S. Extracellular Vesicles as Biomarkers Carriers in Bladder Cancer: Diagnosis, Surveillance, and Treatment. *Int. J. Mol. Sci.* **2021**, *22*, 2744. [CrossRef]
268. Li, Z.; Jiang, P.; Li, J.; Peng, M.; Zhao, X.; Zhang, X.; Chen, K.; Zhang, Y.; Liu, H.; Gan, L.; et al. Tumor-derived exosomal lnc-Sox2ot promotes EMT and stemness by acting as a ceRNA in pancreatic ductal adenocarcinoma. *Oncogene* **2018**, *37*, 3822–3838. [CrossRef]
269. Huang, X.; Yuan, T.; Liang, M.; Du, M.; Xia, S.; Dittmar, R.; Wang, D.; See, W.; Costello, B.A.; Quevedo, F.; et al. Exosomal miR-1290 and miR-375 as Prognostic Markers in Castration-resistant Prostate Cancer. *Eur. Urol.* **2015**, *67*, 33–41. [CrossRef] [PubMed]
270. Buscail, E.; Alix-Panabières, C.; Quincy, P.; Cauvin, T.; Chauvet, A.; Degrandi, O.; Caumont, C.; Verdon, S.; Lamrissi, I.; Moranvillier, I.; et al. High Clinical Value of Liquid Biopsy to Detect Circulating Tumor Cells and Tumor Exosomes in Pancreatic Ductal Adenocarcinoma Patients Eligible for Up-Front Surgery. *Cancers* **2019**, *11*, 1656. [CrossRef]
271. Takahasi, K.; Iinuma, H.; Wada, K.; Minezaki, S.; Kawamura, S.; Kainuma, M.; Ikeda, Y.; Shibuya, M.; Miura, F.; Sano, K. Usefulness of exosome-encapsulated microRNA-451a as a minimally invasive biomarker for prediction of recurrence and prognosis in pancreatic ductal adenocarcinoma. *J. Hepato Biliary Pancreat. Sci.* **2017**, *25*, 155–161. [CrossRef]
272. Kawamura, S.; Iinuma, H.; Wada, K.; Takahashi, K.; Minezaki, S.; Kainuma, M.; Shibuya, M.; Miura, F.; Sano, K. Exosome-encapsulated microRNA-4525, microRNA-451a and microRNA-21 in portal vein blood is a high-sensitive liquid biomarker for the selection of high-risk pancreatic ductal adenocarcinoma patients. *J. Hepato Biliary Pancreat. Sci.* **2019**, *26*, 63–72. [CrossRef]
273. ClinicalTrials.gov. 2021. Available online: https://clinicaltrials.gov/ct2/results?term=Exosomes+cancer&age_v=&gndr=&type=&rslt=&Search=Apply (accessed on 3 June 2021).

Review

Trends in Research on Exosomes in Cancer Progression and Anticancer Therapy

Dona Sinha [1,*]**, Sraddhya Roy** [1]**, Priyanka Saha** [1]**, Nabanita Chatterjee** [1] **and Anupam Bishayee** [2,*]

[1] Department of Receptor Biology and Tumour Metastasis, Chittaranjan National Cancer Institute, Kolkata 700 026, India; sraddhya1@gmail.com (S.R.); poojasaha.saha79@gmail.com (P.S.); nabanita.chatterje@yahoo.com (N.C.)

[2] Lake Erie College of Osteopathic Medicine, Bradenton, FL 34211, USA

[*] Correspondence: dona.sinha@cnci.org.in or donasinha2012@gmail.com (D.S.); abishayee@lecom.edu or abishayee@gmail.com (A.B.)

Simple Summary: Intensive research in the field of cancer biology has discovered a unique mode of interplay between cells via extracellular bioactive vesicles called exosomes. Exosomes serve as intermediators among cells via their cargoes that, in turn, contribute in the progression of cancer. They are ubiquitously present in all body fluids as they are secreted from both normal and tumor cells. These minuscules exhibit multiple unique properties that facilitate their migration to distant locations and modulate the microenvironment for progression of cancer. This review summarizes the multifarious role of exosomes in various aspects of cancer research with its pros and cons. It discusses biogenesis of exosomes, their functional role in cancer metastasis, both protumorigenic and antitumorigenic, and also their applications in anticancer therapy.

Abstract: Exosomes, the endosome-derived bilayered extracellular nanovesicles with their contribution in many aspects of cancer biology, have become one of the prime foci of research. Exosomes derived from various cells carry cargoes similar to their originator cells and their mode of generation is different compared to other extracellular vesicles. This review has tried to cover all aspects of exosome biogenesis, including cargo, Rab-dependent and Rab-independent secretion of endosomes and exosomal internalization. The bioactive molecules of the tumor-derived exosomes, by virtue of their ubiquitous presence and small size, can migrate to distal parts and propagate oncogenic signaling and epigenetic regulation, modulate tumor microenvironment and facilitate immune escape, tumor progression and drug resistance responsible for cancer progression. Strategies improvised against tumor-derived exosomes include suppression of exosome uptake, modulation of exosomal cargo and removal of exosomes. Apart from the protumorigenic role, exosomal cargoes have been selectively manipulated for diagnosis, immune therapy, vaccine development, RNA therapy, stem cell therapy, drug delivery and reversal of chemoresistance against cancer. However, several challenges, including in-depth knowledge of exosome biogenesis and protein sorting, perfect and pure isolation of exosomes, large-scale production, better loading efficiency, and targeted delivery of exosomes, have to be confronted before the successful implementation of exosomes becomes possible for the diagnosis and therapy of cancer.

Keywords: tumor-derived exosomes; exosomal cargoes; protumorigenic effect; drug resistance; anticancer therapy

Citation: Sinha, D.; Roy, S.; Saha, P.; Chatterjee, N.; Bishayee, A. Trends in Research on Exosomes in Cancer Progression and Anticancer Therapy. *Cancers* **2021**, *13*, 326. https://doi.org/10.3390/cancers13020326

Received: 23 November 2020
Accepted: 14 January 2021
Published: 17 January 2021

Publisher's Note: MDPI stays neutral with regard to jurisdictional claims in published maps and institutional affiliations.

1. Introduction

Exosomes are bilayered endosomal nanovesicles, first discovered in 1983, as transferrin conjugated vesicles (50 nm) released by reticulocytes [1]. Due to the increasing interest of scientists in exosome biology, a consensus guideline was proposed by board members of International Society of Extracellular Vesicles under "minimal experimental requirements for definition of extracellular vesicles and their functions" (MISEV2014) which was later updated in 2018 (MISEV2018). The guidelines advocated norms for nomenclature, isolation, separation, characterization, functional studies, and reporting requirements for proper identification of and experimentation with extracellular vesicles and exosomes [2,3]. Exosomes are generally formed by inward budding of late endosomes, also known as multivesicular bodies (MVBs). Intraluminal vesicles (ILVs) of MVBs engulf a variety of biomolecules which are released into extracellular space as exosomes. Exosomes are anucleated particles naturally released by cells, surrounded by lipid bilayer and are not capable of replication. Exosomes are identified by size (30–200 nm) and surface markers, such as membrane-associated proteins, e.g., lysosome-associated membrane glycoprotein 3 (LAMP3)/CD63; intercellular adhesion molecule (ICAM1)/CD81; and tetraspanin membrane protein/CD9. Exosomes are observed in various body fluids, such as blood, plasma, saliva, urine, synovial fluid, amniotic fluid, and breast milk [4,5].

All cellular types (normal and diseased) secrete exosomes, mediating intercellular communications [6]. Exosomes exhibit heterogeneity in size—Exo-Large (90–120 nm), Exo-Small (60–80 nm), and the membrane-less exomere (<50 nm). Exosome-mediated intercellular transfer of specific repertoire of proteins, lipids, RNA and DNA confer physiological and/or pathological functions to the recipient targets. Exosomes regulate physiological functions, such as neuronal communication, immune responses, reproductive activity, cell proliferation homeostasis, maturation and cellular waste disposition. They also contribute in clinical disorders, including inflammation, cancer, cardiovascular diseases, neuronal pathologies and pathogenic infections [5].

Our review deals with exosomal contents, exosome-associated protumorigenic, antitumorigenic effect and therapeutics, unlike other reviews, which discuss combinational roles of all microvesicles in cancer progression [7,8] or have primarily focused on tumor-derived exosomes (TEXs) with little information on therapeutics [9]. In contrast to reviews which have focused on specific exosomal cargoes and therapeutics [10,11], we have envisaged the exosomal contents, the mechanisms influencing cancer progression and their therapeutic implications in cancer management. The inexplicable nature of exosomes has raised concern about their role in the invasion and metastasis of cancer cells, encompassing epithelial-to-mesenchymal transition (EMT), angiogenesis, and immune regulation [12]. Thus, instead of reviewing the isolated impact of exosomes, e.g., evasion of immune surveillance [13] for cancer progression, we have tried to encompass exosome-mediated propagation of oncogenic signaling, epigenetic regulation, modulation of tumor microenvironment (TME) and immune escape, EMT, angiogenesis, metastasis and drug resistance. Considering the clinical applications, the exosomes serve as potent diagnostic and prognostic biomarkers because of their bioavailability, low toxicity and differentiated surface markers [5]. Recent reviews on exosomes have focused on therapeutic efficacy of exosomes by addressing extracellular vesicular interaction with the host immune system [14], constraints and opportunities available with bioengineering of exosomes [15–17], success against multiple cancers [18] and exosome-based drug delivery [19–21]. Anticancer treatments sometimes experience shortfall in their efficacy due to unwanted side effects of the therapeutic agents or shortened shelf-life, but exosomes serve as natural agents to overcome these issues and become a potent therapeutic agent [22]. However, instead of perceiving specific therapeutic potential of exosomes, the present review has tried to decipher the entire repertoire of exosomes, including both protumorigenic and antitumorigenic impact

2. Cargo Composition of Exosomes

Exosomes are rich in enzymes, transcription factors, heat shock proteins (Hsps), major histocompatibility complex (MHC), cytoskeleton components, signal transducers, tetraspanins, lipids, RNAs and DNAs [6,23]. Detailed information about the exosomal components can be accessed via databases, such as ExoCarta [www.exocarta.org], EVpedia [http://evpedia.info] and Vesiclepedia [www.microvesicles.org]. Though exosomes diverge in size and biomolecular inclusions, some common components are observed in all types [5]. Lipid components are cholesterol, sphingomyelin, glycosphingolipids, phosphatidylcholine, phosphatidylserines, phosphatidylethanolamines and saturated fatty acids [4]. RNAs include specific microRNAs (miRNAs), long non-coding RNAs (lncRNAs), vault RNA, Y-RNA, transfer RNAs (tRNAs), ribosomal RNA (rRNA) fragments (such as 28S and 18S rRNA subunits) and messenger RNAs (mRNAs) [24]. Exosomal cargo components also include mitochondrial DNA (mtDNA), single-stranded DNA, double-stranded DNA and retrotransposons [4,6]. Different protein forms include components of the immune system (MHC class I and II molecules, cytokines), endosomal sorting complexes required for the transport (ESCRT) complex, those involved in trafficking (tetraspanins, glycosylphosphatidylinositol-anchored proteins, Rabs, soluble N-ethylmaleimide-sensitive fusion protein attachment protein receptors (SNARES), flotillins, lipid-rafts residents [25] and those involved in carcinogenesis (oncoproteins, tumor suppressor proteins, and transcriptional factors) [4]. The plasma membrane (PM) proteins constitute the vesicle membrane for maintaining composition parity with the cell membrane which helps in sequestration of soluble ligands. Exosomal proteins are involved in (i) antigen presentation, (ii) cell adhesion, (iii) cell structure and motility, (iv) stress regulation, (v) transcription and protein synthesis, and (vi) trafficking and membrane fusion [26]. The structure of exosome with membrane proteins and cargoes have been depicted in Figure 1.

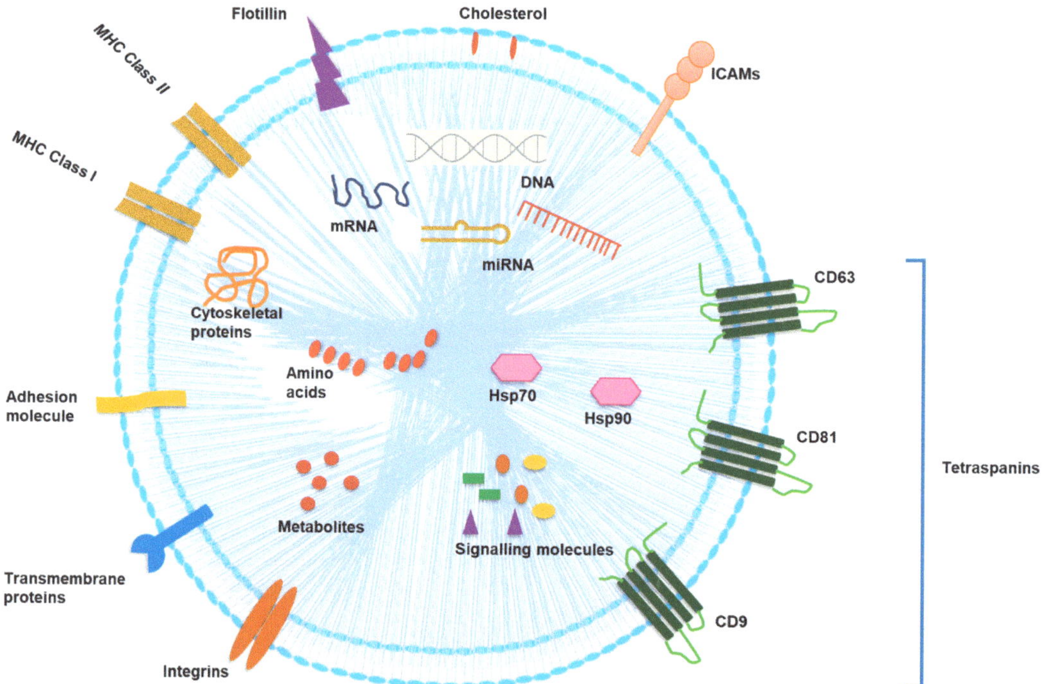

Figure 1. Structure of exosome with membrane proteins and cargoes. Exosomes consist of many constituents of a cell including DNA, RNAs, amino acids, proteins, metabolites, enzymes, lipids (cholesterol) and Hsps along with several cytosolic and cell-surface signaling proteins which are involved in intercellular communications. Exosomal membrane is rich in transmembrane proteins (tetraspanins such as CD81, CD63 and CD9), flotillin, ICAMs, integrins and adhesion molecules. They consist of immune components including MHC class I and class II molecules. Abbreviations: CD, cluster of differentiation; DNA, deoxyribonucleic acids; Hsps, heat shock proteins; ICAMs, intercellular adhesion molecules; MHC, major histocompatibility complex; mRNA, messenger RNA; miRNA, microRNA.

3. Exosome Biogenesis

Endocytosis generates early endosomes via invagination of PM rich in lipid rafts. This internalizes the PM receptors which are either recycled or degraded. The exosome biogenesis involves a complex network of enzymatic actions and signal transductions. Early endosomes mature to MVBs or late endosomes upon internal budding of endosomes, forming ILVs [23]. MVB budding is primed with actin polymerization at PM lipid domains [27,28]. ADP ribosylation factor 6 (ARF6), along with phospholipase D2 (PLD2), converts ILVs into mature MVBs [29]. Heparanase enzyme stimulates the syndecan-syntenin-ALG-2 interacting protein X (ALIX) axis, upregulating exosome formation [30]. ARF6-induced actomyosin contractility and ESCRTs promote ILVs shedding from MVBs as exosomes [31]. The MVBs undergo one of the three type consequences [23,32] mentioned below:

(i) Recycling through the trans-Golgi network (TGN) which may be subdivided into a fast and a slow pathway, considering the duration taken by the specific proteins/lipids from internalization to re-exposure at the cell surface or exocytosis.

(ii) Lysosomal degradation by hydrolytic enzymes which are able to digest complex macromolecules.

(iii) Fusion of MVBs with the cell surface release exosomes via exocytosis. Additional materials may be incorporated to the TGN at any juncture and processed through the canonical secretory pathways.

4. Sorting of Exosomal Cargoes

4.1. ESCRT-Dependent Sorting Pathway

The ESCRT pathway participates in sorting ubiquitinated proteins of exosome, after being internalized within ILVs. The complex includes ESCRT-0, which identifies and processes ubiquitin-dependent cargo inside the vesicles; ESCRT-I and ESCRT-II evoke budding and ESCRT-III causes vesicle scission from endosomal membrane. Other accessory proteins such as ALIX aid in vesicle budding and vacuolar protein sorting associated protein 4 (VPS4) promotes scission [30,33].

4.2. ESCRT-Independent Exosomal Sorting

Ceramide and cholesterol, PLD2, or tetraspanins mediates ESCRT independent sorting machinery. Tetraspanins may promote incorporation of specific cargoes into exosome, e.g., CD9 facilitates encapsulation of metalloproteinase CD10 and CD63. Even the lipid composition and membrane dynamics of the early endosome and MVBs may regulate exosomal cargoes. Ceramide and neutral sphingomyelinase 2 (nSMase2) play a pivotal role in an ESCRT independent process of exosome formation, loading, and release [23]. Podoplanin, a transmembrane glycoprotein, is another regulator of exosome biogenesis and cargo sorting [31].

5. Exocytosis and Secretion of Exosomes

Exocytosis is exosomal secretion into the extracellular matrix (ECM) which is regulated by Rab GTPases, molecular motors, cytoskeletal proteins, SNAREs, intracellular Ca^{2+} levels (increased Ca^{2+} results in increased exosome secretion) and extracellular/intracellular pH gradients [23]. Vesicular SNAREs (v-SNARE) on the MVB bind with the target SNARE (t-SNARE), Syx 5, on the inner surface of the PM for mediating fusion of MVB with the cell membrane [34]. The fusion of exosome with PM occurs at the actin-rich zones of the invadopodia, promoting ECM degradation and metastasis, followed by their exocytosis into extracellular space [34]. Peptidyl arginine deiminases aid exosomal secretion by deaminating actin [35]. A negative feedback mechanism limits excess exosome secretion from the same cells [34].

Rabs Control Endocytic Pathway

The Rab GTPases belong to a large family of highly conserved proteins with 60 members, which regulate vesicular trafficking in eukaryotes. Different Rab forms are involved in endocytic trafficking—Rab4, 5, 9, 11a, 11b, 25 and 35 control recycling [36–39]; Rab5 and 7 cause endosomal maturation [40]; Rab 7 regulates sorting and degradation [41]; Rab 7, 27a and b control secretion of exosomes [42,43] and Rab5 overexpression causes release of exosomal markers [44]. Deregulation of the Rabs perturb the progression of cargo at specific endocytic locations. Rabs also play a crucial role in the regulation of tumor-derived exosomes. Rab11 influenced extrusion of exosome and interaction of MVB with autophagosomes [45] and promoted calcium dependent docking of MVBs to the PM [46] in K562 cells. Rab27A, in association with its GTPase activator, EP164, promoted exosome secretion by A549 lung cancer cells [47]. Rab27A/B are associated with exchange of exosomes between different cells of TME as well as with exosome secretion by macrophages [6]. Various types of Rabs involved in endocytic cargo trafficking have been depicted in Table 1.

Table 1. Different types of Rabs and their function in endocytic trafficking.

Rabs	Effects	Functions	References
Rab27	Secretion of exosomes	Release of markers MHC II, CD63, and CD81 in cancer cells	[32]
Rab7, Rab27a/b		Fusion with plasma membrane	[43]
Rab5, Rab4, Rab35	Recycling	Fast delivery of cargo to the plasma membrane	[36]
Rab5, Rab11a, Rab11b, Rab25		Slow delivery of cargo to the plasma membrane	[37,38]
Rab9		Transportation to TGN	[39]
Rab5, Rab7	Endosome maturation	Release of Rab5	[40]
Rab7	Sorting and degradation	Reduction in pH and acquisition of hydrolytic enzymes	[41]
Rab5 overexpression Note: may be rescued by Rab7	Suppression of release of exosomal markers syndecan, CD63, and ALIX	Inhibition of progression of endocytosed material from early endosomes	[44]

Abbreviations: ALIX, ALG-2 interacting protein X; MHC-II, major histocompatibility complex II; TGN, trans-Golgi network.

6. Exosomal Internalization by Recipient Cells

Exosomes float in the ECM after their release and exosomal surface proteins help in detecting the target cells for their internalization [48]. Exosomes attach to specific target cells by receptor-ligand binding, mediated through integrins, tetraspanins and intercellular adhesion molecules, which then internalizes exosomes (Figure 2) by (i) clathrin/caveolin-mediated endocytosis, (ii) uptake via lipid raft, (iii) macropinocytosis, (iv) direct fusion with the PM and (v) phagocytosis.

Clathrin protein forms a mesh like structure around the exosomes for its internalization. The PM of the recipient cells forms an inward invagination, followed by pinching off the clathrin coated vesicle from the membrane. The exosome empties all its contents in recipient cell's endosomes to perform specific functions [49]. Endocytosis, similar to the clathrin-dependent process, may be also mediated by caveolin-1 whose aggregations in PM form rafts. The invagination of the PM (caveolae) is rich in glycolipids, cholesterol and caveolin 1 [50]. Macropinocytosis involves distortion of PM forming protrusions from the membrane which encompass a region of extracellular fluid and exosomes, thereby internalizing exosomes. This process is Rac1-, actin- and cholesterol-dependent and it requires Na^+/H^+ exchange [51]. LAMP-1, integrins or tetraspanins are involved in the fusion of exosomes with the PM of recipient cells [52,53]. Phagocytosis is similar to macropinocytosis where exosomes are internalized along with some extracellular fluids. This process is followed by both phagocytic cells—like macrophages and dendritic cells (DCs)—and non-phagocytic cells like $\gamma\delta$ T cells [54]. During exosome uptake by soluble signaling, exosomal ligands are cleaved by cytoplasmic proteases and are bound to their respective receptors present on the PM of the recipient cells. In case of juxtacrine signaling, the ligands and receptors need to be in close proximity for efficient ligand–receptor binding [55]. Exosomal tetraspanins (CD9, CD63, CD81 and CD82) regulate cell fission and fusion, target cell selection [42], migration, adhesion, proliferation, and interaction between exosomes and recipient cells [56]. Size distribution in exosomes facilitates their internalization since cells have a propensity for loading smaller exosomes [5]. Oncogenic integrins play a dominant role during internalization of tumor-derived exosomes by recipient cells. Metastasis has been observed to be associated with exosome-integrins, such as $\alpha v \beta 6$ integrin in prostate, $\alpha v \beta 5$ integrins in liver and $\alpha 6 \beta 4$ and $\alpha 6 \beta 1$ integrins in lung [56].

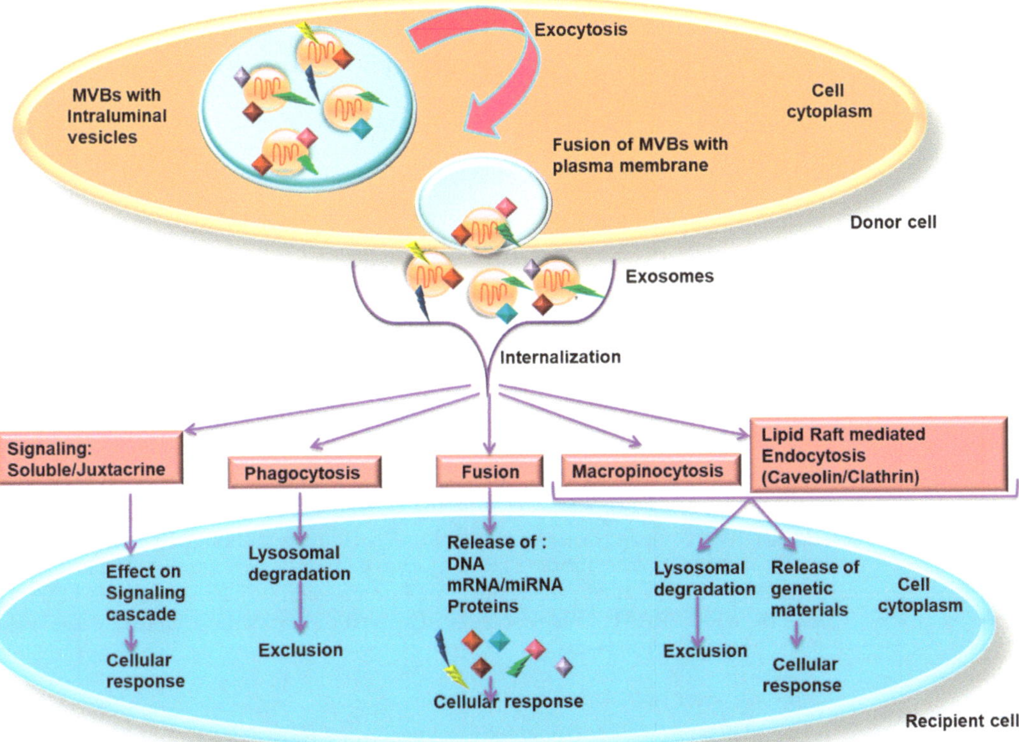

Figure 2. Mechanisms of internalization of exosomes. The exosomes inside the MVBs are extruded out from the donor cells by exocytosis on merging with plasma membrane. The released exosomes are then internalized via different modes: soluble/juxtacrine signaling; phagocytosis; fusion; micropinocytosis and lipid raft mediated endocytosis. The lipid raft mediated endocytosis can be either clathrin or caveolin protein dependent. Exosomes internalized by soluble/juxtacrine signaling affect the signaling cascade of the recipient cell. During phagocytosis, the exosomes undergo degradation, whereas, in the fusion event, genetic material is released that causes cellular response. In macropinocytosis and lipid raft-mediated endocytosis, the exosomes either undergo lysosomal degradation or mediate cellular response. Abbreviations: mRNA, messenger RNA; miRNA, microRNA; MVBs, multivesicular bodies.

7. TEX

The TEXs influence shaping of the TME, tumor progression, invasion and premetastatic niche formation, metastasis, angiogenic switch, and immune escape by paracrine subversion of local and distant microenvironments [57].

7.1. Oncogenic Signaling Involved in Exosomal Trafficking

According to the genometastatic theory, complex biomolecules in exosomes transfer oncogenic traits to target cells. Matrix cells in the TME interact with their oncogenic counterparts through exosomes and mediate tumor evolution and progression. Exosomal cargoes confer oncogenic transformation, EMT, immune surveillance evasion, invasion, and metastatic properties to the recipient cells [58]. Hypoxia and extracellular acidity culminate in greater release of TEXs [58]. Cells having even one oncosuppressor mutation are more prone towards uptake of exosomal oncogenic factors. Mutations leading to upregulated mitogen-activated protein kinase (MAPK) signaling in cancer cells elevated exosomes release [59]. Secretion of exosomes by activated platelets promoted MAPK and phosphoinositide 3-kinase (PI3K)/protein kinase B (Akt)/matrix metalloproteinase (MMP) signaling during cancer progression [31]. Expression of oncogenic RAS in non-tumorigenic epithelial cells promoted secretion of oncoprotein-rich exosomes [60]. Robust expression of oncogenic and truncated forms of epidermal growth factor receptor (EGFR) vIII in glioma cells augmented exosomal secretion and transfer of oncogenic activity to other normal cells [61]. Mutation of liver kinase B1 (STK11), a tumor suppressor, increased exosome secretion in lung cancer [62]. Secretion of exosomal mtDNA induced anaerobic metabolism and dormancy in cancer cells [31].

7.2. Exosomal miRNA-Mediated Cancer Promotion

Breast TEXs, enriched with Dicer, Protein Argonaut 2, and transactivation response element RNA-binding protein, processed precursor miRNAs into mature miRNAs for gene silencing in target cells and induced non-tumorigenic epithelial cells to form tumors [63]. Exosomal miRNAs suppressed cell proliferation by downregulating the C-X-C motif chemokine ligand 12 (CXCL12); exosomal-miR-23b augmented cell quiescence by inhibiting myristoylated alanine-rich C-kinase substrate expression in the metastatic niche [64]; miR-10b molded the TME to promote tumor metastasis [65] of breast cancer (BC) cells. Astrocyte-derived exosomes suppressed phosphatase and tensin homolog (PTEN) by intracellular trafficking of miR-19a in metastatic BC and melanoma brain metastasis models [66]. Release of exosomal miR-1245 from mutant p53 cancer cells reoriented macrophages to transforming growth factor-β (TGF-β)-rich tumor-associated macrophages (TAMs) which, in turn, propagated tumor progression [67]. Exosomal miR-105 and miR-939 in BC and miR-181c in brain cancer dissolved tight junctions, caused vascular leakiness and induced metastasis [31].

7.3. Exosomes and TME

TEXs are well documented for immune suppression by multiple interactions with immune cells of the TME (Figure 3). They hinder helper and cytotoxic T-cell activation and function, activate regulatory T-cell (Tregs), inhibit cytotoxicity of natural killer (NK) cells, augment differentiation of myeloid-derived suppressor cells (MDSCs) and reduce leukocyte adhesion [34]. Exosomes modulate the TME by extracellular signal-regulated kinase (ERK)-mediated cell growth or apoptosis. Interaction of stromal cells and tumor via exosomes inflict dissemination of tight junctions, generating a suitable niche for metastasis [68]. TEXs induced cancer-associated fibroblasts (CAFs) for exosomes' release [69]. The transfer of CAF-derived exosomal cargoes in the form of metabolic intermediates of the tricarboxylic acid cycle to cancer cells promotes neoplastic growth by alteration of glycolysis and glutamine-dependent reductive carboxylation [70]. Exosomes transformed fibroblasts into CAFs in melanoma [71]. CAFs or mesenchymal stem cells (MSCs) derived exosomes maneuvered Wnt signaling-induced migration [68]. Exosomes expressing Fas ligand activated CD8+ T-cell apoptosis [72]. Exosomal αvβ6 integrin inhibited the signal transducer and activator of transcription 1 (STAT1)/MX1/2 signaling in cancer cells and reprogramed monocytes into the M2 phenotype [73]. Exosomal miR-146a-5p from hepatocellular carcinoma (HCC) cells induced M2 polarization [74]. BC cell derived exosomes inhibited NK cells [75] and infiltrated neutrophils into tumors [76]. Melanoma-derived exosomes perturbed maturation of DCs in lymph nodes [77]. However, TEXs can supply antigens to DCs for cross-presentation to cytotoxic T cells [78]. Administration of topotecan/radiation induced the release of exosomal immunostimulatory DNA, which inflicted DC maturation and cytotoxic T cell activation [31]. Programmed death ligand 1 (PD-L1)-positive exosomes positively correlated with head and neck squamous cancer cells (HNSCC) progression in patients and administration of anti-PD-L-1 antibodies inhibited the immunosuppressive function of PD-L1 [79].

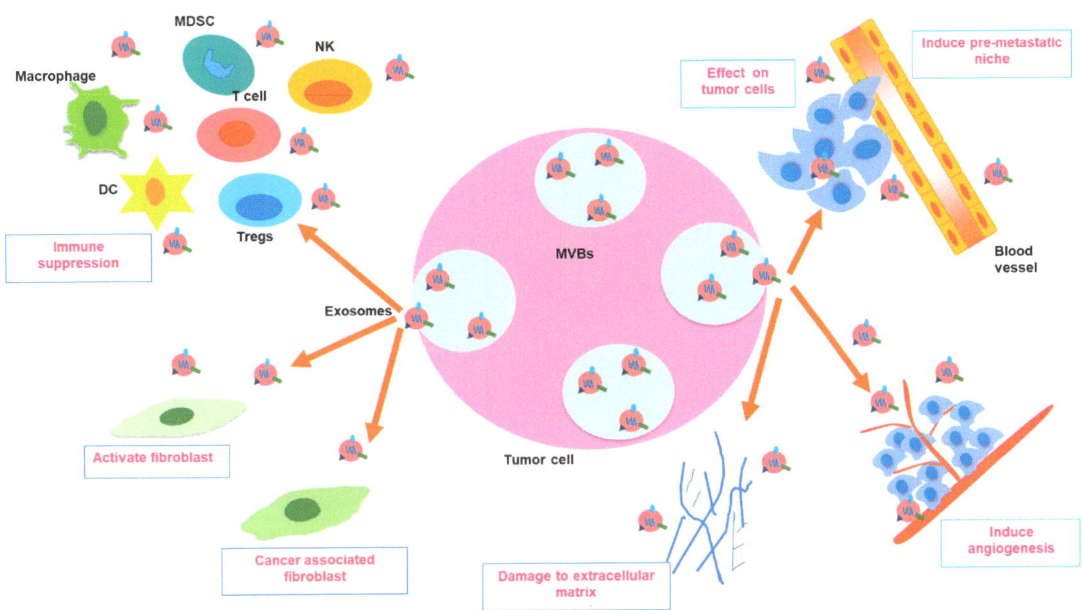

Figure 3. Exosomes in tumor microenvironment. Exosomes secreted from tumor cells containing MVBs exhibited a dynamic signaling between tumor cells and the TME. Exosomes may lead to immune suppression by downregulating macrophages, DC, T cells and NK cells and upregulating immunosuppressive cells like Tregs, MDSCs and TAMs. Exosomes induced differentiation of fibroblasts, activation of CAFs and degradation of ECM, which are associated with TME construction. They are involved in the alteration of ECM, hypoxia-mediated angiogenesis and the formation of pre-metastatic niches that trigger the metastatic escape of tumor cells. Abbreviations: CAFs, cancer-associated fibroblasts; ECM, extracellular matrix; DCs, dendritic cells; MDSCs, myeloid-derived suppressor cells; NK cells, natural killer cells; TAMs, tumor-associated macrophages; TME, tumor microenvironment; Tregs, tumor regulatory cells.

7.4. Impact of Exosomes on EMT, Invasion, Metastasis and Angiogenesis

Exosomal cargoes CD151 and Tspan8 are related with ECM degradation, stromal reprogramming, cell motility and tumor progression [80]. EMT was induced by exosomal miR-663b in bladder cancer [81]; lncRNA SOX2 overlapping transcript (Sox2ot) in pancreatic ductal adenocarcinoma (PDAC cells) [82]; and TGF-β-enriched TEXs in myofibroblasts [83]. Migration of tumor cells was facilitated by the exosome-mediated transfer of αvβ6 in prostate cancer [84]; miR-21 in bladder cancer [85]; TAM derived exosomes in gastric cancer (GC) cells [86]; and lncRNA ubiquitin-fold modifier conjugating enzyme 1 (UFC1) in non-small cell lung carcinoma (NSCLC) [87]. Exosomal lncRNA zinc finger antisense 1 (ZFAS1) induced EMT and migration in GC cells [88]. Metastasis was promoted by exosomal EGFR in GC [89]; MMP1 mRNA in ovarian cancer [90]; miR-25-3p, miR-130b-3p, miR-425-5p in colorectal cancer cells (CRC) [91]; miR-106b in lung cancer [92]; and miR-21 in oesophageal cancer [93]. Cell proliferation and invasion was induced by exosomal miR-1260b in lung adenocarcinoma [94] and miR-222 in PDAC [95]. Angiogenesis and tumor progression were influenced by exosome mediated Wnt4/β-catenin signaling in CRC [96] and by vascular endothelial growth factor A (VEGF-A) enriched exosomes in brain endothelial cells [97]. Tumor progression was augmented by exosomal miRNAs from TP53-mutant cells in colon cancer cells [98] and by exosomal lncRNA ZFAS1 in GC [92]. Exosomal miR-21 reduced apoptosis in GC cells [99], exosomal IL-6 induced metastasis in BC cells [100], exosomal HSP70 induced tumor progression in MSC cells [101], and exosomal TGF-β promoted tumor growth in LAMA84 cells [102]. Various recent studies based on the tumor promoting effect of exosomes have been listed in Table 2.

Table 2. Tumor-promoting effects of exosomal cargoes on recipient cells.

Exosome Donor Cells	Exosomal Cargo	Target Cells	Effects	Mechanisms	References
Human prostate cancer (PC3) cells	Integrin α$_V$β$_6$	Peripheral blood mononuclear cells and THP-1 monocyte cells	↑M2 polarization	↓STAT1/MX1/2 signaling	[73]
		Human prostate cancer DU145 cells	↑Cell adhesion and migration	↑Latency-associated peptide-TGF-β	[84]
HCC (mouse Hepa1-6, H22, and human HepG2, H7402) cells	miR-146a-5p	Mouse RAW264.7 cells, THP-1 cells, mice peritoneal macrophages	↑Pro-inflammatory factors, ↑M2 polarization, ↑T-cell exhaustion by M2 macrophages	↑NF-κB, ↑p-STAT3, ↓p-STAT1	[74]
Human Bladder cancer (T24 and 5637) cells	miR-663b	T24 and 5637 cells	↑Cell proliferation, ↑EMT	↓ERF, ↓E-cadherin, ↑Vimentin	[81]
Human PDAC (Hs 766 T) and metastatic (Hs 766T-L2) cells	lncRNA-Sox2ot	Human PDAC (BxPC-3) cells	↑EMT, ↑stemness, ↑invasion and metastasis	↑Sox-2	[82]
Human bladder cancer (T24) cells	miR-21	Human THP-1 cell-derived macrophages	↑M2 polarization, ↑tumor cell migration and invasion	↓PTEN, ↑PI3K/Akt-STAT3 signaling	[85]
M2 polarized macrophages (TAMs)	Apolipoprotein E	Mouse gastric carcinoma (MFC) cells	↑Cell migration	↑PI3K-Akt signaling	[86]
Human NSCLC (A549 and H1299) cells	lncRNA UFC1	A549 and H1299 cells	↑Cell proliferation, ↑migration, ↑invasion	↓PTEN via EZH2-mediated epigenetic silencing	[87]
Human GC (BGC-823) cells	lncRNA-ZFAS1	Human GC (MKN-28) cells	↑EMT, ↑cell proliferation, ↑migration	↑Cyclin D1, ↑Bcl-2, ↓Bax, ↓E-cad, ↑N-cad, ↑Slug	[88]
Human GC (SGC7901) cells	EGFR	Primary mouse liver cells	↑Cell proliferation, ↑metastasis	↓miR-26a/b, ↑HGF, ↑c-Met	[89]
Human CRC (HCT116) cells	miR-25-3p, miR-130b-3p and miR-425-5p	Macrophages RAW264.7	↑M2 polarization, ↑EMT, ↑liver metastasis	↑CXCL12/CXCR4 axis, ↓PTEN, ↑PI3K-Akt signaling	[91]
Human lung cancer (SPC-A-1 and H1299) cells	miR-106b	SPC-A-1 and H1229 cells	↑Migration and invasion	↓PTEN	[92]
Human esophageal cancer (EC9706) cells	miR-21	EC9706 cells	↑Metastasis	↓PDCD4, ↑MMP2, ↑MMP9	[93]
Human lung adenocarcinoma (H1299) cells	miR-1260b	Human A549 cells	↑Cell invasion, ↑cell proliferation, ↑drug resistance	↑Wnt/β-catenin signaling, ↓sFRP1, ↓Smad4	[94]
Human PDAC (Capan-1 and Hs 766T-L3) cells	miR-222	PDAC (Capan-1 and Hs 766T-L3 cells)	↑Cell invasion, ↑metastasis	↑Akt, ↓PPP2R2A, ↑p-P27	[95]

Table 2. *Cont.*

Exosome Donor Cells	Exosomal Cargo	Target Cells	Effects	Mechanisms	References
Hypoxic human CRC (HT29 and HCT116) cells	Wnt4	Endothelial (HUVECs) and CRC (HT29) cells	↑Proliferation, ↑angiogenesis, ↑migration	↑β-Catenin signaling	[96]
TP53-mutant (HT29) colon cancer cells	miR-1249-5p, miR-6737-5p, and miR-6819-5p	Human colon fibroblasts (CCD-18Co) cells	↑Tumor progression	↓TP53	[98]
Murine bone marrow–derived macrophages	miR-21	Human GC (MFC, MGC-803) cells	↓Apoptosis, ↑resistance to cisplatin	↑PI3K/AKT signalling, ↓PTEN	[99]
Co-culture of THP-1-derived macrophages exposed to apoptotic human BC (MCF-7 or MDA-MB-231) cells	IL-6	Naïve (MCF-7 or MDA-MB-231) cells	↑Proliferation, ↑metastasis	↑p-STAT3, ↑cyclin D1, ↑MMP2, ↑MMP9	[100]
Human lung cancer (A549) cells	HSP70	MSCs extracted from human adipose tissue	Pro-inflammatory MSCs, ↑tumor growth	↑TLR-2/NF-κB signaling, ↑IL-6, ↑IL-8, ↑MCP-1	[101]
Human chronic myeloid leukemia (LAMA84) cells	TGF-β	LAMA84 cells	↑Proliferation, ↓apoptosis, ↑tumor growth	↑SMAD 2/3, ↑Bcl-w, ↑Bcl-xL, ↑survivin, ↓BAD, ↓BAX, ↓PUMA	[102]
Human BC (MCF-7) tamoxifen resistant cells	miR-221/222	Human BC (MCF-7) wild type cells	↑Resistance to tamoxifen	↓P27, ↓ERα,	[103]
Human cisplatin resistant A549 cells	miR-100-5p	Human A549 cells	↑Resistance to cisplatin	↑mTOR	[104]
Gemcitabine treated human PDAC CAFs	Snail and miR-146a	Human pancreatic cancer L3.6pl cells	↑proliferation, ↑resistance to gemcitabine	↑Snail, ↑miR-146a	[105]
Human HER-2-positive BC trastuzumab resistant (SKBR-3 and BT474) cells	lncRNA AFAP1-AS1	SKBR-3 and BT474 cells	↑Resistance to trastuzumab	↑ERBB2	[106]
Tamoxifen resistant BC (LCC2) cells	lncRNA UCA1	ER-positive BC MCF-7 cells	↑Cell viability, ↑resistance to tamoxifen	↓caspase-3	[107]
Human GC (MGC-803 and MKN-45) cisplatin resistant cells	lncRNA HOTTIP	MGC-803 and MKN-45 cells	↑Resistance to cisplatin	↑HMGA1	[108]

Symbols: ↑, upregulated; ↓, downregulated; Abbreviations: AFAP1-AS1, actin filament associated protein1 antisense RNA 1; Akt, protein kinase B; Bad, Bcl-2 associated agonist of cell death; Bax, Bcl-2-associated X protein; Bcl-2, B-cell lymphoma 2; c-Met, Mesenchymal-epithelial transition factor; CXCL12, C-X-C motif chemokine ligand 12; CXCR4, C-X-C chemokine receptor type 4; Erα, estrogen receptor-α; ERF, Ets2-repressor factor; ERBB2, erythroblastic oncogene B; HGF, hepatocyte growth factor; HMGA1, High-mobility group A1; HOTTIP, HOXA transcript at the distal tip; MCP-1, monocyte chemoattractant protein-1; MMP, matrix metalloproteinase; NF-κB, nuclear factor kappa-light-chain-enhancer of activated B cells; PDCD4, programmed cell death 4; PI3K, phosphoinositide 3-kinase; PPP2R2A, protein phosphatase 2 regulatory subunit B alpha; PTEN, phosphatase and tensin homolog; PUMA, p53 upregulated modulator of apoptosis; sFRP, secreted frizzled-related protein 1; STAT, signal transducer and activator of transcription; Sox-2, sex determining region Y-box 2; TGF-β, transforming growth factor-β; TLR-2, toll-like receptor 2; TP53, tumor protein p53.

7.5. Exosomes and Drug Resistance

Exosomes form a physical barrier against drug penetration and confer drug resistance by transfer of cargoes from resistant to sensitive cells [104]. Exosome-mediated drug resistance may be devised through trafficking of non-coding RNAs, drug transporters and neutralization of antibody-based drugs, which has been described in the following sections.

7.5.1. By Trafficking of Non-Coding RNAs

Non-coding RNAs, including miRNAs and lncRNAs, perpetuated drug resistance across an array of cancer cells. Exosomes from M2-macrophage exerted miR-21-mediated upregulation of PI3K/Akt signaling and reduced apoptosis and cisplatin resistance in GC [93]. Exosomal miR-221/222 modulated p27 and ERα for tamoxifen resistance [103] in BC cells. Exosomes derived from cisplatin resistant cells induced resistance in cisplatin sensitive A549 cells in a miR-100-5p-dependent manner [104]. In ovarian cancer cells, exosomal miR-443 induced senescence and resistance against paclitaxel [109]. In prostate cancer, CAF derived exosomes conferred gemcitabine resistance via Snail and miR-146a [105]. Exosomal cargo-lncRNA UCA1 mediated tamoxifen resistance [107] and lncRNA actin filament associated protein1 antisense RNA 1 (AFAP1-AS1) conferred trastuzumab resistance by binding to AU binding factor 1 and translating erythroblastic oncogene B2 (ERBB2) [106] in BC cells. MSC-derived exosomes aided the transfer of lncRNA PSMA3-AS1 to myeloma cells and exerted resistance against proteasome inhibitor [110]. In GC, exosomal lncRNA HoxA transcript at a distal tip (HOTTIP) made sensitive GC cells cisplatin resistant [108].

7.5.2. By Trafficking of Drug Transporters and Neutralizing Antibody-Based Drugs

The exosome-mediated transfer of drug transporter molecules is intimately associated with the spread of drug resistance across diverse cancer forms. Exosomes transported P-glycoprotein (P-gp) from doxorubicin-resistant cells [68] and multidrug resistance protein-1 (MDR-1) from docetaxel-resistant cells [111] to confer drug resistance in sensitive BC cells. Recently, it has been evidenced that exosome-mediated transfer of chloride intracellular channel 1 upregulated P-gp and B cell lymphoma-2 (Bcl-2) and conferred vincristine resistance in GC cell line SGC-7901 [112].

B-cell lymphoma derived exosomes modulated ATP-binding cassette (ABC) transporter A3, carried CD20 antigen which shielded the cancer cells against therapeutic CD20 antibodies and evaded immune surveillance [113]. Exocytosis of TEXs from human epidermal growth factor receptor 2 (HER2) positive BC cells expressed specific decoy molecules and conferred resistance against monoclonal antibody trastuzumab, thus depicting that TEXs are also involved in neutralizing antibody based drugs [114].

8. Strategies against Tumor-Derived Exosomes

There have been, primarily, three approaches for the management of exosomes associated with pathogenesis, as described below.

8.1. Suppression of Exosome Biogenesis and Trafficking

Genetic knockdown of tumor suppressor TSG1 (protein involved with exosome biogenesis and trafficking) reduced Wnt5b-positive exosomes in colon cancer [115]. Suppression of annexin A1 (responsible for membrane contact sites, inward vesiculation and exosome biosynthesis) reduced the number of secreted exosomes in pancreatic cancer cells [116]. Manumycin A was reported to inhibit ESCRT-dependent exosome biogenesis by modulating Ras/Raf/ERK1/2/heterogeneous nuclear ribonucleoprotein H1 axis in prostate cancer cells [117].

Small molecule inhibitor GW4869 against nSMase2 reduced secretion of ceramide enriched exosomes [118] and sensitized breast tumors by inhibition of exosomal PD-L1 [119]. Knockout of nSMase2 reduced exosome secretion, angiogenesis and metastasis in breast tumors [120]. Another inhibitor of lipid metabolism, pantethine, a pantothenic acid (vitamin B_5) derivative, depleted the release of exosomes in MCF-7 variants and increased doxorubicin responsiveness [121]. Genetic silencing of Rab27A/B reduced exosomal secretion by HNSCC and macrophages, thereby minimizing metastasis in BC cells [76] and lung metastasis in melanoma [122]. PRAS40 downregulated Akt, downstream of TGF-β, and mediated antagonistic effects against exosome secretion and chemoresistance in breast and lung cancer cells [123]. WEB2086, a platelet-activating factor receptor (PAFR) antagonist, was shown to reduce gemcitabine-induced exosome release in PAFR-positive pancreatic cancer cells [124]. Other exosome extrusion inhibitors, such as chloramidine, bisindolylmaleimide-I, imipramine, d-pantethine, and calpeptin, and calcium chelators, such as ethylene glycol bis (2-aminoethyl ether) tetra-acetic acid, increased responsiveness toward 5-FU in prostate and BC cells [125]. The inhibition of protease-activated receptor 2 by an anticoagulant, apixaban, which binds to the tissue factor–factor VIIa complex, downregulated the secretion of TF-bearing exosomes from pancreatic cancer cells [126]. Dasitinib inhibited exosome release and beclin-1/Vps34 mediated autophagy in imatinib resistant K562 cells [127]. Reduced exosome secretion by synthetic peptide (constructed with a derivative of the secretion modification region of HIV-1 Nef protein, a N-terminus anchored polyethylene glycol residue and a c-terminus cluster in peptide) [128] and by Docosahexaenoic acid (a polyunsaturated fatty acid) [34] inhibited metastasis and angiogenesis, respectively, in BC cells.

8.2. Depletion of Exosome Uptake

A synthetic nanoparticle, which is a prototype of high-density lipoprotein, was used as an agonist of the scavenger receptor type B-1 (SR-B1) which eliminated cholesterol from lipid rafts and prevented exosome uptake by SR-B1 expressing cancer cells [129]. Other agents, such as heparin sulfate proteoglycans, methyl-β cyclodextrin (molecule used for cholesterol removal from natural and artificial membranes) and dynasore (dynamin inhibitor), have been reported to abrogate exosome endocytosis in cancer cells [130]. Heparin and dynasore attenuated the uptake of multiple myeloma-derived exosome by bone marrow stromal cells and inhibited phosphorylation of STAT1, STAT3, and ERK1/2 signaling pathways [131]. Radiation-derived exosomes made the recipient cancer cells radiation-resistant and aggravated proliferation. Heparin and simvastatin attenuated radiation-derived exosome uptake by recipient cells in in vitro and in vivo models of glioblastoma [132].

8.3. Modulation of Harmful Exosomal Cargo and Inhibition of Exosome Dissemination

Alteration of exosomal cargoes was achieved by viral manipulation or by incorporation of viral proteins/RNA into secreted exosomes [133]. Curcumin culminated the immunosuppressive effect of exosomes in BC by deregulation of the ubiquitin-proteasome system and cargo sorting of ILVs [134]. Subscapular sinus CD169+ macrophages bound with exosomes restricted their interaction with B cells, promoting tumor progression [135]. Exosome release was inhibited by inhibitors like indometacin (COX2 inhibitor) in combination with rapamycin (interfere with MVB biogenesis) in B lymphoma cells, by suppressing ATP-binding cassette sub-family A member 3 expression of the lymphoma cells and induced the cells to undergo complement dependent cytolysis under the effect of drug rituximab [113].

8.4. Removal of Exosomes

A microfluidics-based technology-microscale acoustic standing wave technology facilitates clearance of exosomes from circulation [136]. Innate immune system in co-operation with opsonization effects of complement proteins may be used for elimination of exosomes [137]. Opsonization of exosomal markers CD9 and CD63 by targeting anti CD9 and anti CD63 antibodies elevated exosomes representation to the macrophages, leading to exosomes' elimination, which suppressed lung metastasis in vivo [138]. In colorectal cancer, dimethyl amiloride depleted exosomes, thereby elevating cyclophosphamide efficacy against the cancer cells [139].

9. Cancer Management with Exosomes

Exosomes have emerged as a new arena of clinical interest due to their prospective use in diagnostic applications as potential biomarkers, for carrying specific information of their progenitor cells, as well as for being ideal candidates for liquid biopsy [56].

9.1. Preclinical Studies on Anticancer Potential of Exosomal Cargoes

Uptake of exosomal contents does not always confer procarcinogenic signaling. There are instances where exosomal proteins promoted anticarcinogenic signaling pathways, e.g., exosomal uptake with payload of gastrokine1 suppressed H-Ras/Raf/MEK/ERK-mediated gastric carcinogenesis in gastric epithelial cells [140]. The miR-375 carried by exosomes inhibited cell proliferation and invasive capability in colon cancer cells through

Bcl-2 blocking [141]. Exosomal miR-520b derived from normal fibroblasts cells inhibited proliferation and migration of pancreatic cancer cells [142]. The migratory behavior of lung cancer cells was reduced by exosomal miR-497 through suppression of growth factors, cyclin E1 and VEGF [143]. Exosomal circulating RNA circ-0051443 inhibited tumor progression through apoptosis induction in HCC cells [144]. In BC cells, exosomal miR-100 derived from MSCs inhibited angiogenesis in vitro via modulating mTOR/HIF-1α/VEGF signaling [145].

9.2. Exosomes as Biomarkers

Cancer cells secrete exosomes ten times higher than normal cells, which makes TEXs major potential candidates for liquid biopsy needed for cancer diagnosis and prognosis [57]. The release of exosomes in the extracellular space also aids in cancer diagnosis by examining their increased levels in various body fluids, such as blood, ascites fluid, urine, and saliva [146]. Exosomal DNA represents the entire genome; therefore, liquid biopsies of plasma aid in early detection of cancer-specific mutations. Exosomal CD63 and caveolin-1 served as non-invasive markers of melanoma [121]. Exosomal lncRNA, either with miR-21 or alone, was correlated with tumor classification (III/IV), stage of tumor and lymph node/distant metastasis in many cancer types [5]. Differential expression of exosomal miR-150, miR-155, and miR-1246 in serum of normal individuals and acute myeloid leukemia patients detected minimal residual disease [147]. Phosphatidylserine present on the exosomal surface also serves as a biomarker for diagnosis of early-stage cancer [148]. However, exosomal biomarkers are often overshadowed by highly prevalent complex proteins of the body fluids. Exosome isolation from body fluids follows either of the three methods, namely differential centrifugation coupled with ultracentrifugation, immunoaffinity pull-down, and density gradient separation. Mining of exosomal biomarkers from body fluid of cancer patients has been explored with fluorescence-based analytical techniques, electrochemical aptamer-based detection methods, localized surface plasmon resonance and surface-enhanced Raman scattering [149]. Though exosome biomarker analysis has tremendous translational potential, a gold standard for exosome isolation under clinical settings is yet to be achieved [150]. Since there is no definite consensus for isolation of exosomes, the best suitable body fluid for exosome isolation is also under investigation.

9.3. Role of Exosomes in Immunotherapy and Vaccine Development

DCs and other antigen presenting cells (APCs) derived exosomes are loaded with specific drugs; miRNAs of interest or even exosomes alone are implemented to trigger immune response in the recipient individuals (Figure 4). DC-based exosomes, in therapy, are beneficial as they possess abundant surface lactadherin that helps in efficient exosome uptake [151]. The functional moieties, such as MHC-I, MHC-II, CD40, CD80, CD86 TNF, FasL, TRAIL and natural killer group 2D (NKG2D) ligands on the surface of DC-derived exosomes, facilitate in imparting innate and adaptive antitumor immune response [152]. DC-derived exosomes activated NK cells in NKG2D and interkeukin (IL)-15Rα ligand dependent mode, which restored 50% functionality of NK cells and was implemented as a cell free vaccination strategy [153]. The administration of adjuvants, such as IFN-γ, Toll-like receptor agonists, and polyinosinic: polycyctidylic acid, was explored for production of mature DC-derived exosomes which showed greater potential for activation of Th1 cells [154,155]. Immunogenic cell death was induced by melphalan, an anticancer drug, in multiple myeloma cells by increasing the damage-associated molecular pattern containing exosomes, thus triggering NK cell cytotoxicity [156]. A histone deacetylase inhibitor, MS-275, increased the release of Hsp70 and MHC-I polypeptide-related sequence B (MICB)-rich exosomes which induced NK cytotoxicity and lymphocyte proliferation [157]. Heat shock treatment increasing the immunostimulatory activities of TEXs has been demonstrated in A20 lymphoma/leukemia cells. Heat shock tumor derived exosomes were observed to possess more immune-stimulating activities due to elevated expression of MHC and increased levels of cytokines, such as IL-1β, IL-12p40, and TNF-α [158].

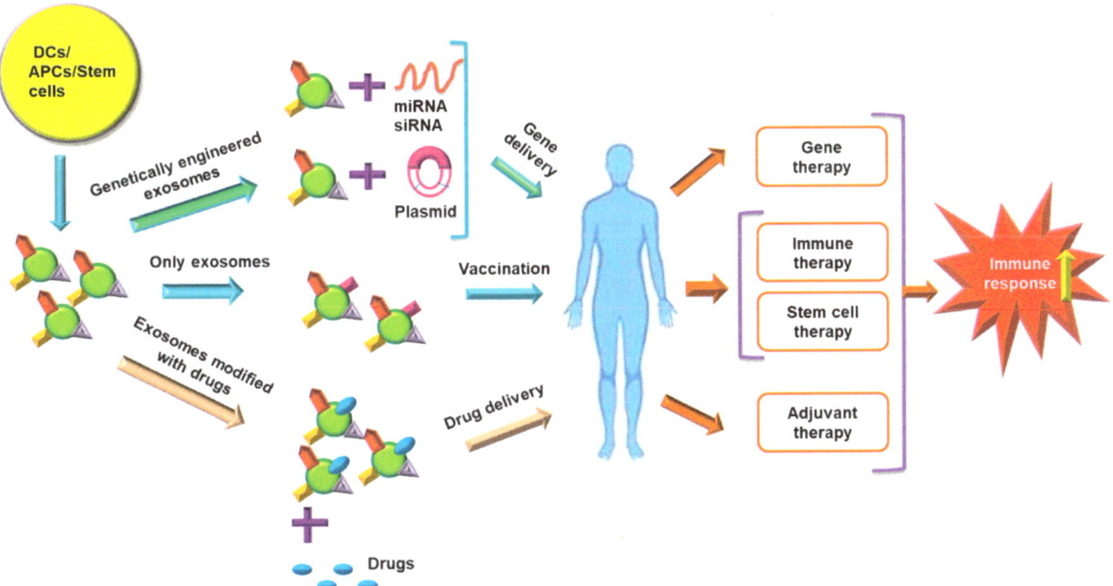

Figure 4. Exosomes in therapeutic approaches. Exosomes derived from DCs, APCs and stem cells can be utilized for immunotherapy, gene therapy, stem cell therapy and adjuvant therapy. Exosome based gene therapy is obtained by genetically engineered exosomes loaded with miRNA, siRNA and plasmids of interest. Stem cell or DC-derived exosomes can be implemented alone as vaccines and confer stem cell-based therapy or immunotherapy. The exosomes can also be utilized for drug delivery vectors by modifying them with drugs of interest. The DC-, APC- and stem cell-derived exosomes administered into the patient help in triggering immune response in combating cancer by targeting and regulating the mechanisms against which the exosomes are implemented. Abbreviations: APCs, antigen presenting cells; DCs, dendritic cells; miRNA, microRNA; siRNA, small interfering RNA.

Exosomes have potential use in vaccine development because the surface-bound proteins on exosomes of APCs, DCs and tumor cells originate from the progenitor cell membranes [5]. Nanoscale immunotherapy treatments with TEX, DC-derived exosomes and ascitic cell-derived exosomes have shown efficacy in stimulating the body's immune system against cancer cells [159]. Ascitic cell-derived exosomes obtained from peritoneal cavity fluid of cancer patients triggered cancer cell lysis via activation of dendritic cells and MHC-1-dependent T cell response. Membrane-bound Hsp70 of TEX exhibited robust priming of T helper cell 1 (Th1)- and NK-mediated antitumor immune response [160]. Chemotherapy accompanied with hyperthermia has evolved as a new treatment mode for cancer involving TEXs. For instance, heat stress has increased the antitumor effect of TEXs derived from doxorubicin-treated MCF-7 cells [161]. DC-derived exosomes control tumor growth by eliciting CD8+ and CD4+ T cell responses [162]. DC-derived exosomes incubated with cancer antigen triggered cancer specific T cell response [163]. Adjuvant-based exosomal vaccines are effective in eliciting immune response. For example, streptavidin-lactadherin protein fused with immunostimulatory biotinylated CpG DNA (adjuvant) after transfection into murine melanoma cells created genetically modified exosomes. These exosomes have the ability to trigger improved antigen presentation to the DCs and other immune cells, contributing to enhanced immune response [164]. DC-derived exosomes have been observed to be more efficient as cell-free vaccines in treating malignancies that respond poorly to immunotherapy. For instance, α-feto protein-rich DC derived exosomes triggered more effective antitumor immune responses and modulated the TME in a HCC mice model [165]. Recently, it was observed that vaccination with TEX-pulsed DC along with cytotoxic drugs specifically targeted immunosuppressive MDSCs in pancreatic cancer cells [166]. DNA vaccines prepared by fusing ovalbumin antigen with lactadherin present on exosomal surface diminished fibrosarcoma, thymoma and melanoma metastasis by activating T lymphocytes [167].

9.4. Exosome-Based RNA Therapy

Exosome-based miRNA therapy exhibited immunosuppressive properties by controlling the gene expressions [19]. An early study reported that exosomes derived from human embryonic kidney cells were effective in regressing tumor growth by delivering miR-let7a in an EGFR-positive BC xenograft model [168]. The MSCs transfected with miR-124a enhanced exosomes carrying the RNA of interest production, which, when implemented against gliomas, reduced the cell viability and targeted FOXA2 that caused accumulation of lipids [169]. Transfer of lncRNA PTEN pseudogene 1 by exosomes derived from normal cells to bladder cancer cells reduced tumor progression in vitro and in vivo [170].

Exosomes also mediated targeted delivery of siRNA, e.g., siRNA transfected into exosomes targeted RAD51 and RAD52 in Hela and fibrosarcoma cells, which inhibited proliferation of the recipient cells [171]. Engineered exosomes containing IL-3 ligand or functional siRNA for BCR-ABL were successfully used against imatinib resistance in chronic myeloid leukemia patients [172]. Exosomes used for trafficking RNA interference (RNAi) mediators counteracted against oncogenic KRAS and improved overall survival in mouse models of pancreatic cancer [173]. Delivery of engineered exosome mediated siRNA inhibited post-operative metastasis of BC, indicating a promising strategy against tumor progression [174]. Successful delivery of antisense miRNA oligonucleotides against miR-21 by electroporating them in exosomal membrane improved the treatment efficacy for glioblastoma by inducing the expression of PTEN and PDCD4, resulting in decreased tumor size [175].

9.5. Exosomes in Stem Cell Therapy

Normal stem cell-derived exosomes are free of tumorigenic factors and are potential candidates for stem cell therapy [176]. MSC-derived exosomes can protect their cargoes from degradation, facilitate easier uptake by recipient cells, elicit low toxicity and immunogenicity, and these exosomes can be modified to enhance cell type-specific targeting and may be a prospective tool for cell-free based therapeutic approaches [177]. Exosomal miR-144 derived from bone marrow derived MSC retarded the spread of NSCLC by targeting cyclin E1 or E2 [178]. Exosomes released from miR-101-3p overexpressing MSCs negatively affected the proliferation and migration of oral cancer cells by targeting the collagen type X α1 chain [179]. MSC-derived exosomes were genetically engineered by loading them with polo-like kinase 1 (PLK-1)-siRNA and were utilized for PLK1 gene silencing in bladder cancer [180]. The primary hurdles of stem cell-based therapy, such as teratoma formation and embolization, are less frequent with exosome-based stem cell therapeutics. Exosomes secreted from induced pluripotent stem cells may exert better therapeutic effects [163].

9.6. Exosomes in Drug Delivery

Normal cell derived exosomes exhibit excellent biodistribution, biocompatibility, low immunogenicity, capacity to cross the blood–brain barrier and high target specificity, which make them potential candidates for drug delivery in cancer [5]. The exosomal surface proteins regulate efficient drug delivery because of their involvement in exosomes uptake by the tumorigenic recipient cells [181]. Exosomes derived from androgen-sensitive human prostate adenocarcinoma cells carrying paclitaxel negatively affect the cancer cells' viability [182]. DC-derived exosomes in BC and macrophage-derived exosomes in lung cancer were loaded with the drugs trastuzumab and paclitaxel, respectively, and successfully delivered to the recipients [183,184]. Moreover, exosomes loaded with doxorubicin conjugated with gold nanoparticles showed anticancer effect against lung cancer cells [185]. Exosomes with A disintegrin and metalloproteinase 15 (ADAM15) expression (A15-Exo) co-delivered with doxorubicin and cholesterol-modified miRNA 159 exhibited anticancer effect in BC cells [186]. Paclitaxel loaded exosomes showed sensitivity towards MDR cancer cells via by-passing P-gp-mediated drug efflux and also inhibited metastasis in a lung cancer xenograft model [187]. Unmodified exosomes encapsulated with doxorubicin reduced tumor proliferation in a mouse mammary carcinoma xenograft model [137]. Exosomal delivery of doxorubicin induced its therapeutic activity in xenograft models of breast and ovarian cancer [188]. Exosomes isolated from engineered immature DCs (expressed Lamp2b fused with αv integrin-specific iRGD peptide (CRGDKGPDC)) loaded with doxorubicin successfully targeted αv integrin-positive breast tumor cells [189]. Exosome encapsulated gemcitabine exhibited anticancer properties in autologous pancreatic cancer cells and in a xenograft model [190].

Phytochemicals, administered via an exosome-mediated drug delivery system, can provide health benefits and anticancer properties [56]. Pancreatic adenocarcinoma cell-derived exosomes aided curcumin in inflicting its anticancer properties among tumor cells [191]. Milk-derived exosomes encapsulated with anthocyanidins exhibited antiproliferative effect in a xenograft lung carcinoma model [192]. Exosomal formulations of black bean extract exhibited pronounced antiproliferative effect in many cancer cells [193]. Exosomal formulations with berry anthocyanidins exhibited anticancer properties in ovarian cancer with enhanced sensitivity in chemoresistant tumors [194]. Exosomal encapsulation of celastrol (a triterpenoid) exhibited antiproliferative effect in lung cancer cells and in a xenograft model [195]. Recent studies on exosomal drug delivery of chemotherapeutic drugs and phytochemicals are listed in Table 3.

9.7. Induction of Chemosensitivity with Exosomes

TEXs impart drug resistance but may also be used for inducing drug sensitivity. Dimethyl amiloride augmented ABC transporter containing exosome secretion revived the cyclophosphamide sensitivity of cancer cells [31]. Downregulation of the GAIP-interacting protein C terminus mediated secretion of ABCG2 drug transporters containing exosomes and suppressed gemicitabine resistance in pancreatic cancer cells [196]. In oral squamous cell carcinoma, exosomal miR-155 increased chemoresistivity in cisplatin-sensitive cancer cells [197]. The exosomes loaded with CRISPR/Cas9 induced apoptosis and cisplatin chemosensitivity in ovarian cancer cells [198]. An increase in apoptosis and chemosensitivity was observed in cisplatin-resistant human gastric adenocarcinoma cells through treatment with si-c-Met containing exosomes derived from human kidney epithelial cell line [199]. Normal intestinal FHC cell-derived exosomes transferred miR-128-3p into oxiplatin resistant CRC cells which induced their chemosensitivity and decreased motility [200]. miR-122-transfected adipose tissue-derived MSCs (AMSCs) released exosomes carrying miR-122 and, when cocultured with hepatocyte carcinoma cells, induced sorafenib chemosensitivity [201]. miR-567 induced chemosensitivity in resistant BC cells towards trastuzumab and blocked autophagy [202]. Exosomal miR-200c induced chemosensitivity towards docetaxel and apoptosis in tongue squamous cell carcinoma [203]. Coculture of miR-199a carrying exosomes derived from AMSCs with HCC cells downregulated mammalian target of rapamycin (mTOR) pathway and induced chemosensitivity towards doxorubicin [204]. Various recent reports on exosome-mediated reversal of chemosensitivity have been listed in Table 4.

Table 3. Exosomes as delivery system for therapeutic implications against cancer.

Exosome Source	Modification of Exosomes with Drugs	Loading Method	Target Cells	Effect	Mechanism	References
			Chemotherapeutic drugs			
Human mammary adenocarcinoma cells (M-CF-7), mouse mammary carcinoma cells (4T1), and human prostate adenocarcinoma cells (PC3)	Doxorubicin	Incubation	4T1 tumor-bearing BALB/c mice	↓Tumor growth, but no significant reduction in tumor growth with exosomes loaded with doxorubicin compared to free drug	-	[137]
Human prostate cancer (LNCaP and PC-3) cells	Paclitaxel	Incubation	LNCaP and PC-3 cells	↑Cytotoxic effect of paclitaxel	-	[182]
Human NSCLC (H1299) cells	Exo-gold nanoparticles-doxorubicin	Incubation	Human NSCLC (H1299 and A549) cells	↑DNA damage, ↑apoptosis	↑caspase-9, ↑ROS	[185]
Human monocyte (THP-1 cells)-derived macrophages	A15-Exo-doxorubicin-cho-miR159	Mixing in triethylamine solution overnight, Incubation	$\alpha v\beta 3+$ and $\alpha v\beta 3-$ human BC (MDA-MB-231 and MCF-7) cells	↓Cell proliferation, ↑apoptosis	↓TCF7, ↓MYC	[186]
			MDA-MB-231 tumor-bearing BALB/c-nu mice	↓Tumor growth, ↑survival of mice	↓TCF7, ↓MYC, ↓Ki67, ↓CD31	
Mouse immature dendritic cells (imDCs)	Doxorubicin	Electroporation	MDA-MB-231 tumor-bearing BALB/c nude mice	↓Tumor growth	-	[189]
Human pancreatic cancer (Panc-1) cells	Gemcitabine	Sonication	Panc-1 and A549 cells	↓Cell viability	-	[190]
			Panc-1 tumor-bearing BALB/c nude mice	↓Tumor volume	↓Alanine aminotransferase, ↓aspartate aminotransferase, ↓TNF-α, ↓IL-6 in exo-gemcitabine group compared to gemcitabine	
Mouse (RAW 264.7) macrophages	Paclitaxel	Sonication	Murine Lewis lung carcinoma cell subline (3LL-M27 cells), Madin-Darby canine kidney (MDCK-WT) and (MDCK-MDR1) cells	↑Drug cytotoxicity, ↑chemosensitization of MDR cells	-	[187]
			8FlmC-FLuc-3LL-M27 tumor bearing C57BL/6 mice	↓Metastasis	-	
Human BC (MDA-MB-231) cells and mouse ovarian cancer (STOSE) cell	Doxorubicin	Electroporation	MDA-MB-231 and STOSE tumor bearing FVB/N mice	↑Doxorubicin efficacy, ↓tumor volume	-	[188]

Table 3. *Cont.*

Exosome Source	Modification of Exosomes with Drugs	Loading Method	Target Cells	Effect	Mechanism	References
			Phytochemicals			
Human pancreatic adenocarcinoma (PANC-1, MIA PaCa-2) cells	Curcumin	Incubation	PANC-1 and MIA PaCa-2 cells	↓Cell viability,	-	[191]
Pooled raw milk from Jersey cows	Anthocyanidins	By mixing in a solution of acetonitrile: ethanol (1:1 *v/v*) and PBS	Human pancreatic cancer (PANC1 and Mia PaCa2), lung cancer (A549 and H1299), colon cancer (HCT116), BC (MDA-MB-231 and MCF7), prostate cancer (PC3 and DU145), and ovarian cancer (OVCA432) cells	↓Cell proliferation, ↓cell survival	↓NF-κB	[192]
			A549 tumor bearing female athymic nude (nu/nu) mice	↓Tumor growth	–	
MCF7, PC3, human liver (HepG2), colon cancer (Caco2) cells	Black bean extract	Electroporation	MCF7, PC3, HepG2 and Caco2 cells	↓Cell viability	–	[193]
Mature bovine milk	Anthocyanidins	By mixing	Human ovarian cancer (A2780, A2780/CP70, OVCA432, and OVCA433) cells	↓Cell survival	-	[194]
			A2780 tumor-bearing female athymic nude mice	↓Tumor volume	-	
Milk from pasture-fed Holstein and Jersey cows	Celastrol	By mixing	Human lung cancer (H1299 and A549) cells	↓Cell survival,	-	[195]
			H1299 and A549 tumor-bearing female athymic nude mice	↓Tumor volume	-	

Symbols: ↑, upregulated; ↓, downregulated; Abbreviations: MYC, master regulator of cell cycle entry and proliferative metabolism; NF-κB, nuclear factor kappa-light-chain-enhancer of activated B cells; ROS, reactive oxygen species; TCF7, transcription factor 7; TNF-α, tumor necrosis factor-α.

Table 4. Reversal of chemoresistance in resistant cancer cells with exosomal cargoes.

Exosome Source	Modification in Exosomal Cargo Content	Target Cells	Effects	Mechanisms	References
Human mesenchymal stem cells (MSCs)	Anti-miR-9	Glioblastoma (U87 and T98G) cells	↑Apoptosis ↑chemosensitivity towards temozolomide	↑Caspase-3 ↓P-gp ↓MDR1	[151]
Human kidney epithelial (HEK293T) cells	si-c-Met	Human gastric adenocarcinoma (SGC7901and SGC7901/DDP) cells	↑Apoptosis ↑chemosensitivity towards cisplatin	↓c-Met gene	[199]
Normal intestinal foetal human cells (FHC)	miR-128-3p	Human oxiplatin resistant colorectal cancer (HCT116OxR) cells	↑Oxiplatin accumulation ↑apoptosis ↓proliferation ↓self-renewal	↓Bmi1 ↓MRP5 ↓N-cadherin ↑E-cadherin	[200]
Human adipose tissue derived mesenchymal stem cells (AMSCs)	miR-122	Human HCC (HepG2, Huh7) cells	↑Apoptosis ↑cell cycle arrest ↑chemosensitivity towards sorafenib	↑G0/G1 arrest ↓CCNG1 ↓ADAM10 ↑Caspase-3 ↑Bax	[201]
Human normal breast epithelial (MCF 10A) cells	miR-567	Human trastuzumab resistant BC (SKBR-3/TR and BT474/TR) cells	↑Chemosensitivity towards trastuzumab ↑autophagy	↓ATG5 ↑p62 ↓LC3-11	[202]
Human normal tongue epithelial (NTECs) cells	miR-200c	Docetaxel resistant hepatic stellate cells (HSC-3DR) cells	↑Chemosensitivity towards docetaxel ↑apoptosis	↓TUBB3 ↓PPP2R1B	[203]
Human adipose tissue derived mesenchymal stem cells (AMSCs)	miR-199a	Human colorectal cancer (CRC) (Huh7, SMMC-7721, PLC/PRF/5) cells	↑Chemosensitivity towards doxorubicin	↓mTOR	[204]

Symbols: ↑, upregulated; ↓, downregulated; Abbreviations: ADAM10, A disintegrin and metalloproteinase 10; ATG5, autophagy related 5 protein; Bax, Bcl-2-associated X protein; BC, breast cancer; c-MET, mesenchymal epithelial transition factor; CCNG1, Cyclin G1; LC3, microtubule associated protein PIA/IB-light chain 3-I; MDR1, multidrug resistance protein-1; MRP5, multidrug resistant associated protein 5; mTOR, mammalian target of rapamycin; P-gp, P-glycoprotein; PPP2R1B, protein phosphatase 2 scaffold subunit 1β; TUBB3, class III β-tubulin gene.

9.8. Exosomes in Clinical Trials

According to the National Institutes of Health website, a large number of clinical trials are being conducted with exosomes (Table 5). In a study, plant exosomes were modified to deliver curcumin in colon cancer patients (ClinicalTrials.gov Identifier: NCT01294072). Phase I and II clinical trials with DC-derived exosomes indicated activation of T cell- and NK cell-based immune responses in NSCLC patients [154]. A phase II clinical trial (ClinicalTrials.gov Identifier: NCT01159288) on NSCLC observed that exosomes derived from TLR4L-or interferon-γ (IFN-γ)-matured DCs enriched with MHC I- and MHC II-restricted cancer antigens as maintenance immunotherapy subsequent to first-line chemotherapy [205]. A study on HER2-positive BC patients measured HER2-HER3 dimer expression in exosomes (ClinicalTrials.gov Identifier: NCT04288141). Another trial led to a therapeutic analysis on cancer-derived exosomes via treatments with lovastatin and vildagliptin in thyroid cancer patients (ClinicalTrials.gov Identifier: NCT02862470). Characterization of exosomal non-coding RNAs was carried out in cholangiocarcinoma patients (ClinicalTrials.gov Identifier: NCT03102268). Another study reported exosome-mediated intercellular signaling in pancreatic cancer (ClinicalTrials.gov Identifier: NCT02393703). In metastatic pancreatic adenocarcinoma, exosomes with KrasG12D siRNA were used to treat pancreatic cancer with KrasG12D mutation (ClinicalTrials.gov Identifier: NCT03608631). In head and neck cancer, the effects of metformin hydrochloride on cytokines and exosomes were investigated (ClinicalTrials.gov Identifier: NCT03109873). A phase I clinical trial

(ClinicalTrials.gov Identifier: NCT01668849) investigated the ability of plant exosomes to prevent oral mucositis induced by combined chemotherapy and radiation in head and neck cancer patients. However, more clinical trials are needed with modified exosomes which may exhibit anticancer effect.

Table 5. Clinical trials on exosomes.

Trial No. (ClinicalTrials.gov Identifier:)	Study Type	Cancer Type	Study Perspective	Study Design	Status
NCT01294072	Phase I	Colon cancer	Interventional	Investigation of the ability of plant-derived exosomes to deliver curcumin	Active, not recruiting
NCT01159288	Phase II	Non-small cell lung cancer	Interventional	Trial of a vaccination with exosomes derived from dendritic cell loaded with tumor antigen	Completed
NCT04288141	Observational	Early HER2-positive BC, Metastatic HER2-positive BC	Prospective	Assessment of HER2-HER3 dimer expression in exosomes from HER2-positive patients receiving HER2 targeted therapies	Recruiting
NCT02862470	Observational	Anaplastic thyroid cancer, Follicular thyroid cancer	Prospective	Analysis of cancer-derived exosomes via lovastatin and vildagliptin treatments and prognostic study through urine exosomal markers	Active, not recruiting
NCT03102268	Observational	Cholangiocarcinoma	Prospective	Characterization of exosomal non-coding RNAs from cholangiocarcinoma patients before anticancer therapies	Unknown
NCT02393703	Observational	Pancreatic cancer	Prospective	Investigation of exosome mediated disease recurrence	Active, not recruiting
NCT03608631	Phase I	Metastatic pancreatic adenocarcinoma, Pancreatic ductal adenocarcinoma	Interventional	Study of the mesenchymal stromal cells-derived exosomes with KrasG12D siRNA (iExosomes) for pancreatic cancer patients having KrasG12D	Not yet recruiting
NCT03109873	Early phase I	Head and neck cancer	Randomized	Assessment of the effect of metformin hydrochloride on cytokines and exosomes in cancer patients	Completed
NCT01668849	Phase I	Head and neck cancer	Interventional	Investigation of the ability of plant-derived exosomes to prevent oral mucositis induced by combined chemotherapy and radiation	Active, not recruiting

10. Current Limitations and Challenges

Exosomes mediate intercellular communication and play significant roles in both physiological and pathological processes. A new hypothesis suggested that the target cells inhibit the incoming signals by forming exosome dimers based on the particle size, zeta potential and/or ligand–receptor pairs which facilitates cancer metastasis, cancer immunoregulation, intraocular pressure homoeostasis, tissue regeneration and many others [206].

Exosomes released by normal and malignant cells are endowed with heterogeneity and pleiotropic physiological and pathological effects. Inhibition of the release of TEXs may have both anti-carcinogenic and pro-carcinogenic effects. The majority of the exosome released inhibitors are not cancer-specific and also affect normal cells. Therefore, inhibition of exosome release may act as a double-edged sword which should be carefully manipulated for minimal adverse effects [34].

Isolation of pure and specific exosomes is limited by technical constraints, the availability of suitable biomarkers for specific exosomes, and expensive technologies [5]. A major hurdle in the execution of liquid biopsy is isolation of exosomes by an economic user-friendly tool. Protein contaminated and heterogenous exosome pool is obtained using ultracentrifugation. Asymmetric flow field-flow fractionation, though a prospective tool, needs technical expertise and requires a huge amount of initial sample. Other exosome isolation methods like microfluidic devices, sucrose gradients, size exclusion chromatography, and affinity-based exosome isolation kits are accompanied with both advantages and disadvantages like lack of robustness and specificity [31]. A perfect exosome isolation method should be robust, reproducible, specific, economic and user friendly as a diagnostic tool.

Detailed research of exosome biogenesis, functional diversity of exosomes and the identification of cancer specific biomarkers may be effective for exosome-based therapeutic approaches with minimum adverse effects [34]. Determination of exosomal cargo sorting and releasing mechanisms holds great potential for the development of various applications in cancer research [31].

Normally, less than 1 µg of exosomal protein is yielded from 1 mL of culture medium, whereas the majority of studies have reported 10–100 µg of exosomal protein as an effective dose for in vivo models [163]. The introduction of exosome-mimetic vesicles (100–200 nm in diameter) has conquered exosomal limitations like low loading efficiency and low yields. These nanovesicles have been used for delivery of chemotherapeutic drugs [204,205] and RNAi [207] to target cancer cells. Hybrid nanocarriers formed by the fusion of exosomes with liposomes changed the exogenous lipid composition and was effective in the delivery of chemotherapeutic drugs [208].

11. Conclusions

It may be deciphered that the intercellular communication via exosomes is evident throughout cancer progression. Apart from cancer pathogenesis, exosome biology heralds the future arena of non-invasive diagnostic tools for cancer management, especially in the spheres of liquid biopsy, immunotherapy and vaccine development, RNA therapy, stem cell therapy, drug delivery, and reversal of chemoresistance. Preclinical studies have undoubtedly proven the immense potential of exosomes in cancer therapeutics, but a number of clinical trials have failed to achieve this success. These inconsistent results indicate major challenges including in-depth knowledge of exosome biogenesis and protein sorting, perfect and pure isolation of exosomes, large scale production, better loading efficiency targeted delivery of exosomes. These hurdles have to be confronted before successful implementation of exosomes for the diagnosis and therapy of cancer. This review has attempted to envisage the implication of exosomes in cancer pathogenesis and cancer therapeutics along with the current limitations so that researchers may be made aware of the existing lacunae with regard to exosomes in their use against cancer. This knowledge may help scientists to improvise innovative technologies for successful translation of the exosome-mediated diagnosis and treatment of malignant neoplasms.

Author Contributions: Conceptualization and writing, D.S.; writing and figure preparation, S.R., P.S. and N.C.; editing and suggestion of improvements, A.B. All authors have read and agreed to the published version of the manuscript.

Funding: This research received no external source of funding.

Institutional Review Board Statement: Not applicable.

Informed Consent Statement: Not applicable.

Data Availability Statement: Not applicable.

Acknowledgments: D.S., S.R., P.S., and N.C. are deeply grateful to Jayanta Chakrabarti, Director, Chittaranjan National Cancer Institute, Kolkata, for providing the infrastructural facilities. P.S. is thankful to Jayanta Chakrabarti for granting an institutional fellowship. S.R. is indebted to the University Grants Commission, New Delhi, India, for providing a fellowship.

Conflicts of Interest: The authors declare no conflict of interest.

Abbreviations

ABC	ATP-binding cassette;
ABCA3	ATP-binding cassette sub-family A member 3;
ADAM10	A disintegrin and metalloproteinase 10;
AFAP1-AS1	actin filament associated protein1 antisense RNA 1;
Akt	protein kinase B;
ALIX	ALG-2 interacting protein X;
AMSCs	adipose tissue-derived MSCs;
APC	antigen presenting cell;
ARF6	ADP ribosylation factor 6;
ATG5	autophagy related 5 protein;
CAFs	cancer-associated fibroblasts;
CCNG1	cyclin G1;
CRC	colorectal cancer cells;
CXCR4	C-X-C chemokine receptor type 4;
DCs	dendritic cells;
ECM	extracellular matrix;
EGFR	epidermal growth factor receptor;
EMT	epithelial mesenchymal transition;
ERBB2	erythroblastic oncogene B2;
ERF	Ets2-repressor factor;
ERK	extracellular signal-regulated kinase;
ERα	estrogen receptor-α;
ESCRT	endosomal sorting complexes required for the transport;
GC	gastric cancer;
HCC	hepatocellular carcinoma;
HER2	human epidermal growth factor receptor 2;
HGF	hepatocyte growth factor;
HNSCC	head and neck squamous cell carcinoma;
HOTTIP	HoxA transcript at a distal tip;
Hsps	heat shock proteins;
ICAM	intercellular adhesion molecule;
IL-6	interleukin-6;
ILV	intraluminal vesicles;
LAMP-1	lysosome-associated membrane glycoprotein-1 LncRNAs
LncRNAs	long non-coding RNAs;
MAPK	mitogen activated protein kinase;
MDR-1	multidrug resistance protein-1;
MDSCs	myeloid-derived suppressor cells;
MHC	major histocompatibility complex;
miRNAs	microRNA;
MMP	matrix metalloproteinase;
mRNA	messenger RNA;
MSCs	mesenchymal stem cells;
mtDNA	mitochondrial DNA;
MVB	multivesicular bodies;
NF-κB	nuclear factor κ-light-chain-enhancer of activated B cells;
NK	natural killer cells;
NKG2D	natural killer group 2D;

NSCLC	non-small cell lung carcinoma;
PAFR	a platelet-activating factor receptor;
PDAC	pancreatic ductal adenocarcinoma;
PDCD4	programmed cell death 4;
PD-L1	programmed death ligand 1;
P-gp	P-glycoprotein;
PI3K	phosphoinositide 3-kinase;
PLD2	phospholipase D2;
PLK-1	polo-like kinase 1;
PM	plasma membrane;
PPP2R1B	protein phosphatase 2 scaffold subunit 1β;
PTEN	phosphatase and tensin homolog;
rRNA	ribosomal RNA;
SNARES	soluble NSF attachment protein receptors;
Sox2ot	SOX2 overlapping transcript;
SR-B1	scavenger receptor type B-1;
STAT1	signal transducer and activator of transcription 1;
TAMs	tumor-associated macrophages;
TEX	tumor derived exosomes;
TGF-β	transforming growth factor-β;
TGN	trans-Golgi network;
TLR-2	toll like receptor-2;
TME	tumor microenvironment;
Tregs	T regulatory cells;
TUBB3	class III β-tubulin gene;
UFC1	Ubiquitin-fold modifier conjugating enzyme 1;
VEGF-A	vascular endothelial growth factor A;
Vps4	vacuolar protein sorting associated protein 4;
ZFAS1	zinc finger antisense 1.

References

1. Harding, C.; Stahl, P. Transferrin Recycling in Reticulocytes: pH and Iron Are Important Determinants of Ligand Binding and Processing. *Biochem. Biophys. Res. Commun.* **1983**, *113*, 650–658. [CrossRef]
2. Lötvall, J.; Hill, A.F.; Hochberg, F.; Buzás, E.I.; Vizio, D.D.; Gardiner, C.; Gho, Y.S.; Kurochkin, I.V.; Mathivanan, S.; Quesenberry, P.; et al. Minimal experimental requirements for definition of extracellular vesicles and their functions: A position statement from the International Society for Extracellular Vesicles. *J. Extracell. Vesicles* **2014**, *3*, 26913. [CrossRef]
3. Théry, C.; Witwer, K.W.; Aikawa, E.; Alcaraz, M.J.; Anderson, J.D.; Andriantsitohaina, R.; Antoniou, A.; Arab, T.; Archer, F.; Atkin-Smith, G.K.; et al. Minimal information for studies of extracellular vesicles 2018 (MISEV2018): A position statement of the International Society for Extracellular Vesicles and update of the MISEV2014 guidelines. *J. Extracell. Vesicles* **2018**, *7*, 1535750. [CrossRef]
4. Zhang, C.; Ji, Q.; Yang, Y.; Li, Q.; Wang, Z. Exosome: Function and role in cancer metastasis and drug resistance. *Technol. Cancer Res. Treat.* **2018**, *17*. [CrossRef]
5. He, C.; Zheng, S.; Luo, Y.; Wang, B. Exosome theranostics: Biology and translational medicine. *Theranostics* **2018**, *8*, 237–255. [CrossRef]
6. McAndrews, K.M.; Kalluri, R. Mechanisms associated with biogenesis of exosomes in cancer. *Mol. Cancer* **2019**, *18*, 52. [CrossRef]
7. Jurj, A.; Zanoaga, O.; Braicu, C.; Lazar, V.; Tomuleasa, C.; Irimie, A.; Berindan-neagoe, I. A comprehensive picture of extracellular vesicles and their contents. Molecular transfer to cancer cells. *Cancers* **2020**, *12*, 298. [CrossRef]
8. Lucchetti, D.; Tenore, C.R.; Colella, F.; Sgambato, A. Extracellular Vesicles and Cancer: A Focus on Metabolism, Cytokines, and Immunity. *Cancers* **2020**, *12*, 171. [CrossRef]
9. Tang, Z.; Li, D.; Hou, S.; Zhu, X. The cancer exosomes: Clinical implications, applications and challenges. *Int. J. Cancer* **2020**, *146*, 2946–2959. [CrossRef]
10. Mohammadi, S.; Yousefi, F.; Shabaninejad, Z.; Movahedpour, A.; Mahjoubin Tehran, M.; Shafiee, A.; Moradizarmehri, S.; Hajighadimi, S.; Savardashtaki, A.; Mirzaei, H. Exosomes and cancer: From oncogenic roles to therapeutic applications. *IUBMB Life* **2020**, *72*, 724–748. [CrossRef]
11. Parayath, N.N.; Padmakumar, S.; Amiji, M.M. Extracellular vesicle-mediated nucleic acid transfer and reprogramming in the tumor microenvironment. *Cancer Lett.* **2020**, *482*, 33–43. [CrossRef]
12. Wee, I.; Syn, N.; Sethi, G.; Goh, B.C.; Wang, L. Role of tumor-derived exosomes in cancer metastasis. *Biochim. Biophys. Acta Rev. Cancer* **2019**, *1871*, 12–19. [CrossRef]
13. Raimondo, S.; Pucci, M.; Alessandro, R.; Fontana, S. Extracellular vesicles and tumor-immune escape: Biological functions and clinical perspectives. *Int. J. Mol. Sci.* **2020**, *21*, 2286. [CrossRef]
14. Gilligan, K.E.; Dwyer, R.M. Extracellular Vesicles for Cancer Therapy: Impact of Host Immune Response. *Cells* **2020**, *9*, 224. [CrossRef]

15. Li, X.; Corbett, A.L.; Taatizadeh, E.; Tasnim, N.; Little, J.P.; Garnis, C.; Daugaard, M.; Guns, E.; Hoorfar, M.; Li, I.T.S. Challenges and opportunities in exosome research—Perspectives from biology, engineering, and cancer therapy. *APL Bioeng.* **2019**, *3*, 011503. [CrossRef]
16. Srivastava, A.; Amreddy, N.; Pareek, V.; Chinnappan, M.; Ahmed, R.; Mehta, M.; Razaq, M.; Munshi, A.; Ramesh, R. Progress in extracellular vesicle biology and their application in cancer medicine. *Wiley Interdiscip. Rev. Nanomed. Nanobiotechnol.* **2020**, *12*, e1621. [CrossRef]
17. Taghikhani, A.; Farzaneh, F.; Sharifzad, F.; Mardpour, S.; Ebrahimi, M.; Hassan, Z.M. Engineered Tumor-Derived Extracellular Vesicles: Potentials in Cancer Immunotherapy. *Front. Immunol.* **2020**, *11*, 221. [CrossRef]
18. Stevic, I.; Buescher, G.; Ricklefs, F.L. Monitoring Therapy Efficiency in Cancer through Extracellular Vesicles. *Cells* **2020**, *9*, 130. [CrossRef]
19. Pullan, J.E.; Confeld, M.I.; Osborn, J.K.; Kim, J.; Sarkar, K.; Mallik, S. Exosomes as Drug Carriers for Cancer Therapy. *Mol. Pharm.* **2019**, *16*, 1789–1798. [CrossRef]
20. Yong, T.; Wang, D.; Li, X.; Yan, Y.; Hu, J.; Gan, L.; Yang, X. Extracellular vesicles for tumor targeting delivery based on five features principle. *J. Control. Release* **2020**, *322*, 555–565. [CrossRef]
21. Zhao, X.; Wu, D.; Ma, X.; Wang, J.; Hou, W.; Zhang, W. Exosomes as drug carriers for cancer therapy and challenges regarding exosome uptake. *Biomed. Pharmacother.* **2020**, *128*, 110237. [CrossRef]
22. Pirisinu, M.; Pham, T.C.; Zhang, D.X.; Hong, T.N.; Nguyen, L.T.; Le, M.T. Extracellular vesicles as natural therapeutic agents and innate drug delivery systems for cancer treatment: Recent advances, current obstacles, and challenges for clinical translation. *Semin. Cancer Biol.* **2020**. [CrossRef]
23. Milane, L.; Singh, A.; Mattheolabakis, G.; Suresh, M.; Amiji, M.M. Exosome Mediated Communication within the Tumor Microenvironment. *J. Control. Release* **2015**, *10*, 278–294. [CrossRef]
24. Jenjaroenpun, P.; Kremenska, Y.; Nair, V.M.; Kremenskoy, M.; Joseph, B.; Kurochkin, I.V. Characterization of RNA in exosomes secreted by human breast cancer cell lines using next-generation sequencing. *PeerJ* **2013**, *1*, e201. [CrossRef]
25. Li, S.P.; Lin, Z.X.; Jiang, X.Y.; Yu, X.Y. Exosomal cargo-loading and synthetic exosome-mimics as potential therapeutic tools. *Acta Pharmacol. Sin.* **2018**, *39*, 542–551. [CrossRef]
26. Zhang, Y.; Liu, Y.; Liu, H.; Tang, W.H. Exosomes: Biogenesis, biologic function and clinical potential. *Cell Biosci.* **2019**, *9*, 19. [CrossRef]
27. Souza-Schorey, C.; Schorey, J.S. Regulation and mechanisms of extracellular vesicle biogenesis and secretion. *Essays Biochem.* **2018**, *62*, 125–133. [CrossRef]
28. Antonyak, M.A.; Cerione, R.A. Microvesicles as mediators of intercellular communication in cancer. *Methods Mol. Biol.* **2014**, *1165*, 147–173. [CrossRef]
29. Ghossoub, R.; Lembo, F.; Rubio, A.; Gaillard, C.B.; Bouchet, J.; Vitale, N.; Slavík, J.; Machala, M.; Zimmermann, P. Syntenin-ALIX exosome biogenesis and budding into multivesicular bodies are controlled by ARF6 and PLD2. *Nat. Commun.* **2014**, *5*, 3477. [CrossRef]
30. Kowal, J.; Tkach, M.; Théry, C. Biogenesis and secretion of exosomes. *Curr. Opin. Cell Biol.* **2014**, *29*, 116–125. [CrossRef]
31. Wortzel, I.; Dror, S.; Kenific, C.M.; Lyden, D. Exosome-Mediated Metastasis: Communication from a Distance. *Dev. Cell* **2019**, *49*, 347–360. [CrossRef]
32. Alenquer, M.; Amorim, M.J. Exosome biogenesis, regulation, and function in viral infection. *Viruses* **2015**, *7*, 5066–5083. [CrossRef]
33. Stoorvogel, W. Resolving sorting mechanisms into exosomes. *Cell Res.* **2015**, *25*, 531–532. [CrossRef]
34. Moloudizargari, M.; Asghari, M.H.; Abdollahi, M. Modifying exosome release in cancer therapy: How can it help? *Pharmacol. Res.* **2018**, *134*, 246–256. [CrossRef]
35. Lange, S.; Gallagher, M.; Kholia, S.; Kosgodage, U.S.; Hristova, M.; Hardy, J.; Inal, J.M. Peptidylarginine deiminases—Roles in cancer and neurodegeneration and possible avenues for therapeutic intervention via modulation of exosome and microvesicle (EMV) release? *Int. J. Mol. Sci.* **2017**, *18*, 1196. [CrossRef]
36. Mohrmann, K.; Leijendekker, R.; Gerez, L.; Van Sluijs, P. Der rab4 regulates transport to the apical plasma membrane in Madin-Darby canine kidney cells. *J. Biol. Chem.* **2002**, *277*, 10474–10481. [CrossRef]
37. Schlierf, B.; Fey, G.H.; Hauber, J.; Hocke, G.M.; Rosorius, O. Rab11b is essential for recycling of transferrin to the plasma membrane. *Exp. Cell Res.* **2000**, *259*, 257–265. [CrossRef]
38. Casanova, J.E.; Wang, X.; Kumar, R.; Bhartur, S.G.; Navarre, J.; Woodrum, J.E.; Altschuler, Y.; Ray, G.S.; Goldenring, J.R. Association of Rab25 and Rab11a with the apical recycling system of polarized Madin-Darby Canine Kidney cells. *Mol. Biol. Cell* **1999**, *10*, 47–61. [CrossRef]
39. Lombardi, D.; Soldati, T.; Riederer, M.A.; Goda, Y.; Zerial, M.; Pfeffer, S.R. Rab9 functions in transport between late endosomes and the trans Golgi network. *EMBO J.* **1993**, *12*, 677–682. [CrossRef]
40. Guerra, F.; Bucci, C. Multiple Roles of the Small GTPase Rab7. *Cells* **2016**, *5*, 34. [CrossRef]
41. Vitelli, R.; Santillo, M.; Lattero, D.; Chiariello, M.; Bifulco, M.; Bruni, C.B.; Bucci, C. Role of the small GTPase RAB7 in the late endocytic pathway. *J. Biol. Chem.* **1997**, *272*, 4391–4397. [CrossRef]
42. Chen, J.; Chopp, M. Exosome therapy for stroke. *Stroke* **2018**, *49*, 1083–1090. [CrossRef]
43. Ostrowski, M.; Carmo, N.B.; Krumeich, S.; Fanget, I.; Raposo, G.; Savina, A.; Moita, C.F.; Schauer, K.; Hume, A.N.; Freitas, R.P.; et al. Rab27a and Rab27b control different steps of the exosome secretion pathway. *Nat. Cell Biol.* **2010**, *12*, 19–30. [CrossRef]
44. Baietti, M.F.; Zhang, Z.; Mortier, E.; Melchior, A.; Degeest, G.; Geeraerts, A.; Ivarsson, Y.; Depoortere, F.; Coomans, C.; Vermeiren, E.; et al. Syndecan-syntenin-ALIX regulates the biogenesis of exosomes. *Nat. Cell Biol.* **2012**, *14*, 677–685. [CrossRef]
45. Savina, A.; Vidal, M.; Colombo, M.I. The exosome pathway in K562 cells is regulated by Rab11. *J. Cell Sci.* **2002**, *115*, 2505–2515.

46. Savina, A.; Fader, C.M.; Damiani, M.T.; Colombo, M.I. Rab11 promotes docking and fusion of multivesicular bodies in a calcium-dependent manner. *Traffic* **2006**, *6*, 131–143. [CrossRef]
47. Li, W.; Hu, Y.; Jiang, T.; Han, Y.; Han, G.; Chen, J.; Li, X. Rab27A regulates exosome secretion from lung adenocarcinoma cells A549: Involvement of EPI64. *APMIS* **2014**, *122*, 1080–1087. [CrossRef]
48. Heusermann, W.; Hean, J.; Trojer, D.; Steib, E.; von Bueren, S.; Graff-Meyer, A.; Genoud, C.; Martin, K.; Pizzato, N.; Voshol, J.; et al. Exosomes surf on filopodia to enter cells at endocytic hot spots, traffic within endosomes, and are targeted to the ER. *J. Cell Biol.* **2016**, *213*, 173–184. [CrossRef]
49. Tian, T.; Zhu, Y.L.; Zhou, Y.Y.; Liang, G.F.; Wang, Y.Y.; Hu, F.H.; Xiao, Z.D. Exosome uptake through clathrin-mediated endocytosis and macropinocytosis and mediating miR-21 delivery. *J. Biol. Chem.* **2014**, *289*, 22258–22267. [CrossRef]
50. Nabi, I.R.; Le, P.U. Caveolae/raft-dependent endocytosis. *J. Cell Biol.* **2003**, *161*, 673–677. [CrossRef]
51. Gonda, A.; Kabagwira, J.; Senthil, G.N.; Wall, N.R. Internalization of exosomes through receptor-mediated endocytosis. *Mol. Cancer Res.* **2019**, *17*, 337–347. [CrossRef]
52. Joshi, B.S.; de Beer, M.A.; Giepmans, B.N.G.; Zuhorn, I.S. Endocytosis of Extracellular Vesicles and Release of Their Cargo from Endosomes. *ACS Nano* **2020**, *14*, 4444–4455. [CrossRef]
53. Andreu, Z.; Yáñez-Mó, M. Tetraspanins in extracellular vesicle formation and function. *Front. Immunol.* **2014**, *5*, 442. [CrossRef]
54. Wu, Y.; Wu, W.; Wong, W.M.; Ward, E.; Thrasher, A.J.; Goldblatt, D.; Osman, M.; Digard, P.; Canaday, D.H.; Gustafsson, K. Human γδ T Cells: A Lymphoid Lineage Cell Capable of Professional Phagocytosis. *J. Immunol.* **2009**, *183*, 5622–5629. [CrossRef]
55. McKelvey, K.J.; Powell, K.L.; Ashton, A.W.; Morris, J.M.; McCracken, S.A. Exosomes: Mechanisms of Uptake. *J. Circ. Biomark.* **2015**, *4*, 7. [CrossRef]
56. Farooqi, A.A.; Desai, N.N.; Qureshi, M.Z.; Librelotto, D.R.N.; Gasparri, M.L.; Bishayee, A.; Nabavi, S.M.; Curti, V.; Daglia, M. Exosome biogenesis, bioactivities and functions as new delivery systems of natural compounds. *Biotechnol. Adv.* **2018**, *36*, 328–334. [CrossRef]
57. Datta, A.; Kim, H.; McGee, L.; Johnson, A.E.; Talwar, S.; Marugan, J.; Southall, N.; Hu, X.; Lal, M.; Mondal, D.; et al. High-Throughput screening identified selective inhibitors of exosome biogenesis and secretion: A drug repurposing strategy for advanced cancer. *Sci. Rep.* **2018**, *8*, 8161. [CrossRef]
58. Logozzi, M.; Spugnini, E.; Mizzoni, D.; Di Raimo, R.; Fais, S. Extracellular acidity and increased exosome release as key phenotypes of malignant tumors. *Cancer Metastasis Rev.* **2019**, *38*, 93–101. [CrossRef]
59. Agarwal, K.; Saji, M.; Lazaroff, S.M.; Palmer, A.F.; Ringel, M.D.; Paulaitis, M.E. Analysis of exosome release as a cellular response to MAPK pathway inhibition. *Langmuir* **2015**, *31*, 5440–5448. [CrossRef]
60. Lee, T.H.; Chennakrishnaiah, S.; Audemard, E.; Montermini, L.; Meehan, B.; Rak, J. Oncogenic ras-driven cancer cell vesiculation leads to emission of double-stranded DNA capable of interacting with target cells. *Biochem. Biophys. Res. Commun.* **2014**, *451*, 295–301. [CrossRef]
61. Al-Nedawi, K.; Meehan, B.; Micallef, J.; Lhotak, V.; May, L.; Guha, A.; Rak, J. Intercellular transfer of the oncogenic receptor EGFRvIII by microvesicles derived from tumour cells. *Nat. Cell Biol.* **2008**, *10*, 619–624. [CrossRef]
62. Zhang, C.; Xiao, X.; Chen, M.; Aldharee, H.; Chen, Y.; Long, W. Liver kinase B1 restoration promotes exosome secretion and motility of lung cancer cells. *Oncol. Rep.* **2018**, *39*, 376–382. [CrossRef]
63. Melo, S.A.; Sugimoto, H.; O'Connell, J.T.; Kato, N.; Villanueva, A.; Vidal, A.; Qiu, L.; Vitkin, E.; Perelman, L.T.; Melo, C.A.; et al. Cancer Exosomes Perform Cell-Independent MicroRNA Biogenesis and Promote Tumorigenesis. *Cancer Cell* **2014**, *26*, 707–721. [CrossRef]
64. Jia, Y.; Chen, Y.; Wang, Q.; Jayasinghe, U.; Luo, X.; Wei, Q.; Wang, J.; Xiong, H.; Chen, C.; Xu, B.; et al. Exosome: Emerging biomarker in breast cancer. *Oncotarget* **2017**, *8*, 41717–41733. [CrossRef]
65. Singh, R.; Pochampally, R.; Watabe, K.; Lu, Z.; Mo, Y.Y. Exosome-mediated transfer of miR-10b promotes cell invasion in breast cancer. *Mol. Cancer* **2014**, *13*, 256. [CrossRef]
66. Zhang, L.; Zhang, S.; Yao, J.; Lowery, F.J.; Zhang, Q.; Huang, W.C.; Li, P.; Li, M.; Wang, X.; Zhang, C.; et al. Microenvironment-induced PTEN loss by exosomal microRNA primes brain metastasis outgrowth. *Nature* **2015**, *527*, 100–104. [CrossRef]
67. Cooks, T.; Pateras, I.S.; Jenkins, L.M.; Patel, K.M.; Robles, A.I.; Morris, J.; Forshew, T.; Appella, E.; Gorgoulis, V.G.; Harris, C.C. Mutant p53 cancers reprogram macrophages to tumor supporting macrophages via exosomal miR-1246. *Nat. Commun.* **2018**, *9*, 771. [CrossRef]
68. Zhang, H.D.; Jiang, L.H.; Hou, J.C.; Zhong, S.L.; Zhu, L.P.; Wang, D.D.; Zhou, S.Y.; Yang, S.J.; Wang, J.Y.; Zhang, Q.; et al. Exosome: A novel mediator in drug resistance of cancer cells. *Epigenomics* **2018**, *10*, 1499–1509. [CrossRef]
69. Yang, X.; Li, Y.; Zou, L.; Zhu, Z. Role of Exosomes in Crosstalk Between Cancer-Associated Fibroblasts and Cancer Cells. *Front. Oncol.* **2019**, *9*, 356. [CrossRef]
70. Zhao, H.; Yang, L.; Baddour, J.; Achreja, A.; Bernard, V.; Moss, T.; Marini, J.C.; Tudawe, T.; Seviour, E.G.; San Lucas, F.A.; et al. Tumor microenvironment derived exosomes pleiotropically modulate cancer cell metabolism. *Elife* **2016**, *5*, e10250. [CrossRef]
71. Hu, T.; Hu, J. Melanoma-derived exosomes induce reprogramming fibroblasts into cancer-associated fibroblasts via Gm26809 delivery. *Cell Cycle* **2019**, *18*, 3085–3094. [CrossRef]
72. Abusamra, A.J.; Zhong, Z.; Zheng, X.; Li, M.; Ichim, T.E.; Chin, J.L.; Min, W.P. Tumor exosomes expressing Fas ligand mediate CD8+ T-cell apoptosis. *Blood Cells Mol. Dis.* **2005**, *35*, 169–173. [CrossRef]
73. Lu, H.; Bowler, N.; Harshyne, L.A.; Craig Hooper, D.; Krishn, S.R.; Kurtoglu, S.; Fedele, C.; Liu, Q.; Tang, H.Y.; Kossenkov, A.V.; et al. Exosomal αvβ6 integrin is required for monocyte M2 polarization in prostate cancer. *Matrix Biol.* **2018**, *70*, 20–35. [CrossRef]

74. Yin, C.; Han, Q.; Xu, D.; Zheng, B.; Zhao, X.; Zhang, J. SALL4-mediated upregulation of exosomal miR-146a-5p drives T-cell exhaustion by M2 tumor-associated macrophages in HCC. *Oncoimmunology* **2019**, *8*, 1601479. [CrossRef]

75. Liu, C.; Yu, S.; Zinn, K.; Wang, J.; Zhang, L.; Jia, Y.; Kappes, J.C.; Barnes, S.; Kimberly, R.P.; Grizzle, W.E.; et al. Murine Mammary Carcinoma Exosomes Promote Tumor Growth by Suppression of NK Cell Function. *J. Immunol.* **2006**, *176*, 1375–1385. [CrossRef]

76. Bobrie, A.; Krumeich, S.; Reyal, F.; Recchi, C.; Moita, L.F.; Seabra, M.C.; Ostrowski, M.; Théry, C. Rab27a supports exosome-dependent and -independent mechanisms that modify the tumor microenvironment and can promote tumor progression. *Cancer Res.* **2012**, *72*, 4920–4930. [CrossRef]

77. Maus, R.L.G.; Jakub, J.W.; Nevala, W.K.; Christensen, T.A.; Noble-Orcutt, K.; Sachs, Z.; Hieken, T.J.; Markovic, S.N. Human melanoma-derived extracellular vesicles regulate dendritic cell maturation. *Front. Immunol.* **2017**, *8*, 358. [CrossRef]

78. Wolfers, J.; Lozier, A.; Raposo, G.; Regnault, A.; Théry, C.; Masurier, C.; Flament, C.; Pouzieux, S.; Faure, F.; Tursz, T.; et al. Tumor-derived exosomes are a source of shared tumor rejection antigens for CTL cross-priming. *Nat. Med.* **2001**, *7*, 297–303. [CrossRef]

79. Theodoraki, M.N.; Yerneni, S.S.; Hoffmann, T.K.; Gooding, W.E.; Whiteside, T.L. Clinical significance of PD-L1 þ exosomes in plasma of head and neck cancer patients. *Clin. Cancer Res.* **2018**, *24*, 896–905. [CrossRef]

80. Yue, S.; Mu, W.; Erb, U.; Zöller, M. The tetraspanins CD151 and Tspan8 are essential exosome components for the crosstalk between cancer initiating cells and their surrounding. *Oncotarget* **2015**, *6*, 2366–2384. [CrossRef]

81. Yin, X.; Zheng, X.; Liu, M.; Wang, D.; Sun, H.; Qiu, Y.; Chen, J.; Shi, B. Exosomal miR-663b targets Ets2-repressor factor to promote proliferation and the epithelial–mesenchymal transition of bladder cancer cells. *Cell Biol. Int.* **2020**, *44*, 958–965. [CrossRef]

82. Li, Z.; Jiang, P.; Li, J.; Peng, M.; Zhao, X.; Zhang, X.; Chen, K.; Zhang, Y.; Liu, H.; Gan, L.; et al. Tumor-derived exosomal lnc-Sox2ot promotes EMT and stemness by acting as a ceRNA in pancreatic ductal adenocarcinoma. *Oncogene* **2018**, *37*, 3822–3838. [CrossRef]

83. Suchorska, W.M.; Lach, M.S. The role of exosomes in tumor progression and metastasis (Review). *Oncol. Rep.* **2016**, *35*, 1237–1244. [CrossRef]

84. Fedele, C.; Singh, A.; Zerlanko, B.J.; Iozzo, R.V.; Languino, L.R. The $\alpha v \beta 6$ integrin is transferred intercellularly via exosomes. *J. Biol. Chem.* **2015**, *290*, 4545–4551. [CrossRef]

85. Lin, F.; Yin, H.B.; Li, X.Y.; Zhu, G.M.; He, W.Y.; Gou, X. Bladder cancer cell-secreted exosomal miR-21 activates the PI3K/AKT pathway in macrophages to promote cancer progression. *Int. J. Oncol.* **2020**, *56*, 151–164. [CrossRef]

86. Zheng, P.; Luo, Q.; Wang, W.; Li, J.; Wang, T.; Wang, P.; Chen, L.; Zhang, P.; Chen, H.; Liu, Y.; et al. Tumor-associated macrophages-derived exosomes promote the migration of gastric cancer cells by transfer of functional Apolipoprotein E. *Cell Death Dis.* **2018**, *9*, 434. [CrossRef]

87. Zang, X.; Gu, J.; Zhang, J.; Shi, H.; Hou, S.; Xu, X.; Chen, Y.; Zhang, Y.; Mao, F.; Qian, H.; et al. Exosome-transmitted lncRNA UFC1 promotes non-small-cell lung cancer progression by EZH2-mediated epigenetic silencing of PTEN expression. *Cell Death Dis.* **2020**, *11*, 215. [CrossRef]

88. Pan, L.; Liang, W.; Fu, M.; Huang, Z.H.; Li, X.; Zhang, W.; Zhang, P.; Qian, H.; Jiang, P.C.; Xu, W.R.; et al. Exosomes-mediated transfer of long noncoding RNA ZFAS1 promotes gastric cancer progression. *J. Cancer Res. Clin. Oncol.* **2017**, *143*, 991–1004. [CrossRef]

89. Zhang, H.; Deng, T.; Liu, R.; Bai, M.; Zhou, L.; Wang, X.; Li, S.; Wang, X.; Yang, H.; Li, J.; et al. Exosome-delivered EGFR regulates liver microenvironment to promote gastric cancer liver metastasis. *Nat. Commun.* **2017**, *8*, 15016. [CrossRef]

90. Yokoi, A.; Yoshioka, Y.; Yamamoto, Y.; Ishikawa, M.; Ikeda, S.I.; Kato, T.; Kiyono, T.; Takeshita, F.; Kajiyama, H.; Kikkawa, F.; et al. Malignant extracellular vesicles carrying MMP1 mRNA facilitate peritoneal dissemination in ovarian cancer. *Nat. Commun.* **2017**, *8*, 14470. [CrossRef]

91. Wang, D.; Wang, X.; Si, M.; Yang, J.; Sun, S.; Wu, H.; Cui, S.; Qu, X.; Yu, X. Exosome-encapsulated miRNAs contribute to CXCL12/CXCR4-induced liver metastasis of colorectal cancer by enhancing M2 polarization of macrophages. *Cancer Lett.* **2020**, *474*, 36–52. [CrossRef]

92. Sun, S.; Chen, H.; Xu, C.; Zhang, Y.; Zhang, Q.; Chen, L.; Ding, Q.; Deng, Z. Exosomal miR-106b serves as a novel marker for lung cancer and promotes cancer metastasis via targeting PTEN. *Life Sci.* **2020**, *244*, 117297. [CrossRef]

93. Liao, J.; Liu, R.; Shi, Y.J.; Yin, L.H.; Pu, Y.P. Exosome-shuttling microRNA-21 promotes cell migration and invasion-targeting PDCD4 in esophageal cancer. *Int. J. Oncol.* **2016**, *48*, 2567–2579. [CrossRef]

94. Xia, Y.; Wei, K.; Hu, L.Q.; Zhou, C.R.; Lu, Z.B.; Zhan, G.S.; Pan, X.L.; Pan, C.F.; Wang, J.; Wen, W.; et al. Exosome-mediated transfer of miR-1260b promotes cell invasion through Wnt/β–catenin signaling pathway in lung adenocarcinoma. *J. Cell. Physiol.* **2020**, *235*, 6843–6853. [CrossRef]

95. Li, Z.; Tao, Y.; Wang, X.; Jiang, P.; Li, J.; Peng, M.; Zhang, X.; Chen, K.; Liu, H.; Zhen, P.; et al. Tumor-secreted exosomal miR-222 promotes tumor progression via regulating P27 expression and re-localization in pancreatic cancer. *Cell. Physiol. Biochem.* **2018**, *51*, 610–629. [CrossRef]

96. Huang, Z.; Feng, Y. Exosomes derived from hypoxic colorectal cancer cells promote angiogenesis through Wnt4-Induced β-catenin signaling in endothelial cells. *Oncol. Res.* **2017**, *25*, 651–661. [CrossRef]

97. Treps, L.; Perret, R.; Edmond, S.; Ricard, D.; Gavard, J. Glioblastoma stem-like cells secrete the pro-angiogenic VEGF-A factor in extracellular vesicles. *J. Extracell. Vesicles* **2017**, *6*, 1359479. [CrossRef]

98. Yoshii, S.; Hayashi, Y.; Iijima, H.; Inoue, T.; Kimura, K.; Sakatani, A.; Nagai, K.; Fujinaga, T.; Hiyama, S.; Kodama, T.; et al. Exosomal microRNAs derived from colon cancer cells promote tumor progression by suppressing fibroblast TP53 expression. *Cancer Sci.* **2019**, *110*, 2396–2407. [CrossRef]

99. Zheng, P.; Chen, L.; Yuan, X.; Luo, Q.; Liu, Y.; Xie, G.; Ma, Y.; Shen, L. Exosomal transfer of tumor-associated macrophage-derived miR-21 confers cisplatin resistance in gastric cancer cells. *J. Exp. Clin. Cancer Res.* **2017**, *36*, 53. [CrossRef]

100. Yu, X.; Zhang, Q.; Zhang, X.; Han, Q.; Li, H.; Mao, Y.; Wang, X.; Guo, H.; Irwin, D.M.; Niu, G.; et al. Exosomes from macrophages exposed to apoptotic breast cancer cells promote breast cancer proliferation and metastasis. *J. Cancer* **2019**, *10*, 2892–2906. [CrossRef]

101. Li, X.; Wang, S.; Zhu, R.; Li, H.; Han, Q.; Zhao, R.C. Lung tumor exosomes induce a pro-inflammatory phenotype in mesenchymal stem cells via NFκB-TLR signaling pathway. *J. Hematol. Oncol.* **2016**, *9*, 42. [CrossRef]

102. Raimondo, S.; Saieva, L.; Corrado, C.; Fontana, S.; Flugy, A.; Rizzo, A.; De Leo, G.; Alessandro, R. Chronic myeloid leukemia-derived exosomes promote tumor growth through an autocrine mechanism. *Cell Commun. Signal.* **2015**, *13*, 8. [CrossRef]

103. Wei, Y.; Lai, X.; Yu, S.; Chen, S.; Ma, Y.; Zhang, Y.; Li, H.; Zhu, X.; Yao, L.; Zhang, J. Exosomal miR-221/222 enhances tamoxifen resistance in recipient ER-positive breast cancer cells. *Breast Cancer Res. Treat.* **2014**, *147*, 423–431. [CrossRef]

104. Qin, X.; Yu, S.; Zhou, L.; Shi, M.; Hu, Y.; Xu, X.; Shen, B.; Liu, S.; Yan, D.; Feng, J. Cisplatin-resistant lung cancer cell–derived exosomes increase cisplatin resistance of recipient cells in exosomal miR-100–5p-dependent manner. *Int. J. Nanomed.* **2017**, *12*, 3721–3733. [CrossRef]

105. Richards, K.E.; Zeleniak, A.E.; Fishel, M.L.; Wu, J.; Littlepage, L.E.; Hill, R. Cancer-associated fibroblast exosomes regulate survival and proliferation of pancreatic cancer cells. *Oncogene* **2017**, *36*, 1770–1778. [CrossRef]

106. Han, M.; Gu, Y.; Lu, P.; Li, J.; Cao, H.; Li, X.; Qian, X.; Yu, C.; Yang, Y.; Yang, X.; et al. Exosome-mediated lncRNA AFAP1-AS1 promotes trastuzumab resistance through binding with AUF1 and activating ERBB2 translation. *Mol. Cancer* **2020**, *19*, 26. [CrossRef]

107. Xu, C.G.; Yang, M.F.; Ren, Y.Q.; Wu, C.H.; Wang, L.Q. Exosomes mediated transfer of lncRNA UCA1 results in increased tamoxifen resistance in breast cancer cells. *Eur. Rev. Med. Pharmacol. Sci.* **2016**, *20*, 4362–4368.

108. Wang, J.; Lv, B.; Su, Y.; Wang, X.; Bu, J.; Yao, L. Exosome-mediated transfer of lncRNA HOTTIP promotes cisplatin resistance in gastric cancer cells by regulating HMGA1/miR-218 axis. *OncoTargets Ther.* **2019**, *12*, 11325–11338. [CrossRef]

109. Weiner-Gorzel, K.; Dempsey, E.; Milewska, M.; Mcgoldrick, A.; Toh, V.; Walsh, A.; Lindsay, S.; Gubbins, L.; Cannon, A.; Sharpe, D.; et al. Overexpression of the microRNA miR-433 promotes resistance to paclitaxel through the induction of cellular senescence in ovarian cancer cells. *Cancer Med.* **2015**, *4*, 745–758. [CrossRef]

110. Xu, H.; Han, H.; Song, S.; Yi, N.; Qian, C.; Qiu, Y.; Zhou, W.; Hong, Y.; Zhuang, W.; Li, Z.; et al. Exosome-transmitted PSMA3 and PSMA3-AS1 promote proteasome inhibitor resistance in multiple myeloma. *Clin. Cancer Res.* **2019**, *25*, 1923–1935. [CrossRef]

111. Giallombardo, M.; Taverna, S.; Alessandro, R.; Hong, D.; Rolfo, C. Exosome-mediated drug resistance in cancer: The near future is here. *Ther. Adv. Med. Oncol.* **2016**, *8*, 320–322. [CrossRef] [PubMed]

112. Zhao, K.; Wang, Z.; Li, X.; Liu, J.L.; Tian, L.; Chen, J.Q. Exosome-mediated transfer of CLIC1 contributes to the vincristine-resistance in gastric cancer. *Mol. Cell. Biochem.* **2019**, *462*, 97–105. [CrossRef] [PubMed]

113. Aung, T.; Chapuy, B.; Vogel, D.; Wenzel, D.; Oppermann, M.; Lahmann, M.; Weinhage, T.; Menck, K.; Hupfeld, T.; Koch, R.; et al. Exosomal evasion of humoral immunotherapy in aggressive B-cell lymphoma modulated by ATP-binding cassette transporter A3. *Proc. Natl. Acad. Sci. USA* **2011**, *108*, 15336–15341. [CrossRef] [PubMed]

114. Ciravolo, V.; Huber, V.; Ghedini, G.C.; Venturelli, E.; Bianchi, F.; Campiglio, M.; Morelli, D.; Villa, A.; Mina, P. Della Menard, S.; et al. Potential Role of Exosomes in Countering Trastuzumab-Based Therapy. *J. Cell. Physiol.* **2012**, *227*, 658–667. [CrossRef] [PubMed]

115. Harada, T.; Yamamoto, H.; Kishida, S.; Kishida, M.; Awada, C.; Takao, T.; Kikuchi, A. Wnt5b-associated exosomes promote cancer cell migration and proliferation. *Cancer Sci.* **2017**, *108*, 42–52. [CrossRef] [PubMed]

116. Pessolano, E.; Belvedere, R.; Bizzarro, V.; Franco, P.; De Marco, I.; Porta, A.; Tosco, A.; Parente, L.; Perretti, M.; Petrella, A. Annexin A1 may induce pancreatic cancer progression as a key player of extracellular vesicles effects as evidenced in the in vitro MIA PaCa-2 model system. *Int. J. Mol. Sci.* **2018**, *19*, 3878. [CrossRef] [PubMed]

117. Datta, A.; Kim, H.; Lal, M.; McGee, L.; Johnson, A.; Moustafa, A.A.; Jones, J.C.; Mondal, D.; Ferrer, M.; Abdel-Mageed, A.B. Manumycin A suppresses exosome biogenesis and secretion via targeted inhibition of Ras/Raf/ERK1/2 signaling and hnRNP H1 in castration-resistant prostate cancer cells. *Cancer Lett.* **2017**, *408*, 73–81. [CrossRef] [PubMed]

118. Trajkovic, K.; Hsu, C.; Chiantia, S.; Rajendran, L.; Wenzel, D.; Wieland, F.; Schwille, P.; Brügger, B.; Simons, M. Ceramide triggers budding of exosome vesicles into multivesicular endosomes. *Science* **2008**, *319*, 1244–1247. [CrossRef]

119. Yang, Y.; Li, C.W.; Chan, L.C.; Wei, Y.; Hsu, J.M.; Xia, W.; Cha, J.H.; Hou, J.; Hsu, J.L.; Sun, L.; et al. Exosomal PD-L1 harbors active defense function to suppress t cell killing of breast cancer cells and promote tumor growth. *Cell Res.* **2018**, *28*, 862–864. [CrossRef]

120. Kosaka, N.; Iguchi, H.; Hagiwara, K.; Yoshioka, Y.; Takeshita, F.; Ochiya, T. Neutral sphingomyelinase 2 (nSMase2)-dependent exosomal transfer of angiogenic micrornas regulate cancer cell metastasis. *J. Biol. Chem.* **2013**, *288*, 10849–10859. [CrossRef]

121. Roseblade, A.; Luk, F.; Ung, A.; Bebawy, M. Targeting Microparticle Biogenesis: A Novel Approach to the Circumvention of Cancer Multidrug Resistance. *Curr. Cancer Drug Targets* **2015**, *15*, 205–214. [CrossRef] [PubMed]

122. Peinado, H.; Alečković, M.; Lavotshkin, S.; Matei, I.; Costa-Silva, B.; Moreno-Bueno, G.; Hergueta-Redondo, M.; Williams, C.; García-Santos, G.; Ghajar, C.; et al. Melanoma exosomes educate bone marrow progenitor cells. *Nat. Med.* **2013**, *18*, 883–891. [CrossRef] [PubMed]

123. Guo, J.; Jayaprakash, P.; Dan, J.; Wise, P.; Jang, G.-B.; Liang, C.; Chen, M.; Woodley, D.T.; Fabbri, M.; Li, W. PRAS40 Connects Microenvironmental Stress Signaling to Exosome-Mediated Secretion. *Mol. Cell. Biol.* **2017**, *37*, e00171-17. [CrossRef] [PubMed]

124. Thyagarajan, A.; Kadam, S.M.; Liu, L.; Kelly, L.E.; Rapp, C.M.; Chen, Y.; Sahu, R.P. Gemcitabine induces microvesicle particle release in a platelet-activating factor-receptor-dependent manner via modulation of the MAPK pathway in pancreatic cancer cells. *Int. J. Mol. Sci.* **2019**, *20*, 32. [CrossRef]

125. Kosgodage, U.S.; Trindade, R.P.; Thompson, P.R.; Inal, J.M. Chloramidine / Bisindolylmaleimide-I-Mediated Inhibition of Exosome and Microvesicle Release and Enhanced Efficacy of Cancer Chemotherapy. *Int. J. Mol. Sci.* **2017**, *18*, 1007. [CrossRef]
126. Featherby, S.; Madkhali, Y.; Maraveyas, A.; Ettelaie, C. Apixaban suppresses the release of TF-positive microvesicles and restrains cancer cell proliferation through directly inhibiting TF-fVIIa activity. *Thromb. Haemost.* **2019**, *119*, 1419–1432. [CrossRef]
127. Liu, J.; Zhang, Y.; Liu, A.; Wang, J.; Li, L.; Chen, X.; Gao, X. Distinct Dasatinib-Induced Mechanisms of Apoptotic Response and Exosome Release in Imatinib-Resistant Human Chronic Myeloid Leukemia Cells. *Int. J. Mol. Sci.* **2016**, *17*, 531. [CrossRef]
128. Huang, M.B.; Gonzalez, R.R.; Lillard, J.; Bond, V.C. Secretion modification region-derived peptide blocks exosome release and mediates cell cycle arrest in breast cancer cells. *Oncotarget* **2017**, *8*, 11302–11315. [CrossRef]
129. Plebanek, M.P.; Mutharasan, R.K.; Volpert, O.; Matov, A.; Gatlin, J.C.; Thaxton, C.S. Nanoparticle targeting and cholesterol flux through scavenger receptor type B-1 inhibits cellular exosome uptake. *Sci. Rep.* **2015**, *5*, 15724. [CrossRef]
130. Bastos, N.; Ruivo, C.F.; da Silva, S.; Melo, S.A. Exosomes in cancer: Use them or target them? *Semin. Cell Dev. Biol.* **2018**, *78*, 13–21. [CrossRef]
131. Zheng, Y.; Tu, C.; Zhang, J.; Wang, J. Inhibition of multiple myeloma-derived exosomes uptake suppresses the functional response in bone marrow stromal cell. *Int. J. Oncol.* **2019**, *54*, 1061–1070. [CrossRef] [PubMed]
132. Mrowczynski, O.D.; Madhankumar, A.B.; Sundstrom, J.M.; Zhao, Y.; Kawasawa, Y.I.; Slagle-Webb, B.; Mau, C.; Payne, R.A.; Rizk, E.B.; Zacharia, B.E.; et al. Exosomes impact survival to radiation exposure in cell line models of nervous system cancer. *Oncotarget* **2018**, *9*, 36083–36101. [CrossRef] [PubMed]
133. Sampey, G.C.; Meyering, S.S.; Asad Zadeh, M.; Saifuddin, M.; Hakami, R.M.; Kashanchi, F. Exosomes and their role in CNS viral infections. *J. Neurovirol.* **2014**, *20*, 199–208. [CrossRef] [PubMed]
134. Zhang, H.G.; Kim, H.; Liu, C.; Yu, S.; Wang, J.; Grizzle, W.E.; Kimberly, R.P.; Barnes, S. Curcumin reverses breast tumor exosomes mediated immune suppression of NK cell tumor cytotoxicity. *Biochim. Biophys. Acta Mol. Cell Res.* **2007**, *1773*, 1116–1123. [CrossRef] [PubMed]
135. Pucci, F.; Garris, C.; Lai, C.P.; Newton, A.; Pfirschke, C.; Engblom, C.; Alvarez, D.; Sprachman, M.; Evavold, C.; Magnuson, A.; et al. SCS macrophages suppress melanoma by restricting tumor-derived vesicle-B cell interactions. *Science* **2016**, *352*, 242–246. [CrossRef]
136. Evander, M.; Gidlöf, O.; Olde, B.; Erlinge, D.; Laurell, T. Non-contact acoustic capture of microparticles from small plasma volumes. *Lab Chip* **2015**, *15*, 2588–2596. [CrossRef]
137. Smyth, T.; Kullberg, M.; Malik, N.; Smith-Jones, P.; Graner, M.W.; Anchordoquy, T.J. Biodistribution and delivery efficiency of unmodified tumor-derived exosomes. *J. Control. Release* **2015**, *199*, 145–155. [CrossRef]
138. Nishida-Aoki, N.; Tominaga, N.; Takeshita, F.; Sonoda, H.; Yoshioka, Y.; Ochiya, T. Disruption of Circulating Extracellular Vesicles as a Novel Therapeutic Strategy against Cancer Metastasis. *Mol. Ther.* **2017**, *25*, 181–191. [CrossRef]
139. Bttpdjbufe, F.; Gspn, T.Q.; Chalmin, F.; Ladoire, S.; Mignot, G.; Vincent, J.; Bruchard, M.; Boireau, W.; Rouleau, A.; Simon, B.; et al. Membrane-associated Hsp72 from tumor-derived exosomes mediates STAT3-dependent immunosuppressive function of mouse and human myeloid-derived suppressor cells Fanny. *J. Clin. Investig.* **2010**, *120*, 457–471. [CrossRef]
140. Yoon, J.H.; Ashktorab, H.; Smoot, D.T.; Nam, S.W.; Hur, H.; Park, W.S. Uptake and tumor-suppressive pathways of exosome-associated GKN1 protein in gastric epithelial cells. *Gastric Cancer* **2020**, *23*, 848–862. [CrossRef]
141. Zaharie, F.; Muresan, M.S.; Petrushev, B.; Berce, C.; Gafencu, G.A.; Selicean, S.; Jurj, A.; Cojocneanu-Petric, R.; Lisencu, C.I.; Pop, L.A.; et al. Exosome-carried microRNA-375 inhibits cell progression and dissemination via Bcl-2 blocking in colon cancer. *J. Gastrointest. Liver Dis.* **2015**, *24*, 435–443. [CrossRef] [PubMed]
142. Shi, H.; Li, H.; Zhen, T.; Dong, Y.; Pei, X.; Zhang, X. The Potential Therapeutic Role of Exosomal MicroRNA-520b Derived from Normal Fibroblasts in Pancreatic Cancer. *Mol. Ther. Nucleic Acids* **2020**, *20*, 373–384. [CrossRef] [PubMed]
143. Jeong, K.; Yu, Y.J.; You, J.Y.; Rhee, W.J.; Kim, J.A. Exosome-mediated microRNA-497 delivery for anti-cancer therapy in a microfluidic 3D lung cancer model. *Lab Chip* **2020**, *20*, 548–557. [CrossRef] [PubMed]
144. Chen, W.; Quan, Y.; Fan, S.; Wang, H.; Liang, J.; Huang, L.; Chen, L.; Liu, Q.; He, P.; Ye, Y. Exosome-transmitted circular RNA hsa_circ_0051443 suppresses hepatocellular carcinoma progression. *Cancer Lett.* **2020**, *475*, 119–128. [CrossRef]
145. Pakravan, K.; Babashah, S.; Sadeghizadeh, M.; Mowla, S.J.; Mossahebi-Mohammadi, M.; Ataei, F.; Dana, N.; Javan, M. MicroRNA-100 shuttled by mesenchymal stem cell-derived exosomes suppresses in vitro angiogenesis through modulating the mTOR/HIF-1α/VEGF signaling axis in breast cancer cells. *Cell. Oncol.* **2017**, *40*, 457–470. [CrossRef]
146. Simpson, R.J.; Lim, J.W.E.; Moritz, R.L.; Mathivanan, S. Exosomes: Proteomic insights and diagnostic potential. *Expert Rev. Proteomics* **2009**, *6*, 267–283. [CrossRef]
147. Hornick, N.I.; Huan, J.; Doron, B.; Goloviznina, N.A.; Lapidus, J.; Chang, B.H.; Kurre, P. Serum Exosome MicroRNA as a minimally-invasive early biomarker of AML. *Sci. Rep.* **2015**, *5*, 11295. [CrossRef]
148. Sharma, R.; Huang, X.; Brekken, R.A.; Schroit, A.J. Detection of phosphatidylserine-positive exosomes for the diagnosis of early-stage malignancies. *Br. J. Cancer* **2017**, *117*, 545–552. [CrossRef]
149. Bu, H.; He, D.; He, X.; Wang, K. Exosomes: Isolation, Analysis, and Applications in Cancer Detection and Therapy. *ChemBioChem* **2019**, *20*, 451–461. [CrossRef]
150. Boukouris, S.; Mathivanan, S. Exosomes in bodily fluids are a highly stable resource of disease biomarkers. *Proteomics Clin. Appl.* **2015**, *9*, 358–367. [CrossRef]

151. Munoz, J.L.; Bliss, S.A.; Greco, S.J.; Ramkissoon, S.H.; Ligon, K.L.; Rameshwar, P. Delivery of functional anti-miR-9 by mesenchymal stem cell-derived exosomes to glioblastoma multiforme cells conferred chemosensitivity. *Mol. Ther. Nucleic Acids* **2013**, *2*, e126. [CrossRef] [PubMed]

152. Syn, N.L.; Wang, L.; Chow, E.K.H.; Lim, C.T.; Goh, B.C. Exosomes in Cancer Nanomedicine and Immunotherapy: Prospects and Challenges. *Trends Biotechnol.* **2017**, *35*, 665–676. [CrossRef] [PubMed]

153. Viaud, S.; Terme, M.; Flament, C.; Taieb, J.; André, F.; Novault, S.; Escudier, B.; Robert, C.; Caillat-Zucman, S.; Tursz, T.; et al. Dendritic cell-derived exosomes promote natural killer cell activation and proliferation: A role for NKG2D ligands and IL-15Rα. *PLoS ONE* **2009**, *4*, e4942. [CrossRef] [PubMed]

154. Pitt, J.M.; André, F.; Amigorena, S.; Soria, J.C.; Eggermont, A.; Kroemer, G.; Zitvogel, L. Dendritic cell-derived exosomes for cancer therapy. *J. Clin. Investig.* **2016**, *126*, 1224–1232. [CrossRef]

155. Viaud, S.; Ploix, S.; Lapierre, V.; Théry, C.; Commere, P.H.; Tramalloni, D.; Gorrichon, K.; Virault-Rocroy, P.; Tursz, T.; Lantz, O.; et al. Updated technology to produce highly immunogenic dendritic cell-derived exosomes of clinical grade: A critical role of interferon-γ. *J. Immunother.* **2011**, *34*, 65–75. [CrossRef] [PubMed]

156. Vulpis, E.; Cecere, F.; Molfetta, R.; Soriani, A.; Fionda, C.; Peruzzi, G.; Caracciolo, G.; Palchetti, S.; Masuelli, L.; Simonelli, L.; et al. Genotoxic stress modulates the release of exosomes from multiple myeloma cells capable of activating NK cell cytokine production: Role of HSP70/TLR2/NF-kB axis. *Oncoimmunology* **2017**, *6*, e1279372. [CrossRef] [PubMed]

157. Xiao, W.; Dong, W.; Zhang, C.; Saren, G.; Geng, P.; Zhao, H.; Li, Q.; Zhu, J.; Li, G.; Zhang, S.; et al. Effects of the epigenetic drug MS-275 on the release and function of exosome-related immune molecules in hepatocellular carcinoma cells. *Eur. J. Med. Res.* **2013**, *18*, 61. [CrossRef] [PubMed]

158. Chen, W.; Wang, J.; Shao, C.; Liu, S.; Yu, Y.; Wang, Q.; Cao, X. Efficient induction of antitumor T cell immunity by exosomes derived from heat-shocked lymphoma cells. *Eur. J. Immunol.* **2006**, *36*, 1598–1607. [CrossRef] [PubMed]

159. Tan, A.; de la Peña, H.; Seifalian, A.M. The application of exosomes as a nanoscale cancer vaccine. *Int. J. Nanomed.* **2010**, *5*, 889–900.

160. Xie, Y.; Bai, O.; Zhang, H.; Yuan, J.; Zong, S.; Chibbar, R.; Slattery, K.; Qureshi, M.; Wei, Y.; Deng, Y.; et al. Membrane-bound HSP70-engineered myeloma cell-derived exosomes stimulate more efficient CD8+ CTL- and NK-mediated antitumour immunity than exosomes released from heat-shocked tumour cells expressing cytoplasmic HSP70. *J. Cell. Mol. Med.* **2010**, *14*, 2655–2666. [CrossRef]

161. Yang, Y.; Chen, Y.; Zhang, F.; Zhao, Q.; Zhong, H. Increased anti-tumour activity by exosomes derived from doxorubicin-treated tumour cells via heat stress. *Int. J. Hyperth.* **2015**, *31*, 498–506. [CrossRef] [PubMed]

162. Hao, S.; Bai, O.; Li, F.; Yuan, J.; Laferte, S.; Xiang, J. Mature dendritic cells pulsed with exosomes stimulate efficient cytotoxic T-lymphocyte responses and antitumour immunity. *Immunology* **2007**, *120*, 90–102. [CrossRef] [PubMed]

163. Yamashita, T.; Takahashi, Y.; Takakura, Y. Possibility of exosome-based therapeutics and challenges in production of exosomes eligible for therapeutic application. *Biol. Pharm. Bull.* **2018**, *41*, 835–842. [CrossRef] [PubMed]

164. Morishita, M.; Takahashi, Y.; Matsumoto, A.; Nishikawa, M.; Takakura, Y. Exosome-based tumor antigens–adjuvant co-delivery utilizing genetically engineered tumor cell-derived exosomes with immunostimulatory CpG DNA. *Biomaterials* **2016**, *111*, 55–65. [CrossRef] [PubMed]

165. Lu, Z.; Zuo, B.; Jing, R.; Gao, X.; Rao, Q.; Liu, Z.; Qi, H.; Guo, H.; Yin, H.F. Dendritic cell-derived exosomes elicit tumor regression in autochthonous hepatocellular carcinoma mouse models. *J. Hepatol.* **2017**, *67*, 739–748. [CrossRef]

166. Xiao, L.; Erb, U.; Zhao, K.; Hackert, T.; Zöller, M. Efficacy of vaccination with tumor-exosome loaded dendritic cells combined with cytotoxic drug treatment in pancreatic cancer. *Oncoimmunology* **2017**, *6*. [CrossRef]

167. Sedlik, C.; Vigneron, J.; Torrieri-Dramard, L.; Pitoiset, F.; Denizeau, J.; Chesneau, C.; de la Rochere, P.; Lantz, O.; Thery, C.; Bellier, B. Different immunogenicity but similar antitumor efficacy of two DNA vaccines coding for an antigen secreted in different membrane vesicle-associated forms. *J. Extracell. Vesicles* **2014**, *3*. [CrossRef]

168. Ohno, S.I.; Takanashi, M.; Sudo, K.; Ueda, S.; Ishikawa, A.; Matsuyama, N.; Fujita, K.; Mizutani, T.; Ohgi, T.; Ochiya, T.; et al. Systemically injected exosomes targeted to EGFR deliver antitumor microrna to breast cancer cells. *Mol. Ther.* **2013**, *21*, 185–191. [CrossRef]

169. Lang, F.M.; Hossain, A.; Gumin, J.; Momin, E.N.; Shimizu, Y.; Ledbetter, D.; Shahar, T.; Yamashita, S.; Parker Kerrigan, B.; Fueyo, J.; et al. Mesenchymal stem cells as natural biofactories for exosomes carrying miR-124a in the treatment of gliomas. *Neuro-Oncology* **2018**, *20*, 380–390. [CrossRef]

170. Zheng, R.; Du, M.; Wang, X.; Xu, W.; Liang, J.; Wang, W.; Lv, Q.; Qin, C.; Chu, H.; Wang, M.; et al. Exosome-transmitted long non-coding RNA PTENP1 suppresses bladder cancer progression. *Mol. Cancer* **2018**, *17*, 143. [CrossRef]

171. Shtam, T.A.; Kovalev, R.A.; Varfolomeeva, E.Y.; Makarov, E.M.; Kil, Y.V.; Filatov, M.V. Exosomes are natural carriers of exogenous siRNA to human cells in vitro. *Cell Commun. Signal.* **2013**, *11*, 88. [CrossRef] [PubMed]

172. Bellavia, D.; Raimondo, S.; Calabrese, G.; Forte, S.; Cristaldi, M.; Patinella, A.; Memeo, L.; Manno, M.; Raccosta, S.; Diana, P.; et al. Interleukin 3- receptor targeted exosomes inhibit in vitro and in vivo chronic myelogenous Leukemia cell growth. *Theranostics* **2017**, *7*, 1333–1345. [CrossRef] [PubMed]

173. Kamerkar, S.; LeBleu, V.; Sugimoto, H.; Yang, S.; Ruivo, C.; Melo, S.A.; Lee, J.; Kalluri, R. Exosomes Facilitate Therapeutic Targeting of Oncogenic Kras in Pancreatic Cancer. *Nature* **2017**, *546*, 498–503. [CrossRef] [PubMed]

174. Zhao, L.; Gu, C.; Gan, Y.; Shao, L.; Chen, H.; Zhu, H. Exosome-mediated siRNA delivery to suppress postoperative breast cancer metastasis. *J. Control. Release* **2020**, *318*, 1–15. [CrossRef] [PubMed]

175. Kim, G.; Kim, M.; Lee, Y.; Byun, J.W.; Hwang, D.W.; Lee, M. Systemic delivery of microRNA-21 antisense oligonucleotides to the brain using T7-peptide decorated exosomes. *J. Control. Release* **2020**, *317*, 273–281. [CrossRef]

176. Fatima, F.; Nawaz, M. Stem cell-derived exosomes: Roles in stromal remodeling, tumor progression, and cancer immunotherapy. *Chin. J. Cancer* **2015**, *34*, 541–553. [CrossRef]

177. Mendt, M.; Rezvani, K.; Shpall, E. Mesenchymal stem cell-derived exosomes for clinical use. *Bone Marrow Transplant.* **2019**, *54*, 789–792. [CrossRef]

178. Liang, Y.; Zhang, D.; Li, L.; Xin, T.; Zhao, Y.; Ma, R.; Du, J. Exosomal microRNA-144 from bone marrow-derived mesenchymal stem cells inhibits the progression of non-small cell lung cancer by targeting CCNE1 and CCNE2. *Stem Cell Res. Ther.* **2020**, *11*, 87. [CrossRef]

179. Xie, C.; Du, L.Y.; Guo, F.; Li, X.; Cheng, B. Exosomes derived from microRNA-101-3p-overexpressing human bone marrow mesenchymal stem cells suppress oral cancer cell proliferation, invasion, and migration. *Mol. Cell. Biochem.* **2019**, *458*, 11–26. [CrossRef]

180. Greco, K.A.; Franzen, C.A.; Foreman, K.E.; Flanigan, R.C.; Kuo, P.C.; Gupta, G.N. PLK-1 Silencing in Bladder Cancer by siRNA Delivered with Exosomes. *Urology* **2016**, *91*, 241.e1–241.e7. [CrossRef]

181. Hu, Q.; Su, H.; Li, J.; Lyon, C.; Tang, W.; Wan, M.; Hu, T.Y. Clinical applications of exosome membrane proteins. *Precis. Clin. Med.* **2020**, *3*, 54–66. [CrossRef] [PubMed]

182. Saari, H.; Lázaro-Ibáñez, E.; Viitala, T.; Vuorimaa-Laukkanen, E.; Siljander, P.; Yliperttula, M. Microvesicle- and exosome-mediated drug delivery enhances the cytotoxicity of Paclitaxel in autologous prostate cancer cells. *J. Control. Release* **2015**, *220*, 727–737. [CrossRef] [PubMed]

183. Wang, L.; Xie, Y.; Ahmed, K.A.; Ahmed, S.; Sami, A.; Chibbar, R.; Xu, Q.; Kane, S.E.; Hao, S.; Mulligan, S.J.; et al. Exosomal pMHC-I complex targets T cell-based vaccine to directly stimulate CTL responses leading to antitumor immunity in transgenic FVBneuN and HLA-A2/HER2 mice and eradicating trastuzumab-resistant tumor in athymic nude mice. *Breast Cancer Res. Treat.* **2013**, *140*, 273–284. [CrossRef] [PubMed]

184. Kim, M.S.; Haney, M.J.; Zhao, Y.; Yuan, D.; Deygen, I.; Klyachko, N.L.; Kabanov, A.V.; Batrakova, E.V. Engineering macrophage-derived exosomes for targeted paclitaxel delivery to pulmonary metastases: In vitro and in vivo evaluations. *Nanomed. Nanotechnol. Biol. Med.* **2018**, *14*, 195–204. [CrossRef] [PubMed]

185. Srivastava, A.; Amreddy, N.; Babu, A.; Panneerselvam, J.; Mehta, M.; Muralidharan, R.; Chen, A.; Zhao, Y.D.; Razaq, M.; Riedinger, N.; et al. Nanosomes carrying doxorubicin exhibit potent anticancer activity against human lung cancer cells. *Sci. Rep.* **2016**, *6*, 38541. [CrossRef]

186. Gong, C.; Tian, J.; Wang, Z.; Gao, Y.; Wu, X.; Ding, X.; Qiang, L.; Li, G.; Han, Z.; Yuan, Y.; et al. Functional exosome-mediated co-delivery of doxorubicin and hydrophobically modified microRNA 159 for triple-negative breast cancer therapy. *J. Nanobiotechnol.* **2019**, *17*, 93. [CrossRef]

187. Kim, M.S.; Haney, M.J.; Zhao, Y.; Mahajan, V.; Deygen, I.; Klyachko, N.L.; Inskoe, E.; Piroyan, A.; Sokolsky, M.; Hingtgen, S.D.; et al. Development of Exosome-encapsulated Paclitaxel to Overcome MDR in Cancer cells. *Nanomedicine* **2017**, *12*, 655–664. [CrossRef]

188. Hadla, M.; Palazzolo, S.; Corona, G.; Caligiuri, I.; Canzonieri, V.; Toffoli, G.; Rizzolio, F. Exosomes increase the therapeutic index of doxorubicin in breast and ovarian cancer mouse models. *Nanomedicine* **2016**, *11*, 2431–2441. [CrossRef]

189. Tian, Y.; Li, S.; Song, J.; Ji, T.; Zhu, M.; Anderson, G.J.; Wei, J.; Nie, G. A doxorubicin delivery platform using engineered natural membrane vesicle exosomes for targeted tumor therapy. *Biomaterials* **2014**, *35*, 2383–2390. [CrossRef]

190. Li, Y.J.; Wu, J.Y.; Wang, J.M.; Hu, X.B.; Cai, J.X.; Xiang, D.X. Gemcitabine loaded autologous exosomes for effective and safe chemotherapy of pancreatic cancer. *Acta Biomater.* **2020**, *101*, 519–530. [CrossRef]

191. Osterman, C.J.D.; Lynch, J.C.; Leaf, P.; Gonda, A.; Bennit, H.R.F.; Griffiths, D.; Wall, N.R. Curcumin modulates pancreatic adenocarcinoma cell-derived exosomal function. *PLoS ONE* **2015**, *10*, e0132845. [CrossRef] [PubMed]

192. Munagala, R.; Aqil, F.; Jeyabalan, J.; Agrawal, A.K.; Mudd, A.M.; Kyakulaga, A.H.; Singh, I.P.; Vadhanam, M.V.; Gupta, R.C. Exosomal formulation of anthocyanidins against multiple cancer types. *Cancer Lett.* **2017**, *393*, 94–102. [CrossRef] [PubMed]

193. Donoso-Quezada, J.; Guajardo-Flores, D.; González-Valdez, J. Exosomes as nanocarriers for the delivery of bioactive compounds from black bean extract with antiproliferative activity in cancer cell lines. *Mater. Today Proc.* **2019**, *13*, 362–369. [CrossRef]

194. Aqil, F.; Jeyabalan, J.; Agrawal, A.K.; Kyakulaga, A.H.; Munagala, R.; Parker, L.; Gupta, R.C. Exosomal delivery of berry anthocyanidins for the management of ovarian cancer. *Food Funct.* **2017**, *8*, 4100–4107. [CrossRef]

195. Aqil, F.; Kausar, H.; Agrawal, A.K.; Jeyabalan, J.; Kyakulaga, A.H.; Munagala, R.; Gupta, R. Exosomal formulation enhances therapeutic response of celastrol against lung cancer. *Exp. Mol. Pathol.* **2016**, *101*, 12–21. [CrossRef]

196. Bhattacharya, S.; Pal, K.; Sharma, A.K.; Dutta, S.K.; Lau, J.S.; Yan, I.K.; Wang, E.; Elkhanany, A.; Alkharfy, K.M.; Sanyal, A.; et al. GAIP interacting protein C-Terminus regulates autophagy and exosome biogenesis of pancreatic cancer through metabolic pathways. *PLoS ONE* **2014**, *9*, e114409. [CrossRef]

197. Kirave, P.; Gondaliya, P.; Kulkarni, B.; Rawal, R.; Garg, R.; Jain, A.; Kalia, K. Exosome mediated miR-155 delivery confers cisplatin chemoresistance in oral cancer cells via epithelial-mesenchymal transition. *Oncotarget* **2020**, *11*, 1157–1171. [CrossRef]

198. Kim, S.M.; Yang, Y.; Oh, S.J.; Hong, Y.; Seo, M.; Jang, M. Cancer-derived exosomes as a delivery platform of CRISPR/Cas9 confer cancer cell tropism-dependent targeting. *J. Control. Release* **2017**, *266*, 8–16. [CrossRef]

199. Zhang, Q.; Zhang, H.; Ning, T.; Liu, D.; Deng, T.; Liu, R.; Bai, M.; Zhu, K.; Li, J.; Fan, Q.; et al. Exosome-Delivered c-Met siRNA Could Reverse Chemoresistance to Cisplatin in Gastric Cancer. *Int. J. Nanomed.* **2020**, *15*, 2323–2335. [CrossRef]

200. Liu, T.; Zhang, X.; Du, L.; Wang, Y.; Liu, X.; Tian, H.; Wang, L.; Li, P.; Zhao, Y.; Duan, W.; et al. Exosome-transmitted miR-128-3p increase chemosensivity of oxaliplatin-resistant colorectal cancer. *Mol. Cancer* **2019**, *18*, 43. [CrossRef]

201. Lou, G.; Song, X.; Yang, F.; Wu, S.; Wang, J.; Chen, Z.; Liu, Y. Exosomes derived from MIR-122-modified adipose tissue-derived MSCs increase chemosensitivity of hepatocellular carcinoma. *J. Hematol. Oncol.* **2015**, *8*, 122. [CrossRef] [PubMed]

202. Han, M.; Hu, J.; Lu, P.; Cao, H.; Yu, C.; Li, X.; Qian, X.; Yang, X.; Yang, Y.; Han, N.; et al. Exosome-transmitted miR-567 reverses trastuzumab resistance by inhibiting ATG5 in breast cancer. *Cell Death Dis.* **2020**, *11*, 43. [CrossRef] [PubMed]

203. Cui, J.; Wang, H.; Zhang, X.; Sun, X.; Zhang, J.; Ma, J. Exosomal miR-200c suppresses chemoresistance of docetaxel in tongue squamous cell carcinoma by suppressing TUBB3 and PPP2R1B. *Aging (Albany N. Y.)* **2020**, *12*, 6756–6773. [CrossRef]

204. Lou, G.; Chen, L.; Xia, C.; Wang, W.; Qi, J.; Li, A.; Zhao, L.; Chen, Z.; Zheng, M.; Liu, Y. MiR-199a-modified exosomes from adipose tissue-derived mesenchymal stem cells improve hepatocellular carcinoma chemosensitivity through mTOR pathway. *J. Exp. Clin. Cancer Res.* **2020**, *39*, 4. [CrossRef] [PubMed]

205. Besse, B.; Charrier, M.; Lapierre, V.; Dansin, E.; Lantz, O.; Planchard, D.; Le Chevalier, T.; Livartoski, A.; Barlesi, F.; Laplanche, A.; et al. Dendritic cell-derived exosomes as maintenance immunotherapy after first line chemotherapy in NSCLC. *Oncoimmunology* **2016**, *5*, 1071008. [CrossRef] [PubMed]

206. Beit-Yannai, E.; Tabak, S.; Stamer, W.D. Physical exosome:exosome interactions. *J. Cell. Mol. Med.* **2018**, *22*, 2001–2006. [CrossRef] [PubMed]

207. Lunavat, T.R.; Jang, S.C.; Nilsson, L.; Park, H.T.; Repiska, G.; Lässer, C.; Nilsson, J.A.; Gho, Y.S.; Lötvall, J. RNAi delivery by exosome-mimetic nanovesicles—Implications for targeting c-Myc in cancer. *Biomaterials* **2016**, *102*, 231–238. [CrossRef]

208. Sato, Y.T.; Umezaki, K.; Sawada, S.; Mukai, S.A.; Sasaki, Y.; Harada, N.; Shiku, H.; Akiyoshi, K. Engineering hybrid exosomes by membrane fusion with liposomes. *Sci. Rep.* **2016**, *6*, 21933. [CrossRef]

Article

Targeted Oral Delivery of Paclitaxel Using Colostrum-Derived Exosomes

Raghuram Kandimalla [1,2,†], Farrukh Aqil [1,3,†], Sara S. Alhakeem [4,‡], Jeyaprakash Jeyabalan [5], Neha Tyagi [1,2], Ashish Agrawal [1,4,§], Jun Yan [1,6], Wendy Spencer [5], Subbarao Bondada [4] and Ramesh C. Gupta [1,2,5,*]

[1] James Graham Brown Cancer Center, University of Louisville, Louisville, KY 40202, USA; raghuram.kandimalla@louisville.edu (R.K.); farrukh.aqil@louisville.edu (F.A.); neha.tyagi@louisville.edu (N.T.); ashish.phe@iitbhu.ac.in (A.A.); jun.yan@louisville.edu (J.Y.)
[2] Department of Pharmacology and Toxicology, University of Louisville, Louisville, KY 40202, USA
[3] Department of Medicine, University of Louisville, Louisville, KY 40202, USA
[4] Department of Microbiology, Immunology & Molecular Genetics, University of Kentucky, Lexington, KY 40536, USA; sara.alhakeem@bioagilytix.com (S.S.A.); Subbarao.Bondada@uky.edu (S.B.)
[5] 3P Biotechnologies, Inc., Louisville, KY 40202, USA; jp3pbiotech@gmail.com (J.J.); wendyspencer3p@gmail.com (W.S.)
[6] Department of Surgery, University of Louisville, Louisville, KY 40202, USA
[*] Correspondence: rcgupta3P@gmail.com or rcgupta@louisville.edu; Tel.: +502-852-3684; Fax: +502-852-3842
[†] Equal First Authors.
[‡] Current address: BioAgilytix, 2300 Englert Drive, Durham, NC 27713, USA.
[§] Current address: Department of Pharmaceutical Engineering and Technology, Indian Institute of Technology, Varanasi 221005, India.

Citation: Kandimalla, R.; Aqil, F.; Alhakeem, S.S.; Jeyabalan, J.; Tyagi, N.; Agrawal, A.; Yan, J.; Spencer, W.; Bondada, S.; Gupta, R.C. Targeted Oral Delivery of Paclitaxel Using Colostrum-Derived Exosomes. *Cancers* **2021**, *13*, 3700. https://doi.org/ 10.3390/cancers13153700

Academic Editor: Maxim V. Berezovski

Received: 17 June 2021
Accepted: 17 July 2021
Published: 23 July 2021

Publisher's Note: MDPI stays neutral with regard to jurisdictional claims in published maps and institutional affiliations.

Simple Summary: Paclitaxel (PAC) is a widely used antitumor agent in the treatment of various early-stage and advanced cancers, including lung cancer. While efficacious, solvent-based PAC generally is not well tolerated and is associated with severe side effects. To overcome such limitations, naturally occurring nanocarriers such as exosomes are attracting great interest. In this paper, we show that tumor-targeted oral formulation of PAC, using bovine colostrum-derived exosomes, not only enhance therapeutic efficacy against orthotopic lung cancer but also mitigate or eliminate systemic and immunotoxicity of the conventional i.v. dosing. These data will leverage the advantages of bovine colostrum exosomes to advance the exosome-mediated targeted oral delivery of PAC as a therapeutic alternative to current therapies.

Abstract: Lung cancer is the leading cause of cancer-related deaths worldwide. Non-small-cell lung cancer (NSCLC) is the most common type accounting for 84% of all lung cancers. Paclitaxel (PAC) is a widely used drug in the treatment of a broad spectrum of human cancers, including lung. While efficacious, PAC generally is not well tolerated and its limitations include low aqueous solubility, and significant toxicity. To overcome the dose-related toxicity of solvent-based PAC, we utilized bovine colostrum-derived exosomes as a delivery vehicle for PAC for the treatment of lung cancer. Colostrum provided higher yield of exosomes and could be loaded with higher amount of PAC compared to mature milk. Exosomal formulation of PAC (ExoPAC) showed higher antiproliferative activity and inhibition of colony formation against A549 cells compared with PAC alone, and also showed antiproliferative activity against a drug-resistant variant of A549. To further enhance its efficacy, exosomes were attached with a tumor-targeting ligand, folic acid (FA). FA-ExoPAC given orally showed significant inhibition (>50%) of subcutaneous tumor xenograft while similar doses of PAC showed insignificant inhibition. In the orthotopic lung cancer model, oral dosing of FA-ExoPAC achieved greater efficacy (55% growth inhibition) than traditional i.v. PAC (24–32% growth inhibition) and similar efficacy as i.v. Abraxane (59% growth inhibition). The FA-ExoPAC given i.v. exceeded the therapeutic efficacy of Abraxane (76% growth inhibition). Finally, wild-type animals treated with *p.o.* ExoPAC did not show gross, systemic or immunotoxicity. Solvent-based PAC caused immunotoxicity which was either reduced or completely mitigated by its exosomal formulations. These studies show that a tumor-targeted oral formulation of PAC (FA-ExoPAC) significantly improved the overall efficacy and safety profile while providing a user-friendly, cost-effective alternative to bolus i.v. PAC and i.v. Abraxane.

Keywords: colostrum exosomes; paclitaxel; drug delivery; lung cancer; immunotoxicity assessment

1. Introduction

Cancer is the second leading cause of death worldwide. In 2020, there were an estimated 1.8 million new cancer cases diagnosed and 606,520 cancer deaths in the United States [1]. More people in the U.S. (135,760) are expected to die of lung cancer in 2021 than prostate, breast and colon cancer combined [2]. Lung cancer remains the leading cause of cancer-related deaths in the United States and worldwide. Non-small cell-lung cancer (NSCLC) is relatively insensitive to chemotherapy and accounts for about 85% of all lung cancer cases. Regrettably, over 80% of all patients diagnosed with NSCLC die eventually due to the disease within five years [1,3]. Despite treatment with platinum-based chemotherapy, new molecularly-targeted therapies and immunotherapies, the overall survival benefit for NSCLC remains modest.

Paclitaxel (PAC) is the first- or second-line chemotherapy for the treatment of various cancers, including lung cancer and exhibits both anti-proliferative and apoptotic effects against cancer cells. Mechanistically, PAC interferes with the normal function of cellular microtubule growth by binding to the β-subunits of the tubulin and locking the microtubules preventing further cell division. Tubulins are the building blocks of microtubules, which play a major role in the migration of chromosomes during anaphase of the cell division [4]. However, the utility and clinical application of PAC has been hindered due to its poor aqueous solubility requiring formulation in the organic solvent Cremophor EL (CrEL) and its dose-related toxicity. For these reasons, the delivery of PAC is associated with substantial challenges. While the use of polyoxyethylated castor oil also known as CrEL and ethanol (50:50) overcomes the solubility problem, this solvent-based approach is associated with severe side effects [5,6]; therefore, PAC formulations are infused over several hours to reduce the effect of bolus dose.

To overcome these solvent-based limitations, several nanoparticle systems have been reported for the delivery of PAC. Abraxane® is an FDA-approved nanoformulation of PAC bound to human serum albumin that was developed to improve the toxicity profile of solvent-based PAC. In a phase III clinical trial, Abraxane was shown to enhance the therapeutic efficacy and pharmacokinetics compared to PAC given in CrEL [7]. However, the i.v. infusion of the Abraxane was reported to lower the blood cell count. Besides toxicity concerns, i.v. administration requires medical assistance, which, in turn, substantially increases the medical costs, besides patient suffering for a long duration.

To overcome these unfavorable physicochemical and pharmacokinetic properties of PAC, several additional delivery approaches have been attempted [8]. Toxicity limitations of solvent-based carriers can be overcome by using nanovesicles derived from natural sources such as milk [9,10]. Further, oral dosing of the chemotherapeutic achieved using these nanovesicles has many advantages such as flexibility of timing and location of administration, flexibility of drug exposure, reduction of the use of healthcare resources and a better quality of life [11,12]. Oral chemotherapy is also good for the metronomic (anti-angiogenic) chemotherapy [13], as it maintains a low serum level of the chemotherapeutic for a longer time than parenteral routes.

Exosomes (Exo) or small extracellular vesicles (sEVs), as the terminology is being debated [14], are biogenic nanocarriers (30–150 nm) with the lipid bilayer and have significant role in cell-to-cell communications. Exosomes are released from essentially all cell types and are present in all bodily fluids like blood, urine, saliva, amniotic fluid, lymphatic fluid and milk etc. [15,16]. Unlike other nanoparticulate systems, exosomes possess special proteins in their membrane surface proteins that may help in the endocytosis, which, in turn, promotes the delivery of tethered content [17,18]. We have previously demonstrated the utility of bovine milk as a source of exosomes for the delivery of small-molecule drugs [19–22] and siRNA [23,24] and for the oral delivery of PAC to inhibit subcutaneous lung tumor xenografts [19]. Biocompatibility, cost-effectiveness and abundance are some of the hallmarks that make milk exosomes a potentially commercially viable option as a nanodrug carrier.

In this study, we used exosomes isolated from a standardized source of bovine colostrum powder obtained from the early lactation period as a delivery vehicle for PAC (ExoPAC). Colostrum powder provides higher yields of exosomes than mature milk. We have shown an overexpression of FRα and RFC in H1299 and A549 lung cancer cells; the overexpression of the folate receptors was more pronounced in tumor tissue versus normal lung tissue (100-fold overexpression) [24]. Here, exosomes, functionalized with folic acid (FA) to target tumor cells, are embedded with PAC (FA-ExoPAC), and the therapeutic efficacy of the formulation was compared with Abraxane for lung tumors grown in a tumor microenvironment. We show that FA-ExoPAC given orally surpassed efficacy of solvent-based PAC and matched efficacy of Abraxane; whereas, i.v. FA-ExoPAC significantly exceeded the efficacy of Abraxane. ExoPAC formulations lacked gross, systemic and immunotoxicity in wild-type mice.

2. Materials and Methods

2.1. Chemicals and Reagents

PAC was procured from LC laboratories, Woburn, MA, USA. XenoLight D-Luciferin, potassium salt was purchased from PerkinElmer, (Waltham, MA, USA). BCA Protein Assay Kit was procured from ThermoFisher Scientific (Waltham, MA, USA) and folic acid was purchased from Sigma-Aldrich (St. Louis, MO, USA). All other chemicals were of analytical grade.

2.2. Isolation of Exosomes

Exosomes were isolated from colostrum powder (Immunodynamics, Inc., Fennimore, WI, USA). Briefly, colostrum powder was rehydrated in deionized water achieving a final concentration of 5% w/v, and exosomes were isolated by sequential centrifugations (13,000× g, 30 min; 65,000× g, 60 min; and 135,000× g, 2 h, as described [22], followed by removal of residual non-exosomal protein by ultrafiltration. After completion of ultracentrifugation, the supernatant containing free drug was discarded and ExoPAC pellet was washed with PBS. The exosome pellet was suspended in PBS (pH 7.4) and sterilized using 0.22 μM filter. The yield of exosomes was measured by means of exosomal protein concentration by a standard BCA protein assay kit. The exosome suspension ($\leq$6 mg/mL) was stored at $-80\,°C$.

2.3. Exosome Characterization

The particle size, polydispersity index (PDI) and zeta potential of the exosomes were determined by Zetasizer (Malvern Instruments Ltd., Malvern, Worcestershire, UK). Particle numbers per milligram of exosomal protein were measured by nanoparticle tracking analyzer (NonoView, Particle Matrix Inc., Grayslake, IL, USA). Samples were analyzed in triplicates. The size of exosomes was confirmed by atomic force microscopy (AFM) as described [22].

2.4. FA-Functionalization of Exosomes for Tumor Targeting

We functionalized exosomes with FA, a known tumor-targeting ligand. To stabilize the interaction of FA with exosomal proteins in vivo, we attached FA covalently by using activated FA. Activated FA was prepared using standard EDC (1-ethyl-3-(-3-dimethyl aminopropyl) carbodiimide hydrochloride) and NHS (N-hydroxysuccinimide esters). Free FA was removed using ultrafiltration. The degree of functionalization was achieved by varying FA concentration, and FA loading was determined by releasing the FA from the formulation in the presence of NaOH, followed by recovery of the exosomes. The FA and exosomal proteins were measured by spectrophotometry and BCA assay, respectively, and percent FA loading was calculated.

2.5. Loading of PAC on Exosomes

PAC was loaded onto the exosomes as described by us previously [19], except that exosomes used were derived from colostrum powder, the ratio of exosomes to PAC was reduced and harvesting time of ExoPAC formulation by ultracentrifuge was reduced to achieve higher drug loading. Briefly, PAC (dissolved in ethanol: acetonitrile; 1:1 v/v) was mixed with exosomes (6 mg/mL in PBS), keeping the solvent concentration $\leq$10%. The reaction mixture was incubated at room temperature for 30 min. The unbound PAC was removed by centrifugation (10,000$\times$ g for 10 min) and the exosomal PAC (ExoPAC) was collected by ultracentrifugation (135,000$\times$ g for 90 min). The resulting pellet was suspended in PBS and filter-sterilized. The ExoPAC solution ($\leq$6 mg/mL) was stored at -80 °C.

2.6. Determination of PAC Loading

PAC loading was determined by analyzing the PAC and Exo concentrations using ultra-performance liquid chromatography (UPLC) and BCA protein estimation kit, respectively, as described [19]. Briefly, 50 μL of the ExoPAC formulation was added to 950 μL of acetonitrile to extract the PAC and precipitate the Exo protein. The reaction mixture was then centrifuged (10,000$\times$ g for 10 min) to separate the pellet. Supernatant was collected separately to analyze PAC. Protein pellet was suspended in PBS and its concentration was determined by BCA.

2.7. UPLC Analysis

UPLC Shim-Pack XR-ODS II reverse-phase column (Shimadzu; 150 $\times$ 3.0 mm i.d., 2.2 μm) was used for the analysis of PAC. Acetonitrile and water were used as a mobile phase with 0.75 mL/min flow rate. In a linear gradient elution, the concentration of acetonitrile was increased from 5 to 60% (from 1.3 to 5.1 min), to 80% (from 5.1 to 7.7. min) and 100% at 10 min and maintained till 10.9 min; the concentration was then reduced to 5% at 12 min. PAC was detected by using PDA-UV detector at 227 nm and concentration was calculated against the standard curve of PAC.

2.8. Mechanism of Drug Loading

Proteins in exosomes show intrinsic fluorescence due to the presence of aromatic residues of tryptophan, tyrosine and phenylalanine. This property was utilized to determine the fluorescent quenching of exosomes due to hydrophobic interaction with different concentration of PAC, as reported for human serum albumin [25]. Briefly, exosomes alone (6 mg/mL) and PAC-loaded exosomes in PBS were analyzed for fluorescent signals at excitation and emission wavelengths of 280 nm and 320 nm, respectively, using a SpectraMax Spectrofluorometer. The reduction in the fluorescent signals in the presence of PAC was calculated and suggestive that the strong hydrophobic interactions play a crucial role in drug loading onto exosomes.

2.9. Cell Lines and Maintenance

Human lung cancer cell lines A549 were obtained from American Type Culture Collection (Manasa, VA, USA) and taxol-resistant A549TR cells were provided by Dr. Bruce Zetter of Children's Hospital Boston, Harvard Medical School (Boston, MA, USA). Bioware® Brite Cell Line A549 Red-FLuc was procured from PerkinElmer, USA. Cells were cultured in RPMI (Gibco, Waltham, MA, USA) supplemented with 10% FBS and antibiotics (penicillin/streptomycin) at 37 °C in 5% CO_2. No antibiotic solution was supplied to the culture media.

2.10. In Vitro Antiproliferative Activity

The effect of PAC and its exosomal formulation on cell viability was measured using the MTT assay. Briefly, A549 and A549TR cells were plated in 96-well plates at an initial density of 3×10^3 cells per well and treated with Exo, PAC or ExoPAC and incubated for 72 h. The cell survival was determined by MTT assay, as described [26]. Briefly, A549-LUC cells (3×10^3 cells/well) were plated in 96-well white plates. Cells were treated with PAC and ExoPAC at different concentrations for 72 h. Culture media were replaced with fresh media containing luciferin (150 μg/mL). The luminescence intensity was measured using a SpectraMax spectrophotometer.

2.11. Colony-Forming Assay

Taxol-sensitive (A549) and taxol-resistant (A549TR) cells were seeded into 6-well tissue culture plates at a density of 500 cells/well, as described [26]. The cells were treated with PAC or ExoPAC at different concentrations for 24 h. The drug-containing medium was discarded and replaced with a fresh drug-free medium. After 10 d, the plates were washed with sterile PBS, and the cells were fixed using methanol/acetic acid solution (3:1) for 5 min and stained with 0.5% crystal violet (in methanol) for 15 min. The crystal violet solution was carefully removed, the cells were rinsed with water and air dried at room temperature. The number of colonies in each well was counted manually.

2.12. Animal Studies

All animals were maintained according to the Institutional Animal Care and Use Committee guidelines (IACUC).

2.12.1. Lung Cancer Subcutaneous Xenograft

Female athymic nude (nu/nu) mice (5–6 weeks old) were procured from Harlan (Indianapolis, IN, USA) and used to assess the antitumor efficacy. Lung tumor xenografts were produced by subcutaneously injecting human lung A549 cells (2.5×10^6), in serum-free media mixed with Matrigel matrix (Becton Dickinson, Bedford, MA, USA), in the left flank of the mice. Animals were provided purified AIN93M diet and water ad libitum. Once the average tumor size reached about 100 mm^3, mice were randomized into four groups ($n = 10$) and provided with oral doses of PBS, PAC, ExoPAC and FA-ExoPAC, three times a week. The PAC doses in all the regimens were kept equal (6 mg/kg). Tumor size, animal weights, diet intake, and overall animal health were monitored weekly. After 7 weeks of treatment, the animals were euthanized and select tissues were collected for further analysis.

2.12.2. Lung Cancer Orthotopic Xenograft
Pilot Study

For the orthotopic lung tumor model, we first performed a pilot study to establish the effect of doses and time on tumor growth, before initiating the tumor inhibition study. After acclimation, female NOD/SCID mice (4–5-week old) were randomized into three groups ($n = 4$) and inoculated with Bioware® Brite A549-Red-Fluc cells (1×10^6, 2×10^6, and 4×10^6 cells) in 50 μL of Matrigel mixed in serum-free media ($1 \cdot 1$; v/v) via intrathoracic injection using 30-gauge needles [27,28]; an untreated group served as control. Luciferase expressions were monitored for tumor growth twice a week. The luciferase signals were detected 15 min post-intraperitoneal injection of luciferin (120 mg/kg) by using Advanced Molecular Imager, AMI1000.

Tumor Inhibition Study (Low Dose)

For the tumor inhibition study, groups of female NOD/SCID mice were inoculated with A549-Red-Fluc cells (2×10^6 cells) via intrathoracic injection, as described above. After 10 days, when the luminescence intensity reached approximately 6×10^6 photons, animals were randomized ($n = 10$) and treated with i.v. PAC, i.v. Abraxane, p.o. ExoPAC, i.v. FA-ExoPAC or p.o. FA-ExoPAC. The i.v. doses of all regimens having PAC (6 mg/kg) were given once a week, whereas oral doses were given three times a week. The PAC and Abraxane were given i.v. to mimic the clinical scenario. Two additional groups were treated with Exo and FA-Exo. The exosome concentration in all the formulations was 50 mg/kg.

Tumor Inhibition—Higher Dose

This study was patterned after the low dose study and the animals were randomized ($n = 10$) and treated with PAC, Abraxane, ExoPAC and FA-ExoPAC given orally or intravenously, as described for the low dose study, except PAC was given initially at 4 mg/kg for three weeks, then switched to 8 mg/kg in all the regimens; the frequency of dosing and the exosome concentration was same as in the low-dose study. Bodyweight gains, diet intake, and overall animal physical health were monitored weekly. At euthanasia, various tissues were collected and imaged ex vivo. Lung, liver, and tumor tissues were collected and stored at $-80\,^\circ$C for marker analysis.

2.12.3. Toxicity Study

Female C57BL/6 mice (5–6 weeks old) were purchased from Charles River Laboratories. Animals were randomized in six groups ($n = 5$) and treated with vehicle, Exo (60 mg/kg/week; oral), FA-Exo (60 mg/kg/week; oral), PAC (6 mg/kg/week; i.p.), ExoPAC (9 mg/kg/week; p.o.) and FA-ExoPAC (9 and 18 mg/kg/week; p.o.). The exosome concentration was kept constant to 60 mg/week in all the exosomal formulation-treated groups. The drug was given three times a week in all the treatment groups and continued for four weeks. Bodyweight, physical mobility and food intake were monitored twice a week throughout the study. After four weeks of treatments, animals were euthanized by CO_2 asphyxiation. At the time of euthanasia, blood and the major organs were weighed and collected for further analysis. The spleen and femur bone were collected in fresh media to harvest the spleen and bone marrow cells.

Systemic Toxicity

Blood was collected at the time of euthanasia and hematological parameters were analyzed using whole blood by the CellDyn 3500 hematology analyzer (Abbott laboratories, Santa Clara, CA, USA). Serum was used to analyze the liver and kidney function enzymes, as described [23]. Electrolyte analysis was done by using an ion-selective electrode while other biochemical parameters were analyzed spectrophotometrically using AU640 Chemistry Immuno Analyzer (Beckman Coulter, Inc., Brea, CA, USA). Spleen and bone marrow were used for immune toxicity studies described below.

Immune Cell Analysis

Immune cell quantification was performed by staining single-cell suspensions of spleen cells with fluorescent dye-coupled antibodies to CD19 for B cells, CD5 (total T-cells), CD4, and CD8 for T-cell subsets, F4/80, CD11b, Gr-1, to identify macrophages and neutrophils, CD11c for dendritic cells, NK1.1 and CD49b for natural killer cells.

Bone marrow stem and progenitor cells were identified by negative staining for lineage-specific markers using biotin-labelled antibodies to B220, CD11b, Gr-1, CD5, CD8, Ter-119 and APC-Cy7 coupled streptavidin and positive staining for Sca-1 and c-kit. The cells stained with fluoresceinated antibodies were analyzed using an LSR II flow cytometer (BD BioSciences) and the data were analyzed using FlowJo software. All statistical analyses were performed by two-way ANOVA using group A as the control group.

For cytokine assays, spleen cells were cultured with lipopolysaccharide (LPS) or anti-CD3 antibody for 24 h in Iscove's DMEM (IMDM) in the presence of 10% fetal calf serum. The culture supernatants were analyzed for IL-6, IL-10, IL-2 and γ-interferon using specific reagents obtained from R&D Biosystems (Minneapolis, MN, USA) and multiples reagent from MesoScale using respective recombinant cytokines as standards. Data are presented as percent control where control is the average value for spleen cells from PBS-treated mice.

2.13. Statistical Analysis

Statistical analysis was performed with GraphPad Prism statistical software (version 4.03; La Jolla, CA, USA) using two-way ANOVA followed by a Bonferroni post-test for xenograft studies. Data in the xenograft studies are expressed as mean $\pm$ standard error of mean (SEM) (n = 10). Statistical significance of differences in immune cell numbers, cytokine assays and proliferation responses between various treatments was evaluated by an unpaired Student's t-test. Values of $p < 0.05$ were considered statistically significant.

3. Results

3.1. Exosome Isolation and Characterization

Colostrum derived exosomes are lipid bilayer nanovesicles and their diameters vary from 30–150 nm. Exosome suspension was homogenous with an average particle size of 59 $\pm$ 1.1 nm, PDI of 0.3 $\pm$ 0.1, and zeta potential of -30.2 ± 0.1 mV, as determined by Zetasizer and NanoView (Figure 1A); these analyses were performed after removing PBS from the exosomes by ultrafiltration (300,000 MWCO spin filter) since the presence of PBS increased zeta potential of the particles. Zeta view analysis of exosomes showed 0.5–1.0 $\times 10^{14}$ particles per mg of exosomal proteins. The size was confirmed with AFM (Figure 1B). Exosomes isolated from colostrum showed hallmark protein markers such as CD81, Tsg101, Alix and the anti-phagocytic protein, CD47, as described elsewhere [24].

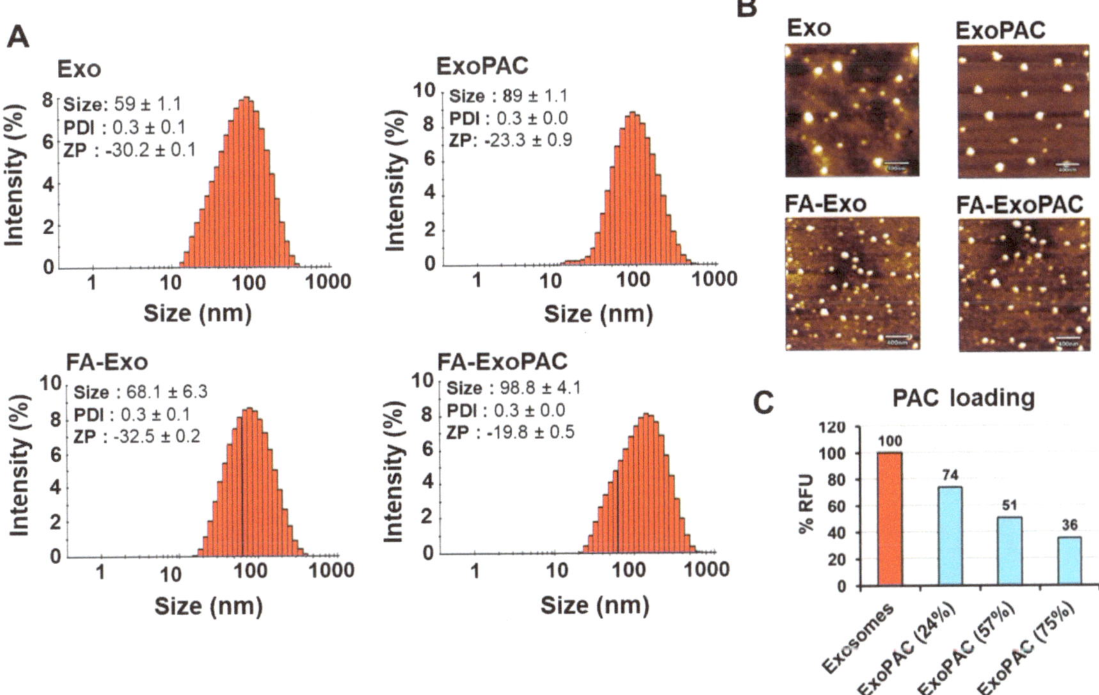

Figure 1. Characterization and drug loading of colostrum-derived exosomes. Size, polydispersity index (PDI), and zeta potential (ZP) of exosomes, FA-Exo, ExoPAC and FA-ExoPAC, analyzed by Zetasizer. Data represent mean $\pm$ SD from three preparations (**A**). Analysis of exosomes and ExoPAC by atomic force microscopy (AFM) after diluting with deionized water up to 10 μg/mL. For measurement, samples were placed on a silica wafer and air-dried for 30 min. AFM in tapping mode and aluminum-coated silicon probes were used for imaging (**B**). The bar diagram shows the quenching of autofluorescence from the exosomes following PAC loading (**C**). Higher quenching of fluorescence in the presence of higher drug load suggests a hydrophobic interaction of drug with exosomal proteins.

3.2. Drug Loading and FA Functionalization

Exosomes were functionalized by covalently attaching FA first, followed by loading of PAC. The PAC was loaded with simple mixing of drug solution with FA-functionalized exosome. The size of exosomes was only slightly increased (68 ± 6.3 nm from 59 ± 1.1) after FA conjugation. However, PAC loading increased the size modestly for both exosomes (89 ± 1.1 from 59 ± 1.1) and FA-exosomes (98.8 ± 4.1 from 68.1 ± 6.3). The zeta potential of FA-ExoPAC (−19.8 ± 0.5) was increased compared with exosomes (−30.2 ± 0.1) and ExoPAC (−23.3 ± 0.9) (Figure 1A,B).

3.3. Mechanistic Understanding of Drug Loading in Exosomes

We utilized quenching of intrinsic fluorescence of surface-bound exosomal proteins to determine if the PAC was surface-bound. We observed a dose-dependent decrease in fluorescence with an increase in PAC loading to exosomes. The percent fluorescence quenching was correlated with the PAC load—a drug load of 24%, 57% and 75% resulted in 26%, 49% and 64% fluorescence quenching, respectively (Figure 1C). These data clearly suggest that at least part of the drug is sequestered in the hydrophobic domains of surface-bound exosomal proteins; however, we cannot rule out that part of the drug is in the lipid bilayer and/or lumen of the exosomes.

3.4. ExoPAC Inhibits Growth of Both Drug-Sensitive and Drug-Resistant Lung Cancer Cells

The antiproliferative effects of PAC, ExoPAC and FA-ExoPAC were determined against drug-sensitive and drug-resistant human lung cancer cells and compared with albumin-bound PAC (Abraxane). PAC and its exosomal formulations showed a dose-dependent cell growth inhibition against A549 cells. FA-ExoPAC, however, showed a twofold reduction in the IC50 values compared to PAC; the IC50 of Abraxane was similar to PAC (Figure 2A). Exosome alone (Figure 2A) and FA-Exo (data not shown) demonstrated about 20% inhibition of A549 cells. To determine if the exosomal formulation could chemosensitize the drug-resistant cells, we tested all the formulations against taxol-resistant A549TR lung cancer cells. PAC did not show any inhibition of the resistant cells up to 200 nM. However, the data indicated that FA-ExoPAC was able to inhibit the growth of the drug-resistant cells dose-dependently with the IC50 values of 12.5 nM. ExoPAC and Abraxane showed a similar effect. (Figure 2B).

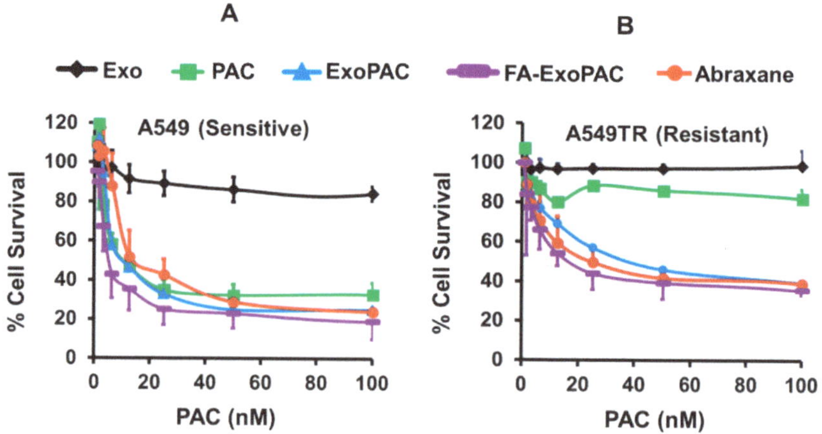

Figure 2. ExoPAC inhibits proliferation of drug-sensitive and drug-resistant cells. (**A**) Drug-sensitive (A549) and drug-resistant (A549TR) cells were treated with Exo, ExoPAC and FA-ExoPAC and compared with Abraxane. Antiproliferative activity was determined by MTT assay after 72 h. Exosomal PAC dose-dependently inhibited the proliferation of drug-sensitive A549 (**A**) and drug-resistant A549TR cells (**B**).

3.5. Colony Formation Assay

To validate the observed antiproliferative effects, we investigated the potential effects of PAC and ExoPAC on the replicative ability of drug-sensitive (A549) and its drug-resistant variant (A549TR) using colony formation assay (Figure 3A,B). Similar to the MTT data, while PAC showed 65% inhibition of colony formation at 6.25 nM, ExoPAC had over 90% inhibition at the same dose. Interestingly, the effect of PAC (25–100 nM) was minimal on the resistant cells, while ExoPAC had significant inhibition starting from 25 nM. As expected, ExoPAC showed dose-dependent inhibition of colony formation against both A549 (Figure 3A; Supplementary Figure S1) and A549TR (Figure 3B; Supplementary Figure S2) cells; ExoPAC inhibited colony formation in both cell lines greater than PAC alone.

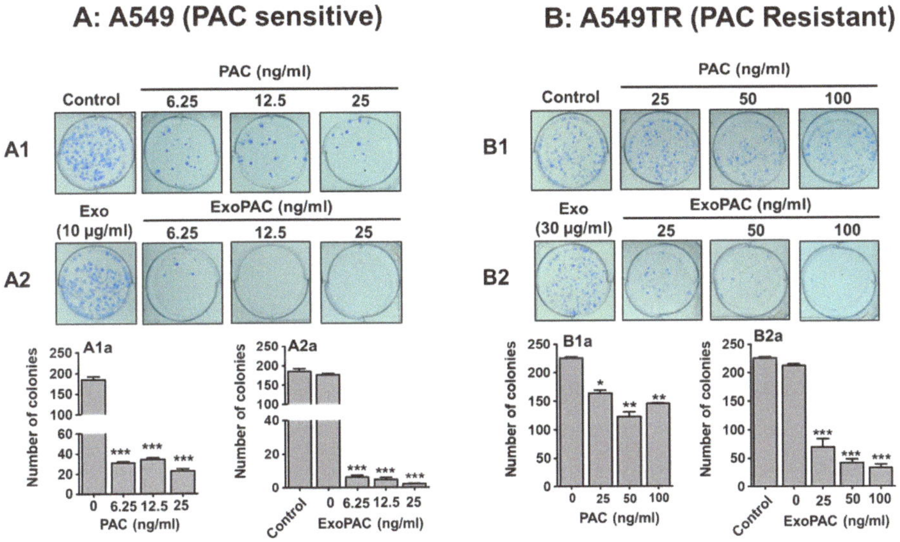

Figure 3. Exosomal PAC inhibited colony formation in NSCLC cells. Representative images showing the colony formation assay in drug-sensitive A549 (**A**) and drug-resistant A549TR (**B**) cells. Lung cancer cells were seeded (500 cells/well) in a six-well plate and incubated with different concentrations of PAC and ExoPAC. After 10 days, developed colonies were fixed, stained and counted manually. While PAC was effective, only against drug-sensitive cells (**A1,B1**), ExoPAC shows dose-dependent inhibition of colony formation of both sensitive and resistant cells (**A2,B2**). Statistical analysis was performed using the Student's *t*-test. *, $p < 0.05$; **, $p < 0.01$; ***, $p < 0.001$.

3.6. Antitumor Efficacy Following Oral Administration of ExoPAC

3.6.1. Subcutaneous Lung Tumor Xenografts

We determined the antitumor efficacy of ExoPAC and FA-ExoPAC using athymic nude mice bearing subcutaneous A549 xenografts and compared them with PAC. There was no difference in the body weight or diet consumption, suggesting no gross toxicity due to PAC or ExoPAC. Compared to untreated control, PAC (6 mg/kg) showed about 30% but statistically insignificant inhibition of the tumor growth. However, ExoPAC (6 mg/kg PAC and 50 mg exosomal proteins/kg) showed a significant (45%; $p < 0.05$) growth inhibition at the end of the study (Figure 4). FA-ExoPAC at the same dose was even more effective (54%; $p < 0.05$) with the growth inhibition occurring as early as five weeks after the treatment (Figure 4).

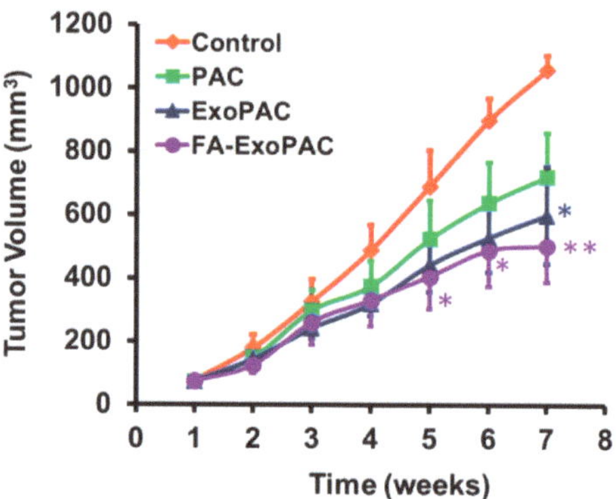

Figure 4. Antitumor activity against subcutaneous xenografts. Following inoculation with A549 cells, nude mice were treated with oral gavage three times a week with PAC (6 mg/kg bw), ExoPAC and FA-ExoPAC (6 mg PAC and 50 mg Exo protein/kg bw). Data represent average ± SD of means (*n* = 8). Statistical analysis was performed using the Student's *t*-test. * $p < 0.05$; ** $p < 0.01$.

3.6.2. Orthotopic Lung Tumor Xenografts

We first established an orthotopic lung tumor model in a pilot study using Bioware® Brite A549 Red-FLuc lung cancer cells. Live animal imaging showed the bioluminescence signals from lung tumors (Figure 5A); tumor growth was dose- and time-dependent (Figure 5A). The tumors could be detected as early as 10–12 days after tumor cell inoculation, with nearly exponential growth. After four weeks, animals inoculated with different cell numbers showed a dose-dependent increase in bioluminescence signals. Based on the tumor growth and expected tumor size, we used 2×10^6 cells in efficacy studies.

In an efficacy study, two doses of PAC (6 mg/kg and 8 mg/kg) were tested. Figure 5B shows the data from the low-dose study; i.v. PAC showed 32% inhibition of tumor growth whereas *p.o.* FA-ExoPAC showed a somewhat higher growth inhibition (39%), although the difference was not statistically significant; the nonfunctionalized formulation (ExoPAC) showed only a slight inhibition (18%). However, FA-ExoPAC administered i.v. resulted in significantly higher growth inhibition (70%; $p < 0.001$) matching the efficacy of i.v. Abraxane (62%; $p < 0.001$).

The data presented in Figure 5(C1,C2) demonstrates significant tumor growth inhibition in the following order: i.v. FA-ExoPAC (76%; $p < 0.001$) > p.o. FA-ExoPAC (55%, $p < 0.001$) ≈ i.v. Abraxane (59%, $p < 0.001$) > p.o. ExoPAC (36%, $p < 0.05$) > i.v. PAC (24%) > p.o. FA-Exo (9%).

Clearly, oral FA-ExoPAC far exceeded the efficacy of i.v. PAC and, in fact, matched the efficacy elicited by i.v. Abraxane; i.v. FA-ExoPAC exceeded the efficacy of i.v. Abraxane. Dose-optimization studies are warranted to identify the most efficacious oral doses and frequency of FA-ExoPAC formulation. At the completion of the study, we observed that animals treated with i.v. FA-ExoPAC (6 and 8 mg/kg) did not exhibit any mortality, while about 42% of the animals died in solvent-based i.v. PAC and 30% in control groups. Importantly, ExoPAC-treated animals also showed a significantly improved overall health index compared to PAC or untreated controls (Supplementary Figure S3).

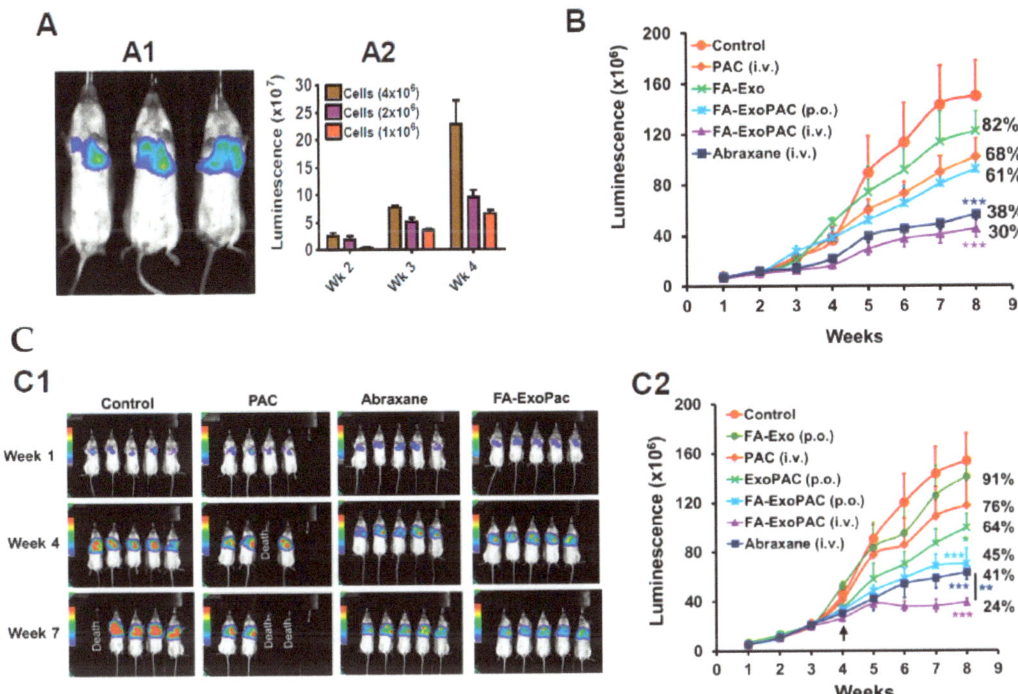

Figure 5. Antitumor activity against orthotopic xenografts. Detection of orthotopic lung cancer using bioluminescent A549-Red-luc cells. (**A**) A1: image of live animals after 27 d of inoculation with 2×10^6 cells (3 mice/group). A2: Mean of bioluminescence signals. (**B**) Inhibition of A549 orthotopic lung tumors in NOD Scid female mice (*n* = 10) by i.v. paclitaxel (PAC), i.v. Abraxane, and *p.o.* ExoPAC and *p.o.* FA-ExoPAC (three doses weekly), all given at 6 mg/kg. FA-Exo was used as control for FA-ExoPAC. (**C**) Inhibition of A549 orthotopic lung tumors in NOD Scid female mice (*n* = 10) by i.v. PAC, Abraxane, FA-ExoPAC (once weekly) and *p.o.* ExoPAC and FA-ExoPAC (three doses weekly), all given at 4 mg/kg until three wks, then switched to 8 mg/kg, as indicated by an arrow in C2. Representative images of animals at different time points in the indicated treatment groups (C1) and time-dependent tumor inhibition (C2). Statistical analysis was performed using the Student's *t*-test. *, *p* < 0.05; **, *p* < 0.01; ***, *p* < 0.001.

3.7. Assessment of Toxicity Due to PAC and ExoPAC

3.7.1. Systemic Toxicity

For the analysis of the potential toxicity of PAC, exosomes and ExoPAC, wild-type C57BL/6 mice were treated for 28 days and assessed for gross and systemic toxicity. We observed no difference in the body weight, diet intake and physical wellness of treated versus control animals. We analyzed the levels of liver enzymes (aspartate transaminase, alanine aminotransferase, alkaline phosphatase, gamma-glutamyl transpeptidase, amylase, and lipase) in the serum of animals treated either with PAC, exosomes or ExoPAC. PAC significantly changed the levels of amylase and total bilirubin. These effects were not evident with ExoPAC suggestive of hepatoprotective role when PAC was embedded in exosomes. Similarly, toxicity caused by PAC in kidney function tests and hemopoietic parameters was mitigated by its formulations in exosome (Tables 1 and 2).

Table 1. Effect on biochemical profile (systemic toxicity) following 28 days exposure to Exo, PAC and ExoPAC in C57BL/6 mice.

Parameter	Control	Exo	PAC	ExoPAC	FA-ExoPAC
		Liver Profile			
AST (SGOT)	400 ± 220	343 ± 102	381 ± 80	314 ± 73	423 ± 83
ALT (SGPT)	39.7 ± 20.0	56.0 ± 19.5	51.8 ± 20.2	42.0 ± 4.9	53.2 ± 17.6
Alk Phosphatase	62.8 ± 62.2	89.2 ± 17.4	10.5 ± 4.9	14.0 ± 9.9	63.0 ± 49.5
GGT	1.0 ± 0.0	1.0 ± 0.0	1.0 ± 0.0	1.0 ± 0.0	1.0 ± 0.0
Amylase	378 ± 164	523 ± 17	987 ± 606 *	525 ± 28	476 ± 80
CPK	1044 ± 499	1488 ± 596	1231 ± 541	1030 ± 393	1313 ± 278
Total Bilirubin	0.6 ± 0.4	0.1 ± 0.0 *	0.1 ± 0.0 *	0.5 ± 0.3 [#]	0.6 ± 0.3 [##]
		Kidney Function Test			
BUN	16.6 ± 4.1	19.6 ± 0.9	12.6 ± 3.1	16.8 ± 3.8	15.4 ± 3.1
Creatinine	0.2 ± 0.0	0.2 ± 0.0	0.2 ± 0.0	0.2 ± 0.0	0.2 ± 0.0
BUN/CreatRatio	83.1 ± 20.5	98.0 ± 4.5	63.0 ± 15.7	84.0 ± 19.2	77.0 ± 15.7
Phosphorus	13.0 ± 4.8	18.5 ± 4.6	10.4 ± 0.6	10.1 ± 0.8	10.1 ± 0.8
Calcium	7.5 ± 1.7	11.0 ± 0.8 ***	8.7 ± 0.6	8.4 ± 0.5	8.1 ± 0.4
Magnesium	5.0 ± 2.0	4.2 ± 0.3	3.8 ± 0.8	3.9 ± 0.4	4.2 ± 0.5
Sodium	120.5 ± 23.5	150.2 ± 5.7 *	137.2 ± 6.3	135.8 ± 6.3	131.6 ± 5.9
Potassium	17.5 ± 15.5	9.2 ± 0.8	9.0 ± 2.2	10.4 ± 1.3	12.5 ± 1.9 [#]
NA/K Ratio	10.6 ± 5.6	16.4 ± 1.8 *	16.0 ± 2.6	13.6 ± 1.9	10.8 ± 2.5 [#]
Chloride	115.4 ± 13.9	111.8 ± 8.0	121.8 ± 3.8	123.2 ± 3.8	117.6 ± 5.9
Total Protein	7.9 ± 3.7	6.2 ± 0.5	5.6 ± 1.3	6.4 ± 0.6	6.7 ± 0.6
Albumin	4.9 ± 2.1	3.7 ± 0.3	2.9 ± 0.8	3.8 ± 0.6	4.2 ± 0.5 [#]
Globulin	2.8 ± 1.0	2.4 ± 0.3	2.7 ± 0.6	2.7 ± 0.3	2.5 ± 0.4
A/G Ratio	1.8 ± 0.4	1.6 ± 0.2	1.1 ± 0.1 **	1.5 ± 0.4	1.7 ± 0.3 [##]
Cholesterol	343 ± 317	142 ± 20	148 ± 44	157 ± 31	204 ± 47
Triglyceride	60.6 ± 29.8	84.6 ± 14.3	93.8 ± 16.9 *	84.0 ± 9.9	64.4 ± 10.4 [#]
Glucose	147.1 ± 40.4	89.8 ± 66.8	109.2 ± 17.6	154.0 ± 7.0 [#]	159.6 ± 26.8 [##]

Female C57BL/6 mice (5–6 weeks old) were provided control diet (AIN 93M) and water ad libitum and treated with colostrum-derived exosomes (60 mg/kg, b. wt.) by oral gavage, i.p. PAC (8 mg/kg) and ExoPAC and FA-ExoPAC (8 mg/kg PAC and 60 mg/kg exosome) for 28 days, three times a week. At euthanasia, blood was collected and analyzed using an automated AU640 Chemistry Analyzer by Antech diagnostics. Data represent average $\pm$ SD of four animals. Statistical analysis was performed by the Student *t*-test. Asterisks represent comparison to control while [#] represents a comparison to PAC group. *, *p*-value <0.05; **, $p < 0.01$; ***, $p < 0.001$; [#], *p*-value <0.05; [##], < 0.01.

Table 2. Effect on hematological parameters (systemic toxicity) following 28-day exposure to exosomes, PAC and ExoPAC in C57BL/6 mice.

Parameter	Control	Exo	PAC	ExoPAC	FA-ExoPAC
WBC	7.4 ± 1.6	4.3 ± 1.9 **	7.7 ± 0.9	5.3 ± 1.7 *,[#]	4.6 ± 2.0 *,[#]
RBC	9.2 ± 0.4	9.0 ± 0.4	7.8 ± 0.5 ***	8.0 ± 2.3	8.5 ± 0.9
HGB	14.5 ± 0.6	14.7 ± 0.5	12.8 ± 0.4 ***	12.4 ± 4.3	13.3 ± 1.7
HCT	45.7 ± 1.4	46.5 ± 2.1	37.2 ± 2.6 ***	38.6 ± 11.9	40.4 ± 4.2 **
MCV	49.3 ± 1.7	51.3 ± 0.5	47.4 ± 1.1	48.0 ± 1.9	47.2 ± 0.8 *
MCH	15.8 ± 0.9	16.3 ± 0.2	16.5 ± 0.7	15.1 ± 1.7	15.6 ± 0.9
MCHC	32.0 ± 2.1	31.8 ± 0.5	34.6 ± 1.9	31.6 ± 2.8	33.0 ± 1.9
Platelet Count	856 ± 111	829 ± 145	891 ± 131	674 ± 260	570 ± 202 **,[#]
Neutrophils	9.7 ± 3.8	13.5 ± 6.1	30 ± 11.6 ***	8.8 ± 2.8 [##]	10.8 ± 2.4 [##]
Bands	0.0 ± 0.0	0.0 ± 0.0	0.0 ± 0.0	0.0 ± 0.0	0.0 ± 0.0
Lymphocytes	87.1 ± 3.1	82.3 ± 4.2	66 ± 12.6 ***	87.6 ± 3.0 [##]	84.8 ± 3.8 [##]
Monocytes	1.1 ± 1.3	3.0 ± 2.3	1.0 ± 0.0	1.0 ± 0.0	1.0 ± 0.0
Eosinophils	2.0 ± 1.2	1.3 ± 0.5	3.0 ± 1.2	2.6 ± 0.9	3.4 ± 1.7
Basophils	0.0 ± 0.0	0.0 ± 0.0	0.0 ± 0.0	0.0 ± 0.0	0.0 ± 0.0
Absolute Neutrophils	688.1 ± 208.5	561.3 ± 314.7	2345 ± 1026 **	453.2 ± 170.1 [##]	479.6 ± 221.1 [##]
Absolute Lymphocytes	6519 ± 1563	3546 ± 1686	5008 ± 1082	4664 ± 1573	3942 ± 1731 *
Absolute Monocytes	44.0 ± 17.0	115.8 ± 79.4	76.6 ± 8.7	53.0 ± 16.8 [#]	46.0 ± 19.5 [#]
Absolute Eosinophils	152.0 ± 101.4	52.3 ± 24.8 *	231 ± 105	130.0 ± 37	132.0 ± 22

Female wild-type C57BL/6 mice (5–6 weeks old) were provided control diet (AIN 93M) and water ad libitum and treated with colostrum-derived exosomes (60 mg/kg, b. wt.) by oral gavage, i.p. PAC (8 mg/kg) and ExoPAC and FA-ExoPAC (8 mg/Kg PAC and 60 mg/kg exosome) for 28 days, three times a week. At euthanasia, blood was collected and analyzed using an automated AU640R Chemistry Analyzer by Antech diagnostics. Data represent average $\pm$ SD of four animals. Statistical analysis was performed by the Student *t*-test. *, $p < 0.05$; **, $p < 0.01$; ***, $p < 0.001$ in comparison to control group. [#], $p < 0.05$ and [##], $p < 0.01$ in comparison to PAC group.

3.7.2. Immunotoxicity

Single cell suspensions of splenic and bone marrow cells were prepared and viable cells were quantified by trypan blue exclusion. No significant differences were found in the total cell numbers of splenic as well as bone marrow cells (Figure 6A,B) upon any of the treatments. Next, splenic cells were stained with multiple fluorochrome-conjugated antibodies specific to B cells (CD19), total T-cells (CD5), T-cell subsets (CD4, CD8) and were analyzed by flow cytometry. Overall percentages of the different lymphocyte subsets were not significantly different between control and ExoPAC or FA-ExoPAC treatment groups. Interestingly there were small reductions in total CD5+ and CD8+ T-cells in the spleen after treatment with PAC alone but these reductions were abrogated when PAC was provided as ExoPAC or FA-ExoPAC. On the other hand (Figure 7B). PAC caused a significant increase ($p < 0.001$) of macrophages (F4/80), neutrophils (CD11b+Gr-1+) and dendritic cells (CD11c+) compared to control, which were mitigated by the use of exosomal formulations. Natural killer cells (NK1.1+) were unaffected irrespective of treatment (Figure 7C). However, in the analysis of bone marrow cells, there was no difference in neutrophils (CD11b+Gr-1+) and B cells (B220) by any treatment (Figure 7C).

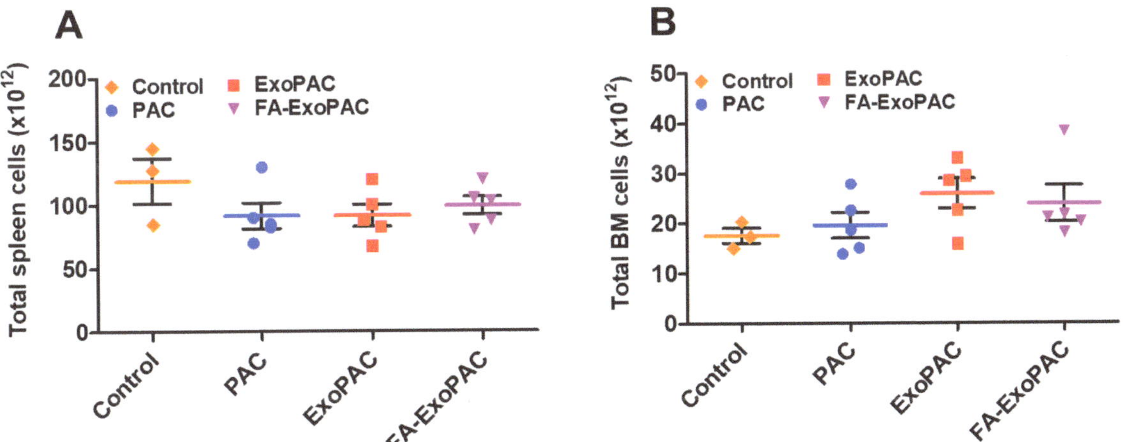

Figure 6. Potential immunotoxicity of PAC and FA-ExoPAC. Female C57BL/6 mice were treated with Exo, PAC and FA-ExoPAC for four weeks. At euthanasia, spleen and bone marrow cells were collected. Live splenic (**A**) and bone marrow (**B**) cell counts were performed by trypan blue exclusion.

Stem and progenitor (LSK or LIN-Sca-1+cKit+) cells were highly increased by treatment with PAC ($p < 0.001$), whereas this increase was not found with ExoPAC formulations suggesting that exosomal formulation could mitigate the toxicity associated with PAC (Figure 7D). Splenic cells were induced to proliferate by treating with different stimulants for 72 h, which is an important requirement for effective immune response to tumor cells or pathogens. Cells were pulsed with 3H-thymidine for 4 h, then, cells were harvested and the incorporation of radioactivity was quantified using a beta-plate counter. PAC treatment showed a lower T-cell proliferation and a lower T-cell independent B-cell proliferation induced by LPS ($p < 0.01$). However, exosomal formulation mitigated these adverse effects. There was no significant difference in the αCD40 treatment, which represents a T-cell-dependent B-cell proliferation response (Figure 7E).

In order to assess cytokine response, splenic cells were treated with different stimulants for 24 h. Sups were collected and MesoScale analysis (V-PLEX of 6 cytokines) was performed. LPS was used for B-cell and macrophage response and αCD3+αCD28 was used for T-cell response (Figure 8). We observed increase in almost all cytokines (except IL-2, which was decreased) in response to PAC treatment, which was mitigated with the exosome formulations suggesting protection of PAC-induced immunotoxicity by exosomal formulations (Figure 8).

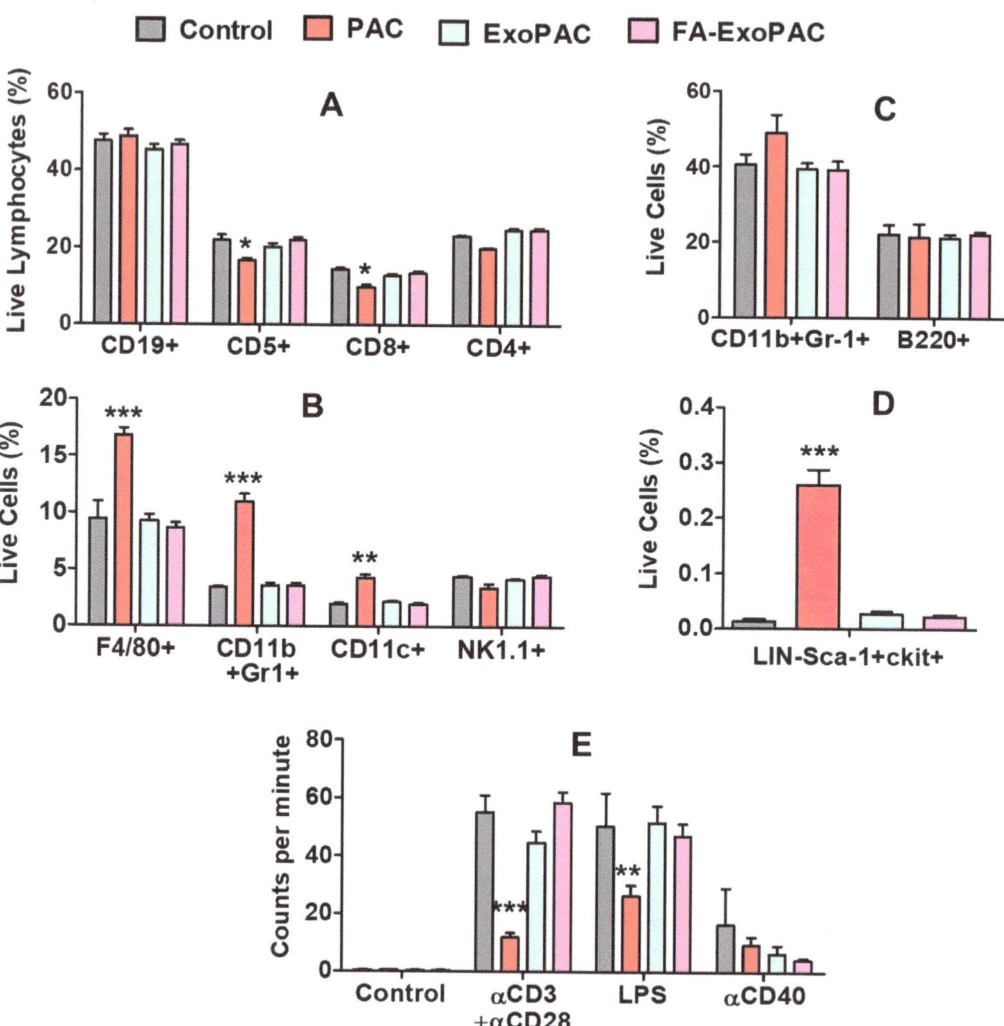

Figure 7. Potential immunotoxicity of PAC and FA-ExoPAC. Female C57BL/6 mice were treated with Exo, PAC and FA-ExoPAC for four weeks. At euthanasia, spleen and bone marrow cells were collected. Splenic cells were stained with multiple fluorochrome-conjugated antibodies. Samples were run using LSRII cytometer and data was analyzed by FlowJo software. Effect of treatment was analyzed on B cells, total T-cells, T helper cells, and cytotoxic T-cells (**A**), macrophages, neutrophils and dendritic cells (**B**), B cell (**C**), %LSK cells (**D**) and T-cell proliferation (**E**). All statistical analysis was performed by two-way ANOVA and compared with untreated control. *, $p < 0.05$; **, $p < 0.01$; ***, $p < 0.001$.

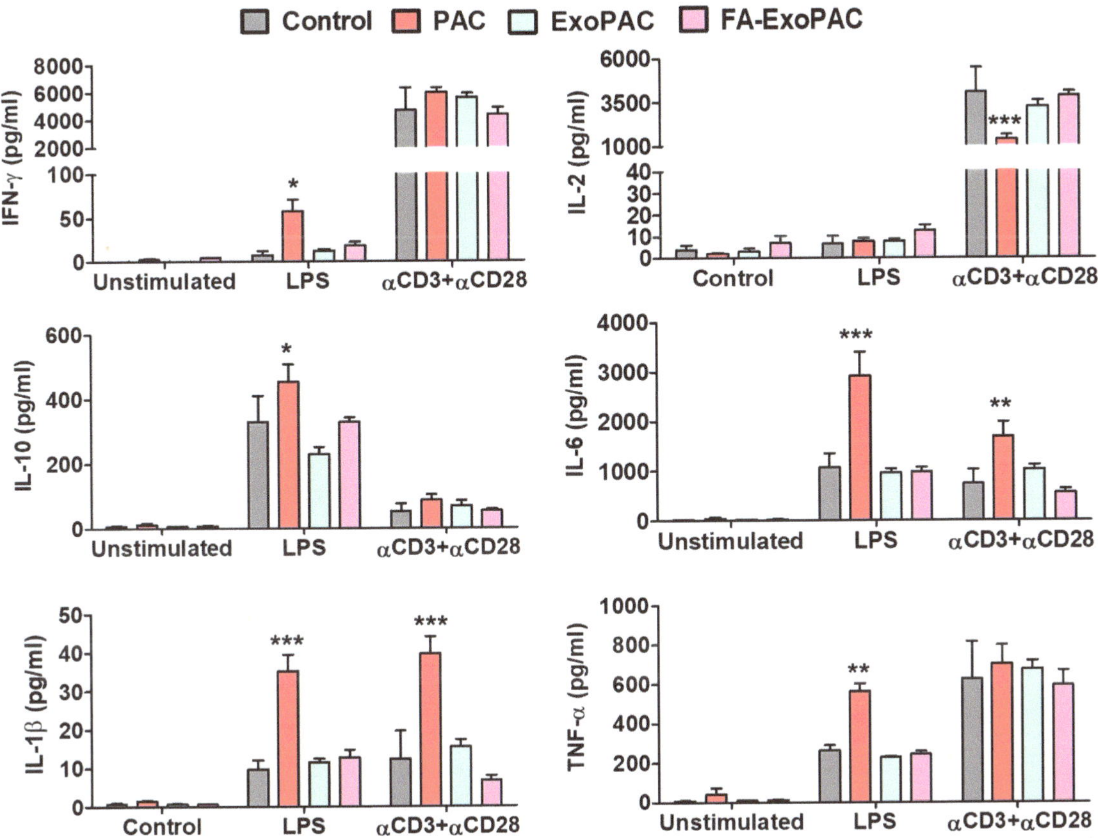

Figure 8. Potential immunotoxicity of PAC and FA-ExoPAC. Female C57BL/6 mice were treated with Exo, PAC and FA-ExoPAC for four weeks. At euthanasia, spleen and bone marrow cells were collected. Splenic cells were treated with the different stimulant for 24 h. Sups were collected and MesoScale analysis (V-PLEX of 6 cytokines) was performed. LPS was used for B-cell response and αCD3+αCD28 for T-cell response. All statistical analysis was performed by two-way ANOVA and compared with untreated control. *, $p < 0.05$; **, $p < 0.01$; ***, $p < 0.001$.

4. Discussion

PAC is an antineoplastic chemotherapeutic drug that is routinely used as the first- or second-line chemotherapeutic in the treatment of a broad spectrum of human cancers, including lung cancer. As with many other chemo drugs, PAC exhibits poor oral bioavailability; hence, it is administered intravenously. To increase bioavailability, several drug delivery formulations of PAC have been developed, including nanoparticle albumin-bound (Abraxane®), liposomal (Lipusu®), polymeric micelles (Genexol® PM), polymeric-drug conjugates (Xyotax™/OPAXIO) and an injection concentrate for nanodispersion (Taclantis™/Bevetex®), as reviewed by Chor et al. [8]. Clinical translatability of these nanoformulations was impeded due to various factors like toxicity, scalability and cost. In addition to cremophor-based PAC, Abraxane is the only formulation of PAC approved by the FDA to date while the remaining formulations, also as i.v. therapeutics, are currently in clinical trials at various stages. In a randomized multinational phase 3 study (NCT02594371) lead by Athenex Inc., an oral formulation of PAC and Encequidar was evaluated in women with metastatic breast cancer. This combination therapeutic showed improved progression-free survival and overall survival compared to i.v. PAC in breast cancer patients [29]. Encequidar, although not systemically absorbed, is an inhibitor of multidrug resistance efflux pump P-glycoprotein that increases the oral bioavailability of PAC by preventing the efflux of PAC from intestinal epithelial cells in the GI tract. While oral PAC/encequidar carried less risk of neuropathy and alopecia compared to i.v. PAC, higher risk for GI and neutropenia adverse events was found [30]. These results exemplify the potential of oral PAC for the treatment of cancer while mitigating, in-part, toxicity of bolus i.v. dosing, however, the oral PAC formulation used in these clinical studies lacks specificity.

The primary objective of this study is to develop a tumor-targeted oral formulation of PAC (FA-ExoPAC) to improve the overall efficacy and safety profile while providing a user-friendly, cost-effective alternative to bolus i.v. PAC. Exosomes provide a nontoxic scalable and cost-effective approach to drug delivery. Exosomes are biogenic nanocarriers and have been shown by us and others to deliver both small and macromolecules to the tumor site [19,21–24]. Most of the current exosomal delivery technologies rely on harvesting exosomes from cells grown in high-density bioreactors. Mendt et al. [31] reported production of $10–15 \times 10^{12}$ exosomes per bioreactor culture. In comparison, exosomes isolated from milk or colostrum are abundant in a readily available source, bovine milk, which dominates commercial production and is estimated to be 85% of worldwide milk consumption [32]. Bovine milk contains abundant exosomes (239 ± 9.6 mg exosomal protein or 33×10^{16} particles/L) [22]. The abundance of exosomes is further increased in bovine colostrum. Clearly, bovine milk/colostrum contains several orders of magnitude higher amounts of exosomes/L than media from high-density bioreactors.

The generally recognized safety of colostrum powder combined with high exosome yield makes it a biocompatible source for cost-effective, large-scale production of exosomes. In these studies, bovine colostrum powder derived exosomes showed small uniform distribution of size, approximating 60 nm, which only slightly increased (13%) following the covalent attachment of the tumor-targeting ligand FA while a modest increase (approx. 30%) in exosomal size was observed following the loading of PAC. This modest increase in exosomal size and our observed fluorescent quenching of exosome surface-bound proteins due to hydrophobic interaction with increasing concentrations of PAC could be attributed to sequestration of PAC by the hydrophobic domains of the exosomal surface-bound proteins; however, we cannot rule out that part of the drug is in the lipid bilayer and/or lumen of the exosomes.

Cancer cells develop resistance to PAC through several mechanisms like initiating the efflux pump, DNA mutations and changes in microtubule dynamics. In our in vitro antiproliferative and colony-forming studies, we noted while PAC is effective against the drug-sensitive cells, its exosomal formulations showed significant activity even against drug-resistant cells, suggesting a role for exosomes in preventing the efflux of PAC. The present study suggests that lung cancer drug resistance towards PAC could be avoided by using exosomal formulation. We have previously shown that the growth of normal epidermal keratinocytes (HEKn) and Beas-2B epithelial cells was unaffected by milk exosomes [21,22]. In the present study we showed that ExoPAC did not disturb the immune homeostasis which was affected by PAC alone in altering some immune cell subsets. Also, ExoPAC, unlike PAC did not alter the cytokine production or growth response of various immune cells further attesting to near absence of any immunotoxicity.

Tumor-targeted drug delivery approaches have attracted extensive attention due to their ability to achieve higher drug accumulation in tumor site and reduce off-target effects. Under physiological conditions, reduced form of folates transports into the cells via reduced folate carrier (RFC) through an anion-exchange mechanism. After entering the cell, folate plays a crucial role in biosynthesis of building blocks of DNA synthesis, methylation and repair [33]. The other form of folate entry into the cell is through folate receptor (FR). There are four isoforms of FR (FRα, FRβ, FRγ and FRδ) identified in humans. In cancer cells, the expression of FRα covers the entire cell surface due to loss of its polarized cellular location. FRα, the target of FA, is present at low levels in normal tissues but it is overexpressed in majority of NSCLCs and in lung adenocarcinomas [34]. Our data show >100-fold higher expression levels in lung tumors versus normal lung. We also showed that FA-Exo-AF750 resulted in significantly higher tumor accumulation of exosomes compared with nonfunctionalized Exo-AF750 [24]. Thus, the limited expression and restricted distribution pattern of this receptor make it attractive for targeting lung tumors [35]. Once FA or RFC binds to the FR receptor, the total complex enters into the cells via the process of endocytosis. In this context, FA serves as a feasible option to direct ExoPAC to cancer cells. In this study, FA was covalently attached to exosomes to enhance the specificity of ExoPAC. Since FA is not retained in the kidneys, no significant toxicities have been observed in rodent models or humans with FRα-targeted agents [35–37].

Oral dosing of chemotherapeutics offers many advantages—including flexibility of timing, location of administration, flexibility of drug exposure, reduction of the use of the healthcare resources for in-patient and ambulatory-patient care services, and a better quality of life [11,12]. However, due to the poor GI absorption and hepatic first-pass effect, bolus doses of PAC and other chemotherapeutics are required for efficacy and likely contribute to overall toxicity.

It is evident from our previous study that ExoPAC using exosomes from bovine milk exhibits enhanced anticancer response versus free PAC against lung cancer cells and efficiently inhibits the lung cancer subcutaneous xenografts [19]. In this study, first we show that given orally, ExoPAC and FA-ExoPAC, using exosomes isolated from bovine colostrum powder, demonstrate much higher activity compared to free PAC in subcutaneous lung cancer xenografts, followed by efficacy studies using an orthotopic model to mimic relevant tumor microenvironment. Our previous studies demonstrate that exosomes maintain their integrity in gastrointestinal pH and the release of PAC was consistent at wide ranges of pH (5, 5.8 and 6.8) resembling physiological conditions of the body, suggesting that ExoPAC formulations are stable in the harsh environment of GI and produce higher activity compared to free PAC.

As per our previous studies, colostrum exosomes express CD47 protein marker along with other hallmark exosomal proteins on their surface, which enhances the circulatory half-life of the exosomal drug formulations [24]. Further, FA functionalization on exosomes leads to trafficking to tumor site due to the presence of folate receptors (FR-α) and reduced folate receptors (RFC) on tumor cells. At the tumor site, exosomes are internalized in the cancer cells through several mechanisms like endocytosis, phagocytosis, micropinocytosis and/or fusion with cellular plasma membrane [38–41]. After entering the cell cytoplasm, exosomes directly release their payloads or undergo lysosomal digestion to release the drug contents [42]. In this study, we postulated that FA-ExoPAC releases PAC inside the cancer cells either through direct release or by the lysozyme-mediated digestion, which was clearly demonstrated by its in vivo antitumor response against subcutaneous and orthotropic lung cancer xenografts.

The orthotopic xenograft models represent a clinically relevant tumor model with respect to the tumor's primary site, microenvironment and metastasis [43,44]. These models are further improved with the advent of imaging techniques, which help in the measurement of internally implanted orthotopic tumors. In this study, using an orthotopic model, we demonstrate that FA-ExoPAC given either i.v. or p.o. has much greater efficacy compared to PAC i.v. and that for the higher dose study, FA-ExoPAC given orally produced efficacy similar to i.v. Abraxane. Noteworthy is that when used i.v., FA-ExoPAC produced significantly higher antitumor activity compared to Abraxane, suggesting the potential of exosome-based drug formulation for the management of cancer. The enhanced activity of FA-ExoPAC could be due to tumor targeting, slow release of PAC from the exosomes and intrinsic ability of exosomes to inhibit the cancer cells [21].

Clinical translatability of new drugs or nanoformulations are often limited due to toxicity concerns at various stages of drug discovery and development. While efficacious, PAC is generally not well tolerated and its limitations include low solubility, and significant toxicity associated with both the drug and the solvent (Cremophor EL), including hypersensitivity reactions, bronchospasms, hypotension, hematological toxicity, peripheral sensory neuropathy, myalgia, arthralgia and alopecia [8,45]. We have demonstrated FA-ExoPAC clearly enhanced therapeutic efficacy of PAC diminishing the dose-related toxicity issues. Our previous toxicity study reports establish milk exosomes as nontoxic and nonimmunogenic [19]. The present study using colostrum exosomes further supports lack of PAC-related gross, systemic and immune toxicity concerns when used in an exosomal formulation.

5. Conclusions

In summary, our findings have potential clinical implications for the management of advanced non-small cell lung cancer and potentially other cancers routinely treated with PAC. We showed that: (i) an abundance of exosomes in standardized bovine colostrum powder displays high PAC loading, and enhanced tumor targeting with FA-functionalized exosomes; (ii) PAC in exosomal formulation exhibits strong activity against both drug-sensitive (A549) and drug-resistant (A549TR) lung cancer; (iii) ExoPAC administered orally inhibit both subcutaneous and orthotopic lung tumors, and the efficacy is enhanced when FA-functionalized exosomes are used; (iv) p.o. FA-ExoPAC surpassed the efficacy of i.v. PAC and matched efficacy of i.v. Abraxane in one study; (v) i.v. FA-ExoPAC exceeded efficacy of i.v. Abraxane, the only FDA-approved albumin-bound nanoformulation of PAC; and (vi) FA-ExoPAC minimally perturbed the immune homeostasis of the host, thus eliminating potential adverse effects of PAC on the immune system. Together, these data provide a strong rationale for the development of oral exosomal formulations of PAC as a therapeutic alternative to current therapies.

Supplementary Materials: The following are available online at https://www.mdpi.com/article/10.3390/cancers13153700/s1, Figure S1: Inhibition of A549 colony formation by PAC and ExoPAC, Figure S2: Inhibition of resistant A549TR colony formation by PAC and ExoPAC, Figure S3: Health index of the animals treated with PAC and ExoPAC.

Author Contributions: Conceptualization, R.C.G., S.B. and F.A.; methodology, R.K., S.S.A., N.T., J.Y. and F.A.; validation, R.K., S.S.A. and F.A.; investigation, R.K., J.J., S.S.A., N.T., A.A., J.Y. and F.A.; resources, R.C.G. and W.S.; writing—original draft preparation, R.K. and F.A.; writing—review and editing, R.C.G., S.B. and W.S.; supervision, R.C.G., S.B. and F.A.; funding acquisition, R.C.G. and W.S. All authors have read and agreed to the published version of the manuscript.

Funding: This work was supported by the NCI grant R44-CA-221487 and, in part, the Agnes Brown Duggan Endowment (to R.C.G.).

Institutional Review Board Statement: All animals were maintained according to the Institutional Animal Care and Use Committee (IACUC) guidelines (IACUC # 17041; approval date: 1 June 2017) of the University of Louisville, Louisville, KY, USA.

Informed Consent Statement: Not applicable.

Data Availability Statement: Not applicable.

Acknowledgments: This work was supported by the NCI grant R44-CA-221487 awarded to 3P Biotechnologies, Inc., Louisville, KY and, in part, the Agnes Brown Duggan Endowment (to R.C.G.). Sarah Wilcher is thankfully acknowledged for administration of test regimens. We are grateful to Mark Burton of Immunodynamics, Inc. (Fennimore, WI, USA) for providing us bovine colostrum powder.

Conflicts of Interest: The authors declare no conflict of interest. Ramesh C. Gupta holds positions both at the University of Louisville and 3P Biotechnologies. The authors have filed a U.S. patent on part of the results reported in this paper.

References

1. ACS. *A.C.S. Cancer Facts & Figures*; American Cancer Society: Atlanta, GA, USA, 2020.
2. Siegel, R.L.; Miller, K.D.; Fuchs, H.E.; Jemal, A. Cancer Statistics, 2021. *CA Cancer J. Clin.* **2021**, *71*, 7–33. [CrossRef]
3. Siegel, R.; Ma, J.; Zou, Z.; Jemal, A. Cancer statistics, 2014. *CA Cancer J. Clin.* **2014**, *64*, 9–29. [CrossRef] [PubMed]
4. Horwitz, S.B. Taxol (paclitaxel): Mechanisms of action. *Ann. Oncol.* **1994**, *5*, S3–S6. [PubMed]

5. Constantinides, P.P.; Lambert, K.J.; Tustian, A.K.; Schneider, B.; Lalji, S.; Ma, W.; Wentzel, B.; Kessler, D.; Worah, D.; Quay, S.C. Formulation development and antitumor activity of a filter-sterilizable emulsion of paclitaxel. *Pharm. Res.* **2000**, *17*, 175–182. [CrossRef]

6. Merisko-Liversidge, E.; Sarpotdar, P.; Bruno, J.; Hajj, S.; Wei, L.; Peltier, N.; Rake, J.; Shaw, J.M.; Pugh, S.; Polin, L.; et al. Formulation and antitumor activity evaluation of nanocrystalline suspensions of poorly soluble anticancer drugs. *Pharm. Res.* **1996**, *13*, 272–278. [CrossRef] [PubMed]

7. Palumbo, R.; Sottotetti, F.; Trifiro, G.; Piazza, E.; Ferzi, A.; Gambaro, A.; Spinapolice, E.G.; Pozzi, E.; Tagliaferri, B.; Teragni, C.; et al. Nanoparticle albumin-bound paclitaxel (nab-paclitaxel) as second-line chemotherapy in HER2-negative, taxane-pretreated metastatic breast cancer patients: Prospective evaluation of activity, safety, and quality of life. *Drug Des. Dev. Ther.* **2015**, *9*, 2189–2199. [CrossRef]

8. Chou, P.L.; Huang, Y.P.; Cheng, M.H.; Rau, K.M.; Fang, Y.P. Improvement of Paclitaxel-Associated Adverse Reactions (ADRs) via the Use of Nano-Based Drug Delivery Systems: A Systematic Review and Network Meta-Analysis. *Int. J. Nanomed.* **2020**, *15*, 1731–1743. [CrossRef]

9. Kooijmans, S.A.; Vader, P.; van Dommelen, S.M.; van Solinge, W.W.; Schiffelers, R.M. Exosome mimetics: A novel class of drug delivery systems. *Int. J. Nanomed.* **2012**, *7*, 1525–1541. [CrossRef]

10. Lakhal, S.; Wood, M.J. Exosome nanotechnology: An emerging paradigm shift in drug delivery: Exploitation of exosome nanovesicles for systemic in vivo delivery of RNAi heralds new horizons for drug delivery across biological barriers. *Bioessays* **2011**, *33*, 737–741. [CrossRef]

11. Aisner, J. Overview of the changing paradigm in cancer treatment: Oral chemotherapy. *Am. J. Health Syst. Pharm.* **2007**, *64*, S4–S7. [CrossRef]

12. Joo, K.M.; Park, K.; Kong, D.S.; Song, S.Y.; Kim, M.H.; Lee, G.S.; Kim, M.S.; Nam, D.H. Oral paclitaxel chemotherapy for brain tumors: Ideal combination treatment of paclitaxel and P-glycoprotein inhibitor. *Oncol. Rep.* **2008**, *19*, 17–23. [CrossRef]

13. Browder, T.; Butterfield, C.E.; Kraling, B.M.; Shi, B.; Marshall, B.; O'Reilly, M.S.; Folkman, J. Antiangiogenic scheduling of chemotherapy improves efficacy against experimental drug-resistant cancer. *Cancer Res.* **2000**, *60*, 1878–1886.

14. Thery, C.; Witwer, K.W.; Aikawa, E.; Alcaraz, M.J.; Anderson, J.D.; Andriantsitohaina, R.; Antoniou, A.; Arab, T.; Archer, F.; Atkin-Smith, G.K.; et al. Minimal information for studies of extracellular vesicles 2018 (MISEV2018): A position statement of the International Society for Extracellular Vesicles and update of the MISEV2014 guidelines. *J. Extracell. Vesicles* **2018**, *7*, 1535750. [CrossRef]

15. Bobrie, A.; Colombo, M.; Raposo, G.; Thery, C. Exosome secretion: Molecular mechanisms and roles in immune responses. *Traffic* **2011**, *12*, 1659–1668. [CrossRef]

16. Kandimalla, R.; Aqil, F.; Tyagi, N.; Gupta, R. Milk exosomes: A biogenic nanocarrier for small molecules and macromolecules to combat cancer. *Am. J. Reprod. Immunol.* **2021**, *85*, e13349. [CrossRef]

17. Kamerkar, S.; LeBleu, V.S.; Sugimoto, H.; Yang, S.; Ruivo, C.F.; Melo, S.A.; Lee, J.J.; Kalluri, R. Exosomes facilitate therapeutic targeting of oncogenic KRAS in pancreatic cancer. *Nature* **2017**, *546*, 498–503. [CrossRef] [PubMed]

18. Vader, P.; Mol, E.A.; Pasterkamp, G.; Schiffelers, R.M. Extracellular vesicles for drug delivery. *Adv. Drug Deliv. Rev.* **2016**, *106*, 148–156. [CrossRef] [PubMed]

19. Agrawal, A.K.; Aqil, F.; Jeyabalan, J.; Spencer, W.A.; Beck, J.; Gachuki, B.W.; Alhakeem, S.S.; Oben, K.; Munagala, R.; Bondada, S.; et al. Milk-derived exosomes for oral delivery of paclitaxel. *Nanomed. Nanotechnol. Biol. Med.* **2017**, *13*, 1627–1636. [CrossRef] [PubMed]

20. Aqil, F.; Jeyabalan, J.; Agrawal, A.K.; Kyakulaga, A.H.; Munagala, R.; Parker, L.; Gupta, R.C. Exosomal delivery of berry anthocyanidins for the management of ovarian cancer. *Food Funct.* **2017**, *8*, 4100–4107. [CrossRef] [PubMed]

21. Munagala, R.; Aqil, F.; Jeyabalan, J.; Agrawal, A.K.; Mudd, A.M.; Kyakulaga, A.H.; Singh, I.P.; Vadhanam, M.V.; Gupta, R.C. Exosomal formulation of anthocyanidins against multiple cancer types. *Cancer Lett.* **2017**, *393*, 94–102. [CrossRef] [PubMed]

22. Aqil, F.; Kausar, H.; Agrawal, A.K.; Jeyabalan, J.; Kyakulaga, A.H.; Munagala, R.; Gupta, R. Exosomal formulation enhances therapeutic response of celastrol against lung cancer. *Exp. Mol. Pathol.* **2016**, *101*, 12–21. [CrossRef] [PubMed]

23. Aqil, F.; Munagala, R.; Jeyabalan, J.; Agrawal, A.K.; Kyakulaga, A.H.; Wilcher, S.A.; Gupta, R.C. Milk exosomes—Natural nanoparticles for siRNA delivery. *Cancer Lett.* **2019**, *449*, 186–195. [CrossRef] [PubMed]

24. Munagala, R.; Aqil, F.; Jeyabalan, J.; Kandimalla, R.; Wallen, M.; Tyagi, N.; Wilcher, S.; Yan, J.; Schultz, D.J.; Spencer, W.; et al. Exosome-mediated delivery of RNA and DNA for gene therapy. *Cancer Lett.* **2021**, *505*, 58–72. [CrossRef]

25. Yang, X.; Ye, Z.; Yuan, Y.; Zheng, Z.; Shi, J.; Ying, Y.; Huang, P. Insights into the binding of paclitaxel to human serum albumin: Multispectroscopic studies. *Lumin. J. Biol. Chem. Lumin.* **2013**, *28*, 427–434. [CrossRef] [PubMed]

26. Kyakulaga, A.H.; Aqil, F.; Munagala, R.; Gupta, R.C. Synergistic combinations of paclitaxel and withaferin A against human non-small cell lung cancer cells. *Oncotarget* **2020**, *11*, 1399–1416. [CrossRef] [PubMed]

27. Fan, T.W.; Lane, A.N.; Higashi, R.M.; Yan, J. Stable isotope resolved metabolomics of lung cancer in a SCID mouse model. *Metab. Off. J. Metab. Soc.* **2011**, *7*, 257–269. [CrossRef] [PubMed]

28. Zhong, W.; Hansen, R.; Li, B.; Cai, Y.; Salvador, C.; Moore, G.D.; Yan, J. Effect of yeast-derived beta-glucan in conjunction with bevacizumab for the treatment of human lung adenocarcinoma in subcutaneous and orthotopic xenograft models. *J. Immunother.* **2009**, *32*, 703–712. [CrossRef]

29. Umanzor, G.; Rugo, H.S.; Barrios, F.J.; Vasallo, R.H.; Chivalan, M.A.; Bejarano, S.; Ramirez, J.R.; Fein, L.; Kowalyszyn, R.D.; Cutler, D.L.; et al. Abstract PD1-08: Oral Paclitaxel and Encequidar (oPac + E) versus IV paclitaxel (IVPac) in the Treatment of Metastatic Breast Cancer (mBC) Patients (Study KX-ORAX-001): Progression Free Survival (PFS) and Overall Survival (OS) Updates. In Proceedings of the San Antonio Breast Cancer Symposium, San Antonio, TX, USA, 8–11 December 2020.

30. Rugo, H.S.; Umanzor, G.; Barrios, F.J.; Vasallo, R.H.; Chivalan, M.A.; Bejarano, S.; Ramirez, J.R.; Fein, L.; Kowalyszyn, R.D.; Cutler, D.L.; et al. Lower Rates of Neuropathy with Oral Paclitaxel and Encequidar (oPac + E) Compared to IV Paclitaxel (IVPac) in Treatment of Metastatic Breast Cancer (mBC): Study KX-ORAX-001. In Proceedings of the San Antonio Breast Cancer Symposium, San Antonio, TX, USA, 8–11 December 2020; p. Abstract PS13-06.

31. Mendt, M.; Kamerkar, S.; Sugimoto, H.; McAndrews, K.M.; Wu, C.C.; Gagea, M.; Yang, S.; Blanko, E.V.R.; Peng, Q.; Ma, X.; et al. Generation and testing of clinical-grade exosomes for pancreatic cancer. *JCI Insight.* **2018**, *3*. [CrossRef] [PubMed]

32. Gerosa, S.; Skoet, J. *Milk Availability: Trends in Production and Demand and Medium-Term Outlook*; ESA Working Paper No. 12-0; FAO: Rome, Italy, 2012; pp. 1–38.

33. Ledermann, J.A.; Canevari, S.; Thigpen, T. Targeting the folate receptor: Diagnostic and therapeutic approaches to personalize cancer treatments. *Ann. Oncol.* **2015**, *26*, 2034–2043. [CrossRef] [PubMed]

34. Shi, H.; Guo, J.; Li, C.; Wang, Z. A current review of folate receptor alpha as a potential tumor target in non-small-cell lung cancer. *Drug Des. Dev. Ther.* **2015**, *9*, 4989–4996. [CrossRef]

35. Cheung, A.; Bax, H.J.; Josephs, D.H.; Ilieva, K.M.; Pellizzari, G.; Opzoomer, J.; Bloomfield, J.; Fittall, M.; Grigoriadis, A.; Figini, M.; et al. Targeting folate receptor alpha for cancer treatment. *Oncotarget* **2016**, *7*, 52553–52574. [CrossRef] [PubMed]

36. Low, P.S.; Kularatne, S.A. Folate-targeted therapeutic and imaging agents for cancer. *Curr. Opin. Chem. Biol.* **2009**, *13*, 256–262. [CrossRef]

37. Sega, E.I.; Low, P.S. Tumor detection using folate receptor-targeted imaging agents. *Cancer Metastasis Rev.* **2008**, *27*, 655–664. [CrossRef]

38. Morelli, A.E.; Larregina, A.T.; Shufesky, W.J.; Sullivan, M.L.; Stolz, D.B.; Papworth, G.D.; Zahorchak, A.F.; Logar, A.J.; Wang, Z.; Watkins, S.C.; et al. Endocytosis, intracellular sorting, and processing of exosomes by dendritic cells. *Blood* **2004**, *104*, 3257–3266. [CrossRef]

39. Aqil, F.; Munagala, R.; Jeyabalan, J.; Agrawal, A.K.; Gupta, R. Exosomes for the Enhanced Tissue Bioavailability and Efficacy of Curcumin. *AAPS J.* **2017**, *19*, 1691–1702. [CrossRef]

40. Parolini, I.; Federici, C.; Raggi, C.; Lugini, L.; Palleschi, S.; De Milito, A.; Coscia, C.; Iessi, E.; Logozzi, M.; Molinari, A.; et al. Microenvironmental pH is a key factor for exosome traffic in tumor cells. *J. Biol. Chem.* **2009**, *284*, 34211–34222. [CrossRef]

41. Feng, D.; Zhao, W.L.; Ye, Y.Y.; Bai, X.C.; Liu, R.Q.; Chang, L.F.; Zhou, Q.; Sui, S.F. Cellular internalization of exosomes occurs through phagocytosis. *Traffic* **2010**, *11*, 675–687. [CrossRef] [PubMed]

42. Kalluri, R.; LeBleu, V.S. The biology, function, and biomedical applications of exosomes. *Science* **2020**, *367*. [CrossRef]

43. Killion, J.J.; Radinsky, R.; Fidler, I.J. Orthotopic models are necessary to predict therapy of transplantable tumors in mice. *Cancer Metastasis Rev.* **1998**, *17*, 279–284. [CrossRef] [PubMed]

44. Kerbel, R.S. Human tumor xenografts as predictive preclinical models for anticancer drug activity in humans: Better than commonly perceived-but they can be improved. *Cancer Biol. Ther.* **2003**, *2*, S134–S139. [CrossRef]

45. Fauzee, N.J.S.; Dong, Z.; Wang, Y.L. Taxanes: Promising Anti-Cancer Drugs. *Asian Pac. J. Cancer Prev.* **2011**, *12*, 837–851. [PubMed]

 *cancers*

Review

Extracellular Vesicles and Their Current Role in Cancer Immunotherapy

Carla Giacobino [1], Marta Canta [1], Cristina Fornaguera [2], Salvador Borrós [2] and Valentina Cauda [1,*]

[1] Department of Applied Science and Technology, Politecnico di Torino, Corso Duca degli Abruzzi 24, 10129 Turin, Italy; s255203@studenti.polito.it (C.G.); marta.canta@polito.it (M.C.)

[2] Grup d'Enginyeria de Materials (Gemat), Institut Químic de Sarrià (IQS), Universitat Ramon Llull (URL), Via Augusta 390, 08017 Barcelona, Spain; cristina.fornaguera@iqs.url.edu (C.F.); salvador.borros@iqs.url.edu (S.B.)

* Correspondence: valentina.cauda@polito.it

Simple Summary: In recent years, immunotherapy has shown great advancement, becoming a powerful tool to combat cancer. In this context, the use of biologically derived vesicles has also acquired importance for cancer immunotherapy. Extracellular vesicles are thus proposed to transport molecules able to trigger an immune response and thus fight cancer cells. As a particular immunotherapeutic approach, a new technique also consists in the exploitation of extracellular vesicles as new cancer vaccines. The present review provides basic notions on cancer immunotherapy and describes several clinical trials in which therapeutic anticancer vaccines are tested. In particular, the potential of extracellular vesicles-based therapeutic vaccines in the treatment of cancer patients is highlighted, even with advanced stage-cancer. A focus on the clinical studies, already completed or still in progress, is offered and a systematic collection and reorganization of the present literature on this topic is proposed to the reader.

Citation: Giacobino, C.; Canta, M.; Fornaguera, C.; Borrós, S.; Cauda, V. Extracellular Vesicles and Their Current Role in Cancer Immunotherapy. *Cancers* **2021**, *13*, 2280. https://doi.org/10.3390/cancers13092280

Academic Editors: Farrukh Aqil and Ramesh Chandra Gupta

Received: 12 April 2021
Accepted: 7 May 2021
Published: 10 May 2021

Publisher's Note: MDPI stays neutral with regard to jurisdictional claims in published maps and institutional affiliations.

Abstract: Extracellular vesicles (EVs) are natural particles formed by the lipid bilayer and released from almost all cell types to the extracellular environment both under physiological conditions and in presence of a disease. EVs are involved in many biological processes including intercellular communication, acting as natural carriers in the transfer of various biomolecules such as DNA, various RNA types, proteins and different phospholipids. Thanks to their transfer and targeting abilities, they can be employed in drug and gene delivery and have been proposed for the treatment of different diseases, including cancer. Recently, the use of EVs as biological carriers has also been extended to cancer immunotherapy. This new technique of cancer treatment involves the use of EVs to transport molecules capable of triggering an immune response to damage cancer cells. Several studies have analyzed the possibility of using EVs in new cancer vaccines, which represent a particular form of immunotherapy. In the literature there are only few publications that systematically group and collectively discuss these studies. Therefore, the purpose of this review is to illustrate and give a partial reorganization to what has been produced in the literature so far. We provide basic notions on cancer immunotherapy and describe some clinical trials in which therapeutic cancer vaccines are tested. We thus focus attention on the potential of EV-based therapeutic vaccines in the treatment of cancer patients, overviewing the clinically relevant trials, completed or still in progress, which open up new perspectives in the fight against cancer.

Keywords: extracellular vesicles; cancer vaccine; immunotherapy; drug delivery; gene delivery; surface funzionalization; nanoparticles

1. Extracellular Vesicles: An Introduction

According to the latest literature reports, extracellular vesicles (EVs) are particles with a spherical shape and delimited by a phospholipid bilayer. They are produced by several cell types according to different physiological processes and even pathological conditions and released into the extracellular environment [1]. EVs can be thus extracted from fluids of biological origin, such as blood, saliva, urine, amniotic fluid, breast milk, cerebrospinal fluid, and synovial fluid [2]. Although EVs were initially considered to be part of the waste management of the cells [3], many advancements have been achieved so far to unravel their mechanisms of the transfer of cell-derived biomolecules (i.e., DNA, various RNA types, proteins, lipids and metabolites), characterizing the communication between different cells and tissues [4]. For this reason, in recent years, more and more attention has been paid to extracellular vesicles and many research and literature reviews have emerged so far [1–6]. Furthermore, EVs are also involved in processes such as angiogenesis, coagulation, cell survival, waste management, immunomodulation and inflammation [5]. The International Society for Extracellular Vesicles (ISEV) suggests the term "extracellular vesicle" to define particles delimited by a lipid bilayer, naturally released by cells and unable to replicate. The lipid bilayer membrane of EVs protects their load from enzymatic degradation during the transfer from donor to recipient cells [3]. Packaging also allows cargo to be stored more efficiently and to deliver it at dedicated target cells by modifying the vesicles with cell type-specific adhesion receptors. For this reason, new clinical applications of EVs for therapy or as biomarkers reporting about healthy and diseased conditions are under development, rising increasing interest in both the research and clinical community [6]. However, due to their heterogeneity, at the moment the definition of EVs subclasses cannot be based on their specific markers expressed at EVs surface. Anyhow, it is possible to classify them into three groups, according to their size and biogenesis' mechanism: microvesicles, apoptotic bodies and exosomes [7]. Figure 1 illustrates the different EVs biogenesis mechanisms and an example of exosome composition.

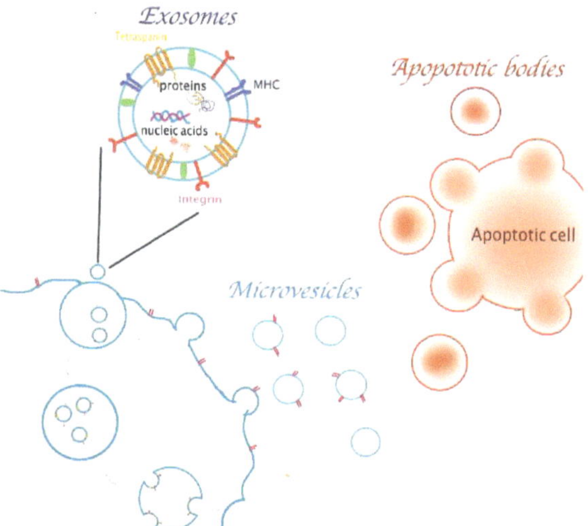

Figure 1. Extracellular vesicles biogenesis and a specific focus on exosome composition. Exosomes originate intracellularly as intraluminal vesicles of the multivesicular bodies; microvesicles originate from the outwards budding and fission of plasma membrane; apoptotic bodies are caused by the fragmentation of the apoptotic cells. The exosomes membrane is composed by different kinds of lipids and proteins and they can carry inside cytosol-derived molecules, such as proteins and nucleic acids.

Microvesicles (MVs) are heterogeneous cell-derived membrane vesicles that protrude from the surface of cells in a highly regulated process and are then released to the extracellular environment [8]. They are large vesicles with a diameter from 100–1000 nm, detected in many bodily fluids, such as blood, urine, synovial fluid, and many others, under both physiological and pathological conditions. Moreover, elevated MV concentrations have been also observed in atherosclerotic plaques and tumor tissues [9]. In general, MVs are the result of the budding of the cell membrane from which they take shape through ARF6-mediated (ADP-ribosylation factor 6) rearrangement of the actin cytoskeleton. The mechanisms of formation and release of MVs remain only partially understood. Microvesicles biogenesis also involves the trafficking of cargo molecules towards the cell membrane and the redistribution of phospholipids of the plasma membrane. The specific function of microvesicles is determined by the composition of their molecular cargo, which, from what described above, is in turn clearly dependent upon the progenitor cell type and condition, as well as upon the microenvironment and the triggering stimuli which preceded the MVs release. Generally, microvesicles transport membrane-derived receptors, cytokines, chemokines, proteins involved in cellular signaling, lipids, carbohydrates [2], and nucleic acids, such as DNA and various types of RNA, including mRNAs, microRNAs (miRNAs), and small-interfering RNAs (siRNAs) [9]. When microvesicles are released into the extracellular environment, their cargo can be released and used to modify or alter the extracellular milieu. In other cases, MVs can dock to recipient cells and be internalized via endocytosis or fusion mechanisms, or even activate a signaling cascade mediated by the cell receptors [8]. Therefore, MVs are also fundamental in modifying the extracellular environment and signaling among cells, in targeting specific recipient cells, as well as in assisting the cell invasion and metastasis towards tissues using a cell-independent matrix proteolysis mechanism, or even in transferring specific receptors to recipient cells, thus enabling new cell signaling in cells initially missing the receptor. In the end, the MVs, while taking part in cellular communication, affect processes such as blood coagulation, thrombosis, angiogenesis, immunomodulation and inflammation [2].

Apoptotic bodies (ApoBDs) are vesicles bound to cell membrane, ranging from 50–5000 nm in diameter, which are then released from cells undergoing apoptosis [4]. Apoptosis is a physiologically-programmed cell death that does not induce inflammatory responses. It is commonly appearing in multicellular organisms as it consists in a homeostatic mechanism for controlling the population of cells in a tissue and has a key role during the processes of development and aging. The splitting of cellular content through the membrane blebbing determines the formation of distinct membrane-delimited vesicles: the apoptotic bodies. These apoptotic bodies are then engulfed by phagocytes for final degradation [10]. This process is immunologically silent, which is a typical characteristic of apoptosis. Different cell types can use different mechanisms to disassemble and thus to undergo in apoptosis. Thus, apoptotic bodies can vary their content and biomolecules such as glycosylated proteins, chromatin, large amounts of RNA, nuclear fragments, and even intact organelles (mitochondria) can be included. Despite this diversity in the apoptotic mechanism depending to the cell type, recent reports have shown that ApoBDs are also involved in the formation and conditioning of the tumor microenvironment and metastatic niche. It has been understood that this mechanism lies in the active role of ApoBDs to transfer biomolecules towards specific "target" cells [11]. Furthermore, ApoBDs have been also reported to show a consistent procoagulant effect towards cancer cells, which in turn would contribute to the prothrombotic state and immune regulation. Actually, the formation process of apoptotic body is closely related to cell death and clearance, and ultimately in the intercellular communication [12]. These mechanisms have direct implications in the regulation of immune system, which is particularly relevant in cancer therapy. Therefore, ApoBDs can promote an important role in anticancer immunity and further studies are needed to clearly demonstrate their role and further use in medicine.

Exosomes are vesicles with a diameter of 40–100 nm and derive from cell secretion upon multivesicular bodies (MVBs) formation [13]. Exosomes are released by almost all cell types and, as the other above-mentioned EVs, are present in many different bodily fluids. Exosome also have a spherical shape and a nanosized dimension, and their phospholipid bilayer membrane is composed of different phospholipids and proteins types (including transport proteins, heat shock proteins, tetraspanins) [14], which in turn derive from the cell of origin. In particular, according to recent databases (ExoCarta and EVPedia), it results that the protein composition of exosomes is somehow defined and include both conserved and cell-specific proteins. Interestingly, the tetraspanins CD9, CD63, CD37, CD81, or CD82 are typically present in the exosomes membrane and thus used as biomarkers for exosome identification [15].

Exosomes were identified in extracellular space for the first time in late 1980s. They were initially considered as cellular waste or as by-products of cell homeostasis [16]. Currently, these extracellular vesicles are considered functional vehicles, because they are able to deliver molecular cargoes to target cells and reprogram the behavior of recipient cells even located far from exosome release site. Generally, exosomes contain proteins, DNA, mRNA, miRNA, lipids, and this molecular composition directly derives from the parent cell, reflecting clearly the signature of the multiple physiological roles or pathological state of the progenitor cell. It is broadly recognized that exosomes play a major role in intercellular communication, but they also take part in many biological processes, including antigen presentation in immune responses, coagulation, inflammation, maturation of erythrocytes, and angiogenesis [14]. It has been recently discovered the role of exosomes in tumor progression and metastasis formation, in particular in pre-conditioning of the metastatic niche and tumor microenvironment. In this sense, exosomes are responsible of transferring bioactive molecules from the primary tumor site to other cells and tissue, both in the local and distant microenvironments [17].

In the recent years, the use of vesicles as biological carriers has also been extended to cancer immunotherapy [18]. This kind of cancer treatment involves the use of extracellular vesicles to transport molecules capable of triggering an immune response to damage cancer cells. In particular, different literature reports have studied the possibility of using extracellular vesicles as new cancer vaccines, which represent a particular form of immunotherapy. However, in the literature there are only few publications that systematically gather and collectively discuss these studies. A recent review [19] highlights specifically the role of exosome, a subclass of EVs, as mediators of immune regulation of both lymphoid and myeloid cells in cancer. However, in that review and others present in the literature, the re-engineering of EVs to become therapeutic players for immunotherapy against cancer is poorly highlighted.

Therefore, the purpose and the novelty of our review is to illustrate and give a reorganization to what has been produced in the literature so far [20], related to both the research and the clinical studies made on therapeutic anticancer vaccines based on extracellular vesicles, completed or still in progress.

The review begins by a defining extracellular vesicles and highlighting importance concepts related to their therapeutic potential. In view of many and prominent reviews present in the literature to date [1–6,14,16,17], here we skip on the detailed the discussion about the EVs nature, origin and types. For the same reason, the description of cargo loading techniques for EVs are only briefly introduced, being recently reported already [21]. Some engineering techniques, aimed at introducing surface markers for cell targeting, are however reported here with the aim to give a proper outline of the main concepts related to extracellular vesicles modifications and re-engineering. We specifically provide the reader with basic notions about cancer immunotherapy and show several examples, including clinical trials, in which therapeutic anticancer vaccines are involved. We then focus on the potential role of EVs as therapeutic vaccines in the treatment of cancer patients, even with advanced stage-cancers, focusing on clinical trials. Thus, the novel perspective of the present review is not only the discussion of the interaction between EVs and immunology, but to report on the biological applications of re-engineered EVs as cancer vaccines.

2. EVs for Therapeutic and Drug Delivery Purposes

EVs have various advantages such low immunogenicity, toxicity and targeting ability. Furthermore, the use of EVs can overcome some limitations encountered in conventional nanoparticulate systems used for drug delivery, such as liposomes, which are vesicular structures prepared from lipids in the laboratory and widely used as drug carriers. Unlike liposomes, exosomes-based delivery systems pass through main biological barriers, such as the blood brain barrier (BBB) [22,23], evade the lysosomal degradation and transport cargoes into the cytoplasm [24]. Therefore, they are ideal candidates for constructing novel therapeutic delivery nanosystems.

Thanks to new technologies enabling for isolating microvesicles and their ability to deliver and transfer biomolecules such as nucleic acids, also microvesicles can be exploited for targeted and therapeutic drug delivery, which can be further improved by engineering the microvesicles. The first study on the delivery of mRNA and proteins through MVs for cancer therapeutics has indeed shown that such genetically engineered vesicles are viable delivery vehicles. In particular the authors showed the ability to deliver suicide mRNA genes to cancerous schwannoma cells [25]. MVs were isolated from cells which stably expressed the suicide gene of interest and a protein–cytosine deaminase (CD) fused to uracil phosphoribosyltransferase (UPRT), which constitutes a valid prodrug-activating combination. Such MVs were then directly injected into schwannoma tumor of an orthotopic mouse model in combination the prodrug (5-fluorocytosine, 5-FC), which is then transformed within the tumor cell to 5-fluorouracil (5-FU), an anticancer agent. Therefore, the combination treatment of CD-UPRT mRNA/protein via MVs and 5-FU led to regression of these tumors. Recently, a group of researchers from Michigan State University (USA) demonstrated the efficacy of using MVs as a delivery vehicle in cancer treatment in a breast cancer mouse model [26]. They loaded MVs with engineered minicircle plasmid DNA encoding a thymidine kinase fusion protein. Such protein is then able to activate two prodrugs, ganciclovir and CB1954, against breast cancer cells. Then, they detected that the efficiency of delivery of MVs loaded with the engineered DNA was 14-times greater than the efficacy given by MVs loaded with regular plasmids. In addition, minicircle-loaded MVs were even more successful in destroying cancer cells. These outcomes confirm that gene delivery via MVs enables an effective drug delivery, so it can be considered an alternative to chemotherapy and, thanks to the high compatibility with the human body, unwanted immune responses can also be reduced.

The natural role of exosomes in cell communication, as well as their unique characteristics, enable them to be ideal candidates as drug delivery systems, promoting their application in both drugs and biomolecules delivery. Indeed, other than having a low immunogenicity and toxicity, these vesicles are able to transport a great variety of substances both in their core or associated to their membrane.

Exosomes are secreted by many different cell types, each influencing their nature and biological role. From the proteomic analysis it emerges that other than the ubiquitous proteins (f.i. tetraspanin, alix, TSG101), exosomes are composed by cell type-specific proteins inherited by their cellular source that condition their biological activities. For instance, exosomes released by mature reticulocytes show elevated levels of transferrin receptors, that is lost by these cells during their maturation. Similarly, the aquaporin proteins, involved in the water transport, are enriched in exosomes derived from kidney cells [24,27].

Other two important factors that promote the application of exosomes as new delivery systems are the amenability to modifications, thus to enhance exosomal targeting capability and the scalability of the process. Generally, it is possible to introduce modifications to the exosomal membrane proteins which are responsible for cell targeting. However, exosomes derived from specific cell source have been investigated to solve the problem of cell targeting or to produce a desirable therapeutic effect. Some exosomes released by tumor cells are reported harbouring an intrinsic targeting activity versus the tumor site that could be exploited in clinical applications. Other surface specific proteins, called tumor associated antigens, could be transferred to dendritic cells (DC), thus used to promote an immune reaction against cancer cells [28].

Therefore, the delivery ability of these kind of vesicles has already opened new scenarios for the development of new exosomes-based therapeutic agents and also for their application in cancer immunotherapy.

Cancer immunotherapy is defined as "the artificial stimulation of the immune system to treat cancer" [29]. This cancer treatment approach is based on the idea to use the immune system to fight and destroy the cancer cells, exploiting the natural immune mechanisms already normally used by the immune system to defeat diseases. Indeed, EVs could play a key role in immune regulation of cancer development through the use of those EVs released by immune cells or cancer cells. There is actually a controversy on the most suitable exosomes donor cells to be used, due to their putative involvement in immune system activation against cancer but also as immunesupressors, as reoprted already for glioblastoma, pancreatic and ovarian cancers [30–32]. In addition, cancer derived exosomes arise also some controversy due to their origin from malignant cells. However, upon engineering, there is no doubt on their beneficial use given that the antibodies produced by the immune system could be trained to bind to the tumor's antigens, identifying in this way the cancer cells and promoting their elimination. However, at the same time, tumor-derived exosomes could produce undesired effects, such as the inhibition or the killing of the cytotoxic T cells, used by the immune system to destroy its enemies. As a result, the release of exosomes by tumors may allow them to evade the "immunosurveillance" and interfere with cancer immunotherapy [33,34]. The potential risk in using tumor exosomes and the possibility in aggravating the pathological condition of the patients, induced the scientific community to explore new exosomes sources. Exosomes derived from plants or agricultural products, such as milk, have been proved to be a most safe and really scalable options to produce exosomes for clinical applications [35]. However, in the context of cancer immunotherapy they are not the best choice. In fact, unlike those derived from tumors, these are unable to stimulate the immune system for the cancer treatment.

As a consequence, immune cell-derived exosomes have been investigated, in particular those derived from DC cells [36,37]. These cells are main components of the immune systems and act as antigen presenting cells to the T effector cells [38]. Consequently, DC-derived exosomes contain antigen presenting molecules, adhesion molecules and costimulatory molecules, that are the necessary equipment required for generating powerful immune responses and thus for exosome-based vaccines. This new technique for the therapeutic administration of the vaccines is based on the production of exosomes, engineered with the vaccine antigen of interest, to induce a powerful cytotoxic T cell (CTL) mediated immune response against a large number of tumors and viral antigens. Other than the DC cells, another attractive healthy exosomes' source are the Mesenchymal Stem Cells (MSC). These cells are able to self-renewal and are multipotent since they can differentiate in different kind of cells and also own many useful characteristics [39]. They could be isolated from many different sources (f.i. bone marrow, placenta, umbilical cord), they are easily expanded In vitro to produce large amounts of exosomes and they harbour immunosuppressive properties that are demonstrated transferable to their exosomes [40]. The only main limitation associated to these cells is that they are not immortal, so the number of produced exosomes is limited by the number of their replications. Yeo et al. [27] investigated the immortalization of human embryonic stem cells (hESC-MSC) with the oncogene *MYC* to avoid this limitation, allowing the growth of this kind of cells In vitro for even long periods. Similar immortalization approaches have been reporte so far, proposing a feasible manufacturing method for therapeutic EVs [41,42].

Another, yet to be developed, application of exosomes, could be in enhancing the performance of CAR-T cell therapies, given the natural role of exosomes to activate T cells through antigen presentation. Although some studies previously reported the possible potent antitumor effect of exosomes isolated from CAR-T cells [43], bibliography is still scarce and many efforts will be required to achieve this specific application transference to clinical use.

3. Cargo-Loading Methods of EVs

The unique structure of EV membrane, constituted by a phospholipid bilayer with a hydrophobic space within the bilayer and a hydrophilic surface, allows a great variety of molecules to be loaded into the EVs [3,21]. Hydrophobic and hydrophilic molecules can be loaded into EVs, including anticancer drugs, miRNA, siRNA, DNA and proteins additional to their physiological molecular composition [35,44,45]. Recently, EVs have been proposed to carry nanoparticles (NPs) as well [46–49]. By loading NPs into EVs, it is possible to overcome problems such as particle aggregation, degradation and rapid clearance, which often occur in the use of nanoparticles.

Therapeutic cargos are incorporated into EVs by following two main loading approaches: exogenous (or direct loading) and endogenous loading approaches [3,21]. By exogenous methods, therapeutic cargos are loaded into EVs after their isolation. In addition, these techniques are subclassified into passive and active cargo-loading methods. Passive loading refers to a simple method wherein cargo is passively loaded into EVs without any external interventions. Instead, active loading uses different techniques that force EVs to load the cargo. Cargo-loading methods of EVs have been illustrated in other more detailed reviews [1,3,21] and will thus be no further discussed here.

Endogenous loading refers to (a) genetic engineering of donor cells to constitutively produce exosomes loaded with the Active Pharmaceutical Ingredients (API) of interest, or to (b) transient donor cells transfection to achieve the release of loaded exosomes. In particular, some genetic engineering techniques are used for surface functionalization and are briefly described below in Section 4.

4. EVs Surface Functionalization: An Overview

The functionalization of the EVs surface is carried out to improve targeting abilities, biodistribution and therapeutic applications of EVs. However, the use of this approach needs the development of protocols with a strict control of the experimental conditions (f.i. temperature, pressure, solvents, salt concentrations) to preserve the exosomes integrity and functions. Indeed, undesired effects for therapeutic applications, such as vesicles aggregation due to inadequate reaction conditions, have been reported in the literature so far [50].

4.1. Post-Isolation Methods

Between the methods used for the EVs surface modifications there are the covalent and non-covalent chemical modifications [1]. The non-covalent approaches provide membrane modifications through the use of gentle reactions, such as the electrostatic interactions or the hydrophobic insertions [50]. For this purpose, Nakase et al. [51] used Lipofectamine, a commercial transfection reagent containing cationic lipids, to modify the EVs charge and promote their interaction with the plasma membrane. Indeed, the adsorption of the Lipofectamine on the EVs surface confers them a positive charge helping their interaction with the negative domains of the target cell membrane. In addition, the functionalization with cationic lipids was used to recruit the negatively charged fusogenic peptide GALA. This peptide promoted the EVs escape from the endosomal compartments, essential for many drug delivery applications [52]. Therefore, as it is evident, the functionalization based on electrostatic interaction combined with the application of the GALA peptide can be used to improve cellular uptake and also enhance the exosomal escape and content release in the cytosolic space.

The covalent methods, on the contrary, need the formation of new chemical bonds at the EVs surface. A widely diffused method is the click chemistry, that adds molecules on the EVs surface through a cicloaddiction reaction. This method can be easily performed and is highly versatile in terms of reaction conditions, such as pH and temperature ranges [53]. Furthermore, a study demonstrated that click chemistry did not alter exosome size and functionality [54]. Another common approach of surface functionalization of EVs, based on covalent bonding, is the PEGylation, to achieve stealth surfaces, as performed everywhere for different kinds of nanosystems. In fact, the addition of the polyethylene glycol (PEG) on the EV surfaces forms a corona that is effective in reducing their immunogenicity. A study by Kooijmans et al. [55] showed that this surface modification significantly increases the EV circulation half-life in mice because it reduces EVs recognition by the mononuclear phagocyte system (MPS) avoiding plasma protein opsonization. However, the presence of this PEG corona around the EVs surface is also been associated to a reduction in EVs interaction with the plasma membrane, thus a reduction of cellular uptake. To overcome these limitations, many authors functionalize the PEG with targeting ligands, directed to receptors overexpressed by the target cells [55,56].

For example, Kim et al. [57] loaded exosome with paclitaxel (PTX), an anti-cancer agent widely used against lung cancer, to be delivered to pulmonary metastases. They developed a specific procedure based on sonication and incubation to add PEG, and a similar procedure to incorporate a vector moiety with Aminoethylanisamide AA, a ligand of the sigma receptor, overexpressed by lung cells [58]. Experimental results showed that the AA-PEG-exoPTX formulation can be easily internalized into target cancer cells, other than a high loading capacity and strong anticancer effects compared to the unfunctionalized exosomes.

4.2. Genetic Engineering of Parental Cells for Surface Functionalization

The EVs surface can be functionalized with ligands not only through EV post-isolation methods such as click chemistry, but also through genetic engineering, wherein the cells that will produce the EVs are induced to express the protein or peptide of interest. This approach was employed by Alvarez-Erviti et al. [22] that used exosomes loaded with small interfering RNAs (siRNAs) to achieve knockdown of β-site APP-cleaving enzyme 1 (BACE 1), a therapeutic target for Alzheimer's disease [59]. In particular, to obtain a selective targeting of this EVs versus the mouse brain, they functionalized the EVs with the targeting peptide obtained from Rabies Virus Glycoprotein (RVG), able to bind the nicotinic acetyl choline receptor (AchR) highly expressed by the cells at the blood brain barrier level [13]. This functionalization was obtained attaching this peptide to the Lamp2b protein expressed on the EVs surface, transfecting the cell with the plasmid codifying for this construct. Moreover, researchers observed a strong knockdown of BACE 1 and an important reduction of amyloid plaques components, which are in turn associated to Alzheimer's pathology. In a recent study, Yang et al. [60] have found a similar approach to systematically deliver nerve growth factor (NGF) into ischemic cortex for the treatment of stroke, in a photothrombotic ischemia model. NGF has a primary role in the growth, as well as the maintenance, proliferation, and survival of nerve cells [61]. The exosomes loaded with the mRNA coding for the NGF protein were functionalized on their surface with the peptide RVG, thus obtaining NGF@ExoRVG. The in vivo administration of such modified exosomes led to an effective delivery of the NGF mRNA to the cells of the ischemic cortex and in its consequent protein translation. Interestingly, the treatment with these exosomes was associated to a reduced inflammation mediated by the M2 microglia: a process which mediates the immune response in the CNS.

All these results suggest that the exosomes engineering could be a useful tool for the production of exosome-based delivery vehicles, providing new options for target therapies.

5. Cancer Immunotherapy

Immune system is a defensive apparatus that the human body uses to fight illness. It is constituted by a complex "surveillance network" made up of several highly specialized organs and cells, shared by the lymphatic vessels, and located in various parts of the body. All of them cooperate, each with a specific role, to defend the organism and keep it healthy. The immune system has the main role to trace the "foreign" substances present in the organism. Therefore, whenever substance not normally present in the organism enters the body triggers an alarm and causes an immune response, aimed at destroying it. Unfortunately, cancer can commonly escape the immune system's natural defences, allowing cancer cells to continue growing. Immunotherapy is specialized in the development of novel anti-cancer therapies by understanding and making use of immune pathways. This innovative cancer treatment stimulates the natural immune system to fight cancer by finding and attacking cancer cells. There are many types of immunotherapy including monoclonal antibodies, oncolytic virus therapy, adoptive cell therapies and cancer vaccines.

Monoclonal antibodies are highly specific molecules produced in laboratory from identical immune cells and engineered to work as substitute antibodies that target only a single site (epitope) on a single antigen in order to obtain a strong immune response against cancer cells [62]. The monoclonal antibody recognizes the presence of a specific receptor on the surface of the tumor cell and binds to it. In this way, it induces the immune system to attack and destroy cancer cells while minimizing the damage to healthy cells. It can also induce cancer cells to self-destruct or can block the receptor preventing it from binding to a different protein that stimulates the cancer growth. Some examples of commercialized monoclonal antibodies for cancer treatment or maintenance cancer therapy are trastuzumab and pertuzumab for breast cancer and rituximab for non-Hodgkins lymphoma.

The goal of oncolytic virus therapy is to use viruses to destroy tumors; in particular, the oncolytic viruses (OVs) are able to replicate in cancer cells (while not in healthy ones) producing a lysis of the tumor mass. At the same time, OVs can stimulate the immune system to implement a strong and lasting response against the tumor itself. Unfortunately, it is necessary to consider that this kind of therapy could be hindered by the possible occurrence of an anti-viral response due to the recognition of the virus as a pathogen. Therefore, many researchers are currently looking for a solution that allows a right balance between anti-tumor and anti-viral immunity to make OV therapy as successful as possible [63].

Adoptive cell therapy requires that the patient's autologous T cells be genetically engineered in the laboratory so that they express a receptor, called chimeric antigen receptor (CAR) specific for a tumor antigen. T cells are taken from a patient's blood, then the gene for CAR expression is added to the T cells in the laboratory. After ex vivo cell expansion, CAR T cells are administered to the patient by infusion, then they bind the antigen on the cancer cells and kill them [64].

Lastly, cancer vaccines are designed to stimulate the immune system to recognize cancer cells in the patient and thus fight cancer more effectively [65].

The next section will investigate cancer vaccines more in detail and in particular the role of extracellular vesicles as delivery vehicles to enhance and amplify the effect of the immune response against tumor cells.

6. What Is a Cancer Vaccine?

Vaccines are generally prophylactic agents which are administered to healthy people to achieve long-term immunity against a virus and to prevent the consequent onset of the disease. However, cancer vaccines are different from prophylactic antiviral vaccines as they include not only (i) prophylactic vaccines used for cancer prevention, but also (ii) therapeutic vaccines [66]. Actually, currently existing prophylactic cancer vaccines only apply to virus-induced malignancies, such as liver cancer caused by the hepatitis B virus or genital cancers caused by the human papilloma virus (HPV). In contrast, therapeutic vaccines are designed for individuals with an existing disease. There are two aspects to take into account. First, cancer is often a silent disease and for this reason, at the time of diagnosis, it is already out of the control of the immune system and therefore already established in the body. Secondly, the vast majority of tumors that are classically treated with surgery, chemotherapy or radiotherapy definitively disappears. In some patients, however, the tumor may relapse and become increasingly resistant to treatment [67–69]. This happens because in some cases a certain number of cancer cells can escape treatment and remain in the body, even if the anticancer therapy was initially very effective. These cells are able to reproduce the tumor even years after its first appearance, and conse-quently the disease becomes more difficult to eradicate with classic treatments. In this context, recent clinical trials propose the combined or sequential use of different therapeutic strategies to create a "multimodal" therapeutic approach [70]. For example, therapeutic cancer vaccines are administered after complete surgical removal of the tumor, followed or not by chemotherapy, in lung cancer or cutaneous melanoma patients (NCT00530634, NCT04245514, NCT02211131, NCT04330430). The main goal of therapeutic vaccines is to elicit strong antigen-specific T cell responses, particularly CD8+ cytotoxic T lymphocytes (CTLs) mediated responses, with the assistance of suitable adjuvants which enhance the immune response [71]. CTLs are primarily responsible for the recognition and suppression of tumor cells: they recognize tumor-associated antigenic epitopes that are expressed by human leukocyte antigen (HLA) class I molecules on tumor cells, then attack, proliferate and cause cancer cell lysis [72]. The making of a cancer vaccine is based on the fundamental choice of the antigen together with its characteristics. The ideal antigen must be expressed only by tumor cells and not by normal cells; it must be present on all tumor cells in such a way that the cancer does not escape immune attack due to antigen downregulation; it must be highly immunogenic [73]. Antigens meeting all of these criteria do not exist, but they can be classified as (i) tumor-associated antigens (TAAs) and (ii) tumor-specific antigens (TSAs).

TAAs are self-antigens that can be expressed by both tumor cells and normal cells. However, the use of these antigens presents a major obstacle as the T lymphocytes that bind to TAAs are typically cleared by immune tolerance mechanisms. Immune tolerance is an important means by which growing tumors manipulate the tumor microenvironment with the aim of preventing elimination by the host immune system. Tolerance is the result of the so-called immune checkpoints. These pathways include the presence of protein receptors on the surface of immune system cells, which, if bound to specific ligands, prevent the immune system from attacking cells indiscriminately. Often, these specific ligands are expressed by tumor cells. In this way, cancer cells are able to trigger inhibitory signals that make immune cells inert or tolerant. For example, the binding between the checkpoint proteins PD-1 on the surface of T cells and the PD-L1 inhibitory receptors on tumor cells not only prevents T cells from attacking malignant cells in the body, but also inhibits T cells proliferation [74,75]. Therefore, to be effective, a cancer vaccine based on TAAs must be able to contrast and interrupt the tolerance mechanisms described above. Moreover, many vaccine clinical studies have revealed that the triggered immune response does not have significant efficacy. Indeed, Hollingsworth and Jansen [73] explain that these vaccines allow a very low rate of activation and proliferation of CD 8 T cells, compared to the amounts typically obtained with antiviral vaccines.

Conversely, TSAs, or tumor neoantigens, are truly tumor-specific as they are expressed only by cancer cells, highly immunogenic and dependent on tumor type [76]. Most importantly, these immunogenic neoantigens are the result of hotspot mutations which occur in numerous cancer patients so they are unique to each patient. For this reason, the design of a cancer vaccine based on single neoantigens requires a patient-specific approach. It starts with the sequencing of the tumor genome, then the mutations are identified and at the end, through computerized algorithms, the neoantigens are designed, possibly with experimental confirmation [73].

Antigens to be employed in therapeutic vaccines can be given to patients in the form of peptide-restricted epitopes, proteins, heat shock proteins transporting immunogenic peptides, recombinant viruses, autologous or allogeneic tumor cells, or DNA constructs. The main routes of administration are the direct injection, the coupling to immunostimulatory adjuvants and the ex-vivo loading of antigen presenting cells (APCs), typically dendritic cells (DCs) [72,77].

A peptide-based vaccine consists of immunogenic restricted epitopes, usually from tumor-specific or tumor-associated antigens, typically conjugated to a carrier protein [78]. The binding with the carrier protein helps enhance immune response thanks to its immunogenic properties and its characteristic of increasing the half-life of the epitope. Cancer cells can be distinguished from healthy cells because most of them are capable of overexpressing proteins or are characterized by mutations of those same proteins. Therefore, any gene product that has a mutated form with respect to healthy cells can be a potential target for the vaccine, just like the HER2 gene and its modified form HER2/neu that is one of the most studied oncogenes in cancer. HER2/neu is an epidermal growth factor (EGF) receptor. Human EGF is a protein naturally produced by the human body, which when attached to another protein, such as HER2 or CerbB2, stimulates the multiplication of cancer cells. About 30% of breast cancers develop along with the amplification of the HER2/neu gene or the overexpression of its protein product [79]. Its overexpression also takes place in other cancers such as stomach, ovarian, and colorectal cancers and in aggressive forms of pancreatic cancer, uterine cancer, and non-small cell lung cancer. Tumors characterized by HER2 overexpression are known as HER2-positive and can be highly aggressive [80]. Several peptide-based vaccines derived from the HER2 receptor have been developed in the last few years. NeuVax^{TM} is a 9-aminoacid peptide resulting from the combination of the extracellular domain of HER2 with granulocyte-macrophage colony-stimulating factor (GM-CSF). The vaccine stimulates specific CD8+ cytotoxic T lymphocytes to destroy HER2 positive cancer cells by promoting the lysis of breast cancer cells. Actually, Brossart et al. demonstrated that CTLs are also able to lyse other types of cancer cells expressing the HER2/neu oncogene, such as in the case of colon carcinoma and renal cell carcinoma [81]. Initial clinical trials showed that NeuVax^{TM} was well tolerated by patients but had no significant effect in preventing recurrence (NCT01479244). Currently, two Phase II clinical trials are examining NeuVax^{TM} treatment combined with trastuzumab in HER2-positive breast cancer to evaluate the real effectiveness of a combination therapy (NCT01570036, NCT02297698). Trastuzumab works by interfering with one of the ways in which breast cancer cells grow and divide. In particular, it attaches to the HER2 protein, thereby preventing the human growth factor in the epidermis from reaching the neoplastic cells and, consequently, preventing their division and growth. Trastuzumab also acts as a stimulator of the immune cells to improve their killing action against cancer cells. The main objectives of these two trials are disease-free survival at 24 months and 36 months and invasive disease-free survival from the beginning of therapy until the end of the study (5 years). Only NCT01570036 has been completed so far, by also achieving promising results in terms of rate of survival: disease-free survival as percentage of participants who survived at 24 months was 89.8%, at 36 months was 86.7%.

GP2 is another 9-amino acid peptide derived from HER2. Likewise, for this vaccine the preliminary In vitro tests have confirmed the ability to induce CTLs immune responses. Clinical studies have also shown that it is well tolerated and have also found an increase in HER2-specific CTLs. These are just two of the peptide-based vaccines designed specifically for breast cancer treatment [78]. It is also important to mention another peptide-based vaccine, HerVaxx. It consists of three peptides isolated from HER2 extracellular domain and linked to another carrier protein that is diphtheria toxin [82]. In the last few months a Phase 2 HerVaxx study was updated to evaluate the overall survival and the progression-free survival and to measure the efficacy, safety and immune response in 68 patients with metastatic gastric cancer overexpressing the HER-2 protein (NCT02795988).

In the examples described so far, the vaccine is directly injected to the patient. As already mentioned, however, it is possible to administer the vaccine also through the ex vivo loading of DCs. DCs are specialized antigen presenting cells (APCs) fundamental to the proper behavior of the immune system because they play as mediators between innate and adaptive immune responses. In particular, they act as sentinels throughout the organism and are able to recognize antigens. At the same time, through the processing of the antigenic material and the so-called "antigen presentation", they allow the activation of the immune response implemented by the T lymphocytes. More precisely, DCs activate T lymphocytes through major histocompatibility complex (MHC) signaling [83]. These properties led to many attempts in the development of DC-based vaccines, achieving also promising results in individuals with advanced-stage cancers [84]. In 2010, the FDA approved the sipuleucel-T (PROVENGE$^{®}$) for metastatic castration-resistant prostate cancer (mCRPC) [85]. Sipuleucel-T is an autologous DC vaccine based on enriched blood APCs cultured with a recombinant protein derived from the combination of prostatic acid phosphatase (PAP) with granulocyte-macrophage colony-stimulating factor (GM-CSF), that are, respectively, an antigen expressed in prostate cancer tissue and an immune cell activator [86]. Sipuleucel-T is a personalized therapy: DCs are directly taken from the peripheral blood mononuclear cells (PBMC) of the patient. Then, after ex vivo loading of PAP and activation, the vaccine is administered to the patient by infusion. In 2010, a Phase III clinical trial enrolled 512 mCRPC patients to receive sipuleucel-T (NCT00065442). The study primarily aimed at evaluate the overall survival compared to a placebo: it resulted 25.8 versus 21.7 months. Despite this positive outcome, the complex formulation and consequently the high cost of production of sipuleucel-T have hindered a more extensive diffusion. Nonetheless, sipuleucel-T has given a contribution to the development of numerous other DC-based vaccines. As various kind of DC vaccines are currently undergoing clinical trials for different cancer disease, many therapeutic candidates are envisioned, and the most relevant ones are listed in Table 1.

Table 1. Representative selection of the currently active clinical trials investigating dendritic cell-based cancer vaccines.

Condition	Treatment	Clinical Phase	NCT Identifier
Breast cancer	HER-2 pulsed DC Vaccine	Phase I	NCT02063724
Brain tumors	Autologous DCs pulsed with CSC Lysate	Phase I	NCT02010606
Prostate cancer	Autologous DCs loaded with mRNA from Primary prostate cancer tissue + hTERT + survivin	Phase I/II	NCT01197625
Sarcoma/soft tissue Sarcoma/bone sarcoma	DC vaccine + tumor lysate + imiquimod	Phase I	NCT01803152
Brain Metastases	Personalized cellular vaccine: tumor antigen mRNA-pulsed autologous DCs	Phase I	NCT02808416
Newly diagnosed glioblastoma	AV-GBM-1: autologous DCs loaded with autologous tumor antigens derived from self-renewing TICs	Phase II	NCT03400917
Multiple myeloma	ASCT + DC myeloma fusion vaccine + MAb CT-011 (pidilizumab)	Phase II	NCT01067287
AML	DC AML fusion vaccine	Phase II	NCT01096602
Advanced breast cancer	DCs co-cultured with CIK cells + capecitabine monotherapy	Phase II	NCT02491697

Legend: DC = dendritic cell, CSC = cancer stem cell, hTERT = human telomerase reverse transcriptase, TIC = tumor-initiating cell, ASCT = autologous stem cell transplantation, Mab = monoclonal antibody AML = acute myelogenous leukemia, CIK cells = cytokine-induced killer cells.

These studies highlight that these new strategies can become real possibilities in the fight against cancer, therefore they must be increasingly explored, as they can seriously improve and enhance the capabilities of existing cancer drugs.

7. EVs in Anti-Tumor Immunotherapy

The capabilities of the DCs as powerful and versatile APCs make them suitable to be the vehicles in cancer vaccines and anti-tumor immunotherapy. However, many drawbacks hinder their use in clinical treatments. DCs are a heterogeneous cellular population comprising various subtypes having different functional properties. Depending on the subset and on the received stimuli, DCs can display different capacities for antigen presentation, migration, and cytokine secretion. In particular, they can induce different T cell behaviors, by polarizing them into effector or tolerogenic cells [87]. Because of this heterogeneity they can thus promote either antitumor activity or regulation of immune tolerance, which is known to be a very limiting factor in vaccine success [88]. For example, some soluble immunosuppressive cytokines produced by tumor cells can convert immature DCs into tolerogenic DCs. These converted dendritic cells activate regulatory T (Treg) cells. Treg cells inhibit or downregulate proliferation of T cells such as CD4+ and CD8+, by promoting tumor proliferation [18]. Another limitation in the use of DCs is also due to their difficult storage aimed at maintaining their efficacy even for long periods of time. Finally, applying such therapies to a broad population is expensive and needs to meet rigorous quality control parameters. The exploitation of DC-derived exosomes (Dex) has proven to be a possible solution to the problems encountered in DC-based immunotherapy. Dex are characterized by unique molecular composition that allows them to maintain the immunostimulatory abilities of DCs. Indeed, Dex contain MHC-I and MHC-II molecules, which can respectively stimulate cytotoxic and helper T cells (T_h cells), together with costimulatory (CD86, CD40) and adhesion molecules (ICAMs), which can elicit strong immune responses toward cancer cells [89]. Furthermore, the lipid composition of exosomal membrane allows to storage Dex at −80 °C for more than 6 months maintaining high stability. Finally, compared to other types of cancer vaccines, cell-free treatment such as Dex-based treatment may be less vulnerable to immunotolerance and other immunomodulatory mechanisms that usually occur in tumors [90]. Some important preclinical studies carried out to evaluate the immunogenicity of Dex and their possibility of use in the production of new therapeutic vaccines are collected in Table 2.

Table 2. Preclinical studies evaluating Dex immunogenicity for cancer vaccines.

Authors	Method	Main Outcomes	Refs
Théry C. et al.	In vitro	Dex can transfer functional peptide-loaded MHC class I and II complexes to DCs.	[91]
André F. et al.	In vitro and in vivo	Dex harbouring MHC class I/peptide complexes require DC for efficient priming of CTLs.	[92]
	In vivo	Dex mimic the capacity of mature DCs to initiate peptide-specific CD8+ T cell responses.	
Segura E. et al.	In vitro	Dex from immature DCs (imDC) and mature DCs (mDC) have different protein composition due to maturation signals. MHC class I molecules are up-regulated in mDC and reduced in mature exosomes. Molecules stimulating CD4 T cells are up-regulated in mDC and mature exosomes.	[93]
Sprent J.	In vitro	Peptide-pulsed Dex are immunogenic for CD8+ T cells also in the absence of APCs.	[94]
	In vivo	Peptide-loaded Dex induce high proliferative responses and CTLs induction, so priming CD8+ T cells.	
Viaud S. et al.	In vivo	Dex administration promotes proliferation, activation and cytotoxicity of NK cells.	[95]
	In vitro	Human Dex harbouring IL-15Rα lead to NK cell proliferation and IFNγ production	

Many Phase I clinical trials performed on various cancers have shown exosome-based vaccines to be safe, so exosomes could effectively represent a new approach for cancer immunotherapy. In particular, a Phase I clinical study investigated not only the feasibility and efficacy, but also the safety of administering autologous Dex loaded with peptides of melanoma-associated antigen (MAGE) to individuals with advanced NSCLC [96]. MAGE gene-derived peptides are widely expressed in many tumor and are able to stimulate antigen-specific CTL responses against MAGE-expressing cancer cells, resulting in tumor cell lysis. Therefore, thanks to their effectiveness in preventing and treating diverse cancers, MAGE has been used as a target for cancer immunotherapies [97]. In this Phase I trial, Morse et al. [96] isolated Dex by ultracentrifugation from peripheral blood mononuclear cells and loaded them with MAGE-A3, -A4, -A10, and MAGE-3DPO4 peptides. MAGE peptides were loaded directly into Dex after the purification step or indirectly into cultured DCs. The three patient groups in the clinical study were given the same Dex dose of 1.3×10^{13} MHC class II molecules in a volume of 3 mL once a week for 4 weeks, varying both the peptide loading method and the peptide concentration in each cohort. Dex therapy administered to 9 patients resulted non-toxic, tolerated and without the appearance of autoimmune responses. By contrast, In vitro immunologic analysis detected an increase in T cell activity in only one of tested patients, probably due to T_{reg}-mediated suppression of immune cells. In 2/3 patients who had analyzable samples, an increase in T_{reg} was observed after the conclusion of Dex therapy. Other possible explanations were related to not well performing or low-sensitive assays, insufficient antigen presentation, or the absence of circulating antigen-specific T cells. Actually, it was hypothesized that might be due to activation of natural killer (NK) cells, that are cytotoxic lymphocytes involved in the innate immune system. NK cells are named "natural" because they recognize tumor or infected cells without the need of antigen-specific cell surface receptors or any preparatory activation and kill them by producing cytokines [98]. Confirming the hypothesis, Morse et al. [96] observed an increased activity of NK cells after immunization in two over four of the analysed patients, but in general, the clinical observations confirmed the successful immunization of the treated patients. Finally, the main clinical result was a very good disease stability, in both two patients with disease progression and in two initially stable patients. A similar study used MAGE peptides as cancer vaccine and was carried out in 15 patients with stage III/IV malignant melanoma [99]. Again, MAGE peptides were loaded directly into autologous DC-derived exosomes or indirectly into cultured DCs and administered to patients every week for 4 weeks. All patients underwent assessment of tumor status at 2 weeks after the fourth exosome vaccination. Vaccination was well-tolerated and resulted in a therapy response in four patients; in particular, a stabilization of the disease was proved in two patients receiving the highest dosage of directly-loaded exosomes. In contrast, one partial response and a minor response were identified in two patients who were then subjected to a continuation therapy, allowing for stabilization. Unfortunately, as in the previous study by Morse et al., no significant T cell response was observed.

Briefly, although there are clear advantages in Dex-based vaccine applications, not all treated patients show satisfactory responses due to the obstacles described above. Tumor-derived exosomes (Tex) are another type of exosomes investigated for the improvement of antitumor immune responses. Tex contain tumor-associated antigens expressed in the parental tumor cells and major histocompatibility complex (MHC) class I molecules [100]. Thus, Tex could present tumor antigens to DC and induce CD8+ T cell-dependent antitumor immune responses. In addition, tumor cells release a larger number of exosomes compared to healthy cells-derived exosomes. As a result, they are designated to become a new source of tumor antigens and thus a novel type of cell-free cancer vaccine [101,102]. However, previous studies reported that Tex can promote immune escape through different immunosuppressive mechanisms as described in the review by Whiteside et al. [103]. Generally, Tex are involved in the progression, regression and drug resistance of tumors and contribute to the development of metastases because they regulate immune responses by mediating communication between immune cells and cancer cells [104]. Many studies have verified the feasibility and functionality of Tex in activating immune responses against cancer in mouse models. Bu et al. [105] showed that a single dose of L1210 leukaemia-derive exosome-based vaccine not only inhibited tumor formation but also promoted protection against tumor growth in syngeneic mice. To assess the efficacy of the vaccine, the treated mice and the control group, consisting of unvaccinated mice, were stimulated with L1210 tumor cells 2 weeks after vaccination. Regarding tumorigenesis, 87.5% of the vaccinated mice were found to be tumor free, while in all untreated mice the tumor appeared in less than 10 days. Meanwhile, the rate of protection against tumor growth after 60 days was 85% in the vaccinated with a dose of 5 µg and 60% with the dose reduced to 2.5 µg. Finally, with regard to the immune response, CTL activity against L1210 was also observed and was better than the control group. Other studies on leukaemia vaccination proposed Tex as a possible antigen source for DC-based vaccination because of: Tex contain high amounts of tumor-associated antigens; Tex have markers that make easier the uptake by the DCs; the antigen presentation through MHC by DC induce CD8 T cell-dependent antitumor immune responses. Thus, a new perspective is emerging that is to produce cancer vaccines based on DCs loaded with exosomes derived from tumor cells, with the aim of exploiting the advantages deriving from the combined use of Tex and DCs. Indeed, Yao et al. [106] demonstrated that exosome-pulsed DCs induced stronger antitumor immunity than exosomes and DCs alone. Gu et al. have shown that In vitro, in the presence of DCs, it is possible to avoid inhibition of T_h lymphocytes activation by Tex and for this reason they have considered vaccination based on DC loaded with Tex (DC-TEX): they have proposed to load the DCs of patients with Tex circulating in the peripheral blood and to test the vaccine in a myeloid leukaemia mouse model [107]. In the end, they confirmed that Tex-pulsed DCs really increased the survival time of mice and determined a conspicuous activation of CTLs thanks to the combined use of Tex and DCs. Other studies that have evaluated the possibility of using extracellular vesicles as a vehicle in cancer vaccines are shown in Table 3.

Table 3. Preclinical studies investigating the use of EVs in cancer vaccines.

Therapeutic Agent	Condition	Outcome	Refs
Irradiated C6 glioma cell-derived MVs (IR-MVs)	Malignant C6 glioma	In vivo vaccination with IR-MVs promotes antitumor immune response leading to the apoptosis of glioblastoma cells and increases Th cells and CTL infiltration into the tumor.	[108]
DC-derived-exosomes functionalized with costimulatory molecules, MHCs, antigenic Ovalbumin peptide and anti-CTLA-4 antibody (EXO-OVA-mAb)	B16-OVA melanoma tumor model	Exosomes are targeted to T cells in vivo. EXO-OVA-mAb are able to effectively prime T-cell activation and proliferation, In vitro and in vivo. The fraction of memory T cells is increased in mice treated with vaccination. The antitumor efficacy is confirmed by the infiltration of both CD4 + and CD8 + cells and the CTLs/Treg ratio within the tumor site of vaccinated mice.	[109]
Interferon-γ-modified prostate cancer cell-derived exosomes	RM-1 prostate cancer	Vaccine induces macrophages differentiation and the production of antibodies, reduces tumor angiogenesis and metastasis rate, inhibits tumor growth and prolongs survival time of mice with metastatic prostate cancer.	[110]

Table 3. *Cont.*

Therapeutic Agent	Condition	Outcome	Refs
Interferon-γ-modified prostate cancer cell-derived exosomes + IFN-γ-modified RM-1 cell vaccine	RM-1 prostate cancer	Exosomal vaccine improves the T cell response generated by the tumor cell vaccine and downregulates in the expression of IDO1 and PD-L1 immune checkpoints. Combination therapy show the highest tumor-specific cytotoxic activities compared to vaccine monotherapies and tumor growth is significantly suppressed.	[110]
Mature DCs pulsed with ovalbumin protein-pulsed DC-derived exosomes (EXO-pulsed DCs)	B16-OVA melanoma tumor model	EXO-pulsed DCs stimulate CD8+ T-cell proliferation and differentiation into CTL effectors In vitro and in vivo. EXO-pulsed DCs induce stronger immunity against lung tumor metastases and can eradicate established tumors. They also induce strong long-term OVA-specific CD8+ T-cell memory	[111]

As described above, exosomes, and more broadly EVs, are also natural carriers of RNA and can be employed to deliver siRNA in silencing of genes for cancer treatment [112,113]. In addition, silencing of immunosuppressive genes via siRNA combined with immune checkpoint blockade therapy has been found to be a promising practice in new cancer immunotherapy applications [114,115]. As already mentioned, immune checkpoint blockade plays a key role in preventing the interruption of immune responses. Actually, immune checkpoint inhibitors, including monoclonal antibodies against programmed death 1 (PD-1), inhibit immunosuppressive molecules and restores the ability of the CTLs to kill cancer [116]. For all these reasons, Matsuda et al. used extracellular vesicles (EVs) for targeting β-catenin in hepatocellular carcinoma (HCC) via intrahepatic delivery of siRNA [117] β-catenin takes part in the signaling pathway of cell proliferation and often its alteration causes the development of carcinomas. In particular, Wnt/β-catenin pathway activation is associated with poor spontaneous T cell infiltration across most human cancers; thus, it contributes to immune escape [118,119]. By taking into account that responsiveness to anti-PD-1 therapy needs the presence of tumor antigen-specific T cells within tumor tissue, a poor or even absent T-cell infiltration can result in immune deserts and weak response to immunotherapy. Moreover, mutations in gene encoding β-catenin were identified among the most frequent alterations associated to the development of HCC [120,121]. Consequently, in this study, together with the β-catenin siRNA-loaded EVs, also anti-PD-1-based therapy were administered in order to reduce the tumor growth and at the same time, to improve the therapeutic action of immune checkpoint inhibitors. A synthetic model of hepatocellular cancer was induced in mouse livers by the co-expression of c-tyrosine-protein kinase Met (cMET) and mutant β-catenin via hydrodynamic injection (HDI) of DNA and plasmids [122]. The EVs were derived from bovine milk, which is a safe and scalable source of EVs [123]. First, the therapeutic effectiveness of anti-PD-1 was evaluated. Three weeks after HDI, two groups of mice bearing HCC were respectively administered with 250 μg/mouse of anti-PD-1 and with phosphate-buffered saline (PBS) for control measurements, for 2 weeks. The transfection of Gaussia luciferase (g-luc) and its expression level allowed to verify the success of the therapy against tumor growth in terms of relative luminescence units [124]. It was observed a reduced rate of tumor growth over a 6-week period and a very extended survival of the mice which received anti-PD-1 therapy, with a mean of 119 days compared to 96 in the control group. Secondly, to evaluate the efficacy of combined treatment with both therapeutic EVs and anti-PD-1 in targeting β-catenin, four groups of mice received different treatments three weeks after HDI for a period of 2 weeks. The first group received only 250 μg of anti-PD-1, the second group received only EVs (2×10^{12} particles/body), and the third group both anti-PD-1 and EVs. Last group was used as control and did not receive any treatment. Matsuda et al. [117] first demonstrated the In vitro effectiveness of β-catenin siRNA delivery via EVs: they incubated HepG2 cells with siRNA-loaded EVs and then found a decreased expression of β-catenin protein. Subsequently, they evaluated the effect of siRNA delivery in vivo. Tumor growth rate decreased with anti-PD-1, EV or both in the first three weeks after stopping treatment, but later a relapse was noted in 38% and 100% of the first and the second groups, respectively. Finally, the mice treated with both anti-PD-1 and EVs did not display any relapse, on the contrary they showed a considerable decrease in the tumor growth rate (between 3 and 12 weeks) even greater than that with either treatment alone. Therefore, this confirmed that targeting an oncogenic factor can improve the therapeutic effect of anti-PD-1. Ultimately, the combined treatment produced the highest degree of infiltration of CD8+ T cells within the tumor microenvironment. Furthermore, it has been confirmed that inhibition of β-catenin signaling in HCC improve the activation of specific T cells, promoting their infiltration into the tumor microenvironment and preventing CD8+ T-cell exhaustion following an initial response to anti-PD-1 therapy. However, there is still no scientific evidence to elucidate the mechanism used by β-catenin to promote CD8+ infiltration. Table 4 shows the currently active or already completed clinical trials involving EVs in immunotherapy. Figure 2 illustrates the main EVs application in cancer immunotherapy.

Table 4. Collected currently active or completed clinical trials investigating the use of EVs-based immunotherapies.

Condition	Treatment	Year	Clinical Phase	NCT Identifier and References
Advanced NSCLC	Dex loaded with the MAGE tumor antigens	2005	Phase I	[96]
Metastatic melanoma	Autologous exosomes pulsed with MAGE 3 peptides	2005	Phase I	[99]
Colorectal cancer	Ascites-derived exosomes (Aex) in combination with GM-CSF	2008	Phase I	[125]
Melanoma	Human Dex bearing NKG2D ligands	2009	Phase I	[95]
NSCLC	Tumor Antigen-loaded Dex	2010	Phase II	NCT01159288
Unresectable NSCLC	IFN-γ-Dex loaded with MHC class I- and class II-restricted cancer antigens	2015	Phase II	[89]

NSCLC = non-small-cell lung cancer, IFN = interferon γ, GM-CSF = granulocyte–macrophage colony-stimulating factor.

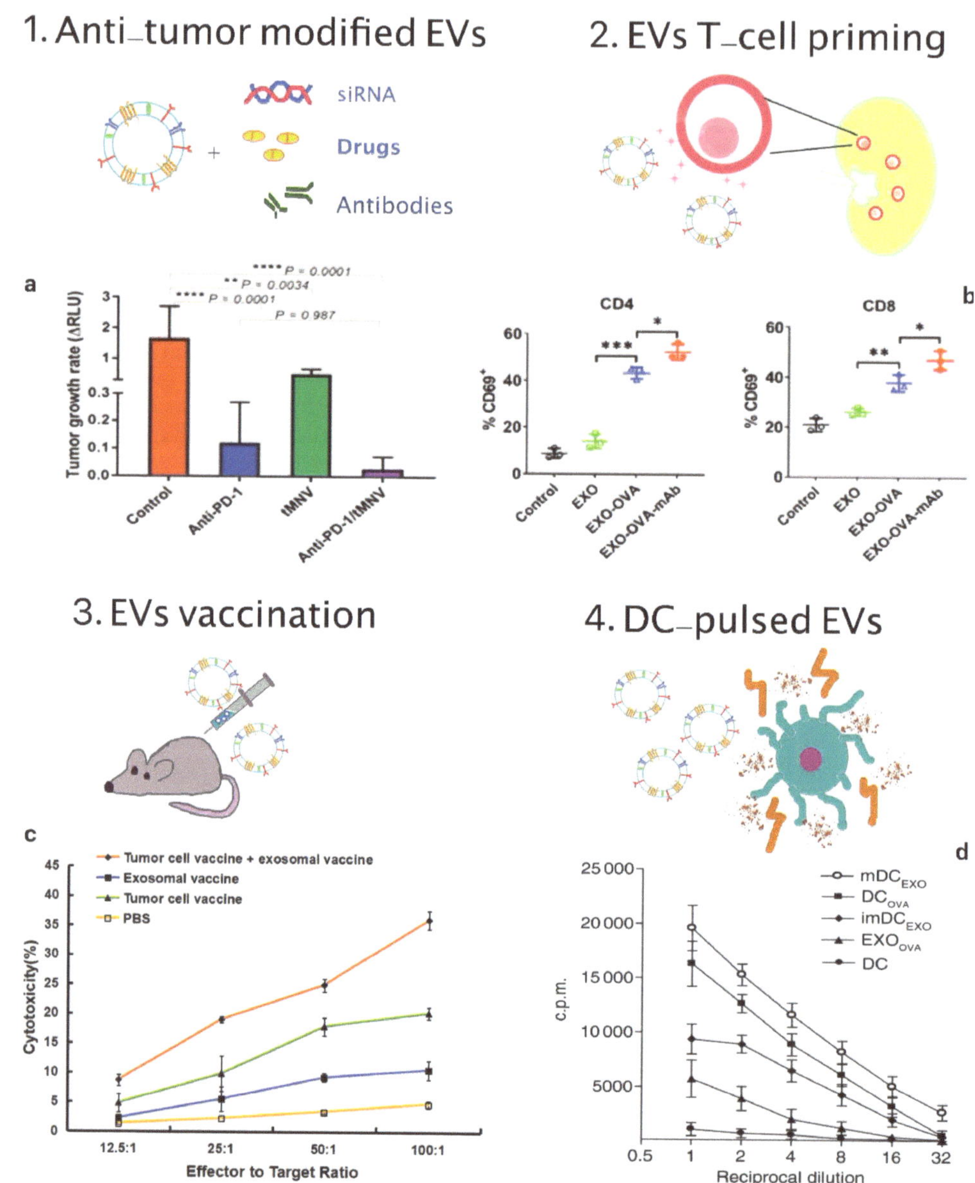

Figure 2. EVs applications in cancer immunotherapy. (**1**) Scheme of EVs as cargos of siRNAs, drugs and monoclonal antibodies and (**a**) therapeutic effect on tumor growth rate of anti-PD-1 and tMNV-directed therapy targeting B-catenin. Reproduced from [117] with the permission of Jhon Wiley and Sons. *p*-Values as indicated, one-way ANOVA analysis. (**2**) Scheme of EVs-mediated T-cell activation and (**b**) data analysis of CD69, a T-cell activation marker on CD4+ and CD8+ T cells following incubation with different exosomes formulations. Reproduced from [109] with permission of Elsevier. * $p < 0.05$, ** $p < 0.01$, *** $p < 0.001$, one-way ANOVA analysis. (**3**) Scheme of EVs vaccination and (**c**) tumor specific cytotoxic activity of the combination therapy involving exosomal vaccine and tumor cell vaccine against prostate cancer cells. Reproduced from [110] with permission of John Wiley and Sons. (**4**) Scheme of DC-pulsed EVs and (**d**) proliferative response of CD8+ T cells co-cultured with EXOOVA (10 μg/mL), DCOVA, mDCEXO and imDCEXO (3×10^4 cells/well), determined by (3H)thymidine uptake assay after two days. Reproduced from [111] with permission of Jhon Wiley and Sons.

8. Conclusions

The research on extracellular vesicles has recently intensified considerably, leading to the achievement of important results in the application EVs for gene and drug delivery. In particular, exosomes have proved to be the ideal candidates to be used as therapeutic delivery vehicles. Based on this, the use of extracellular vesicles has also been extended to the field of cancer immunotherapy. A particular form of immunotherapy is represented by therapeutic anticancer vaccines. Among the different types of cancer vaccines investigated, the most promising are the cancer vaccines based on dendritic cells, as these cells facilitate the triggering of the immune response. Several clinical trials are currently active to evaluate the efficacy of therapeutic vaccines against various forms of cancer. As shown by numerous pre-clinical tests, dendritic cell-derived exosomes can also be used in the production of cancer vaccines. Clinical trials have confirmed the safety and feasibility of exosome-based cancer vaccines; however, some studies have not been fully satisfactory. In fact, vaccines are generally well tolerated by patients undergoing treatment, but often no significant immune response is observed. This indicates that significant progress has been made in building safe delivery vehicles but, at the same time, the clinical efficacy of extracellular vesicles based-cancer vaccines remains to be determined. The exosomes derived from tumor cells have been proposed as an alternative to the previous ones, as they are able to improve the antitumor immune responses. Nevertheless, some studies have shown that exosomes derived from tumor cells could be involved in processes that promote tumor proliferation, therefore, they are not preferable. To overcome these drawbacks, the possibility of producing cancer vaccines based on dendritic cells loaded with exosomes derived from tumor cells has recently emerged, but they are still under investigation. The use of extracellular vesicles in immunotherapy therefore seems to be hindered only by technological problems and not by qualitative ones. Therefore, an ever-increasing effort in this direction could lead to tangible results and above all to a new and innovative way to fight cancer.

Author Contributions: Conceptualization, C.F., S.B. and V.C.; Methodology, C.G.; Investigation, C.G.; Resources, C.F., S.B. and V.C.; Data Curation, M.C. and V.C.; Writing—Original Draft Preparation, C.G. and M.C.; Writing—Review and Editing, C.F. and V.C.; Supervision, C.F., S.B. and V.C.; Project Administration, C.F., S.B. and V.C.; Funding Acquisition, C.F., S.B. and V.C. All authors have read and agreed to the published version of the manuscript.

Funding: Part of this work has received funding from the European Research Council (ERC) under the European Union's Horizon 2020 research and innovation programme (grant agreement no. 678151—Project Acronym 'TROJANANOHORSE'—ERC starting Grant). Part of this work has been also supported by a grant from the Spanish Ministerio de Ciencia, Inovación y Universidades through the Grant: RTI2018-094734-B-C22. S.B and C.F. would like to thank Generalitat de Catalunya for the consolidated Research Group Grant 2017SGR 01559 to GEMAT.

Acknowledgments: Part of this work derives from the Maser Thesis monography of Giacobino Carla, Entitled "Extracellular vesicles and their current role in cancer immunotherapy", dissertation hold on 7 October 2020 at Politecnico di Torino, Turin, Italy.

Conflicts of Interest: There are no conflict to declare.

References

1. Susa, F.; Limongi, T.; Dumontel, B.; Vighetto, V.; Cauda, V. Engineered extracellular vesicles as a reliable tool in cancer nanomedicine. *Cancers* **2019**, *11*, 1979. [CrossRef] [PubMed]
2. Ståhl, A.L.; Johansson, K.; Mossberg, M.; Kahn, R.; Karpman, D. Exosomes and microvesicles in normal physiology, pathophysiology, and renal diseases. *Pediatr. Nephrol.* **2019**, *34*, 11–30. [CrossRef]
3. Sutaria, D.S.; Badawi, M.; Phelps, M.A.; Schmittgen, T.D. Achieving the Promise of Therapeutic Extracellular Vesicles: The Devil is in Details of Therapeutic Loading. *Pharm. Res.* **2017**, *34*, 1053–1066. [CrossRef]
4. Kalra, H.; Drummen, G.P.; Mathivanan, S. Focus on Extracellular Vesicles: Introducing the Next Small Big Thing. *Int. J. Mol. Sci.* **2016**, *17*, 170. [CrossRef]
5. Yuana, Y.; Sturk, A.; Nieuwland, R. Extracellular vesicles in physiological and pathological conditions. *Blood Rev.* **2013**, *27*, 31–39. [CrossRef]

6. Van der Pol, E.; Böing, A.N.; Harrison, P.; Sturk, A.; Nieuwland, R. Classification, functions, and clinical relevance of extracellular vesicles. *Pharmacol. Rev.* **2012**, *64*, 676–705. [CrossRef] [PubMed]
7. Théry, C.; Witwer, K.W.; Aikawa, E.; Alcaraz, M.J.; Anderson, J.D.; Andriantsitohaina, R.; Antoniou, A.; Arab, T.; Archer, F.; Atkin-Smith, G.K.; et al. Minimal information for studies of extracellular vesicles 2018 (MISEV2018): A position statement of the International Society for Extracellular Vesicles and update of the MISEV2014 guidelines. *J. Extracell. Vesicles* **2018**, *7*, 1535750. [CrossRef]
8. Tricarico, C.; Clancy, J.; D'Souza-Schorey, C. Biology and biogenesis of shed microvesicles. *Small GTPases* **2017**, *8*, 220–232. [CrossRef]
9. Chen, Y.; Li, G.; Liu, M.L. Microvesicles as Emerging Biomarkers and Therapeutic Targets in Cardiometabolic Diseases. *Genom. Proteom. Bioinf.* **2018**, *16*, 50–62. [CrossRef]
10. Xu, X.; Lai, Y.; Hua, Z.C. Apoptosis and apoptotic body: Disease message and therapeutic target potentials. *Biosci. Rep.* **2019**, *39*, BSR20180992. [CrossRef] [PubMed]
11. Jurj, A.; Zanoaga, O.; Braicu, C.; Lazar, V.; Tomuleasa, C.; Irimie, A.; Berindan-Neagoe, I. A comprehensive picture of extracellular vesicles and their contents. Molecular transfer to cancer cells. *Cancers* **2020**, *12*, 298. [CrossRef]
12. Battistelli, M.; Falcieri, E. Apoptotic bodies: Particular extracellular vesicles involved in intercellular communication. *Biology* **2020**, *9*, 21. [CrossRef]
13. Van Dommelen, S.M.; Vader, P.; Lakhal, S.; Kooijmans, S.A.A.; Van Solinge, W.W.; Wood, M.J.A.; Schiffelers, R.M. Microvesicles and exosomes: Opportunities for cell-derived membrane vesicles in drug delivery. *J. Control. Release* **2012**, *161*, 635–644. [CrossRef]
14. Ha, D.; Yang, N.; Nadithe, V. Exosomes as therapeutic drug carriers and delivery vehicles across biological membranes: Current perspectives and future challenges. *Acta Pharm. Sin. B* **2016**, *6*, 287–296. [CrossRef] [PubMed]
15. Andreu, Z.; Yáñez-Mó, M. Tetraspanins in extracellular vesicle formation and function. *Front. Immunol.* **2014**, *5*, 442. [CrossRef]
16. Zhang, Y.; Liu, Y.; Liu, H.; Tang, W.H. Exosomes: Biogenesis, biologic function and clinical potential. *Cell Biosci.* **2019**, *9*, 1–18. [CrossRef] [PubMed]
17. Tai, Y.L.; Chen, K.C.; Hsieh, J.T.; Shen, T.L. Exosomes in cancer development and clinical applications. *Cancer Sci.* **2018**, *109*, 2364–2374. [CrossRef] [PubMed]
18. Tian, H.; Li, W. Dendritic cell-derived exosomes for cancer immunotherapy: Hope and challenges. *Ann. Transl. Med.* **2017**, *5*, 3–5. [CrossRef]
19. Kugeratski, F.G.; Kalluri, R. Exosomes as mediators of immune regulation and immunotherapy in cancer. *FEBS J.* **2021**, *288*, 10–35. [CrossRef]
20. Syn, N.L.; Wang, L.; Chow, E.K.; Lim, C.T.; Goh, B.C. Exosomes in Cancer Nanomedicine and Immunotherapy: Prospects and Challenges. *Trends Biotechnol.* **2017**, *35*, 665–676. [CrossRef]
21. Villata, S.; Canta, M.; Cauda, V. EVs and Bioengineering: From Cellular Products to Engineered Nanomachines. *Int. J. Mol. Sci.* **2020**, *21*, 6048. [CrossRef]
22. Alvarez-Erviti, L.; Seow, Y.; Yin, H.; Betts, C.; Lakhal, S.; Wood, M.J.A. Delivery of siRNA to the mouse brain by systemic injection of targeted exosomes. *Nat. Biotechnol.* **2011**, *29*, 341–345. [CrossRef] [PubMed]
23. Yang, T.; Martin, P.; Fogarty, B.; Brown, A.; Schurman, K.; Phipps, R.; Yin, V.P.; Lockman, P.; Bai, S. Exosome delivered anticancer drugs across the blood-brain barrier for brain cancer therapy in Danio rerio. *Pharm. Res.* **2015**, *32*, 2003–2014. [CrossRef] [PubMed]
24. Patil, S.M.; Sawant, S.S.; Kunda, N.K. Exosomes as drug delivery systems: A brief overview and progress update. *Eur. J. Pharm. Biopharm.* **2020**, *154*, 259–269. [CrossRef]
25. Mizrak, A.; Bolukbasi, M.F.; Ozdener, G.B.; Brenner, G.J.; Madlener, S.; Erkan, E.P.; Ströbel, T.; Breakefield, X.O.; Saydam, O. Genetically engineered microvesicles carrying suicide mRNA/protein inhibit schwannoma tumor growth. *Mol. Ther.* **2013**, *21*, 101–108. [CrossRef]
26. Kanada, M.; Kim, B.D.; Hardy, J.W.; Ronald, J.A.; Bachmann, M.H.; Bernard, M.P.; Perez, G.I.; Zarea, A.A.; Ge, T.J.; Withrow, A.; et al. Microvesicle-Mediated Delivery of Minicircle DNA Results in Effective Gene-Directed Enzyme Prodrug Cancer Therapy. *Mol. Cancer Ther.* **2019**, *18*, 2331–2342. [CrossRef] [PubMed]
27. Yeo, R.W.Y.; Lai, R.C.; Zhang, B.; Tan, S.S.; Yin, Y.; Teh, B.J.; Lim, S.K. Mesenchymal stem cell: An efficient mass producer of exosomes for drug delivery. *Adv. Drug Deliv. Rev.* **2013**, *65*, 336–341. [CrossRef] [PubMed]
28. Markov, O.; Oshchepkova, A.; Mironova, N. Immunotherapy Based on Dendritic Cell-Targeted/-Derived Extracellular Vesicles-A Novel Strategy for Enhancement of the Anti-tumor Immune Response. *Front. Pharmacol.* **2019**, *10*, 1152. [CrossRef]
29. American Cancer Society. How Immunotherapy Is Used to Treat Cancer. Available online: https://www.cancer.org/treatment/treatments-and-side-effects/treatment-types/immunotherapy/what-is-immunotherapy.html (accessed on 10 September 2020).
30. Benecke, L.; Coray, M.; Umbricht, S.; Chiang, D.; Figueiró, F.; Muller, L. Exosomes: Small EVs with Large Immunomodulatory Effect in Glioblastoma. *Int. J. Mol. Sci.* **2021**, *22*, 3600. [CrossRef]
31. Batista, I.A.; Melo, S.A. Exosomes and the Future of Immunotherapy in Pancreatic Cancer. *Int. J. Mol. Sci.* **2019**, *20*, 567. [CrossRef]
32. Nawaz, M.; Fatima, F.; Nazarenko, I.; Ekström, K.; Murtaza, I.; Anees, M.; Sultan, A.; Neder, L.; Camussi, G.; Valadi, H.; et al. Extracellular vesicles in ovarian cancer: Applications to tumor biology, immunotherapy and biomarker discovery. *Expert Rev. Proteomics.* **2016**, *13*, 395–409. [CrossRef]
33. Bae, S.; Brumbaugh, J.; Bonavida, B. Exosomes derived from cancerous and non-cancerous cells regulate the anti-tumor response in the tumor microenvironment. *Genes Cancer* **2018**, *9*, 87–100. [CrossRef] [PubMed]

34. Taylor, D.D.; Gercel-Taylor, C. Exosomes/microvesicles: Mediators of cancer-associated immunosuppressive microenvironments. *Semin. Immunopathol.* **2011**, *33*, 441–454. [CrossRef]
35. Luan, X.; Sansanaphongpricha, K.; Myers, I.; Chen, H.; Yuan, H.; Sun, D. Engineering exosomes as refined biological nanoplatforms for drug delivery. *Acta Pharmacol. Sin.* **2017**, *38*, 754–763. [CrossRef]
36. Shenoda, B.B.; Ajit, S.K. Modulation of Immune Responses by Exosomes Derived from Antigen-Presenting Cells. *Clin. Med. Insights Pathol.* **2016**, *9*, 1–8. [CrossRef] [PubMed]
37. Martin-Gayo, E.; Yu, X.G. Role of Dendritic Cells in Natural Immune Control of HIV-1 Infection. *Front. Immunol.* **2019**, *10*, 1306. [CrossRef] [PubMed]
38. Turnis, M.E.; Rooney, C.M. Enhancement of dendritic cells as vaccines for cancer. *Immunotherapy* **2010**, *2*, 847–862. [CrossRef] [PubMed]
39. Jones, E.A.; Yang, X.; Giannoudis, P.; McGonagle, D. Mesenchymal Stem Cells: Discovery in Bone Marrow and Beyond. In *Mesenchymal Stem Cells and Skeletal Regeneration*; Jones, E.A., Yang, X., Giannoudis, P., McGonagle, D., Eds.; Academic Press: Boston, MA, USA, 2013; Chapter 2; pp. 7–13.
40. Marigo, I.; Dazzi, F. The immunomodulatory properties of mesenchymal stem cells. *Semin. Immunopathol.* **2011**, *33*, 593–602. [CrossRef] [PubMed]
41. Chen, T.S.; Arslan, F.; Yin, Y.; Tan, S.S.; Lai, R.C.; Choo, A.B.H.; Padmanabhan, J.; Lee, C.N.; de Kleijn, D.P.; Lim, S.K. Enabling a robust scalable manufacturing process for therapeutic exosomes through oncogenic immortalization of human ESC-derived MSCs. *J. Transl. Med.* **2011**, *9*, 47. [CrossRef]
42. Irfan Maqsood, M.; Matin, M.M.; Bahrami, A.R.; Ghasroldasht, M.M. Immortality of cell lines: Challenges and advantages of establishment. *Cell Biol. Int.* **2013**, *37*, 1038–1045. [CrossRef]
43. Fu, W.; Lei, C.; Liu, S.; Cui, Y.; Wang, C.; Qian, K.; Li, T.; Shen, Y.; Fan, X.; Lin, F. CAR exosomes derived from effector CAR-T cells have potent antitumour effects and low toxicity. *Nat. Commun.* **2019**, *10*, 4355. [CrossRef]
44. Sun, D.; Zhuang, X.; Xiang, X.; Liu, Y.; Zhang, S.; Liu, C.; Barnes, S.; Grizzle, W.; Miller, D.; Zhang, H.G. A novel nanoparticle drug delivery system: The anti-inflammatory activity of curcumin is enhanced when encapsulated in exosomes. *Mol. Ther.* **2010**, *18*, 1606–1614. [CrossRef] [PubMed]
45. Goh, W.J.; Lee, C.K.; Zou, S.; Woon, E.C.; Czarny, B.; Pastorin, G. Doxorubicin-loaded cell-derived nanovesicles: An alternative targeted approach for anti-tumor therapy. *Int. J. Nanomedicine* **2017**, *12*, 2759–2767. [CrossRef]
46. Dumontel, B.; Susa, F.; Limongi, T.; Canta, M.; Racca, L.; Chiodoni, A.; Garno, N.; Chiabotto, G.; Centomo, M.L.; Pignochino, Y.; et al. ZnO nanocrystals shuttled by extracellular vesicles as effective Trojan nano-horses against cancer cells. *Nanomedicine* **2019**, *14*, 2815–2833. [CrossRef] [PubMed]
47. Illes, B.; Hirschle, P.; Barnert, S.; Cauda, V.; Wuttke, S.; Engelke, H. Exosome-coated metal–organic framework nanoparticles: An efficient drug delivery platform. *Chem. Mater.* **2017**, *29*, 8042–8046. [CrossRef]
48. Sancho-Albero, M.; Navascués, N.; Mendoza, G.; Sebastián, V.; Arruebo, M.; Martín-Duque, P.; Santamaria, J. Exosome origin determines cell targeting and the transfer of therapeutic nanoparticles towards target cells. *J. Nanobiotechnology* **2019**, *17*, 1–13. [CrossRef]
49. Piffoux, M.; Silva, A.K.A.; Lugagne, J.-B.; Hersen, P.; Wilhelm, C.; Gazeau, F. Extracellular Vesicle Production Loaded with Nanoparticles and Drugs in a Trade-off between Loading, Yield and Purity: Towards a Personalized Drug Delivery System. *Adv. Biosyst.* **2017**, *1*, 1700044. [CrossRef] [PubMed]
50. Armstrong, J.P.K.; Holme, M.N.; Stevens, M.M. Re-Engineering Extracellular Vesicles as Smart Nanoscale Therapeutics. *ACS Nano* **2017**, *11*, 69–83. [CrossRef] [PubMed]
51. Nakase, I.; Futaki, S. Combined treatment with a pH-sensitive fusogenic peptide and cationic lipids achieves enhanced cytosolic delivery of exosomes. *Sci. Rep.* **2015**, *5*, 10112. [CrossRef]
52. Nakase, I.; Kogure, K.; Harashima, H.; Futaki, S. Application of a fusiogenic peptide GALA for intracellular delivery. *Methods Mol. Biol.* **2011**, *683*, 525–533. [CrossRef]
53. Hein, C.D.; Liu, X.M.; Wang, D. Click Chemistry, A Powerful Tool for Pharmaceutical Sciences. *Pharm. Res.* **2008**, *25*, 2216–2230. [CrossRef]
54. Smyth, T.; Petrova, K.; Payton, N.M.; Persaud, I.; Redzic, J.S.; Graner, M.W.; Smith-Jones, P.; Anchordoquy, T.J. Surface Functionalization of Exosomes Using Click Chemistry. *Bioconjug. Chem.* **2014**, *25*, 1777–1784. [CrossRef] [PubMed]
55. Kooijmans, S.A.A.; Fliervoet, L.A.L.; van der Meel, R.; Fens, M.H.A.M.; Heijnen, H.F.G.; van Bergen en Henegouwen, P.M.P.; Vader, P.; Schiffelers, R.M. PEGylated and targeted extracellular vesicles display enhanced cell specificity and circulation time. *J. Control. Release* **2016**, *224*, 77–85. [CrossRef]
56. Li, S.-D.; Huang, L. Targeted Delivery of Antisense Oligodeoxynucleotide and Small Interference RNA into Lung Cancer Cells. *Mol. Pharm.* **2006**, *3*, 579–588. [CrossRef] [PubMed]
57. Kim, M.S.; Haney, M.J.; Zhao, Y.; Yuan, D.; Deygen, I.; Klyachko, N.L.; Kabanov, A.V.; Batrakova, E.V. Engineering macrophage-derived exosomes for targeted paclitaxel delivery to pulmonary metastases: In vitro and in vivo evaluations. *Nanomedicine: Nanotechnol. Biol. Med.* **2018**, *14*, 195–204. [CrossRef]
58. Kim, M.S.; Haney, M.J.; Zhao, Y.; Mahajan, V.; Deygen, I.; Klyachko, N.L.; Inskoe, E.; Piroyan, A.; Sokolsky, M.; Okolie, O.; et al. Development of exosome-encapsulated paclitaxel to overcome MDR in cancer cells. *Nanomed. Nanotechnol. Biol. Med.* **2016**, *12*, 655–664. [CrossRef] [PubMed]

59. Flintoft, L. Getting RNAi therapies to the brain. *Nat. Rev. Genet.* **2011**, *12*, 296. [CrossRef]
60. Yang, J.; Wu, S.; Hou, L.; Zhu, D.; Yin, S.; Yang, G.; Wang, Y. Therapeutic Effects of Simultaneous Delivery of Nerve Growth Factor mRNA and Protein via Exosomes on Cerebral Ischemia. *Mol. Ther. Nucleic Acids* **2020**, *21*, 512–522. [CrossRef]
61. Aloe, L.; Rocco, M.L.; Balzamino, B.O.; Micera, A. Nerve growth factor: A focus on neuroscience and therapy. *Curr. Neuropharmacol.* **2015**, *13*, 294–303. [CrossRef] [PubMed]
62. Sullivan, M.; Kaur, K.; Pauli, N.; Wilson, P.C. Harnessing the immune system's arsenal: Producing human monoclonal antibodies for therapeutics and investigating immune responses. *F1000 Biol. Rep.* **2011**, *3*, 1–8. [CrossRef]
63. Marelli, G.; Howells, A.; Lemoine, N.R.; Wang, Y. Oncolytic Viral Therapy and the Immune System: A Double-Edged Sword Against Cancer. *Front. Immunol.* **2018**, *9*, 26866. [CrossRef]
64. Miliotou, A.N.; Papadopoulou, L.C. CAR T-cell Therapy: A New Era in Cancer Immunotherapy. *Curr. Pharm. Biotechnol.* **2018**, *19*, 5–18. [CrossRef]
65. Gatti-Mays, M.E.; Redman, J.M.; Collins, J.M.; Bilusic, M. Cancer vaccines: Enhanced immunogenic modulation through therapeutic combinations. *Hum. Vaccin. Immunother.* **2017**, *13*, 2561–2574. [CrossRef]
66. Beatty, P.L.; Finn, O.J. Therapeutic and Prophylactic Cancer Vaccines. In *Encyclopedia of Immunobiology*; Ratcliffe, M.J.H., Ed.; Academic Press: Oxford, UK, 2016; pp. 542–549.
67. Esmatabadi, M.J.D.; Bakhshinejad, B.; Motlagh, F.M.; Babashah, S.; Sadeghizadeh, M. Therapeutic resistance and cancer recurrence mechanisms: Unfolding the story of tumour coming back. *J. Biosci.* **2016**, *41*, 497–506. [CrossRef]
68. Terraneo, N.; Jacob, F.; Dubrovska, A.; Grünberg, J. Novel Therapeutic Strategies for Ovarian Cancer Stem Cells. *Front. Oncol.* **2020**, *10*, 319. [CrossRef]
69. Civenni, G.; Albino, D.; Shinde, D.; Vázquez, R.; Merulla, J.; Kokanovic, A.; Mapelli, S.; Carbone, G.M.; Catapano, C.V. Transcriptional Reprogramming and Novel Therapeutic Approaches for Targeting Prostate Cancer Stem Cells. *Front. Oncol.* **2019**, *9*, 385. [CrossRef] [PubMed]
70. Coventry, B.J. Therapeutic vaccination immunomodulation: Forming the basis of all cancer immunotherapy. *Ther. Adv. Vaccines Immunother.* **2019**, *7*, 1–28. [CrossRef]
71. Ye, Z.L.; Qian, Q.; Jin, H.J.; Qian, Q.J. Cancer vaccine: Learning lessons from immune checkpoint inhibitors. *J. Cancer* **2018**, *9*, 263–282. [CrossRef] [PubMed]
72. Nencioni, A.; Grünebach, F.; Patrone, F.; Brossart, P. Anticancer vaccination strategies. *Ann. Oncol.* **2004**, *15*, iv153–iv160. [CrossRef] [PubMed]
73. Hollingsworth, R.E.; Jansen, K. Turning the corner on therapeutic cancer vaccines. *NPJ Vaccines* **2019**, *4*, 1–10. [CrossRef]
74. Nurieva, R.; Wang, J.; Sahoo, A. T-cell tolerance in cancer. *Immunotherapy* **2013**, *5*, 513–531. [CrossRef] [PubMed]
75. Lin, Z.; Zhang, Y.; Cai, H.; Zhou, F.; Gao, H.; Deng, L.; Li, R. A PD-L1-Based Cancer Vaccine Elicits Antitumor Immunity in a Mouse Melanoma Model. *Mol. Ther. Oncolytics* **2019**, *14*, 222–232. [CrossRef]
76. Jiang, T.; Shi, T.; Zhang, H.; Hu, J.; Song, Y.; Wei, J.; Ren, S.; Zhou, C. Tumor neoantigens: From basic research to clinicalapplications. *J. Hematol. Oncol.* **2019**, *12*, 93. [CrossRef]
77. Kimiz-Gebologlu, I.; Gulce-Iz, S.; Biray-Avci, C. Monoclonal antibodies in cancer immunotherapy. *Mol. Biol. Rep.* **2018**, *45*, 2935–2940. [CrossRef]
78. Malonis, R.J.; Lai, J.R.; Vergnolle, O. Peptide-Based Vaccines: Current Progress and Future Challenges. *Chem. Rev.* **2020**, *120*, 3210–3229. [CrossRef]
79. Pegram, M.D.; Konecny, G.; Slamon, D.J. The molecular and cellular biology of HER2/neu gene amplification/overexpression and the clinical development of herceptin (trastuzumab) therapy for breast cancer. *Cancer Treat. Res.* **2000**, *103*, 57–75. [CrossRef] [PubMed]
80. Jackson, D.O.; Trappey, F.A.; Clifton, G.T.; Vreeland, T.J.; Peace, K.M.; Hale, D.F.; Litton, J.K.; Murray, J.L.; Perez, S.A.; Papamichail, M.; et al. Effects of HLA status and HER2 status on outcomes in breast cancer patients at risk for recurrence—Implications for vaccine trial design. *Clin. Immunol.* **2018**, *195*, 28–35. [CrossRef] [PubMed]
81. Brossart, P.; Stuhler, G.; Flad, T.; Stevanovic, S.; Rammensee, H.G.; Kanz, L.; Brugger, W. Her-2/neu-derived peptides are tumor-associated antigens expressed by human renal cell and colon carcinoma lines and are recognized by in vitro induced specific cytotoxic T lymphocytes. *Cancer Res.* **1998**, *58*, 732–736. [PubMed]
82. Kaumaya, P.T. B-cell epitope peptide cancer vaccines: A new paradigm for combination immunotherapies with novel checkpoint peptide vaccine. *Future Oncol.* **2020**, *16*, 1767–1791. [CrossRef]
83. DeMaria, P.J.; Bilusic, M. Cancer Vaccines. *Hematol. Oncol. Clin. North. Am.* **2019**, *33*, 199–214. [CrossRef]
84. Farkona, S.; Diamandis, E.P.; Blasutig, I.M. Cancer immunotherapy: The beginning of the end of cancer? *BMC Med.* **2016**, *14*, 73. [CrossRef] [PubMed]
85. Cheever, M.A.; Higano, C.S. PROVENGE (Sipuleucel-T) in prostate cancer: The first FDA-approved therapeutic cancer vaccine. *Clin. Cancer Res.* **2011**, *17*, 3520–3526. [CrossRef] [PubMed]
86. Hammerstrom, A.E.; Cauley, D.H.; Atkinson, B.J.; Sharma, P. Cancer immunotherapy: Sipuleucel-T and beyond. *Pharmacotherapy* **2011**, *31*, 813–828. [CrossRef]
87. Calmeiro, J.; Carrascal, M.A.; Tavares, A.R.; Ferreira, D.A.; Gomes, C.; Falcão, A.; Cruz, M.T.; Neves, B.M. Dendritic cell vaccines for cancer immunotherapy: The role of human conventional type 1 dendritic cells. *Pharmaceutics* **2020**, *12*, 158. [CrossRef]

88. Cintolo, J.A.; Datta, J.; Mathew, S.J.; Czerniecki, B.J. Dendritic cell-based vaccines: Barriers and opportunities. *Future Oncol.* **2012**, *8*, 1273–1299. [CrossRef]

89. Besse, B.; Charrier, M.; Lapierre, V.; Dansin, E.; Lantz, O.; Planchard, D.; Le Chevalier, T.; Livartoski, A.; Barlesi, F.; Laplanche, A.; et al. Dendritic cell-derived exosomes as maintenance immunotherapy after first line chemotherapy in NSCLC. *Oncoimmunology* **2016**, *5*, e1071008. [CrossRef]

90. Pitt, J.M.; André, F.; Amigorena, S.; Soria, J.C.; Eggermont, A.; Kroemer, G.; Zitvogel, L. Dendritic cell-derived exosomes for cancer therapy. *J. Clin. Investig.* **2016**, *126*, 1224–1232. [CrossRef]

91. Théry, C.; Duban, L.; Segura, E.; Véron, P.; Lantz, O.; Amigorena, S. Indirect activation of naïve CD4+ T cells by dendritic cell–derived exosomes. *Nat. Immunol.* **2002**, *3*, 1156–1162. [CrossRef]

92. André, F.; Chaput, N.; Schartz, N.E.C.; Flament, C.; Aubert, N.; Bernard, J.; Lemonnier, F.; Raposo, G.; Escudier, B.; Hsu, D.H.; et al. Exosomes as Potent Cell-Free Peptide-Based Vaccine. I. Dendritic Cell-Derived Exosomes Transfer Functional MHC Class I/Peptide Complexes to Dendritic Cells. *J. Immunol.* **2004**, *172*, 2126–2136. [CrossRef] [PubMed]

93. Segura, E.; Amigorena, S.; Théry, C. Mature dendritic cells secrete exosomes with strong ability to induce antigen-specific effector immune responses. *Blood Cells Mol. Dis.* **2005**, *35*, 89–93. [CrossRef]

94. Sprent, J. Direct stimulation of naïve T cells by antigen-presenting cell vesicles. *Blood Cells Mol. Dis.* **2005**, *35*, 17–20. [CrossRef]

95. Viaud, S.; Terme, M.; Flament, C.; Taieb, J.; André, F.; Novault, S.; Escudier, B.; Robert, C.; Caillat-Zucman, S.; Tursz, T.; et al. Dendritic cell-derived exosomes promote natural killer cell activation and proliferation: A role for NKG2D ligands and IL-15Rα. *PLoS ONE* **2009**, *4*, e4942. [CrossRef]

96. Morse, M.A.; Garst, J.; Osada, T.; Khan, S.; Hobeika, A.; Clay, T.M.; Valente, N.; Shreeniwas, R.; Sutton, M.A.; Delcayre, A.; et al. A phase I study of dexosome immunotherapy in patients with advanced non-small cell lung cancer. *J. Transl. Med.* **2005**, *3*, 1–8. [CrossRef] [PubMed]

97. Zhang, X.M.; Zhang, Y.F.; Huang, Y.; Qu, P.; Ma, B.; Si, S.Y.; Li, Z.S.; Li, W.X.; Li, X.; Ge, W.; et al. The anti-tumor immune response induced by a combination of MAGE-3/MAGE-n-derived peptides. *Oncol. Rep.* **2008**, *20*, 245–252. [CrossRef] [PubMed]

98. Vivier, E.; Raulet, D.H.; Moretta, A.; Caligiuri, M.A.; Zitvogel, L.; Lanier, L.L.; Yokoyama, W.M.; Ugolini, S. Innate or adaptive immunity? The example of natural killer cells. *Science* **2011**, *331*, 44–49. [CrossRef] [PubMed]

99. Escudier, B.; Dorval, T.; Chaput, N.; André, F.; Caby, M.P.; Novault, S.; Flament, C.; Leboulaire, C.; Borg, C.; Amigorena, S.; et al. Vaccination of metastatic melanoma patients with autologous dendritic cell (DC) derived-exosomes: Results of the first phase 1 clinical trial. *J. Transl. Med.* **2005**, *3*, 1–13. [CrossRef] [PubMed]

100. Chen, W.; Wang, J.; Shao, C.; Liu, S.; Yu, Y.; Wang, Q.; Cao, X. Efficient induction of antitumor T cell immunity by exosomes derived from heat-shocked lymphoma cells. *Eur. J. Immunol.* **2006**, *36*, 1598–1607. [CrossRef] [PubMed]

101. Shao, Y.; Shen, Y.; Chen, T.; Xu, F.; Chen, X.; Zheng, S. The functions and clinical applications of tumor-derived exosomes. *Oncotarget* **2016**, *7*, 60736–60751. [CrossRef]

102. De la Torre Gomez, C.; Goreham, R.V.; Bech Serra, J.J.; Nann, T.; Kussmann, M. "Exosomics"-A Review of Biophysics, Biology and Biochemistry of Exosomes With a Focus on Human Breast Milk. *Front. Genet.* **2018**, *9*, 92. [CrossRef]

103. Whiteside, T.L.; Mandapathil, M.; Szczepanski, M.; Szajnik, M. Mechanisms of tumor escape from the immune system: Adenosine-producing Treg, exosomes and tumor-associated TLRs. *Bull. Cancer* **2011**, *98*, E25–E31. [CrossRef]

104. Xavier, C.P.R.; Caires, H.R.; Barbosa, M.A.G.; Bergantim, R.; Guimarães, J.E.; Vasconcelos, M.H. The Role of Extracellular Vesicles in the Hallmarks of Cancer and Drug Resistance. *Cells* **2020**, *9*, 1141. [CrossRef]

105. Bu, N.; Li, Q.-L.; Feng, Q.; Sun, B.-Z. Immune protection effect of exosomes against attack of L1210 tumor cells. *Leuk. Lymphoma* **2006**, *47*, 913–918. [CrossRef] [PubMed]

106. Yao, Y.; Wang, C.; Wei, W.; Shen, C.; Deng, X.; Chen, L.; Ma, L.; Hao, S. Dendritic Cells Pulsed with Leukemia Cell-Derived Exosomes More Efficiently Induce Antileukemic Immunities. *PLoS ONE* **2014**, *9*, e91463. [CrossRef]

107. Gu, X.; Erb, U.; Büchler, M.W.; Zöller, M. Improved vaccine efficacy of tumor exosome compared to tumor lysate loaded dendritic cells in mice. *Int. J. Cancer* **2015**, *136*, E74–E84. [CrossRef]

108. Pineda, B.; Sánchez García, F.J.; Olascoaga, N.K.; Pérez de la Cruz, V.; Salazar, A.; Moreno-Jiménez, S.; Pedro, N.H.; Marquez-Navarro, A.; Plata, A.O.; Sotelo, J. Malignant Glioma Therapy by Vaccination with Irradiated C6 Cell-Derived Microvesicles Promotes an Antitumoral Immune Response. *Mol. Ther.* **2019**, *27*, 1612–1620. [CrossRef]

109. Phung, C.D.; Pham, T.T.; Nguyen, H.T.; Nguyen, T.T.; Ou, W.; Jeong, J.H.; Choi, H.G.; Ku, S.K.; Yong, C.S.; Kim, J.O. Anti-CTLA-4 antibody-functionalized dendritic cell-derived exosomes targeting tumor-draining lymph nodes for effective induction of antitumor T-cell responses. *Acta Biomater.* **2020**, *4*, 1–12. [CrossRef] [PubMed]

110. Shi, X.; Sun, J.; Li, H.; Lin, H.; Xie, W.; Li, J.; Tan, W. Antitumor efficacy of interferon-γ-modified exosomal vaccine in prostate cancer. *Prostate* **2020**, *80*, 811–823. [CrossRef]

111. Hao, S.; Bai, O.; Li, F.; Yuan, J.; Laferte, S.; Xiang, J. Mature dendritic cells pulsed with exosomes stimulate efficient cytotoxic T-lymphocyte responses and antitumour immunity. *Immunology* **2007**, *120*, 90–102. [CrossRef]

112. Lu, M.; Xing, H.; Xun, Z.; Yang, T.; Ding, P.; Cai, C.; Wang, D.; Zhao, X. Exosome-based small RNA delivery: Progress and prospects. *Asian J. Pharm. Sci.* **2018**, *13*, 1–11. [CrossRef]

113. Zhang, D.; Lee, H.; Jin, Y. Delivery of Functional Small RNAs via Extracellular Vesicles In Vitro and In Vivo. *Methods Mol. Biol.* **2020**, *2115*, 107–117. [CrossRef]

114. Wu, Y.; Chen, W.; Xu, Z.P.; Gu, W. PD-L1 Distribution and Perspective for Cancer Immunotherapy-Blockade, Knockdown, or Inhibition. *Front. Immunol.* **2019**, *10*, 2022. [CrossRef]

115. Imbert, C.; Montfort, A.; Fraisse, M.; Marcheteau, E.; Gilhodes, J.; Martin, E.; Bertrands, F.; Marcellin, M.; Burlet-Schiltz, O.; Gonzalez de Peredo, A.; et al. Resistance of melanoma to immune checkpoint inhibitors is overcome by targeting the sphingosine kinase-1. *Nat. Commun.* **2020**, *11*, 437. [CrossRef]

116. Wieder, T.; Eigentler, T.; Brenner, E.; Röcken, M. Immune checkpoint blockade therapy. *J. Allergy Clin. Immunol.* **2018**, *142*, 1403–1414. [CrossRef]

117. Matsuda, A.; Ishiguro, K.; Yan, I.K.; Patel, T. Extracellular Vesicle-Based Therapeutic Targeting of β-Catenin to Modulate Anticancer Immune Responses in Hepatocellular Cancer. *Hepatol. Commun.* **2019**, *3*, 525–541. [CrossRef] [PubMed]

118. De Galarreta, M.R.; Bresnahan, E.; Molina-Sánchez, P.; Lindblad, K.E.; Maier, B.; Sia, D.; Puigvehi, M.; Miguela, V.; Casanova-Acebes, M.; Dhainaut, M.; et al. β-Catenin Activation Promotes Immune Escape and Resistance to Anti-PD-1 Therapy in Hepatocellular Carcinoma. *Cancer Discov.* **2019**, *9*, 1124–1141. [CrossRef] [PubMed]

119. Spranger, S.; Gajewski, T.F. A new paradigm for tumor immune escape: β-catenin-driven immune exclusion. *J. Immunother. Cancer* **2015**, *3*, 43. [CrossRef] [PubMed]

120. Petitprez, F.; Meylan, M.; de Reyniès, A.; Sautès-Fridman, C.; Fridman, W.H. The Tumor Microenvironment in the Response to Immune Checkpoint Blockade Therapies. *Front. Immunol.* **2020**, *11*, 784. [CrossRef] [PubMed]

121. Berraondo, P.; Ochoa, M.C.; Olivera, I.; Melero, I. Immune Desertic Landscapes in Hepatocellular Carcinoma Shaped by β-Catenin Activation. *Cancer Discov.* **2019**, *9*, 1003–1005. [CrossRef]

122. Bell, J.B.; Podetz-Pedersen, K.M.; Aronovich, E.L.; Belur, L.R.; McIvor, R.S.; Hackett, P.B. Preferential delivery of the Sleeping Beauty transposon system to livers of mice by hydrodynamic injection. *Nat. Protoc.* **2007**, *2*, 3153–3165. [CrossRef]

123. Munagala, R.; Aqil, F.; Jeyabalan, J.; Gupta, R.C. Bovine milk-derived exosomes for drug delivery. *Cancer Lett.* **2016**, *371*, 48–61. [CrossRef]

124. Subleski, J.J.; Scarzello, A.J.; Alvord, W.G.; Jiang, Q.; Stauffer, J.K.; Kronfli, A.; Saleh, B.; Back, T.; Weiss, J.M.; Wiltrout, R.H. Serum-based tracking of de novo initiated liver cancer progression reveals early immunoregulation and response to therapy. *J. Hepatol.* **2015**, *63*, 1181–1189. [CrossRef] [PubMed]

125. Dai, S.; Wei, D.; Wu, Z.; Zhou, X.; Wei, X.; Huang, H.; Li, G. Phase I clinical trial of autologous ascites-derived exosomes combined with GM-CSF for colorectal cancer. *Mol. Ther.* **2008**, *16*, 782–790. [CrossRef] [PubMed]

MDPI

Review

DC-Derived Exosomes for Cancer Immunotherapy

Yi Yao [1,2], Chunmei Fu [1], Li Zhou [1,2], Qing-Sheng Mi [1,2,*] and Aimin Jiang [1,2,*]

1   Center for Cutaneous Biology and Immunology, Department of Dermatology, Henry Ford Health System, Detroit, MI 48202, USA; yyao1@hfhs.org (Y.Y.); cfu1@hfhs.org (C.F.); lzhou1@hfhs.org (L.Z.)
2   Immunology Program, Henry Ford Cancer Institute, Henry Ford Health System, Detroit, MI 48202, USA
*   Correspondence: qmi1@hfhs.org (Q.-S.M.); ajiang3@hfhs.org (A.J.);
    Tel.: +313-876-1017 (Q.-S.M.); +313-876-7292 (A.J.)

Simple Summary: Dendritic cells (DCs)-based cancer vaccines have not succeeded in generating significant clinical responses despite their capacity to induce host anti-tumor CD8 T cell immunity, and one major hurdle is tumor-mediated immunosuppression. Exosomes are nano-sized inert membrane vesicles derived from the endocytic pathway that play a critical role in intercellular communication. DC-derived exosomes (DCexos) additionally carried MHC class I/II (MHCI/II) often complexed with antigens and co-stimulatory molecules, thus capable of priming antigen-specific CD4 and CD8 T cells. Indeed, vaccines with DCexos have been shown to exhibit better anti-tumor efficacy in eradicating tumors compared to DC vaccines in pre-clinical models. Coupled with their resistance to tumor immunosuppression, DCexo-based cancer vaccines have been heralded as the superior alternative cell-free therapeutic vaccines over DC vaccines, and have now been tested in multiple clinical trials. In this review, current studies of DCexo cancer vaccines as well as potential future directions will be discussed.

Abstract: As the initiators of adaptive immune responses, DCs play a central role in regulating the balance between CD8 T cell immunity versus tolerance to tumor antigens. Exploiting their function to potentiate host anti-tumor immunity, DC-based vaccines have been one of most promising and widely used cancer immunotherapies. However, DC-based cancer vaccines have not achieved the promised success in clinical trials, with one of the major obstacles being tumor-mediated immunosuppression. A recent discovery on the critical role of type 1 conventional DCs (cDC1s) play in cross-priming tumor-specific CD8 T cells and determining the anti-tumor efficacy of cancer immunotherapies, however, has highlighted the need to further develop and refine DC-based vaccines either as monotherapies or in combination with other therapies. DC-derived exosomes (DCexos) have been heralded as a promising alternative to DC-based vaccines, as DCexos are more resistance to tumor-mediated suppression and DCexo vaccines have exhibited better anti-tumor efficacy in pre-clinical animal models. However, DCexo vaccines have only achieved limited clinical efficacy and failed to induce tumor-specific T cell responses in clinical trials. The lack of clinical efficacy might be partly due to the fact that all current clinical trials used peptide-loaded DCexos from monocyte-derived DCs. In this review, we will focus on the perspective of expanding current DCexo research to move DCexo cancer vaccines forward clinically to realize their potential in cancer immunotherapy.

Keywords: dendritic cells; exosomes; vaccines; plasmacytoid DCs; cancer immunotherapy

Citation: Yao, Y.; Fu, C.; Zhou, L.; Mi, Q.-S.; Jiang, A. DC-Derived Exosomes for Cancer Immunotherapy. *Cancers* **2021**, *13*, 3667. https://doi.org/10.3390/cancers13153667

Academic Editors: Farrukh Aqil and Ramesh Chandra Gupta

Received: 14 June 2021
Accepted: 18 July 2021
Published: 21 July 2021

Publisher's Note: MDPI stays neutral with regard to jurisdictional claims in published maps and institutional affiliations.

1. Dendritic Cells, Anti-Tumor Immunity and Dendritic Cell-Based Cancer Vaccines

As the sentinel of the immune system, DCs play a central role in bridging innate and adaptive immune responses [1]. Known as the most potent professional antigen presenting cells (APCs), DCs initiate all adaptive immune responses by uptaking, processing, and presenting antigens including tumor antigens to activate naive antigen-specific CD4 and CD8 T cells [2,3]. Since their identification in 1973 [4], intensive studies have shown that DCs are heterogeneous populations comprising several subsets distinguished by their development, phenotype, localization, and functional specialization [5–9]. DCs originate in bone marrow from progenitors called common myeloid progenitors (CMPs). In the presence of transcription factor Nur77, CMPs differentiate into monocytes which can further differentiate into monocyte DCs (MoDCs) under inflammatory conditions [8]. Alternatively, CMPs differentiate into macrophage/DC progenitors (MDPs) that give rise to common DC progenitors (CDPs), which then differentiate into two major DC subsets: classical/conventional DCs (cDCs) and plasmacytoid DCs (pDCs) [5,7–11]. Murine cDCs consist of two subtypes currently described as cDC1s (CD11b⁻, type 1 cDCs) and cDC2s (CD11b⁺, type 2 cDCs), and their human counterparts are CD141⁺ DCs (also known as BDCA3⁺) and CD1c⁺ DCs (also known as BDCA1⁺), respectively [12]. These two subtypes of cDCs differ in their transcriptional factor dependency, function, and phenotypes. cDC1 cells include lymphoid tissue CD8α⁺ cDC1s and non-lymphoid tissue CD103⁺ cDC1s [13]. cDC1 cells depend on interferon regulatory factor 8 (IRF8) and basic leucine zipper transcriptional factor ATF-like 3 (Batf3) for their development, and are specialized in presenting internalized antigens bound to MHCI to CD8 T cells in a process termed cross-presentation [13,14]. cDC2s depend on interferon regulatory factor 4 (IRF4) and represent a heterogeneous population with enhanced MHCII antigen presentation that preferentially activate CD4 T cells [15–18]. On the other hand, pDCs are a multifunctional population best known for their specialized ability in producing and secreting large amount of type I interferons (IFNIs) [19–21]. pDCs also express high level of IRF8 similar to cDC1s, but depend on the E2-2 transcription factor for their development from CDPs in both mice and humans [22]. Besides MDPs, recent studies have shown that pDCs also develop from lymphoid progenitors with distinct function from their MDP-derived counterparts [23].

Cross-priming, a process which DCs prime CD8 T cells following cross-presentation of exogenous antigens onto MHCI [24,25], plays a critical role in inducing anti-tumor CD8 T cell immunity as well as mediating CD8 T cell tolerance (cross-tolerance) [26–29]. In fact, the ability of DCs to cross-present tumor-associated antigens onto an MHCI molecule to prime CD8 T cells is the foundation of the Cancer-Immunity cycle proposed by Chen and Mellman [30]. Exploiting DCs' function to potentiate host anti-tumor immunity, DC-based vaccines have become one of the most widely-used cancer immunotherapies [7,8,31–34]. However, DC-based vaccines, the vast majority of which make use of monocyte-derived DCs (MoDCs) differentiated in vitro, remain mostly unsuccessful, only resulting in 5–15% objective clinical responses in numerous clinical trials. To date, Provenge (Sipuleucel-T) remains the first and only "DC" cancer vaccine to be approved by FDA for treatment of castration-resistant prostate cancer in 2010 [35]. Despite mostly unsuccessful clinical trials, recent findings that cDC1s play a critical role in cross-priming tumor-specific CD8 T cells and in determining the anti-tumor efficacy of cancer immunotherapies [36–40], has reignited the efforts to further develop/refine DC-based vaccines either as monotherapies or combined therapies. One of the major obstacles for DC-based cancer vaccines is tumor-mediated immunosuppression, often targeting DC function in cross-priming leading to CD8 T cell tolerance (cross-tolerance) instead of immunity [5,10,11,41–43]. Vaccines with DC-derived exosomes (DCexos), which are superior in their resistance to tumor-mediated suppression, bioavailability, and biostability compared to DCs, have demonstrated better anti-tumor efficacy than DC-based vaccines in preclinical trials, and thus have emerged as the promising alternative cell-free therapeutic vaccines that could overcome the obstacles of DC vaccines [44]. This review will examine the results and limitations of clinical trials, recent development on DCexo research and the future of DCexos as viable cancer vaccines. For more detailed review on DC-based cancer vaccines, readers are hereby referred to many recent excellent reviews [6–8,33,34,45].

2. DC-Derived Exosomes and Their Function

All cells release extracellular vesicles (EVs) of different sizes and intracellular origin. These EVs could be broadly classified into three main groups according to their origin and size: the nanosized exosomes, microvesicles (MVs, also referred to as ectosomes), and apoptotic bodies that are constructed by direct sprouting of the cellular membrane in living and dying cells, respectively [46–49]. Apoptotic bodies are large vesicles generated by cells undergoing apoptosis, with 1000–5000 nm in size. MVs are generated through the direct budding or shedding from the plasma membrane by living cells, with a diameter from 50 to 1000 nm. On the other hand, exosomes are a heterogenous group of nano-sized EVs originating from endosomal pathway, with size ranging from 40 to 160 nm (Figure 1). Exosomes are produced in the endosomal compartment by inward budding of limiting endosomal membrane into intraluminal vesicles (ILVs) within the lumen of multivesicular bodies (MVBs, or so-called late endosomes) [48–50]. MVBs are either targeted for lysosomal degradation or they may fuse with the cellular membrane to release these ILVs into the extracellular space as free exosomes (Figure 1). Exosomes can contain membrane proteins, cytosolic and nuclear proteins, extracellular matrix proteins, metabolites, and nucleic acids including mRNA, miRNA, non-coding RNA, and DNA [49] (Figure 1). Although Exosomes were initially presumed to be an alternate route to excrete waste products in order to sustain cellular homeostasis, a seminal study have shown that exosomes carry and transfer mRNA and miRNA between cells [51]. It is well established now that exosomes play significant roles in intercellular communications and material transfer of their cargo [44,52,53]. In addition, exosomes are also no-immunogenic due to their similar membrane composition to the cells, biostable in vitro and in vivo, capable of targeting tissues and penetrating of biological barrier, making them attractive delivery vehicles for genetic material (miRNA, mRNA) and loaded drugs [54,55]. It should be noted that exosomes have great heterogeneity reflective of their size, content, function, and cellular origin, and current phenotypic and functional analyses of exosomes have been performed on only exosome-enriched populations, thus demanding caution when interpreting the data [49,55,56]. For more in-depth reading, several recent publications provided excellent comprehensive review on exosomes [46–49].

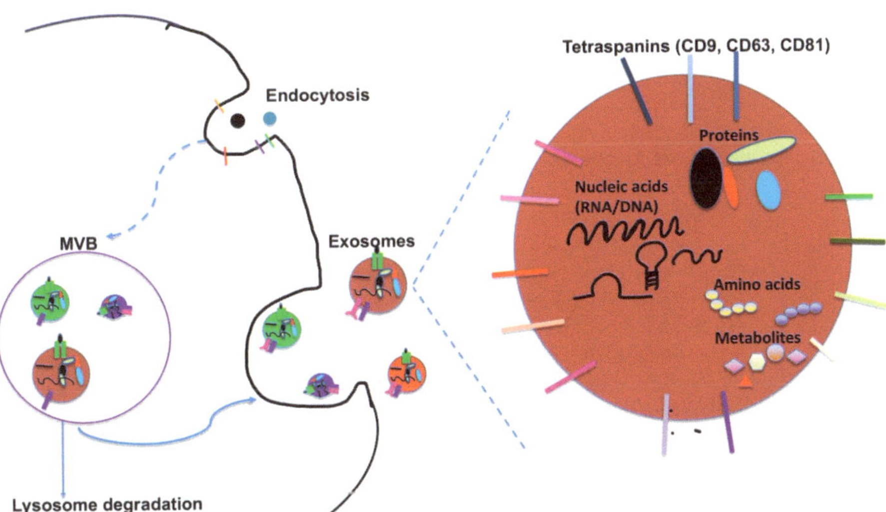

Figure 1. Biogenesis and characteristics of exosomes. Exosomes are produced in the endosomal compartment by inward budding of limiting endosomal membrane into intraluminal vesicles (ILVs) within the lumen of multivesicular bodies (MVBs). MVBs are either targeted for lysosomal degradation or they may fuse with plasma membrane to release these ILVs into the extracellular space as free exosomes. Exosomes are highly heterogenous with size ranging from 40 to 160 nm. Besides expressing an array of receptors on their surface, exosomes carry proteins, metabolites and nucleic acids including mRNA, miRNA, other non-coding RNA and DNA. Tetraspanins (CD9, CD63 and CD81) and other proteins such as Alix and TSG101 are often enriched in exosomes, and are commonly used as markers for exosomes.

DC-derived exosomes (DCexo) additionally carry functional MHCI and MHCII, co-stimulatory molecules (CD80, CD86) and adhesion molecules (ICAM1) involved in antigen uptake and presentation on their surface, which might be the most prominent feature distinct from exosomes produced from other immune cells [44,49,55,57]. As DCs process exogenous antigens in endosomal compartments including MVBs, which in turn fuse with plasma membrane resulting in the release of DCexos, DCexos have been shown to express both functional peptide/MHCI and peptide/MHCII complexes (pMHCI and pMHCII) for priming antigen-specific CD4 and CD8 T cells. It was first reported that exosomes generated from another APC–B cells express functional pMHCII complexes on their surface to activate antigen-specific CD4 T cells [58]. DCexos also carry high level of MHCI molecules, thus affording them the capacity to induce antigen-specific CD8 T cell responses [59]. Indeed, Zitvogel et al. have showed that DCexos from tumor-associated antigens (TAAs)–stimulated DCs prime tumor-specific CD8 T cell responses in vivo, and a single intradermal injection of DCexos achieved better anti-tumor efficacy in eradicating established tumors compared to the injection of DCs [60]. Although DCexos have been shown to directly prime T cells in vitro [59,61], this direct mechanism has been reported to be unable to prime naive T cells and is less likely to play a significant role in vivo [44,62] (Figure 2). Indeed, studies have shown that DCexos prime antigen-specific T cells far more efficiently with bystander DCs through indirect mechanisms. Several exosome membrane proteins, including integrins and ICAMs, facilitate their binding and subsequent internalization by bystander DCs, leading to indirect antigen presentation. One of the indirect antigens is called "cross-dressing", referring to the direct transfer of exosomal pMHC complexes to the bystander DCs following exosome binding to bystander DCs (or other APCs) [44] (Figure 2). The second mechanism involves the processing and presentation of exosomal antigens onto MHC of bystander DCs, following binding and internalization of DCexos (Figure 2). In one scenario, DCexos pMHC$_{Exo}$ complexes could be reprocessed through endosomal pathway in bystander DCs, resulting in the transfer of exosomal antigenic epitopes to bystander DC MHC molecules to be presented as pMHC$_{DC}$ complexes on bystander DCs [44,63] (Figure 2). Alternatively, protein antigens carried by DCexos, which have been shown previously [64], could be processed by bystander DCs, and multiple different epitopes for both CD4 and CD8 T cells (or even B cells) could be presented on MHC$_{DC}$ of bystander DCs (Figure 2). This mechanism might be most relevant for the use of DCexos as cancer vaccines, as studies have shown that only OVA protein-loaded but not peptide-loaded DCexos induced strong (allogeneic) CD8 T cell responses without requiring exosomal MHCI in vivo [65,66].

While DCexo-mediated T cell activation plays a critical function in their potential application in immunotherapy, DCexos also express NK receptors to induce NK cell activation [67] (Figure 2). In addition, DCexos have been shown to exert regulatory functions through exosomal membrane proteins or miRNA [68–70].

The capacity of DCexos to prime T cells—especially antigen-specific CD8 T cells—and their ability to shuttle different biomolecules, including proteins (such as antigens and cytokines) and RNAs to modulate immune responses, coupled with their amenability to modification of their composition and cargos, have presented DCexo-based vaccines as much improved alternative to DC cell-based vaccines [44,71–74]. Additionally, DCexos possess other advantages over DC(cell)-based vaccines. (1) DCexos have a more restricted and controllable molecular composition than DCs, owing to specific sorting and loading mechanisms. (2) DCexos have much longer shelf life than the very short shelf life of DCs. (3) DCexos can reach the proper location on secondary lymphoid organs more efficiently compared to DCs, and could be easily modified to deliver their cargos to specific targeted destinations [44]. (4) DCs are susceptible to tumor-mediated immunosuppression often observed in cancer patients, whereas DCexos being inert vesicles are not or more resistant. (5) DCexos might be more potent than DCs in activating T and NK cells, as DCexos actually present 10–100 times more pMHC complexes on their surface than DCs and have been shown to be enriched of activation ligands for NK cells [67,75]. Indeed, it has been reported that DCexos loaded with tumor antigens achieved better anti-tumor efficacy in eradicating established murine tumors compared to vaccines using DCs in preclinical models [60], thus supporting their clinical application as cancer vaccines [44,57,76–78].

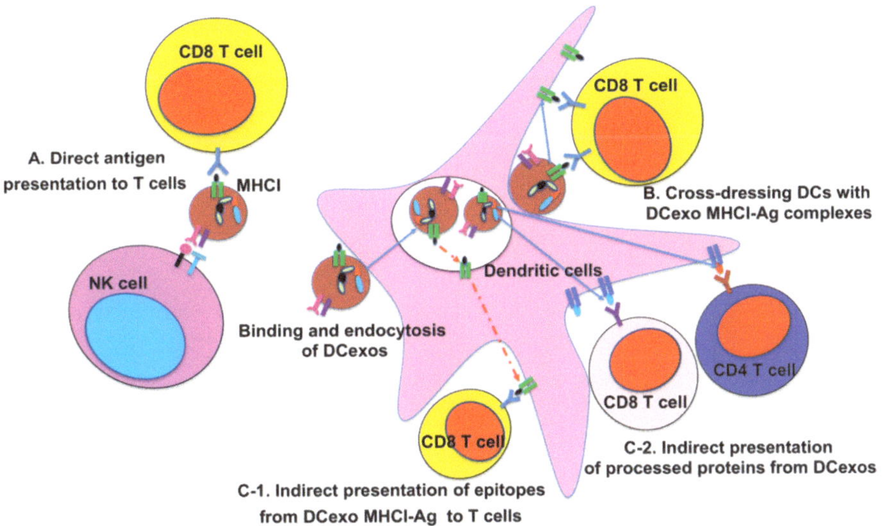

Figure 2. DCexo-mediated antigen presentation to activate T cells. (**A**): The presence of MHC_{Exo}-Ag complexes on the surface of DCexos enables them to activate antigen-specific CD4 and CD8 T cells directly. Only MHCI and CD8 T cells are illustrated. NK cell-expressed ligands (NKG2D-L, IL-15R and BAG6) on DCexos can also activate natural killer (NK) cells directly. (**B–C**): DCexos activate antigen-specific T cells more efficiently indirectly via bystander DCs. (**B**): DCexos can transfer MHC_{Exo}-Ag complexes to the DC plasma membrane, a process termed cross-dressing, leading to activation of antigen-specific T cells. MHC of DCexos and T cells has to be the same while MHC of the bystander DCs is not required. DCexos can also transfer MHC_{Exo}-Ag complexes to tumor cells to be presented to host T cells (not depicted). (**C**): After internalization, (**C-1**) DCexos could transfer antigenic peptides to the MHC_{DC} in bystander DCs. The $pMHC_{DC}$ complexes could be transported to the DC plasma membrane to be presented to T cells. DCexos, bystander DCs and T cells are required to have the same MHC in this mode. (**C-2**) Protein antigens carried by DCexos could be processed by bystander DCs, and multiple and different epitopes for both CD4 and CD8 T cells could be presented on MHC of bystander DCs. Only bystander DCs and T cells need to have the same MHC, allowing DCexos to induce allogeneic T cell responses.

3. DC-Derived Exosomes in Clinical Trials

Three clinical trials using DCexos including two phase I and one phase II clinical trials (see Table 1) have been completed [79–81]. In addition, there was one phase I clinical trial, which treated advanced colorectal carcinoma (CRC) patients with autologous ascites-derived exosomes (ASexos) either alone or in combination with GM-CSF [82]. ASexos were prepared from ascites of the CRC patients, that were enriched for MHCI and MHCII, HSPs (including HSC70, HSP70 and HSP90), CD80 and ICAM1. While ASexos were mainly derived from CRC cells in the ascites, they likely also contained exosomes from immune cells including DCs.

Table 1. Summary of current clinical trials with DC-derived exosomes (DCexos).

Cancer Type	Phase	Exosomes /Antigen	Doses	Patients	Toxicity	Clinical Outcomes
Advanced Non-small cell lung cancer	I	Exosomes were isolated from autologous MoDCs generated in vitro, and loaded with MAGE peptides	once weekly for 4 weeks	13 (9 completed) HLA-A2$^+$ stage IIIb and IV NSCLC patients with tumor expression of MAGE3 or MAGE4	Grade 1–2 toxicity	DTH reactivity against MAGE peptides in 3/9; MAGE-specific T cell responses in 1/3 patients examined; increased NK lytic activity in 2/4 [79].

Table 1. *Cont.*

Cancer Type	Phase	Exosomes /Antigen	Doses	Patients	Toxicity	Clinical Outcomes
MAGE3-expressing advanced melanoma	I	Autologous MoDC-derived exosomes were loaded with MAGE3 peptides	once weekly for 4 weeks	15 stage IIIb and IV, HLA-A1[+], B35[+] or HLA-DPO4[+] patients	Only grade 1 toxicity	No detectable MAGE3-specific CD4 and CD8 T cell responses; restored NKG2D expression and NKG2D-dependent function of NK cells in 7/14 patients; 1/15 partial responses [67,80].
Advanced colorectal cancer	I	Exosomes from patient ascites $\pm$ GM-CSF, ASexos contained CEA with no additional antigen loading.	once weekly for 4 weeks	40 HLA-A2[+]CEA[+] stage III and IV CRC patients	Grade 1–2 toxicity	DTH induction in both groups, and CEA-specific CTL responses were observed in ASexo + GM-CSF group. 1 stable disease and 1 minor response in ASexo + GM-CSF group [82].
Non-small cell lung cancer	II	IFN-γ-matured autologous MoDCs were loaded with both MHCI and MHCII tumor epitopes.	exosome immunization in 1, 2 and 3 week intervals in a maintenance immunotherapy regime	26 (22 HLA-A2[+] stage IIIb and IV NSCLC patients	1/22 grade 3 hepato-toxicity	No detectable induction of antigen-specific T cell responses; increased NKp30-dependent NK cell function; 7 patients (32%) with progression-free survival at 4 months after chemotherapy cessation; no objective tumor response according to RECIST criteria [81].

4. DCexo Phase I Clinical Trials

The first phase I clinical trial with DCexos as cancer vaccines employed DCexos obtained from autologous immature monocyte-derived DCs (MoDCs) (Table 1). Exosomes were isolated by ultracentrifugation and loaded with both MHCI and MHCII melanoma-associated antigen (MAGE) peptide epitopes. An MHCI-restricted cytomegalovirus (CMV) peptide and an MHCII-restricted tetanus toxoid-derived peptide were also loaded onto exosomes. Peptides were loaded either directly onto isolated exosomes, or indirectly by adding peptides into DC culture that produced exosomes, and the antigen-loaded exosomes were then administered into advanced non-small cell lung cancer (NSCLC) patients [79]. A total of 13 HLA-A2[+] patients with pre-treated advanced stage (IIIb and IV) NSCLC were enrolled, and 9 patients completed therapy after receiving 4 exosome doses at weekly interval. Only grade 1–2 toxicity and no autoimmune reactions were observed, suggesting that the exosome vaccine was safe and well-tolerated in patients similar to DC vaccines. One week after the last DCexo injection, three patients of the tested participants, who had not shown MAGE-specific immune responses before DCexo injections, exhibited systemic MAGE-specific immune reactivity as measured by delayed- type hypersensitivity (DTH) response. Increased MAGE-specific T cell responses were only observed in one of five patients examined by enzyme-linked immunospot (ELISPOT) assay. However, no antigen (MAGE)-specific T cell responses were observed in PBMCs by in-vitro assays. The low rate of T cell activation was attributed to potential suppression by regulatory T cells (Tregs, CD4[+]CD25[+] T cells). In two of three patients examined, the percentages of Tregs out of total CD4[+] T cells were increased following DCexo vaccinations. Interestingly, two of four tested samples exhibited increased NK cell lysis activity. Overall, the NSCLC phase I study showed that DCexo vaccines were well-tolerated with an acceptable safety profile, with disease stability observed in two patients who had progressive cancer at diagnosis. In addition, disease stability continued for over 12 months in two of four patients with stable disease [79].

The other DCexo phase I clinical trial enrolled 15 MAGE3[+] advanced (stage IIIb or IV) metastatic melanoma (MM) patients (Table 1). Exosomes were isolated by ultracentrifugation from autologous immature MoDCs, and loaded with both MHCI and MHCII MAGE3 peptide epitopes either directly or indirectly. Similar to the NSCLC trial, an MHCII-restricted tetanus toxoid-derived peptide was also loaded unto DCexos. All patients were administered 4 doses of DCexos at a weekly interval, and evaluation of the vaccine efficacy was conducted two weeks after vaccinations. Of these patients, one patient exhibited a partial response to DCexo immunotherapy. In this patient, a halo of depigmentation around naevi was observed, and the arterial neovasculature disappeared and tumor size reduced. This patient received 4 months of continuation therapy with DCexos, leading to disease stabilization and reduced toxicity. Stabilization of the disease for up to 24 months was also observed in another patient who was given continued DCexo immunotherapy. Overall, this clinical trial resulted in two stable diseases, as well as one minor, one partial, and one mixed response at lymph nodes or skin. Some of these responses were achieved in patients with progressive disease who had formerly been given other cancer therapies. Similar to the NSCLC phase I clinical trial, neither DTH responses or MAGE-specific T cell responses were observed in peripheral blood, although T cell responses against MART1 (melanoma antigen recognized by T cells 1) which were not included in the vaccines, were detected [80]. In contrast, NK cell effector functions were enhanced in peripheral blood of 8/13 patients following DCexo vaccination [80], thus suggesting that augmented NK cell functions in vivo might account for the T cell-independent clinical responses.

As enhanced NK cell activation was observed in the two phase I clinical trials, Viaud S. et al. further examined whether NK cell activation instead of T cells could be responsible for the clinical effects observed in the clinical trial carried out by Escudier B. et al. [67]. Indeed, exosomes generated from immature human DCs express killer cell lectin–like receptor subfamily K, member 1 ligands (NKG2D-L), which can directly interact with NKG2D on NK cells, resulting in their activation [67]. Examining samples from the DCexo clinical trial on MM [80], Viaud S. et al. observed that circulating NK cell numbers were significantly increased after 4 DCexo vaccinations. Moreover, the expression of NKG2D and NK cytotoxicity were restored after vaccinations in 50% of patients who had NK cell function defects at the beginning of the clinical trial [67]. Further studies have shown that DCexo vaccinations induce NK cell proliferation in an IL-15Rα–dependent manner. These findings on DCexo-mediated effects on NK cells are consistent with improved control of tumor metastasis in B16F10-bearing C57BL/6 mice by NK1.1$^+$ cells [67]. Interestingly, exosomes generated from human immature DCs have also been reported to express BCL2-associated athanogene 6 (BAG6, also known as BAT3), a ligand for NKp30 receptors expressed on NK cells [83,84]. Cytokine production of NK cells has been reported to directly correlate with exosomal BAG6 expression levels l [84]. Additionally, DCexo expression of TNF superfamily ligands TNF, FasL, and TRAIL on their surface activate NK cells and stimulate them to secrete IFN-γ [69]. Taken together, the two phase I clinical trials and follow-up studies suggest that DCexos likely possess the capability to activate NK cells to generate anti-tumor immunity.

The third phase I clinical trial that might involve DCexos used ascites-derived exosomes (ASexos) alone or in combination with GM-CSF to treat 40 advanced colorectal carcinoma (CRC) patients [82] (Table 1). Exosomes were prepared from patient ascites, and were shown to be enriched in MHCI and MHCII, CD80, and ICAM1 similar to DCexos. In addition, these ASexos also contained the immunogenic carcinoembryonic antigen (CEA) of CRC. The patients received 4 weekly immunizations. Unlike the other two phase I clinical studies, DTH responses as well as CEA-specific CTL cell responses were observed in patients treated with ASexos plus GM-CSF. A higher level of tumor-associated antigens (TAAs) in ASexos may be responsible for the augmented T cell responses compared to the two aforementioned DCexo phase I trials. Despite the T cell responses, however, no detectable therapeutic responses were observed with the exception of one stable disease and a minor response after ASexos plus GM-CSF treatment. Another drawback for this model is that the majority of the ASexos were likely derived from the CRC cells instead of immune cells including DCs, and tumor-derived exosomes have been shown to be often immunosuppressive and promote tumor growth, metastasis, and development of drug resistance [85,86].

5. DCexo Phase II Clinical Trial

The limited clinical benefit shown by the phase I trials using exosomes from immature MoDCs prompted researchers to design and develop innovative strategies to promote DCexo-induced anti-tumor host immune responses. Based on previous studies showing that DCexos from mature DCs prime T cells more efficiently compared to DCexos from immature DCs [61,87], one strategy is to utilize DCexos originated from matured DCs. A clinical-grade process for producing DCexo vaccines was developed with human DC cultures [88]. Here, IFNγ was employed to stimulate human MoDCs in culture, and subsequently, costimulatory factors and ICAMs were upregulated, resulting in second-generation DCexos (IFNγ DCexos) with increased immunostimulatory capacity [88]. A phase II clinical trial was carried out with these second-generation IFNγ DCexos, aiming to investigate whether maintenance immunotherapy of advanced NSCLC patients using IFNγ–DCexos could increase progression-free survival (PFS) at 4 months following platinum-based chemotherapy [81] (Table 1). Twenty-two advanced HLA-A2$^+$ NSCLC patients who had inoperable (stage IIIb or IV) NSCLC with neutrophils $\geq 1.5 \times 10^9$/L and showed immune responses or disease stabilization following 4 rounds of a first-line platinum-based chemotherapy were eligible to receive IFNγ DCexos [81]. Both MHCI-restricted (MAGE-A1, MAGE-A3, NY-ESO, MART1) and MHCII-restricted (EBV) TAA were used. Patients first received 3 weeks of metronomic oral low-dose cyclophosphamide (CTX) prior to IFNγ DCexo maintenance therapy, based on both preclinical and clinical data showing that this protocol reduces Treg function and induces IFN-γ/IL-17–producing T cells [89–92]. Of these participants, 7 patients (32%) exhibited stable disease after 9 times of IFNγ DCexo vaccinations, and proceeded to receive DCexo therapy every 3 weeks. Unfortunately, a PFS of 50%, the primary end-point of the trial, was not reached, and no objective response was reported in the clinical trial. However, one patient exhibited a long-term disease stabilization, which allowed for surgical removal of the tumor and the eligibility for local adjuvant radiotherapy.

As to immunological readouts, the use of IFNγ DCexos as cancer immunotherapy were again insufficient to induce TAA-specific T cell responses in patients despite loading of multiple epitopes and CTX adjuvant therapy [81]. Thus, the immunostimulatory effects by IFNγ DCexos was likely mediated via augmented NK cell activation through NKp30 signaling. Indeed, increase in NKp30-mediated IFNγ and TNFα production by circulating NK cells was observed upon four IFNγ DCexo vaccinations, although NK cells in these advanced NSCLC patients only exhibited low levels of NKp30. More significantly, this increased NKp30-elicited NK cell activation correlated with longer PFS. In addition, the membrane-associated NKp30 ligand, BAG6, was detected on the membrane of DCexo vaccine preparations and was reported to play a critical role in mediating NK cell activation through a NKp30-dependent manner, thus supporting IFNγ DCexos promote NK cell activation/function through a NKp30-dependent mechanism. Moreover, BAG6 levels correlated with the MHCII concentrations of DCexos and NKp30-dependent NK cell activation. It should be noted that the NKp30-dependent NK activation differs from the finding of the phase I clinical trial on MM where NKG2D/NKG2D-L (and potentially IL-15/IL-15Rα) signaling mediated DCexo-induced NK activation [67,80]. Given that the DCexos employed in the MM clinical trial were not generated from MoDCs matured by IFNγ, which has been shown to upregulate BAG6, NKG2D/NKG2D-L–mediated NK cell activation likely plays a more prominent role instead of NKp30/BAG6 signaling.

Overall, these DCexo phase I and II clinical trials have demonstrated that DCexo vaccines are well-tolerated and safe, and are amenable to large-scale production in clinical settings. While only partial or minor responses were observed in these clinical trials with small number of patients, some patients achieved stabilization of disease.

While we focused our discussion on DCexos, it's worth pointing out that exosomes from tumor cells, mesenchymal stem cells, and other immune cells such as B cells, macrophages, and NK cells, T cells have also been examined for their application in cancer immunotherapy [55,62]. Indeed, tumor cells were likely the main source of the ASexos in one of the clinical trials we mentioned above [82]. Although tumor cell-derived exosomes (Texos) are capable of inducing anti-tumor immune responses, Texos generally seem to exhibit immune-suppressive functions [55,62]. Nevertheless, vaccines with Texos have emerged as promising cancer vaccines, likely due to their enrichment of tumor antigens making them capable of inducing T and B cell responses [93]. One promising approach to counter the suppressive properties of these exosomes is exosome engineering—to modify surface molecules on exosomes to induce tumor cell death or improve targeted delivery of exosomal cargos, to modify exosomal contents to deliver miRNA, mRNA, and cytokines for immune modulation, thus improving their efficacy [72,73]. The application of these exosomes was excellently reviewed recently [47,86,93,94].

6. Conclusions and Future Directions

The three clinical trials of DCexos as cancer vaccines have shown limited clinical efficacy in advanced cancer patients, which could be attributed to weak induction of adaptive immune responses specifically T cell responses in these patients. The poor adaptive immune responses could be potentially due to several factors: (1) the heterogeneity and the limited number of the patients, who had received previous treatments before enrollment; (2) systemic and local immunosuppressive mechanisms often present in these advanced-stage patients; (3) lack of sufficient maturation/adjuvant signals; (4) autologous MoDCs might not be the best DCs to achieve the optimal anti-tumor T cell responses; and (5) T cell antigens employed in these clinical trials might be insufficient to induce tumor antigen-specific T cell responses [44,63].

Given the low clinical efficacy and lack of antigen-specific T cell responses in all clinical trials with DCexos, there is a critical need to develop strategies to augment DCexo functions in generating anti-tumor T cell immunity to improve the anti-tumor efficacy of DCexo vaccines. A number of approaches have been discussed in detail in several reviews recently [44,63,95]. Briefly, the following improvements have been proposed: (1) Based on the phase II trial data on NSCLC [81], DCexo immunotherapy was likely most effective in patients with measurable levels of serum BAG6, which is possibly related to NKp30 functional defects. Thus, selection of patients who showed downregulation or defective functions of NK receptors (particularly NKG2D or NKp30), will likely improve the efficacy of DCexo therapy. The screening of NK receptors in the patients can be achieved now by using high-dimensional immunoprofiling approach such as CyTOF [96]. Along the same line, to generate synergistic immunogenic effects against NK-dependent cancers including gastrointestinal stromal tumors, neuroblastomas, and kidney cancers, DCexo vaccines could be combined with NK-based therapies, such as anti-KIR Ab (anti-killer cell immunoglobulin-like receptor antibody) [97–99]. (2) To counter systemic or local immunosuppressive mechanisms often observed in patients with advanced cancers, DCexo vaccines could be combined with other therapy regimes that reduce tumor-mediated immunosuppression. For example, in the phase II clinical trial DCexo vaccines were combined with CTX treatment [81], which has been shown in preclinical and clinical studies to reduce Treg function and stimulate dual IFN-γ/IL-17–producing T cells [89–92]. Unfortunately, objective responses were not observed in the Phase II NSCLC clinical trial even with the combination treatment, likely due to the advanced stages of NSCLC. However, combination treatments with other immunotherapies including immune checkpoint blockade (ICB) remain promising approaches. (3) To employ additional TLR ligand adjuvants as DC maturation signals, as DCexo-induced anti-tumor immune response directly depends on the degree of maturity of DCs and the type of maturation stimuli. For example, DCexos from DCs treated with poly(I:C) have been shown to be the most efficient in a model of B16-OVA melanoma in vivo compared to other TLR ligands, inducing robust activation of melanoma-specific CD8 T cells in tumor-draining lymph nodes, spleen, and tumor tissues and recruited NK and NKT cells to the tumor site, resulting in drastic inhibition of tumor growth and an increase in survival in tumor-bearing animals [100]. Together with other studies, TLR3 ligand poly(I:C) seems to be a favorable TLR agonist for DC maturation during antigen loading, which significantly increased the potential for DCexo-induced anti-tumor immunity, and could be employed as a promising maturation stimulus to generate DCexos in future clinical trials. (4) In addition, DCexos could be engineered to improve their migration and immunostimulatory capacity. DCexos could be modified to express TNF, FasL, and TRAIL to target tumor cells directly and induce tumor cell apoptosis, and DCexos could be engineered to transfer miRNAs, cytokines, and chemokines, mRNAs that encode relevant neoantigens or regulatory proteins to modulate gene functions in targeted immune cells or cancer cells. Similarly, immortalized DC cell lines, which could bypass the demanding procedure of generating autologous MoDCs on advanced cancer patients, could be amendable to generate DCexos of desired modification. (5) For DC vaccines, one promising approach to overcome the functional limitations of autologous MoDCs used in all three DCexo clinical trials is to use naturally circulating primary DCs (nDCs) [45,101]. Indeed, several clinical trials using naturally circulating DCs including cDC2s and/or pDCs have shown that nDC vaccines are safe and well-tolerated in patients, with the induction of antigen-specific immunity in some patients [102–107]. Conceivably, corresponding exosomes generated from theses DCs could be tested as vaccines. Similarly, exosomes from immortalized DC cell lines such as the human pDC cell line used in GeniusVac-Mel4 clinical trial) would bypass the need of using autologous MoDCs [108]. (6) To expand TAAs beyond the current T cell-restricted epitopes to augment anti-tumor adaptive immune responses, as recent studies have suggested that both B cells and CD4 T cells played critical role in DCexo-induced antigen-specific CD8 T cell responses [64,65]. In addition, the same group has shown that DCexos loaded with protein antigens but not with peptide antigens were capable of inducing allogeneic CD8 T cell responses, leading to the suggestion that allogeneic DCexos should be tested as cancer vaccines [66].

It should be noted that the two major presumed advantages for DCexo vaccines over DC-based vaccines; namely, better anti-tumor efficacy and resistance to immunosuppression, have not been realized in the three DCexo clinical trials [79–81]. While these strategies discussed above will undoubtedly improve on current DCexo-based cancer vaccines, they are unlikely to overcome the hurdles to move DCexo vaccines forward as discussed below. One of the major drawbacks of the DCexo clinical trials is that all three current DCexo clinical trials use peptide-pulsed DCexos from labor- and cost-intensive autologous MoDCs, based on the idea that exosomal pMHC complexes play a critical role in priming T cells. MoDCs were generated from autologous DC culture systems [44], where a leukapheresis is performed on already immunocompromised advanced cancer patients. The patient loses important immune cells, and the cells may be suboptimal. Indeed, ex vivo differentiated MoDCs have been shown to differ from the primary DCs both in phenotypic and transcriptional features and are less efficient in migratory capacity and T cell activation [95]. All three DCexo clinical trials, however, have shown limited clinical efficacy and induced little or no antigen-specific T cell responses, although enhanced NK cell activity was observed, which likely contributed to the improved clinical outcomes [79–81]. Collectively, these data suggest that exosomal pMHC complexes on peptide-pulsed DCexos from autologous MoDCs are likely not sufficient and/or critical to prime T cells in vivo, consistent with recent report showing that protein-loaded DCexos but not peptide-loaded DCexos induced antigen-specific T cell responses in vivo [65]. However, protein-loaded DCexos have not been tested in clinical settings. Given that current DCexo studies are focusing only on peptide- or protein-loaded DCexos from ex vivo differentiated MoDCs [44,109], there is an urgent need to expand our studies on DCexos beyond MoDCs, to develop new approaches to generate DCexos that are able to prime (CD8) T cells and generate anti-tumor immunity in vivo.

Several developments support/demand the expansion of DCexos from other DCs such as other DC subsets and primary DCs. For example, clinical trials with naturally circulating primary DCs including CD1c$^+$ conventional type 2 DCs (cDC2s) and plasmacytoid DCs (pDCs) are well-tolerated and safe in patients with promising results [103,104,107]. A new report on previously unreported pDCexos offers an important and exciting addition to current arsenal of DCexos [110], which we will discuss in more details.

7. Plasmacytoid DC-Derived Exosomes—The New Addition to DCexos

Although pDCs were generally thought to play a tolerogenic role in tumors as accumulation of pDCs in multiple tumors was often associated with poor prognosis [20,21,111–113], immunotherapies with pDCs either alone or in combination with cDCs have shown promising clinical results [45,101,103,114,115]. However, it remains unclear whether pDCs exert their effects directly through their cross-priming or indirectly by regulating other immune cells (i.e., cDCs, regulatory T cells, and NK cells) through pDC activation and subsequent production of IFNI and other cytokines [116,117]. In fact, even the involvement of pDCs in cross-priming in vivo is still under debate [118–121], although pDCs have been shown to be capable of cross-presentation in vitro [122–126]. Moreover, whether pDCs generate exosomes have not been investigated, although earlier studies have shown that exosomes could regulate the functions of pDCs [127–129].

As it remains unclear how pDCs exert their functions in inducing anti-tumor CD8 T cell immunity or promoting tolerance, our group decided to use a pDC-targeted vaccine model to investigate how pDCs achieve cross-priming of antigen-specific CD8 T cells [110]. Previous studies have shown that pDC-targeted anti-Siglec-H and anti-Bst2 antibodies delivered antigens to only pDCs, but not cDCs in vivo [130,131]. As anti-Siglec-H-Ag have been reported to induce CD4 T cell tolerance with or without adjuvants [130], we first employed pDC-targeted anti-Siglec-H-OVA to investigate if pDCs similarly cross-prime CD8 T cells to induce tolerance in vivo. To our surprise, vaccination with anti-Siglec-H-OVA plus CpG as adjuvant resulted in strong cross-priming of OVA-specific CD8 T (OTI) cells and recalled memory CD8 T cell responses [110]. Interestingly, pDC-mediated cross-priming in vivo is dependent on non-targeted cDCs, as depletion of cDCs abrogated effector differentiation of antigen-specific CD8 T cells [110]. Further analysis revealed that while pDCs transferred antigens to cDCs leading to both pDCs and cDCs expressing MHCI-antigen (pMHCI, H-2K^b-SIINFEKL) complexes on their surfaces, only cDCs but not pDCs effectively primed naive OTI cells ex vivo, suggesting that pDCs likely achieve cross-priming by transferring antigens to non-targeted cDCs [110]. Taking advantage of an in vitro culture system where antigens were only accessible to pDCs, we were able to confirm the requirement of bystander cDCs for pDC-mediated cross-priming, showing that cross-presenting pDCs primed naive CD8 T cells by transferring antigens to bystander cDCs [110]. Thus, our data suggest that cross-presenting pDCs transferred antigens to naïve bystander cDCs, thus conferring bystander cDCs the ability to cross-priming CD8 T cells [110].

We next investigated how cross-presenting pDCs transferred the antigens to cDCs. Using multiple approaches, we have further demonstrated that cross-presenting pDCs transferred antigens to bystander cDCs through pDC-derived exosomes (pDCexos). Interestingly, pDCexo-mediated priming of CD8 T cells was dependent on the presence of bystander cDCs, similar to cross-presenting pDCs, suggesting that cross-presenting pDCs achieve cross-priming through a novel mechanism of pDCexo-mediated antigen transfer to cDCs [110]. The pDCexo-mediated antigen transfer to cDCs is not limited to targeting pDCs via Siglec-H. Using both soluble proteins and antigens targeted to pDCs through anti-Bst2, we further showed that pDCs loaded with soluble proteins or antigens delivered through anti-Bst2 generated pDCexos, which similarly induced cDC-dependent cross-priming by transferring antigens to bystander cDCs. Taken together, our data has suggested that pDCs employ an exosome-mediated and cDC-dependent mechanism for cross-priming under multiple settings [110].

8. The Future of DCexos as Cancer Vaccines?

The identification of previously unreported pDCexos not only provides an interesting addition to current DCexo arsenal, but also open up new avenues for expanding DCexo research. As pDCexos function similarly to their counterpart pDCs, it's worthy to explore exosomes from different in-vitro differentiated DCs, as well as primary DCs to determine their potential in cancer vaccines [102–107,132].

As multiple clinical trials of pDC-based vaccines have already reported promising results [103,108,115], it is conceivable that cancer vaccines with pDCexos could combine the advantages of both pDC and DCexo vaccines. Compared to pDCs, pDCexos are more resistant to tumor-mediated immunosuppression and more biostable, and thus might achieve better anti-tumor efficacy. More excitingly, a pDC vaccine clinical trial using a human pDC cell line (GeniusVac-Mel4 clinical trial) has shown promising results [108]. The availability of multiple well-characterized human pDC cell lines, including the one used in GeniusVac-Mel4 clinical trial [108,133–135], will in theory produce pDCexos without quantity limitation at low cost, and eliminate the need of the demanding leukapheresis on vaccine patients often with advanced cancers. Production of pDCexos from these immortalized pDC cell lines will also reduce production time and is more amendable to quality control and scale up. Further studies on these pDCexos are warranted to determine their potential clinical application in cancer immunotherapy.

On the other hand, the use of DC-targeted antibody carrying desired antigens to generate pDCexos also opens up the field beyond the current peptide- or protein-loaded DCexos [44,109]. As cDC1-targeted anti-DEC-205-antigen has been shown to be about 1000 times more efficient in cross-presentation compared to soluble protein antigens [136], it would be interesting to investigate if pDCexos loaded with pDC-targeted antigens are also more efficient in cross-priming than pDCexos loaded with non-targeted protein antigens. Along the same line, our identification of pDCexos loaded with pDC-targeted antigens prompted us to ask if we could similarly generate cDC-derived exosomes (cDCexos) using cDC-targeted antigens such as anti-DEC-205-Ag, and whether such exosomes function more efficiently in cross-priming than cDCexos loaded with peptide antigens or non-targeted protein antigens. Of note, human anti-DEC-205 carrying NY-ESO-1 targeting DEC205-expressing cDCs induced both humoral and NY-ESO-1-specific CD4 and CD8 T cell responses, and achieved partial clinical responses in a phase I clinical trial [137]. Studies are warranted to further investigate how these pDCexos and/or cDCexos loaded with DC-targeted antigens function in vivo to determine if these DCexos are suitable as cancer vaccines.

Another related question raised from our pDCexo study is how pDCexos generated with pDC-targeted antigens transfer antigens to cDCs to achieve CD8 T cell priming. Interestingly, Gabrielsson's group has reported recently that DCexos generated with soluble OVA protein carried intact OVA protein [64], and these OVA-loaded DCexos induced strong allogeneic CD8 T cell responses requiring no exosomal MHC [65,66]. As uptake of pDC-targeted antigens was mediated by receptor-mediated endocytosis similar to soluble OVA protein [138], pDCexos might similarly carry intact antigens to be transferred to bystander cDCs. Indeed, our preliminary data have shown that pDCexos loaded with pDC-targeted anti-Siglec-H-Ag carried the intact anti-Siglec-H-Ag and efficiently primed allogeneic CD8 T cells in vitro and in vivo [139]. Future studies are warrantied to determine if pDCexos could be employed as impersonalized vaccines that are broadly applicable without MHC restriction [63]. Together with the potential of generating pDCexos from available human pDC cell lines, one of which has already been employed in a clinical trial with promising results [108], we would argue that pDCexos might represent the most promising DCexo candidate as cancer vaccines that could overcome the hurdles presented in previous DCexo clinical trials.

While we are focused on the application of exosomes in cancer immunotherapy, exosome-based vaccines have also emerged as good candidates for rapid development of vaccines against infectious diseases due to their increased efficacy and versatility [71]. Cross-talk between their applications in cancers and infectious diseases will likely benefit both fields. Indeed, a casual search will find that at least 7 registered human clinical trials are testing exosomes/EVs as therapeutics for treating severe acute respiratory syndrome coronavirus 2 (SARS-CoV2) (ClinicalTrials.gov Identifier: NCT04276987, ClinicalTrials.gov Identifier: NCT4384445, ClinicalTrials.gov Identifier: NCT04389385, ClinicalTrials.gov Identifier: NCT04491240, ClinicalTrials.gov Identifier: NCT04493242, ClinicalTrials.gov Identifier: NCT04602442 and ClinicalTrials.gov Identifier: NCT04798716). Our recent study has shown that resting primary HPBCs harbor abundant cytoplasmic angiotensin-converting enzyme 2 (ACE2) protein and that circulating exosomes contain ACE2, the surface expression of which is indispensable for SARS-CoV2 infection of circulating monocytes/macrophages [140], highlighting the potential of exosome-based and/or exosome-targeted immunotherapies against COVID-19. Furthermore, exosome-based vaccines might potentially synergize with mRNA-based vaccines, which have shown much success in generating efficient vaccines against SARS-CoV2 [141]. Indeed, mRNA-based vaccines are revolutionizing the field of rapid development of vaccines for emerging pathogens and have reported encouraging data in personalized neoantigen vaccines using mRNAs encoding neoantigens of patients [142,143], although achieving an efficient cytoplasmic delivery of mRNA to target cells remains one major challenge. Exosomes/EVs are known to play a critical role in intercellular communication, shuttling proteins, metabolites and nucleic acids including mRNA, miRNA, non-coding RNA, and DNA [47,49,78]. While exosomes have not been tested as potential vehicles in mRNA vaccination, exosomes have been reported to be excellent vehicles to transport mRNAs and to target the delivery to secondary lymphoid organs efficiently [144]. More importantly, direct application of mRNA or its electroporation into DCs was shown to induce polyclonal antigen-specific CD4 and CD8 T cell responses as well as the production of protective antibodies with the ability to eliminate transformed or infected cells [141]. More than 10 clinical trials on mRNA vaccines using DCs as vehicles have been completed, with similar number of ongoing clinical trials [141]. Given that exosomes are amendable to modification by introducing exogenous cargos into or unto exosomes, either through direct modification or manipulation of the parental cells [71–74], exosomes could serve as suitable vehicles for delivering mRNAs as vaccines. Similarly, DCexos would be easily modified to carry multiple desired mRNAs to augment anti-tumor immune responses and improve the anti-tumor efficacy of mRNA-based cancer vaccines. Thus, combining exosome-based and mRNA-based vaccines might represent a promising strategy to further improve mRNA-based vaccines again infectious diseases and cancer (Figure 3).

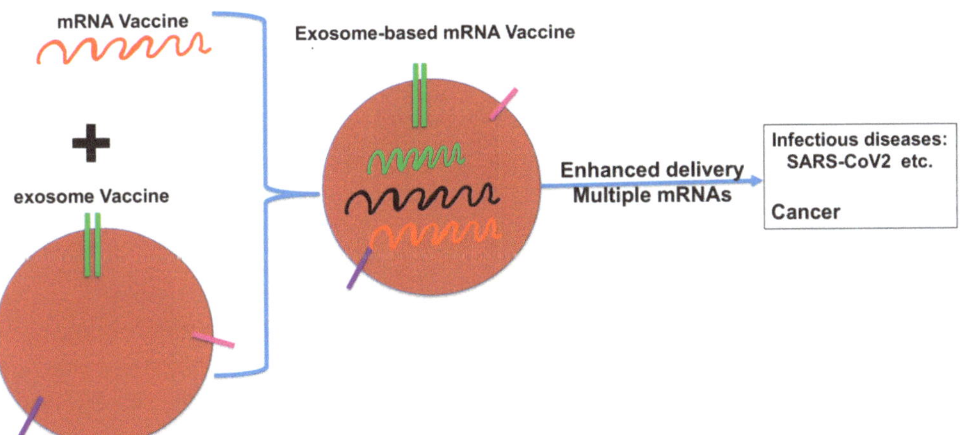

Figure 3. Exosome-based mRNA vaccines for immunotherapy. Combining exosome and mRNA vaccines might represent a promising strategy to further improve mRNA-based vaccines again infectious diseases and cancer: receptors on exosomes could enhance the targeted delivery of mRNAs, and multiple mRNAs could be efficiently delivered.

Despite the great promise of DCexo-based immunotherapy, the advance of DCexos as cancer vaccines has stalled due to the limited clinical efficacy of recent clinical trials [79–81]. However, the expansion of DCexos beyond the peptide-loaded DCexos from MoDCs, coupled with recent advance in exosome-based therapies against COVID-19 and their potential use in combination with mRNA-based vaccines, suggest that the future for exosomes/DCexos as cancer vaccines is very bright.

Author Contributions: Conceptualization, A.J.; writing, review and editing, Y.Y., C.F., L.Z., Q.-S.M. and A.J. All authors have read and agreed to the published version of the manuscript.

Funding: This work was supported by an internal grant from Henry Ford Health System and R01CA198105 (To A.J.).

Data Availability Statement: The data presented in this study is available within the article.

Conflicts of Interest: No potential conflicts of interest were disclosed.

References

1. Banchereau, J.; Steinman, R.M. Dendritic cells and the control of immunity. *Nature* **1998**, *392*, 245–252. [CrossRef]
2. Steinman, R.M.; Hawiger, D.; Nussenzweig, M.C. Tolerogenic dendritic cells. *Annu. Rev. Immunol* **2003**, *21*, 685–711. [CrossRef] [PubMed]
3. Ma, Y.; Shurin, G.V.; Peiyuan, Z.; Shurin, M.R. Dendritic cells in the cancer microenvironment. *J. Cancer* **2013**, *4*, 36–44. [CrossRef]
4. Steinman, R.M.; Cohn, Z.A. Identification of a novel cell type in peripheral lymphoid organs of mice. I. Morphology, quantitation, tissue distribution. *J. Exp. Med.* **1973**, *137*, 1142–1162. [CrossRef]
5. Fu, C.; Jiang, A. Dendritic Cells and CD8 T Cell Immunity in Tumor Microenvironment. *Front. Immunol.* **2018**, *9*, 3059. [CrossRef] [PubMed]
6. Huber, A.; Dammeijer, F.; Aerts, J.; Vroman, H. Current State of Dendritic Cell-Based Immunotherapy: Opportunities for in vitro Antigen Loading of Different DC Subsets? *Front. Immunol.* **2018**, *9*, 2804. [CrossRef]
7. Wculek, S.K.; Cueto, F.J.; Mujal, A.M.; Melero, I.; Krummel, M.F.; Sancho, D. Dendritic cells in cancer immunology and immunotherapy. *Nat. Rev. Immunol.* **2020**, *20*, 7–24. [CrossRef] [PubMed]
8. Gardner, A.; de Mingo Pulido, A.; Ruffell, B. Dendritic Cells and Their Role in Immunotherapy. *Front. Immunol.* **2020**, *11*, 924. [CrossRef] [PubMed]
9. Anderson, D.A., 3rd; Dutertre, C.A.; Ginhoux, F.; Murphy, K.M. Genetic models of human and mouse dendritic cell development and function. *Nat. Rev. Immunol.* **2021**, *21*, 101–115. [CrossRef]
10. Bandola-Simon, J.; Roche, P.A. Dysfunction of antigen processing and presentation by dendritic cells in cancer. *Mol. Immunol.* **2018**. [CrossRef]
11. Chrisikos, T.T.; Zhou, Y.; Slone, N.; Babcock, R.; Watowich, S.S.; Li, H.S. Molecular regulation of dendritic cell development and function in homeostasis, inflammation, and cancer. *Mol. Immunol.* **2019**, *110*, 24–39. [CrossRef] [PubMed]
12. Guilliams, M.; Ginhoux, F.; Jakubzick, C.; Naik, S.H.; Onai, N.; Schraml, B.U.; Segura, E.; Tussiwand, R.; Yona, S. Dendritic cells, monocytes and macrophages: A unified nomenclature based on ontogeny. *Nat. Rev. Immunol.* **2014**, *14*, 571–578. [CrossRef]
13. Gutierrez-Martinez, E.; Planes, R.; Anselmi, G.; Reynolds, M.; Menezes, S.; Adiko, A.C.; Saveanu, L.; Guermonprez, P. Cross-Presentation of Cell-Associated Antigens by MHC Class I in Dendritic Cell Subsets. *Front. Immunol.* **2015**, *6*, 363. [CrossRef]
14. Hildner, K.; Edelson, B.T.; Purtha, W.E.; Diamond, M.; Matsushita, H.; Kohyama, M.; Calderon, B.; Schraml, B.U.; Unanue, E.R.; Diamond, M.S.; et al. Batf3 deficiency reveals a critical role for CD8alpha+ dendritic cells in cytotoxic T cell immunity. *Science* **2008**, *322*, 1097–1100. [CrossRef] [PubMed]
15. Suzuki, S.; Honma, K.; Matsuyama, T.; Suzuki, K.; Toriyama, K.; Akitoyo, I.; Yamamoto, K.; Suematsu, T.; Nakamura, M.; Yui, K.; et al. Critical roles of interferon regulatory factor 4 in CD11bhighCD8alpha- dendritic cell development. *Proc. Natl. Acad. Sci. USA* **2004**, *101*, 8981–8986. [CrossRef]
16. Mildner, A.; Jung, S. Development and function of dendritic cell subsets. *Immunity* **2014**, *40*, 642–656. [CrossRef]
17. Merad, M.; Sathe, P.; Helft, J.; Miller, J.; Mortha, A. The dendritic cell lineage: Ontogeny and function of dendritic cells and their subsets in the steady state and the inflamed setting. *Annu. Rev. Immunol.* **2013**, *31*, 563–604. [CrossRef]
18. Gardner, A.; Ruffell, B. Dendritic Cells and Cancer Immunity. *Trends Immunol.* **2016**, *37*, 855–865. [CrossRef]
19. Reizis, B.; Colonna, M.; Trinchieri, G.; Barrat, F.; Gilliet, M. Plasmacytoid dendritic cells: One-trick ponies or workhorses of the immune system? *Nat. Rev. Immunol.* **2011**, *11*, 558–565. [CrossRef] [PubMed]
20. Swiecki, M.; Colonna, M. The multifaceted biology of plasmacytoid dendritic cells. *Nat. Rev. Immunol.* **2015**, *15*, 471–485. [CrossRef]
21. Mitchell, D.; Chintala, S.; Dey, M. Plasmacytoid dendritic cell in immunity and cancer. *J. Neuroimmunol.* **2018**, *322*, 63–73. [CrossRef]

22. Cisse, B.; Caton, M.L.; Lehner, M.; Maeda, T.; Scheu, S.; Locksley, R.; Holmberg, D.; Zweier, C.; den Hollander, N.S.; Kant, S.G.; et al. Transcription factor E2-2 is an essential and specific regulator of plasmacytoid dendritic cell development. *Cell* **2008**, *135*, 37–48. [CrossRef] [PubMed]

23. Rodrigues, P.F.; Alberti-Servera, L.; Eremin, A.; Grajales-Reyes, G.E.; Ivanek, R.; Tussiwand, R. Distinct progenitor lineages contribute to the heterogeneity of plasmacytoid dendritic cells. *Nat. Immunol.* **2018**, *19*, 711–722. [CrossRef]

24. Bevan, M.J. Cross-priming for a secondary cytotoxic response to minor H antigens with H-2 congenic cells which do not cross-react in the cytotoxic assay. *J. Exp. Med.* **1976**, *143*, 1283–1288. [CrossRef]

25. Bevan, M.J. Minor H antigens introduced on H-2 different stimulating cells cross-react at the cytotoxic T cell level during in vivo priming. *J. Immunol.* **1976**, *117*, 2233–2238.

26. Kurts, C.; Robinson, B.W.; Knolle, P.A. Cross-priming in health and disease. *Nat. Rev. Immunol.* **2010**, *10*, 403–414. [CrossRef] [PubMed]

27. Andersen, B.M.; Ohlfest, J.R. Increasing the efficacy of tumor cell vaccines by enhancing cross priming. *Cancer Lett.* **2012**, *325*, 155–164. [CrossRef] [PubMed]

28. Chen, L.; Fabian, K.L.; Taylor, J.L.; Storkus, W.J. Therapeutic Use of Dendritic Cells to Promote the Extranodal Priming of Anti-Tumor Immunity. *Front. Immunol.* **2013**, *4*, 388. [CrossRef]

29. Fuertes, M.B.; Woo, S.R.; Burnett, B.; Fu, Y.X.; Gajewski, T.F. Type I interferon response and innate immune sensing of cancer. *Trends Immunol.* **2013**, *34*, 67–73. [CrossRef]

30. Chen, D.S.; Mellman, I. Oncology meets immunology: The cancer-immunity cycle. *Immunity* **2013**, *39*, 1–10. [CrossRef]

31. Palucka, K.; Banchereau, J. Dendritic-cell-based therapeutic cancer vaccines. *Immunity* **2013**, *39*, 38–48. [CrossRef] [PubMed]

32. Reardon, D.A.; Mitchell, D.A. The development of dendritic cell vaccine-based immunotherapies for glioblastoma. *Semin Immunopathol.* **2017**, *39*, 225–239. [CrossRef] [PubMed]

33. Saxena, M.; Bhardwaj, N. Re-Emergence of Dendritic Cell Vaccines for Cancer Treatment. *Trends Cancer* **2018**, *4*, 119–137. [CrossRef]

34. Fu, C.; Zhou, L.; Mi, Q.S.; Jiang, A. DC-Based Vaccines for Cancer Immunotherapy. *Vaccines* **2020**, *8*, 706. [CrossRef]

35. Kantoff, P.W.; Higano, C.S.; Shore, N.D.; Berger, E.R.; Small, E.J.; Penson, D.F.; Redfern, C.H.; Ferrari, A.C.; Dreicer, R.; Sims, R.B.; et al. Sipuleucel-T immunotherapy for castration-resistant prostate cancer. *N. Engl. J. Med.* **2010**, *363*, 411–422. [CrossRef]

36. Broz, M.L.; Binnewies, M.; Boldajipour, B.; Nelson, A.E.; Pollack, J.L.; Erle, D.J.; Barczak, A.; Rosenblum, M.D.; Daud, A.; Barber, D.L.; et al. Dissecting the tumor myeloid compartment reveals rare activating antigen-presenting cells critical for T cell immunity. *Cancer Cell* **2014**, *26*, 638–652. [CrossRef]

37. Spranger, S.; Bao, R.; Gajewski, T.F. Melanoma-intrinsic beta-catenin signalling prevents anti-tumour immunity. *Nature* **2015**, *523*, 231–235. [CrossRef] [PubMed]

38. Salmon, H.; Idoyaga, J.; Rahman, A.; Leboeuf, M.; Remark, R.; Jordan, S.; Casanova-Acebes, M.; Khudoynazarova, M.; Agudo, J.; Tung, N.; et al. Expansion and Activation of CD103(+) Dendritic Cell Progenitors at the Tumor Site Enhances Tumor Responses to Therapeutic PD-L1 and BRAF Inhibition. *Immunity* **2016**, *44*, 924–938. [CrossRef] [PubMed]

39. Sanchez-Paulete, A.R.; Cueto, F.J.; Martinez-Lopez, M.; Labiano, S.; Morales-Kastresana, A.; Rodriguez-Ruiz, M.E.; Jure-Kunkel, M.; Azpilikueta, A.; Aznar, M.A.; Quetglas, J.I.; et al. Cancer Immunotherapy with Immunomodulatory Anti-CD137 and Anti-PD-1 Monoclonal Antibodies Requires BATF3-Dependent Dendritic Cells. *Cancer Discov.* **2016**, *6*, 71–79. [CrossRef]

40. Spranger, S.; Dai, D.; Horton, B.; Gajewski, T.F. Tumor-Residing Batf3 Dendritic Cells Are Required for Effector T Cell Trafficking and Adoptive T Cell Therapy. *Cancer Cell* **2017**, *31*, 711–723.e714. [CrossRef]

41. Veglia, F.; Gabrilovich, D.I. Dendritic cells in cancer: The role revisited. *Curr. Opin. Immunol.* **2017**, *45*, 43–51. [CrossRef]

42. Sanchez-Paulete, A.R.; Teijeira, A.; Cueto, F.J.; Garasa, S.; Perez-Gracia, J.L.; Sanchez-Arraez, A.; Sancho, D.; Melero, I. Antigen cross-presentation and T-cell cross-priming in cancer immunology and immunotherapy. *Ann. Oncol.* **2017**, *28*, xii44–xii55. [CrossRef]

43. Jhunjhunwala, S.; Hammer, C.; Delamarre, L. Antigen presentation in cancer: Insights into tumour immunogenicity and immune evasion. *Nat. Rev. Cancer* **2021**, *21*, 298–312. [CrossRef]

44. Pitt, J.M.; Andre, F.; Amigorena, S.; Soria, J.C.; Eggermont, A.; Kroemer, G.; Zitvogel, L. Dendritic cell-derived exosomes for cancer therapy. *J. Clin. Invest.* **2016**, *126*, 1224–1232. [CrossRef] [PubMed]

45. Bol, K.F.; Schreibelt, G.; Rabold, K.; Wculek, S.K.; Schwarze, J.K.; Dzionek, A.; Teijeira, A.; Kandalaft, L.E.; Romero, P.; Coukos, G.; et al. The clinical application of cancer immunotherapy based on naturally circulating dendritic cells. *J. Immunother Cancer* **2019**, *7*, 109. [CrossRef] [PubMed]

46. Gurunathan, S.; Kang, M.H.; Jeyaraj, M.; Qasim, M.; Kim, J.H. Review of the Isolation, Characterization, Biological Function, and Multifarious Therapeutic Approaches of Exosomes. *Cells* **2019**, *8*, 307. [CrossRef]

47. Marar, C.; Starich, B.; Wirtz, D. Extracellular vesicles in immunomodulation and tumor progression. *Nat. Immunol.* **2021**, *22*, 560–570. [CrossRef] [PubMed]

48. Stefanius, K.; Servage, K.; Orth, K. Exosomes in cancer development. *Curr. Opin. Genet. Dev.* **2021**, *66*, 83–92. [CrossRef] [PubMed]

49. Kalluri, R.; LeBleu, V.S. The biology, function, and biomedical applications of exosomes. *Science* **2020**, *367*. [CrossRef]

50. Leone, D.A.; Rees, A.J.; Kain, R. Dendritic cells and routing cargo into exosomes. *Immunol. Cell Biol.* **2018**, *96*, 683–693. [CrossRef]

51. Valadi, H.; Ekstrom, K.; Bossios, A.; Sjostrand, M.; Lee, J.J.; Lotvall, J.O. Exosome-mediated transfer of mRNAs and microRNAs is a novel mechanism of genetic exchange between cells. *Nat. Cell Biol.* **2007**, *9*, 654–659. [CrossRef]

52. Colombo, M.; Raposo, G.; Thery, C. Biogenesis, secretion, and intercellular interactions of exosomes and other extracellular vesicles. *Annu. Rev. Cell Dev. Biol.* **2014**, *30*, 255–289. [CrossRef]
53. Jella, K.K.; Nasti, T.H.; Li, Z.; Malla, S.R.; Buchwald, Z.S.; Khan, M.K. Exosomes, Their Biogenesis and Role in Inter-Cellular Communication, Tumor Microenvironment and Cancer Immunotherapy. *Vaccines* **2018**, *6*, 69. [CrossRef]
54. Moore, C.; Kosgodage, U.; Lange, S.; Inal, J.M. The emerging role of exosome and microvesicle- (EMV-) based cancer therapeutics and immunotherapy. *Int. J. Cancer* **2017**, *141*, 428–436. [CrossRef]
55. Veerman, R.E.; Gucluler Akpinar, G.; Eldh, M.; Gabrielsson, S. Immune Cell-Derived Extracellular Vesicles-Functions and Therapeutic Applications. *Trends Mol. Med.* **2019**. [CrossRef]
56. Jeppesen, D.K.; Fenix, A.M.; Franklin, J.L.; Higginbotham, J.N.; Zhang, Q.; Zimmerman, L.J.; Liebler, D.C.; Ping, J.; Liu, Q.; Evans, R.; et al. Reassessment of Exosome Composition. *Cell* **2019**, *177*, 428–445.e418. [CrossRef] [PubMed]
57. Lindenbergh, M.F.S.; Stoorvogel, W. Antigen Presentation by Extracellular Vesicles from Professional Antigen-Presenting Cells. *Annu. Rev. Immunol.* **2018**, *36*, 435–459. [CrossRef] [PubMed]
58. Raposo, G.; Nijman, H.W.; Stoorvogel, W.; Liejendekker, R.; Harding, C.V.; Melief, C.J.; Geuze, H.J. B lymphocytes secrete antigen-presenting vesicles. *J. Exp. Med.* **1996**, *183*, 1161–1172. [CrossRef] [PubMed]
59. Admyre, C.; Johansson, S.M.; Paulie, S.; Gabrielsson, S. Direct exosome stimulation of peripheral human T cells detected by ELISPOT. *Eur. J. Immunol.* **2006**, *36*, 1772–1781. [CrossRef]
60. Zitvogel, L.; Regnault, A.; Lozier, A.; Wolfers, J.; Flament, C.; Tenza, D.; Ricciardi-Castagnoli, P.; Raposo, G.; Amigorena, S. Eradication of established murine tumors using a novel cell-free vaccine: Dendritic cell-derived exosomes. *Nat. Med.* **1998**, *4*, 594–600. [CrossRef]
61. Segura, E.; Amigorena, S.; Thery, C. Mature dendritic cells secrete exosomes with strong ability to induce antigen-specific effector immune responses. *Blood Cells Mol. Dis.* **2005**, *35*, 89–93. [CrossRef]
62. Robbins, P.D.; Morelli, A.E. Regulation of immune responses by extracellular vesicles. *Nat. Rev. Immunol.* **2014**, *14*, 195–208. [CrossRef]
63. Samuel, M.; Gabrielsson, S. Personalized medicine and back-allogeneic exosomes for cancer immunotherapy. *J. Intern. Med.* **2019**. [CrossRef] [PubMed]
64. Qazi, K.R.; Gehrmann, U.; Domange Jordo, E.; Karlsson, M.C.; Gabrielsson, S. Antigen-loaded exosomes alone induce Th1-type memory through a B-cell-dependent mechanism. *Blood* **2009**, *113*, 2673–2683. [CrossRef] [PubMed]
65. Naslund, T.I.; Gehrmann, U.; Qazi, K.R.; Karlsson, M.C.; Gabrielsson, S. Dendritic cell-derived exosomes need to activate both T and B cells to induce antitumor immunity. *J. Immunol.* **2013**, *190*, 2712–2719. [CrossRef] [PubMed]
66. Hiltbrunner, S.; Larssen, P.; Eldh, M.; Martinez-Bravo, M.J.; Wagner, A.K.; Karlsson, M.C.; Gabrielsson, S. Exosomal cancer immunotherapy is independent of MHC molecules on exosomes. *Oncotarget* **2016**, *7*, 38707–38717. [CrossRef] [PubMed]
67. Viaud, S.; Terme, M.; Flament, C.; Taieb, J.; Andre, F.; Novault, S.; Escudier, B.; Robert, C.; Caillat-Zucman, S.; Tursz, T.; et al. Dendritic cell-derived exosomes promote natural killer cell activation and proliferation: A role for NKG2D ligands and IL-15Ralpha. *PLoS ONE* **2009**, *4*, e4942. [CrossRef]
68. Montecalvo, A.; Larregina, A.T.; Shufesky, W.J.; Stolz, D.B.; Sullivan, M.L.; Karlsson, J.M.; Baty, C.J.; Gibson, G.A.; Erdos, G.; Wang, Z.; et al. Mechanism of transfer of functional microRNAs between mouse dendritic cells via exosomes. *Blood* **2012**, *119*, 756–766. [CrossRef]
69. Munich, S.; Sobo-Vujanovic, A.; Buchser, W.J.; Beer-Stolz, D.; Vujanovic, N.L. Dendritic cell exosomes directly kill tumor cells and activate natural killer cells via TNF superfamily ligands. *Oncoimmunology* **2012**, *1*, 1074–1083. [CrossRef]
70. Alexander, M.; Hu, R.; Runtsch, M.C.; Kagele, D.A.; Mosbruger, T.L.; Tolmachova, T.; Seabra, M.C.; Round, J.L.; Ward, D.M.; O'Connell, R.M. Exosome-delivered microRNAs modulate the inflammatory response to endotoxin. *Nat. Commun.* **2015**, *6*, 7321. [CrossRef]
71. Fernandez-Delgado, I.; Calzada-Fraile, D.; Sanchez-Madrid, F. Immune Regulation by Dendritic Cell Extracellular Vesicles in Cancer Immunotherapy and Vaccines. *Cancers* **2020**, *12*, 3558. [CrossRef] [PubMed]
72. Bell, B.M.; Kirk, I.D.; Hiltbrunner, S.; Gabrielsson, S.; Bultema, J.J. Designer exosomes as next-generation cancer immunotherapy. *Nanomedicine* **2016**, *12*, 163–169. [CrossRef]
73. Gilligan, K.E.; Dwyer, R.M. Engineering Exosomes for Cancer Therapy. *Int. J. Mol. Sci.* **2017**, *18*, 1122. [CrossRef] [PubMed]
74. Nicolini, A.; Ferrari, P.; Biava, P.M. Exosomes and Cell Communication: From Tumour-Derived Exosomes and Their Role in Tumour Progression to the Use of Exosomal Cargo for Cancer Treatment. *Cancers* **2021**, *13*, 822. [CrossRef] [PubMed]
75. Andre, F.; Escudier, B.; Angevin, E.; Tursz, T.; Zitvogel, L. Exosomes for cancer immunotherapy. *Ann. Oncol.* **2004**, *15* (Suppl 4), iv141–iv144. [CrossRef] [PubMed]
76. Kowal, J.; Tkach, M. Dendritic cell extracellular vesicles. *Int. Rev. Cell Mol. Biol.* **2019**, *349*, 213–249. [CrossRef] [PubMed]
77. Zhang, B.; Yin, Y.; Lai, R.C.; Lim, S.K. Immunotherapeutic potential of extracellular vesicles. *Front. Immunol.* **2014**, *5*, 518. [CrossRef]
78. Kugeratski, F.G.; Kalluri, R. Exosomes as mediators of immune regulation and immunotherapy in cancer. *FEBS J.* **2021**, *288*, 10–35. [CrossRef] [PubMed]
79. Morse, M.A.; Garst, J.; Osada, T.; Khan, S.; Hobeika, A.; Clay, T.M.; Valente, N.; Shreeniwas, R.; Sutton, M.A.; Delcayre, A.; et al. A phase I study of dexosome immunotherapy in patients with advanced non-small cell lung cancer. *J. Transl. Med.* **2005**, *3*, 9. [CrossRef] [PubMed]

80. Escudier, B.; Dorval, T.; Chaput, N.; Andre, F.; Caby, M.P.; Novault, S.; Flament, C.; Leboulaire, C.; Borg, C.; Amigorena, S.; et al. Vaccination of metastatic melanoma patients with autologous dendritic cell (DC) derived-exosomes: Results of the first phase I clinical trial. *J. Transl. Med.* **2005**, *3*, 10. [CrossRef]

81. Besse, B.; Charrier, M.; Lapierre, V.; Dansin, E.; Lantz, O.; Planchard, D.; Le Chevalier, T.; Livartoski, A.; Barlesi, F.; Laplanche, A.; et al. Dendritic cell-derived exosomes as maintenance immunotherapy after first line chemotherapy in NSCLC. *Oncoimmunology* **2016**, *5*, e1071008. [CrossRef] [PubMed]

82. Dai, S.; Wei, D.; Wu, Z.; Zhou, X.; Wei, X.; Huang, H.; Li, G. Phase I clinical trial of autologous ascites-derived exosomes combined with GM-CSF for colorectal cancer. *Mol. Ther.* **2008**, *16*, 782–790. [CrossRef] [PubMed]

83. Pogge von Strandmann, E.; Simhadri, V.R.; von Tresckow, B.; Sasse, S.; Reiners, K.S.; Hansen, H.P.; Rothe, A.; Boll, B.; Simhadri, V.L.; Borchmann, P.; et al. Human leukocyte antigen-B-associated transcript 3 is released from tumor cells and engages the NKp30 receptor on natural killer cells. *Immunity* **2007**, *27*, 965–974. [CrossRef]

84. Simhadri, V.R.; Reiners, K.S.; Hansen, H.P.; Topolar, D.; Simhadri, V.L.; Nohroudi, K.; Kufer, T.A.; Engert, A.; Pogge von Strandmann, E. Dendritic cells release HLA-B-associated transcript-3 positive exosomes to regulate natural killer function. *PLoS ONE* **2008**, *3*, e3377. [CrossRef] [PubMed]

85. Mashouri, L.; Yousefi, H.; Aref, A.R.; Ahadi, A.M.; Molaei, F.; Alahari, S.K. Exosomes: Composition, biogenesis, and mechanisms in cancer metastasis and drug resistance. *Mol. Cancer* **2019**, *18*, 75. [CrossRef] [PubMed]

86. Olejarz, W.; Dominiak, A.; Zolnierzak, A.; Kubiak-Tomaszewska, G.; Lorenc, T. Tumor-Derived Exosomes in Immunosuppression and Immunotherapy. *J. Immunol. Res.* **2020**, *2020*, 6272498. [CrossRef]

87. Segura, E.; Nicco, C.; Lombard, B.; Veron, P.; Raposo, G.; Batteux, F.; Amigorena, S.; Thery, C. ICAM-1 on exosomes from mature dendritic cells is critical for efficient naive T-cell priming. *Blood* **2005**, *106*, 216–223. [CrossRef] [PubMed]

88. Viaud, S.; Ploix, S.; Lapierre, V.; Thery, C.; Commere, P.H.; Tramalloni, D.; Gorrichon, K.; Virault-Rocroy, P.; Tursz, T.; Lantz, O.; et al. Updated technology to produce highly immunogenic dendritic cell-derived exosomes of clinical grade: A critical role of interferon-gamma. *J. Immunother.* **2011**, *34*, 65–75. [CrossRef]

89. Taieb, J.; Chaput, N.; Schartz, N.; Roux, S.; Novault, S.; Menard, C.; Ghiringhelli, F.; Terme, M.; Carpentier, A.F.; Darrasse-Jeze, G.; et al. Chemoimmunotherapy of tumors: Cyclophosphamide synergizes with exosome based vaccines. *J. Immunol.* **2006**, *176*, 2722–2729. [CrossRef]

90. Viaud, S.; Flament, C.; Zoubir, M.; Pautier, P.; LeCesne, A.; Ribrag, V.; Soria, J.C.; Marty, V.; Vielh, P.; Robert, C.; et al. Cyclophosphamide induces differentiation of Th17 cells in cancer patients. *Cancer Res.* **2011**, *71*, 661–665. [CrossRef]

91. Viaud, S.; Saccheri, F.; Mignot, G.; Yamazaki, T.; Daillere, R.; Hannani, D.; Enot, D.P.; Pfirschke, C.; Engblom, C.; Pittet, M.J.; et al. The intestinal microbiota modulates the anticancer immune effects of cyclophosphamide. *Science* **2013**, *342*, 971–976. [CrossRef] [PubMed]

92. Ghiringhelli, F.; Menard, C.; Puig, P.E.; Ladoire, S.; Roux, S.; Martin, F.; Solary, E.; Le Cesne, A.; Zitvogel, L.; Chauffert, B. Metronomic cyclophosphamide regimen selectively depletes CD4+CD25+ regulatory T cells and restores T and NK effector functions in end stage cancer patients. *Cancer Immunol. Immunother.* **2007**, *56*, 641–648. [CrossRef]

93. Naseri, M.; Bozorgmehr, M.; Zoller, M.; Ranaei Pirmardan, E.; Madjd, Z. Tumor-derived exosomes: The next generation of promising cell-free vaccines in cancer immunotherapy. *Oncoimmunology* **2020**, *9*, 1779991. [CrossRef] [PubMed]

94. Markov, O.; Oshchepkova, A.; Mironova, N. Immunotherapy Based on Dendritic Cell-Targeted/-Derived Extracellular Vesicles-A Novel Strategy for Enhancement of the Anti-tumor Immune Response. *Front. Pharmacol.* **2019**, *10*, 1152. [CrossRef]

95. Nikfarjam, S.; Rezaie, J.; Kashanchi, F.; Jafari, R. Dexosomes as a cell-free vaccine for cancer immunotherapy. *J. Exp. Clin. Cancer Res.* **2020**, *39*, 258. [CrossRef]

96. Strauss-Albee, D.M.; Fukuyama, J.; Liang, E.C.; Yao, Y.; Jarrell, J.A.; Drake, A.L.; Kinuthia, J.; Montgomery, R.R.; John-Stewart, G.; Holmes, S.; et al. Human NK cell repertoire diversity reflects immune experience and correlates with viral susceptibility. *Sci. Transl. Med.* **2015**, *7*, 297ra115. [CrossRef] [PubMed]

97. Remark, R.; Alifano, M.; Cremer, I.; Lupo, A.; Dieu-Nosjean, M.C.; Riquet, M.; Crozet, L.; Ouakrim, H.; Goc, J.; Cazes, A.; et al. Characteristics and clinical impacts of the immune environments in colorectal and renal cell carcinoma lung metastases: Influence of tumor origin. *Clin. Cancer Res.* **2013**, *19*, 4079–4091. [CrossRef]

98. Rusakiewicz, S.; Semeraro, M.; Sarabi, M.; Desbois, M.; Locher, C.; Mendez, R.; Vimond, N.; Concha, A.; Garrido, F.; Isambert, N.; et al. Immune infiltrates are prognostic factors in localized gastrointestinal stromal tumors. *Cancer Res.* **2013**, *73*, 3499–3510. [CrossRef]

99. Semeraro, M.; Rusakiewicz, S.; Minard-Colin, V.; Delahaye, N.F.; Enot, D.; Vely, F.; Marabelle, A.; Papoular, B.; Piperoglou, C.; Ponzoni, M.; et al. Clinical impact of the NKp30/B7-H6 axis in high-risk neuroblastoma patients. *Sci. Transl. Med.* **2015**, *7*, 283ra255. [CrossRef] [PubMed]

100. Damo, M.; Wilson, D.S.; Simeoni, E.; Hubbell, J.A. TLR-3 stimulation improves anti-tumor immunity elicited by dendritic cell exosome-based vaccines in a murine model of melanoma. *Sci. Rep.* **2015**, *5*, 17622. [CrossRef]

101. Wimmers, F.; Schreibelt, G.; Skold, A.E.; Figdor, C.G.; De Vries, I.J. Paradigm Shift in Dendritic Cell-Based Immunotherapy: From in vitro Generated Monocyte-Derived DCs to Naturally Circulating DC Subsets. *Front. Immunol.* **2014**, *5*, 165. [CrossRef]

102. Davis, I.D.; Chen, Q.; Morris, L.; Quirk, J.; Stanley, M.; Tavarnesi, M.L.; Parente, P.; Cavicchiolo, T.; Hopkins, W.; Jackson, H.; et al. Blood dendritic cells generated with Flt3 ligand and CD40 ligand prime CD8+ T cells efficiently in cancer patients. *J. Immunother.* **2006**, *29*, 499–511. [CrossRef] [PubMed]

103. Tel, J.; Aarntzen, E.H.; Baba, T.; Schreibelt, G.; Schulte, B.M.; Benitez-Ribas, D.; Boerman, O.C.; Croockewit, S.; Oyen, W.J.; van Rossum, M.; et al. Natural human plasmacytoid dendritic cells induce antigen-specific T-cell responses in melanoma patients. *Cancer Res.* **2013**, *73*, 1063–1075. [CrossRef]

104. Prue, R.L.; Vari, F.; Radford, K.J.; Tong, H.; Hardy, M.Y.; D'Rozario, R.; Waterhouse, N.J.; Rossetti, T.; Coleman, R.; Tracey, C.; et al. A phase I clinical trial of CD1c (BDCA-1)+ dendritic cells pulsed with HLA-A*0201 peptides for immunotherapy of metastatic hormone refractory prostate cancer. *J. Immunother.* **2015**, *38*, 71–76. [CrossRef] [PubMed]

105. Schreibelt, G.; Bol, K.F.; Westdorp, H.; Wimmers, F.; Aarntzen, E.H.; Duiveman-de Boer, T.; van de Rakt, M.W.; Scharenborg, N.M.; de Boer, A.J.; Pots, J.M.; et al. Effective Clinical Responses in Metastatic Melanoma Patients after Vaccination with Primary Myeloid Dendritic Cells. *Clin. Cancer Res.* **2016**, *22*, 2155–2166. [CrossRef] [PubMed]

106. Davis, I.D.; Quirk, J.; Morris, L.; Seddon, L.; Tai, T.Y.; Whitty, G.; Cavicchiolo, T.; Ebert, L.; Jackson, H.; Browning, J.; et al. A pilot study of peripheral blood BDCA-1 (CD1c) positive dendritic cells pulsed with NY-ESO-1 ISCOMATRIX adjuvant. *Immunotherapy* **2017**, *9*, 249–259. [CrossRef]

107. Hsu, J.L.; Bryant, C.E.; Papadimitrious, M.S.; Kong, B.; Gasiorowski, R.E.; Orellana, D.; McGuire, H.M.; Groth, B.F.S.; Joshua, D.E.; Ho, P.J.; et al. A blood dendritic cell vaccine for acute myeloid leukemia expands anti-tumor T cell responses at remission. *Oncoimmunology* **2018**, *7*, e1419114. [CrossRef] [PubMed]

108. Charles, J.; Chaperot, L.; Hannani, D.; Bruder Costa, J.; Templier, I.; Trabelsi, S.; Gil, H.; Moisan, A.; Persoons, V.; Hegelhofer, H.; et al. An innovative plasmacytoid dendritic cell line-based cancer vaccine primes and expands antitumor T-cells in melanoma patients in a first-in-human trial. *Oncoimmunology* **2020**, *9*, 1738812. [CrossRef]

109. Tian, H.; Li, W. Dendritic cell-derived exosomes for cancer immunotherapy: Hope and challenges. *Ann. Transl. Med.* **2017**, *5*, 221. [CrossRef] [PubMed]

110. Fu, C.; Peng, P.; Loschko, J.; Feng, L.; Pham, P.; Cui, W.; Lee, K.P.; Krug, A.B.; Jiang, A. Plasmacytoid dendritic cells cross-prime naive CD8 T cells by transferring antigen to conventional dendritic cells through exosomes. *Proc. Natl Acad Sci. USA* **2020**. [CrossRef]

111. Demoulin, S.; Herfs, M.; Delvenne, P.; Hubert, P. Tumor microenvironment converts plasmacytoid dendritic cells into immuno-suppressive/tolerogenic cells: Insight into the molecular mechanisms. *J. Leukoc. Biol.* **2013**, *93*, 343–352. [CrossRef]

112. Aspord, C.; Leccia, M.T.; Charles, J.; Plumas, J. Melanoma hijacks plasmacytoid dendritic cells to promote its own progression. *Oncoimmunology* **2014**, *3*, e27402. [CrossRef] [PubMed]

113. Li, S.; Wu, J.; Zhu, S.; Liu, Y.J.; Chen, J. Disease-Associated Plasmacytoid Dendritic Cells. *Front. Immunol.* **2017**, *8*, 1268. [CrossRef] [PubMed]

114. Aspord, C.; Leccia, M.T.; Salameire, D.; Laurin, D.; Chaperot, L.; Charles, J.; Plumas, J. HLA-A(*)0201(+) plasmacytoid dendritic cells provide a cell-based immunotherapy for melanoma patients. *J. Invest. Dermatol.* **2012**, *132*, 2395–2406. [CrossRef] [PubMed]

115. Westdorp, H.; Creemers, J.H.A.; van Oort, I.M.; Schreibelt, G.; Gorris, M.A.J.; Mehra, N.; Simons, M.; de Goede, A.L.; van Rossum, M.M.; Croockewit, A.J.; et al. Blood-derived dendritic cell vaccinations induce immune responses that correlate with clinical outcome in patients with chemo-naive castration-resistant prostate cancer. *J. Immunother. Cancer* **2019**, *7*, 302. [CrossRef]

116. Thomas, M.; Ponce-Aix, S.; Navarro, A.; Riera-Knorrenschild, J.; Schmidt, M.; Wiegert, E.; Kapp, K.; Wittig, B.; Mauri, C.; Domine Gomez, M.; et al. Immunotherapeutic maintenance treatment with toll-like receptor 9 agonist lefitolimod in patients with extensive-stage small-cell lung cancer: Results from the exploratory, controlled, randomized, international phase II IMPULSE study. *Ann. Oncol.* **2018**, *29*, 2076–2084. [CrossRef]

117. Nierkens, S.; den Brok, M.H.; Garcia, Z.; Togher, S.; Wagenaars, J.; Wassink, M.; Boon, L.; Ruers, T.J.; Figdor, C.G.; Schoenberger, S.P.; et al. Immune adjuvant efficacy of CpG oligonucleotide in cancer treatment is founded specifically upon TLR9 function in plasmacytoid dendritic cells. *Cancer Res.* **2011**, *71*, 6428–6437. [CrossRef]

118. Joffre, O.P.; Segura, E.; Savina, A.; Amigorena, S. Cross-presentation by dendritic cells. *Nat. Rev. Immunol.* **2012**, *12*, 557–569. [CrossRef]

119. Nierkens, S.; Tel, J.; Janssen, E.; Adema, G.J. Antigen cross-presentation by dendritic cell subsets: One general or all sergeants? *Trends Immunol.* **2013**, *34*, 361–370. [CrossRef]

120. Adiko, A.C.; Babdor, J.; Gutierrez-Martinez, E.; Guermonprez, P.; Saveanu, L. Intracellular Transport Routes for MHC I and Their Relevance for Antigen Cross-Presentation. *Front. Immunol.* **2015**, *6*, 335. [CrossRef]

121. Embgenbroich, M.; Burgdorf, S. Current Concepts of Antigen Cross-Presentation. *Front. Immunol.* **2018**, *9*, 1643. [CrossRef]

122. Hoeffel, G.; Ripoche, A.C.; Matheoud, D.; Nascimbeni, M.; Escriou, N.; Lebon, P.; Heshmati, F.; Guillet, J.G.; Gannage, M.; Caillat-Zucman, S.; et al. Antigen crosspresentation by human plasmacytoid dendritic cells. *Immunity* **2007**, *27*, 481–492. [CrossRef]

123. Di Pucchio, T.; Chatterjee, B.; Smed-Sorensen, A.; Clayton, S.; Palazzo, A.; Montes, M.; Xue, Y.; Mellman, I.; Banchereau, J.; Connolly, J.E. Direct proteasome-independent cross-presentation of viral antigen by plasmacytoid dendritic cells on major histocompatibility complex class I. *Nat. Immunol.* **2008**, *9*, 551–557. [CrossRef] [PubMed]

124. Klechevsky, E.; Flamar, A.L.; Cao, Y.; Blanck, J.P.; Liu, M.; O'Bar, A.; Agouna-Deciat, O.; Klucar, P.; Thompson-Snipes, L.; Zurawski, S.; et al. Cross-priming CD8+ T cells by targeting antigens to human dendritic cells through DCIR. *Blood* **2010**, *116*, 1685–1697. [CrossRef] [PubMed]

125. Segura, E.; Durand, M.; Amigorena, S. Similar antigen cross-presentation capacity and phagocytic functions in all freshly isolated human lymphoid organ-resident dendritic cells. *J. Exp. Med.* **2013**, *210*, 1035–1047. [CrossRef]

126. Oberkampf, M.; Guillerey, C.; Mouries, J.; Rosenbaum, P.; Fayolle, C.; Bobard, A.; Savina, A.; Ogier-Denis, E.; Enninga, J.; Amigorena, S.; et al. Mitochondrial reactive oxygen species regulate the induction of CD8(+) T cells by plasmacytoid dendritic cells. *Nat. Commun.* **2018**, *9*, 2241. [CrossRef] [PubMed]

127. Bastos-Amador, P.; Perez-Cabezas, B.; Izquierdo-Useros, N.; Puertas, M.C.; Martinez-Picado, J.; Pujol-Borrell, R.; Naranjo-Gomez, M.; Borras, F.E. Capture of cell-derived microvesicles (exosomes and apoptotic bodies) by human plasmacytoid dendritic cells. *J. Leukoc. Biol.* **2012**, *91*, 751–758. [CrossRef]

128. Bracamonte-Baran, W.; Florentin, J.; Zhou, Y.; Jankowska-Gan, E.; Haynes, W.J.; Zhong, W.; Brennan, T.V.; Dutta, P.; Claas, F.H.J.; van Rood, J.J.; et al. Modification of host dendritic cells by microchimerism-derived extracellular vesicles generates split tolerance. *Proc. Natl. Acad. Sci. USA* **2017**, *114*, 1099–1104. [CrossRef]

129. Salvi, V.; Gianello, V.; Busatto, S.; Bergese, P.; Andreoli, L.; D'Oro, U.; Zingoni, A.; Tincani, A.; Sozzani, S.; Bosisio, D. Exosome-delivered microRNAs promote IFN-alpha secretion by human plasmacytoid DCs via TLR7. *JCI Insight* **2018**, *3*. [CrossRef]

130. Loschko, J.; Heink, S.; Hackl, D.; Dudziak, D.; Reindl, W.; Korn, T.; Krug, A.B. Antigen targeting to plasmacytoid dendritic cells via Siglec-H inhibits Th cell-dependent autoimmunity. *J. Immunol.* **2011**, *187*, 6346–6356. [CrossRef]

131. Loschko, J.; Schlitzer, A.; Dudziak, D.; Drexler, I.; Sandholzer, N.; Bourquin, C.; Reindl, W.; Krug, A.B. Antigen delivery to plasmacytoid dendritic cells via BST2 induces protective T cell-mediated immunity. *J. Immunol.* **2011**, *186*, 6718–6725. [CrossRef]

132. Dubsky, P.; Saito, H.; Leogier, M.; Dantin, C.; Connolly, J.E.; Banchereau, J.; Palucka, A.K. IL-15-induced human DC efficiently prime melanoma-specific naive CD8+ T cells to differentiate into CTL. *Eur. J. Immunol.* **2007**, *37*, 1678–1690. [CrossRef] [PubMed]

133. Maeda, T.; Murata, K.; Fukushima, T.; Sugahara, K.; Tsuruda, K.; Anami, M.; Onimaru, Y.; Tsukasaki, K.; Tomonaga, M.; Moriuchi, R.; et al. A novel plasmacytoid dendritic cell line, CAL-1, established from a patient with blastic natural killer cell lymphoma. *Int. J. Hematol.* **2005**, *81*, 148–154. [CrossRef]

134. Narita, M.; Watanabe, N.; Yamahira, A.; Hashimoto, S.; Tochiki, N.; Saitoh, A.; Kaji, M.; Nakamura, T.; Furukawa, T.; Toba, K.; et al. A leukemic plasmacytoid dendritic cell line, PMDC05, with the ability to secrete IFN-alpha by stimulation via Toll-like receptors and present antigens to naive T cells. *Leuk. Res.* **2009**, *33*, 1224–1232. [CrossRef]

135. Drexler, H.G.; Macleod, R.A. Malignant hematopoietic cell lines: In vitro models for the study of plasmacytoid dendritic cell leukemia. *Leuk. Res.* **2009**, *33*, 1166–1169. [CrossRef] [PubMed]

136. Bonifaz, L.; Bonnyay, D.; Mahnke, K.; Rivera, M.; Nussenzweig, M.C.; Steinman, R.M. Efficient targeting of protein antigen to the dendritic cell receptor DEC-205 in the steady state leads to antigen presentation on major histocompatibility complex class I products and peripheral CD8+ T cell tolerance. *J. Exp. Med.* **2002**, *196*, 1627–1638. [CrossRef] [PubMed]

137. Dhodapkar, M.V.; Sznol, M.; Zhao, B.; Wang, D.; Carvajal, R.D.; Keohan, M.L.; Chuang, E.; Sanborn, R.E.; Lutzky, J.; Powderly, J.; et al. Induction of antigen-specific immunity with a vaccine targeting NY-ESO-1 to the dendritic cell receptor DEC-205. *Sci. Transl. Med.* **2014**, *6*, 232ra251. [CrossRef] [PubMed]

138. Burgdorf, S.; Schuette, V.; Semmling, V.; Hochheiser, K.; Lukacs-Kornek, V.; Knolle, P.A.; Kurts, C. Steady-state cross-presentation of OVA is mannose receptor-dependent but inhibitable by collagen fragments. *Proc. Natl. Acad. Sci. USA* **2010**, *107*, E48–E49, author reply E50-41. [CrossRef]

139. Fu, C.; Yao, Y.; Zhou, L.; Mi, Q.-S.; Cui, W.; Lee, K.P.; Krug, A.B.; Jiang, A. pDC-derived exosomes employ a novel mechanism to induce CD8 T cell responses in vivo. Center for Cutaneous Biology and Immunology, Department of Dermatology, Henry Ford Health System, Detroit, MI, USA, 2021. manuscript in preparation.

140. Yao, Y.; Subedi, K.; Sexton, J.Z.; Liu, T.; Khalasawi, N.; Pretto, C.D.; Wotring, J.W.; Wang, J.; Yin, C.; Jiang, A.; et al. Circulating monocytes co-expressing surface ACE2 and TMPRSS2 upon TLR4/7/8 activation are susceptible to SARS-CoV-2 infection. *Cell Res.* **2021**, submitted.

141. Heine, A.; Juranek, S.; Brossart, P. Clinical and immunological effects of mRNA vaccines in malignant diseases. *Mol. Cancer* **2021**, *20*, 52. [CrossRef]

142. Sahin, U.; Derhovanessian, E.; Miller, M.; Kloke, B.P.; Simon, P.; Lower, M.; Bukur, V.; Tadmor, A.D.; Luxemburger, U.; Schrors, B.; et al. Personalized RNA mutanome vaccines mobilize poly-specific therapeutic immunity against cancer. *Nature* **2017**, *547*, 222–226. [CrossRef] [PubMed]

143. Gebre, M.S.; Brito, L.A.; Tostanoski, L.H.; Edwards, D.K.; Carfi, A.; Barouch, D.H. Novel approaches for vaccine development. *Cell* **2021**, *184*, 1589–1603. [CrossRef] [PubMed]

144. Srinivasan, S.; Vannberg, F.O.; Dixon, J.B. Lymphatic transport of exosomes as a rapid route of information dissemination to the lymph node. *Sci. Rep.* **2016**, *6*, 24436. [CrossRef] [PubMed]

Review

The Role of Extracellular HSP70 in the Function of Tumor-Associated Immune Cells

Manuel Linder and Elke Pogge von Strandmann *

Institute for Tumor Immunology, Clinic for Hematology, Immunology, and Oncology, Philipps University of Marburg, Hans-Meerwein-Strasse 3, 35043 Marburg, Germany; manuel.linder@staff.uni-marburg.de
* Correspondence: poggevon@staff.uni-marburg.de; Tel.: +49-6421-282-1640

Simple Summary: The intracellular heat shock protein 70 (HSP70) is essential for cells to respond to stress, for instance, by refolding damaged proteins or inhibiting apoptosis. However, in cancer, HSP70 is overexpressed and can translocate to the extracellular milieu, where it emerged as an important modulator of tumor-associated immune cells. By targeting the tumor microenvironment (TME) through different mechanisms, extracellular HSP70 can trigger pro- or anti-tumorigenic responses. Therefore, understanding the pathways and their consequences is crucial for therapeutically targeting cancer and its surrounding microenvironment. In this review, we summarize current knowledge on the translocation of extracellular HSP70. We further elucidate its functions within the TME and provide an overview of potential therapeutic options.

Abstract: Extracellular vesicles released by tumor cells (T-EVs) are known to contain danger-associated molecular patterns (DAMPs), which are released in response to cellular stress to alert the immune system to the dangerous cell. Part of this defense mechanism is the heat shock protein 70 (HSP70), and HSP70-positive T-EVs are known to trigger anti-tumor immune responses. Moreover, extracellular HSP70 acts as an immunogen that contributes to the cross-presentation of major histocompatibility complex (MHC) class I molecules. However, the release of DAMPs, including HSP70, may also induce chronic inflammation or suppress immune cell activity, promoting tumor growth. Here, we summarize the current knowledge on soluble, membrane-bound, and EV-associated HSP70 regarding their functions in regulating tumor-associated immune cells in the tumor microenvironment. The molecular mechanisms involved in the translocation of HSP70 to the plasma membrane of tumor cells and its release via exosomes or soluble proteins are summarized. Furthermore, perspectives for immunotherapies aimed to target HSP70 and its receptors for cancer treatment are discussed and presented.

Keywords: HSP70; extracellular vesicles; tumor microenvironment; cancer; immune modulation

Citation: Linder, M.; Pogge von Strandmann, E. The Role of Extracellular HSP70 in the Function of Tumor-Associated Immune Cells. *Cancers* **2021**, *13*, 4721. https://doi.org/10.3390/cancers13184721

Academic Editor: David Wong

Received: 15 July 2021
Accepted: 14 September 2021
Published: 21 September 2021

Publisher's Note: MDPI stays neutral with regard to jurisdictional claims in published maps and institutional affiliations.

1. Introduction

Heat shock protein 70 (HSP70/HSPA1A/HSP72) is a molecular chaperone and belongs to the HSP70 family. It is upregulated upon stress-stimuli, such as heat, exercise, or pathological stress. HSP70 comprises a plethora of crucial housekeeping and chaperone activities, including folding of newly synthesized proteins, refolding or disposal of damaged proteins, preventing protein aggregates, or translocating proteins to different compartments. In addition, HSP70 can directly inhibit apoptosis, thus protecting cells from stress-induced cell death [1]. However, in pathologies like cancer, upregulated HSP70 can lead to disease progression and therapy resistance [2]. The chaperone consists of three main domains, the N-terminal ATPase domain (NBD), the substrate-binding domain (SBD), and a C-terminal domain. The NBD is crucial for binding and hydrolyzing ATP to ADP, thus regulating the chaperone's conformational state. In general, the binding of ATP leads to an open or low-affinity conformation allowing HSP70 to bind its substrates via its SBD.

In the ADP state, the chaperone turns into a closed or high-affinity conformation in which the protein is further protected and processed. Importantly, all functions of HSP70 require different co-chaperones, such as BCl-2-associated anthanogene 3 and 4 (BAG3, BAG4) or carboxy-terminus of HSP70 interacting protein (CHIP), and is mainly mediated by the C-terminal region of the chaperone [1]. Initially, HSP70 was identified as an intracellular protein (iHSP70), but it can translocate to the extracellular milieu (exHSP70) under stressful conditions. This includes physiological stress (e.g., physical activity) and pathological stress (e.g., cancer) and is higher in the latter. Hereby, it is either associated with the plasma membrane (mHSP70), with extracellular vesicles (evHSP70), or secreted as soluble proteins (sHSP70) [2].

Extracellular vesicles (EVs) are nanoparticles secreted by virtually all cells under physiological and pathophysiological conditions. EVs differ in size, origin, and cargo and can be mainly subdivided into ectosomes and exosomes [3]. Ectosomes, also referred to as microvesicles, are generally described as particles with a size of 50–1000 nm, generated by a direct outward budding of the plasma membrane. In contrast, exosomes are small EVs with a size between 40 and 200 nm and are generated along an endosomal pathway by double invagination of the plasma membrane (Box 1). EVs contain a plethora of molecules, including proteins, lipids, and nucleic acids, usually reflecting the cell of origin [3]. In the past, it has been speculated that they function as trash bins to remove unwanted cellular components; however, increasing evidence demonstrates crucial functions as an intercellular communication system [3–5].

Box 1. Biology and function of exosomes

<table><tr><td>

Exosomes are small extracellular vesicles (40–200 nm; small EVs), which are produced by a double invagination of the plasma membrane. In particular, endocytosis leads to the generation of early sorting endosomes (ESE), containing factors from the cell membrane and the extracellular milieu. After maturation to late sorting endosomes (LSE), a second invagination of the membrane occurs, leading to the formation of multivesicular bodies (MVB) containing future exosomes. During MVB formation, cargo is specifically sorted to the vesicles by several distinct mechanisms. The endosomal sorting complexes required for transport (ESCRT) machinery is widely believed to play a role in the sorting process; however, specific loading mechanisms remain largely unknown and need to be elucidated. Typical exosomal markers often include ESCRT components, such as Alix or TSG101. Interestingly, HSP70 is also described as an exosome marker; however, its expression is not specific to the small EVs. The cargo can include various proteins, lipids, or nucleic acids, indicating a multitude of distinct functions. After the release, exosomes can interact with target cells by binding to specific receptors, fusion with the plasma membrane, or endocytosis. Since they are critical players of intercellular communication in physiological and pathological conditions, the research interest significantly increased in the last decade [3,4].

</td></tr></table>

In this review, we summarize the current knowledge on the HSP70 translocation to the extracellular milieu, either in association with the plasma membrane, extracellular vesicles, or as soluble extracellular proteins. Moreover, we outline the functions of exHSP70, especially its ability to modulate the immune system in cancer. We address possible interactions of the chaperone with the tumor microenvironment. Finally, we provide an overview of the potential therapeutic options regarding exHSP70.

2. Translocation of exHSP70

HSP70 has several functions, including folding newly synthesized proteins, regulating protein activity, or preventing aggregation, indicating a cytosolic localization of the chaperone [1]. However, HSP70 has also been found on the plasma membrane, associated with EVs, or secreted as free soluble protein [6–8]. Here, the translocation was shown to occur independently of the physiological state, although stress, such as in cancer, drastically increases the extracellular localization [8,9].

In the classical secretory pathway, proteins synthesized by ribosomes are released into the lumen of the endoplasmic reticulum (ER), where chaperones assist and control the proper protein folding. Correctly folded proteins subsequently enter the Golgi apparatus and are eventually secreted by transport vesicles. This pathway usually requires a short peptide sequence that targets proteins for secretion [10]. HSP70 lacks such a specific signal peptide indicating different mechanisms for its translocation [11].

2.1. Membrane HSP70 (mHSP70)

In accordance, Broquet and colleagues reported an unhindered translocation of HSP70 to the plasma membrane after treatment with the classical secretory pathway inhibitors brefeldin A or monensin [12]. Brefeldin A is an antiviral lactone compound blocking the anterograde transport from the ER to the Golgi. In contrast, monensin is a polyether antibiotic acting as an ionophore, inhibiting the transport from the Golgi apparatus [13]. This study further revealed that mHSP70 preferentially localizes in lipid rafts, which are microdomains enriched in cholesterol, glycosphingolipids, and protein receptors. They typically form in the exoplasmic leaflet of the Golgi apparatus and are also known as detergent-resistant microdomains (DRMs) due to their composition and resistance to non-ionic detergents [14]. Moreover, treatment with the drug methyl-β-cyclodextrin (MβCD), which degrades cholesterol and disrupts DRMs, resulted in a significant decrease in mHSP70, suggesting lipid raft-mediated translocation of HSP70 [12]. In accordance, Hunter and Levin showed HSP70 release in peripheral blood mononuclear cells (PBMCs), which could be blocked by MβCD but not by brefeldin A [15]. However, in contrast to the previous study, translocation was also partially inhibited by monensin. The authors postulated that the Golgi but not the ER might be important for HSP70 secretion or that monensin led to a disruption of the plasma membrane, ultimately inhibiting the transport [15]. It has also been reported that HSP70 can non-covalently bind to the lipid raft component globotriaosylceramide (Gb3), further supporting the role of lipid rafts in the translocation of the chaperone [16,17]. Therefore, HSP70 could be associated with Gb3 and subsequently secreted and recruited to lipid rafts. Another mechanism of membrane integration by HSP70 could be a direct interaction with the plasma membrane, as suggested by its ability, to bind to phosphatidylserine (PS), integrate into membranes, or form ionic channels [7,18–21]. It has been hypothesized that HSP70 translocates to the membrane after oligomerization, where it binds to PS [7]. Spontaneous flipping allows PS to reach the outer membrane layer, leading to the integration of the chaperone. It is further described that the return of PS into the cytosolic layer does not affect the integration of HSP70 [7].

2.2. EV-Associated HSP70 (evHSP70)

In 2005, Lancaster and co-workers presented another mechanism for the translocation of HSP70 [6]. They reported that HSP70 is released by exosomes in PBMCs independently of stress. Neither brefeldin A nor MβCD were able to inhibit the transport, indicating an endosomal-mediated secretion mechanism. It was further postulated that the controversial results of previous studies in PBMCs from Hunter and Levin could be due to an insufficient concentration of the inhibitor MβCD [6,15]. However, distinct exocytosis processes could occur in different and even within the same cell types. It can be hypothesized that translocation of HSP70 into or onto exosomes could eventuate after the chaperones are localized in DRM. Lipid raft-mediated endocytosis could then lead to the formation of endosomes, further maturing into multivesicular bodies (MVB) containing future exosomes. Many other reports have confirmed exosomal secretion of HSP70; however, the mechanism of sorting the chaperone to the vesicles is still elusive [7,22–24]. In general, posttranslational modifications, including ubiquitination, phosphorylation, or sumoylation, have been indicated to play a role in cargo sorting into exosomes [25–28]. Interestingly, HSP70 was found to be ubiquitinated by the co-chaperone CHIP, mainly presumed to mark the protein for degradation [29,30]. However, Jiang and colleagues showed that ubiquitylation of the constitutive isoform HSC70 by CHIP did not lead to an increased degradation [31]. This is in line with a report displaying that crucial amounts of secreted HSP70 in A431 cells are ubiquitinylated [32]. In contrast, acetylation of the HSP70 family member high glucose-regulated protein 78 (GRP78) was shown to inhibit exosomal secretion through interaction with the phosphoinositide-3 kinase VPS34 [33]. Acetylated HSP70 was also reported to bind to VPS34 [34]. Hereby, autophagosomal stress led to disruption of the interaction of histone deacetylase 6 (HDAC6) with the chaperone, resulting in increased acetylation. Interestingly, oligomerization, which was shown to play a role in membrane integration of HSP70, was indicated to be preferentially loaded into exosomes [7,35]. This is supported by studies of Nimmervoll and co-workers, demonstrating mHSP70-mediated clathrin-independent endocytosis, which was dependent on the oligomerization of the chaperone [36,37]. Therefore, HSP70 could be sorted into exosomes after integration into the membrane in an oligomeric form.

2.3. Soluble Extracellular HSP70 (sHSP70)

In addition to mHSP70 and evHSP70, the chaperone can also be found in a free soluble form, which was previously thought to result exclusively from passive release upon cell death [11]. In particular, necrosis rather than apoptosis was assumed to be responsible for the release of sHSP70 [38]. However, cell death accounts only for a minor fraction of sHSP70, and it is mainly released in an active manner [11]. Mambula and Calderwood postulated an endolysosomal route as a mechanism for solubleHSP70 (sHSP70), as the secretion correlated with the lysosomal marker lysosomal-associated membrane protein 1 (LAMP1) and could be inhibited by lysosomotropic compounds [39]. Moreover, they showed that HSP70 could enter lysosomes via ATP-binding cassette (ABC) family transporter proteins [39]. Interestingly, ABC transporters are also expressed on endosomes and exosomes, where one could postulate a possible way for HSP70 loading into exosomes. The secreted chaperone was also found to bind back to the plasma membrane, indicating an alternative route to the lipid raft-mediated pathway for membrane-associated HSP70 [39].

Another way of HSP70 release was demonstrated by Evdonin et al., showing the formation of secretory-like granules upon inhibition of phospholipase C [32]. Interestingly, this secretion could be blocked by brefeldin A, indicating an involvement of the classical secretory pathway [32,40]. This was contradictory to prior studies that demonstrated no inhibition of the secretory pathway by brefeldin A [6,12,15]. However, the authors suggested a time-dependent effect since previous studies evaluated the translocation of HSP70 at least 4 h after inducing stress or treatment with brefeldin A [40]. Therefore, it can be postulated that distinct cells potentially use different mechanisms, possibly depending on the stress level and exposure.

Altogether, it can be concluded that translocation of HSP70 into the extracellular milieu is increased under stress conditions and that several pathways lead to either membrane-, EV-associated, or soluble HSP70 (Figure 1). Still, additional research is needed to further unravel the different mechanisms and their functional consequences.

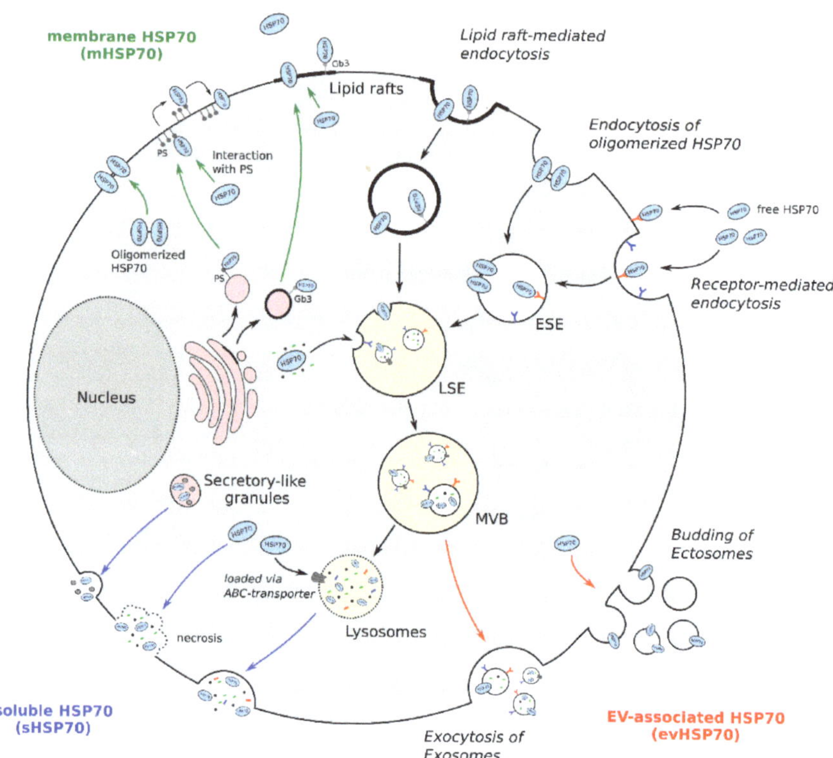

Figure 1. Translocation of extracellular HSP70 (exHSP70). Cellular stress leads to the translocation of HSP70 via different potential mechanisms. In particular, membrane-associated HSP70 (mHSP70, green) can be found in lipid rafts, mainly interacting with globotriaosylceramide 3 (Gb3). Moreover, HSP70 can interact with phosphatidylserine (PS), leading to its integration into the plasma membrane. Oligomerized HSP70 was also reported to integrate into the plasma membrane directly. HSP70 can also be released into or onto extracellular vesicles (evHSP70, red), by either budding of the plasma membrane (ectosomes) or by an endolysosomal pathway (exosomes). Soluble HSP70 (sHSP70, blue) can be secreted either passively during cell necrosis or actively via secretory granules or lysosomes. ESE: early sorting endosomes; LSE: late sorting endosomes; MVB: multivesicular bodies; EV: extracellular vesicles; PS: phosphatidylserine; Gb3: globotriaosylceramide 3; HSP70: heat shock protein 70.

3. Role of exHSP70: Regulation of Immune Responses

The immune system is a complex network of biological processes protecting us from various pathogens and diseases. It is mainly divided into two groups, the innate immune system, which is defined by a non-specific and rapid response, and the adaptive immune system, which can respond and adapt to specific stress stimuli. A typical immune response can be divided into four phases: (I) Recognition of pathogen- or damage-associated molecular patterns (PAMPs/DAMPs) by innate immune cells resulting in phagocytosis, complement activation, and secretion of pro-inflammatory cytokines. (II) Released cytokines, such as IL-1, Il-12, or TNF-α, then trigger an acute inflammatory response helping to control the infection. (III) Meanwhile, antigen-presenting cells (APCs) activate naïve T-helper cells via their MHC class II and co-stimulatory signal molecules. (IV) Activated T-helper cells subsequently initiate the adaptive immune system, including a cell-mediated and humoral immune response [41].

In the case of inflammation, intracellular HSP70 expression is dramatically increased and exerts cytoprotective functions. This is achieved both through classical chaperone functions, such as refolding and repair of proteins, and by direct inhibition of apoptosis [42]. Since excessive apoptosis can lead to severe human inflammatory diseases (HIDs), HSP70 plays a crucial role in balancing the appropriate response to cell stress. However, in cancer, HSP70 is known to be overexpressed, upsetting the balance and increasing proliferation, invasiveness, and resistance of malignant cells [2]. As described in the previous chapter, HSP70 is known to translocate into the extracellular milieu primarily upon stress stimuli. Moreover, exHSP70 was shown to exert pro-inflammatory functions, leading to increased tissue damage, indicating a dual role of HSP70 [43]. In the following section, we provide an overview of the general mechanisms of exHSP70 in immunity and summarize current data of its functions towards immune cells in the tumor microenvironment.

3.1. General Mechanisms of Immunomodulation by exHSP70

The role of HSP70 in immunity is still extensively discussed in the literature, and a plethora of distinct functions and mechanisms are described. It is associated with developing an innate and adaptive immune response, formation of memory cells, and termination of the immune response [2].

To exert specific immunomodulatory functions, proteins need to interact with immune cells. Here, membrane-bound and free HSP70 were both shown to bind to different immune cells, including macrophages, dendritic cells, and natural killer (NK) cells [44–46]. Moreover, in the early 2000s, Asea and colleagues demonstrated that exHSP70 specifically binds to monocytes, increasing the pro-inflammatory cytokines TNF-α, IL-6, and IL-1β [47]. This was postulated to be mediated by two distinct pathways: a CD14-dependent and a CD14-independent mechanism, with both being dependent on intracellular calcium. Consequently, exHSP70 binds monocytes, increasing the intracellular calcium flux and resulting in NF-κB-mediated transcription of pro-inflammatory cytokines [47]. Hence, the authors termed HSP70 a "chaperokine" [48]. This finding led to many publications displaying pro-inflammatory cytokine functions of HSP70 [49–51]. In addition to CD14, Toll-like receptors 2 and 4 (TLR2/TLR4) and CD40 were also described to mediate those effects, mainly via NF-κB and the MAPK pathway [49,51,52]. The fact that HSP70 could be released during necrotic cell death and subsequently induce an inflammatory response by TLR and being increasingly expressed in various inflammatory diseases emphasized its role as a DAMP [38,47,51].

However, shortly after the first publication, Gao and Tsan issued their concerns about the pro-inflammatory functions and postulated that endotoxin contamination and not HSP70 itself could trigger those effects [53]. They showed that highly purified recombinant HSP70 could not induce TNF-α in murine macrophages [53]. Moreover, endotoxin-free HSP70 was not able to mature dendritic cells, as described earlier [38,54]. Further studies from different groups supported the theory that most pro-inflammatory functions could be due to contamination [55–58]. Therefore, results obtained using recombinant HSP70, mainly produced in *Escherichia coli (E. coli)*, need to be interpreted with caution. In addition, studies with purified recombinant proteins are limited since the activity of HSP70-protein complexes is often overlooked.

Fong and co-workers gave another explanation of the contradictory results. They showed that exHSP70 binds to sialic acid-binding immunoglobulin-type lectins (Siglecs), which are membrane proteins expressed on immune cells. In particular, exHSP70 can interact with the anti-inflammatory Siglec-5 as well as the pro-inflammatory Siglec-14. Different expression patterns of Siglecs in specific immune cell populations could therefore explain different inflammatory responses [59].

Interestingly, recent studies introduced another receptor that potentially induces proinflammatory functions [60,61]. Here, the authors showed that exHSP70 binds to the receptor for advanced glycation endproducts (RAGE), leading to extracellular signal-regulated kinases (ERK)-dependent activation of NF-κB. Apart from releasing pro-inflammatory cytokines, activated NF-κB also increased RAGE expression, resulting in a positive feedback loop [61].

In contrast, it was also reported that exHSP70 has rather anti-inflammatory than pro-inflammatory functions [62]. Studies showed that highly purified exHSP70 downregulated TNFα and IL-6 production in monocytes [63]. This was explained by upregulation of the heat shock factor 1 (HSF1), subsequently inhibiting NF-κB activation and directly binding the *TNFα* gene promotor [63,64]. Extracellular HSP70 was also shown to repress LPS-induced cytokines in rats [65]. In addition, in most neurological disorders, the chaperone was described to be primarily anti-inflammatory [66].

The function of HSP70 in the adaptive immune response is less controversial. It is widely accepted that the chaperone plays a significant role in the antigen-presenting process. Firstly, HSP70 can bind antigens either inside or outside of the cell [67]. The complex is then recognized by APCs via the CD91 or scavenger receptors, such as LOX-1 or SREC-1, resulting in endocytosis [68]. Inside of the APCs, HSP70 protects the antigen until it reaches the proteasome. Here, the antigen is released, processed, and transported to MHC class I molecules [69]. This unique cross-presentation finally leads to the activation of CD8+ T-cells. Moreover, the HSP70-antigen complex can be processed in the lysosome leading to the presentation of the antigen on MHC class II molecules, subsequently activating CD4+ T-cells [69]. The presentation itself was shown to be significantly enhanced by the HSP70-antigen complex compared to the antigen alone [70]. Therefore, HSP70 is essential for the transport, uptake, protection, and effective (cross-)presentation of antigens.

Additionally, to the four phases of a typical immune response, the generation of immunological memory is crucial. Generally, the formation of B- and T-memory cells depends on the respective B- and T-cell receptors (BCR/TCR) and the MHC class II antigen complex [41]. In 2005, researchers of the University of Tuebingen showed enhanced activation of CD4+ memory T-cells by HSP70-peptide complexes compared to peptides alone [71]. This activation was shown to be dependent on CD91 and scavenger receptors [72]. Moreover, Wang and colleagues demonstrated a TCR- and MHC class II-independent mechanism of CD4+ memory T-cell formation [73]. Here, stress-induced dendritic cells upregulated intracellular and membrane-associated HSP70, which activates NF-κB via CD40, resulting in increased membrane-bound IL-15 expression. The IL-15 then activates the JAK3 and STAT5 pathway in CD4+ T-cells, upregulating CD40L, which can reactivate the DCs for a positive feedback loop and subsequently lead to the formation of CD4+ memory T-cells [73]. This was the first time researchers could demonstrate the formation of memory cells independent of the antigen. These results were also validated in vivo [74].

The last step of the immune response is the termination, which is vital to prevent tissue damage by excessive inflammation [75]. Overexpression of HSP70 on the surface of immune cells serves as a regulator for the termination by displaying a "death signal" [76]. HSP70 is hereby recognized by γδT killer cells, which subsequently terminate the cells [76]. Moreover, HSP70 was shown to prime γδT killer cells, leading to higher proliferation and killing [77]. This priming could be dependent on TLRs [78].

All in all, HSP70 is crucially involved in many aspects of immunity, including the development of the innate and adaptive immune response, the formation of memory cells, and the termination of the immune response. Hereby, the chaperone exhibit either pro- or anti-inflammatory functions, potentially depending on its location, cell type, and expression level. Several receptors are known to interact with exHSP70, such as CD40, TLR2, TLR4, CD14, RAGE, CD91, as well as different scavenger receptors, primarily exhibiting their function via the NF-κB or MAPK pathway.

3.2. The Role of exHSP70 in Immunomodulation of Cancer

The communication of cancer cells and immune cells is a crucial step in cancer progression. Therefore, understanding the mechanisms and consequences of the interaction is essential. A potential key player of the crosstalk in the tumor microenvironment may be HSP70, which is overexpressed in cancer compared to normal tissue [2,79–82]. The expression was shown to be correlated to tumor grade, therapy resistance, and worse overall survival [80–85]. In particular, HSP70 can inhibit the intrinsic and extrinsic apoptotic pathway and block oncogene-induced senescence, resulting in therapy resistance [2]. The chaperone was also reported to be translocated into the extracellular milieu in cancer, including as a membrane-bound or as an exosome secreted form [82,83,86–89]. Recently, Finkernagel and colleagues were able to identify HSP70 as a major constituent of ovarian cancer EVs and demonstrated a significant correlation of exHSP70 with patient survival [88]. Moreover, in a prospective clinical study of breast and lung cancer, exosomal HSP70 was correlated to metastasis and disease status. Here, the authors suggested HSP70-positive exosomes as a potential biomarker to predict tumor responses and clinical outcomes [82].

One candidate of the pro-tumorigenic effects of exHSP70 could be TLR4 and the subsequent PI3K/Akt pathway engagement. It is described that the pathway leads to IL-10 and galectin-1 production, resulting in an increase of matrix-metalloproteases 2 and 9 (MMP-9/MMP-2), finally enhancing tumor migration [90]. The induction of MMP-2 and MMP-9 by exosomal HSP70 was already reported in mesoangioblasts, where initiation occurred via TLR4 and CD91 in an autocrine fashion [91]. In addition, MMP-9 induction was also reported in monocytes. Extracellular HSP70 stimulated NF-κB and activating protein-1 (AP-1), enabling MMP-9 expression [92]. Interestingly, HSP70 was also shown to increase its own expression via the TLR4 by inactivating glycogen synthase kinase-3β (GSK-3β) via Akt signaling. Inactivation of GSK-3β then stimulated HSF1, finally inducing intracellular HSP70 [93]. It was also reported that tumor-derived exosomal HSP70 activated myeloid-derived suppressor cells (MDSC), enhancing tumor growth [94]. This is due to TLR2 activation and subsequent MyD88-dependent phosphorylation of the signal transducer and activator of transcription 3 (STAT3) [94,95]. STAT3 is known to be critically involved in tumor progression and generation of an immunosuppressive and therefore pro-tumorigenic environment [96,97]. Moreover, activation of TLR2 by exosomal HSP70 led to upregulation of IL-6, iNos, and Arg-1 [95]. Thereby, iNos enhances nitric oxide production, whereas Arg-1 leads to arginine depletion, both inhibiting T-cell proliferation and function [98,99]. MyD88-independent activation of TLR4 by exHSP70 was also described to facilitate cancer growth, potentially through the TRIF pathway [100].

The interaction of exHSP70 with TLR 2/4 was further demonstrated to activate neutrophils [88]. Klink and co-workers showed that activation of those receptors led to the production of reactive oxygen species (ROS) and the release of IL-8. This is associated with cancer progression since ROS can stimulate the expression of vascular endothelial growth factor (VEGF) and activate MMPs, thus enhancing tumor angiogenesis and metastasis [88,101]. Moreover, exHSP70-TLR2 interaction also leads to activation and pro-inflammatory cytokine production in neutrophils [102].

Additionally, the TLR4 pathway was suggested to play a role in chemotherapy resistance [103]. In ovarian cancer, TLR4 activation resulted in MyD88-dependent nuclear localization of NFκB, upregulating the production of IL-6 as well as the chemokines MCP-1 and GRO-α, which are all associated with tumor progression. Furthermore, the Akt pathway was activated, followed by enhanced expression of the anti-apoptotic protein XIAP, thus exhibiting resistance to chemotherapy [103,104]. In accordance, a recent study showed that transfer of exHSP70 via small EVs resulted in therapy resistance of breast cancer by increasing ROS production [105].

Epithelial to mesenchymal transition (EMT) is believed to be a key step of cancer cells to enable tumor invasion and metastasis. Li and colleagues showed that treating cells with tumor-derived exHSP70 resulted in the decrease of E-cadherin and the increase of α-SMA and, therefore, in the induction of EMT [106]. This was mediated by TLR2/4 and subsequent activation of the JNK1/2 and MAPK pathways [107].

Another target of exHSP70 is the receptor RAGE, which was shown to result in a pro-inflammatory response via ERK-dependent NF-κB activation [61]. Interestingly, RAGE was also found to be overexpressed in cancer, where it was correlated to tumor size and cancer stage [108,109]. Additionally, RAGE is expressed on immune cells, such as monocytes or macrophages, further extending possible targets for exHSP70 [110]. It is also described that

RAGE interacts with TLR4 and that this crosstalk leads to MyD88-dependent activation of NF-κB [110,111] (Figure 2).

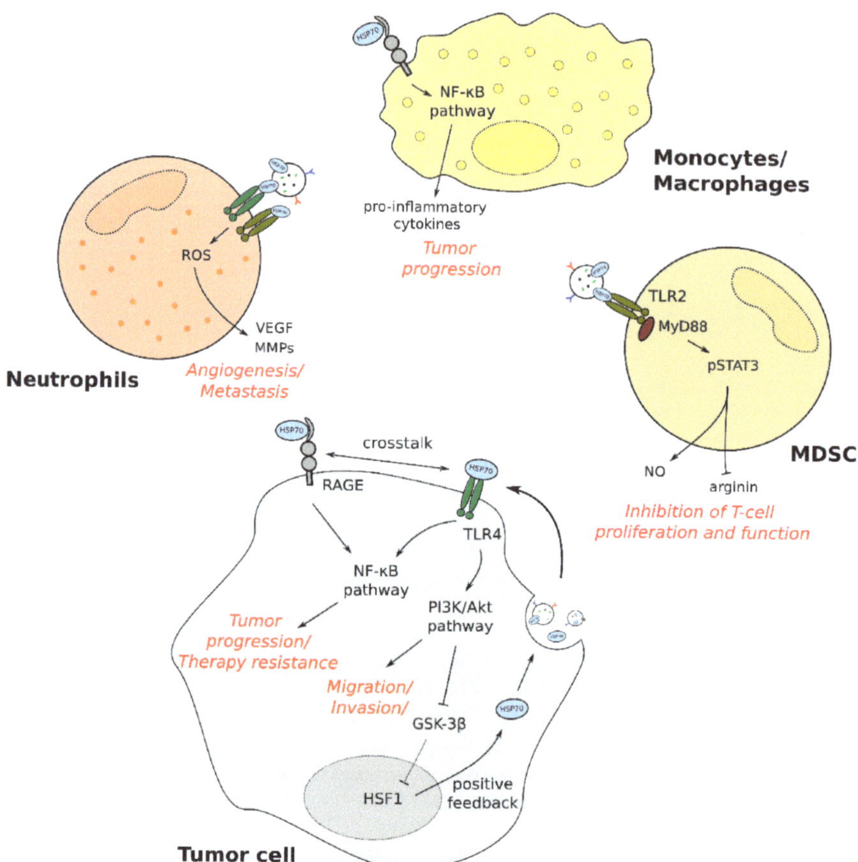

Figure 2. Pro-tumorigenic functions of exHSP70. MDSC: myeloid-derived suppressor cell; RAGE: receptor for advanced glycosylation endproducts; GSK-3β: glycogen synthase kinase-3β; NO: nitride oxide; TLR: Toll-like receptor; NFκB: nuclear factor kappa-light-chain-enhancer of activated B cells; HSF1: heat shock factor 1; VEGF: vasculogenic endothelial growth factor; pSTAT3: phosphorylated signal transducer and activator of transcription 3; ROS: reactive oxygen species; MyD88: myeloid differentiation primary response 88; PI3K: phosphoinositide 3-kinase; Akt: protein kinase B; MMP: matrix metalloproteinases; HSP70: heat shock protein 70.

Extracellular HSP70 can also exert anti-tumorigenic functions, which can be mediated by activating or priming NK-, dendritic- or T-cells (Figure 3). For instance, NK cells specifically interact with mHSP70 on tumor cells, probably via the C-type lectin receptor CD94 [112–114]. Moreover, it was described that exHSP70 could mature dendritic cells and increase the expression of the NK ligand MICA and the co-stimulatory molecules CD86 or CD40. Therefore, exHSP70 stimulates NK cells either directly via CD94 or indirectly via MICA, which binds the NK cell activating receptor NKG2D [87,115]. NK cells can then kill tumor cells in an NKG2D-dependent way or by increasing granzyme B release [115]. Interestingly, granzyme B was found to be incorporated in tumor cells by explicitly binding to mHSP70, which results in perforin-independent apoptosis of the cells [44]. Moreover, Sharapova and colleagues recently reported another possibility of NK cell activation while investigating a novel exHSP70 target receptor [116]. They showed that exHSP70 binds TREM-1 on monocytes, leading to the secretion of TNF-α and INF-

γ. The cytokines then stimulate CD4$^+$ T-cells to secrete IL-2, finally activating NK cells. Moreover, CD8$^+$ T-cells are also activated by this mechanism and can kill tumor cells via FasL/Fas interaction [116]. Interestingly, FasL can trigger the secretion of an HSP70-Tag7 complex in T-cells. This complex further induces tumor cell lysis via the tumor necrosis factor receptor 1 (TNFR1) [117,118].

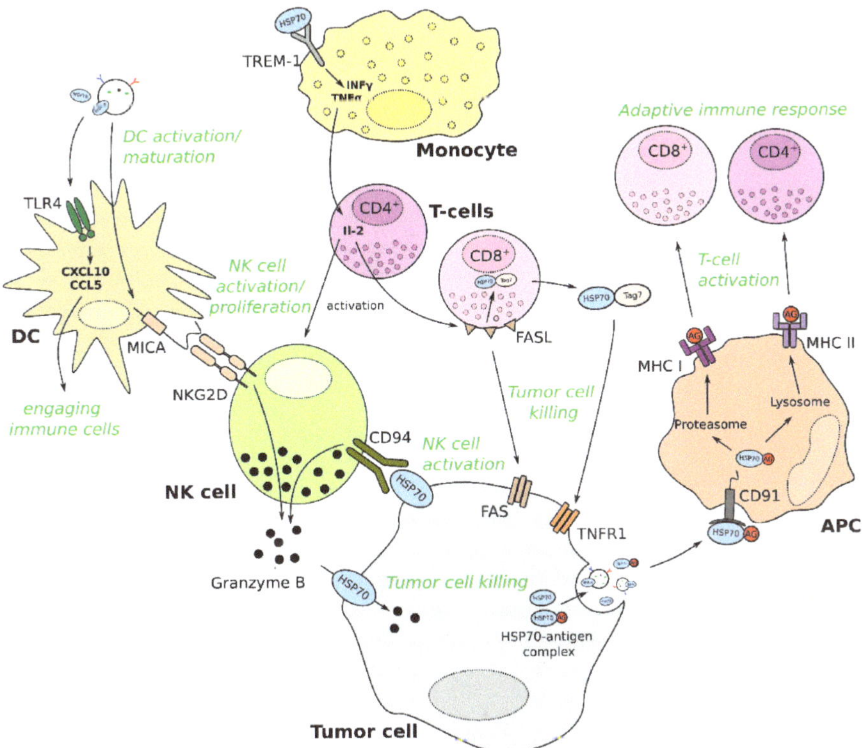

Figure 3. Anti-tumorigenic functions of exHSP70. FASL: Fas ligand; TNFR1: tumor necrosis factor receptor 1; APC: antigen-presenting cell; NK: natural killer; DC: dendritic cell; MHC: major histocompatibility complex; MICA: MHC class I polypeptide–related sequence A; TREM-1: triggering receptor expressed on myeloid cells 1; INF-γ: interferon-gamma; TNF-α: tumor necrosis factor-alpha; IL-2: interleukin-2; Tag7: peptidoglycan recognition protein; CXCL10: C-X-C motif chemokine ligand 10; CCL5: Chemokine (C-C motif) ligand 5; HSP70: heat shock protein 70.

Dendritic cells can also be stimulated by binding to the TLR4. Activation of MyD88- and TRIF-dependent pathways result in the production of chemokines, such as CXCL10 or CCL5, engaging immune cells and inducing anti-tumor immunity [119].

Furthermore, exHSP70 was reported to inhibit the conversion of monocytes to a pro-tumor phenotype, which is potentially due to a diminished expression of pro-tumor cytokines, such as IL-10 and MCP-1 [120].

Finally, as described in the previous chapter, exHSP70 can induce an adaptive immune response by (cross-) presentation of antigens on MHC class I or II molecules. In particular, HSP70-antigen complexes are endocytosed by APCs after binding to CD91 or other scavenger receptors, where the antigen is processed and presented to T-cells [69].

Increasing evidence emphasizes exHSP70 as an important player in mediating immune responses in cancer by either exerting pro- or anti-tumorigenic functions. However, there is still much unknown or controversially described; thus, specific functions and their associated mechanism urgently need to be investigated.

4. Therapeutic Potential of exHSP70

Since intracellular HSP70 is upregulated in cancer and elicits anti-apoptotic functions, it represents an important target for developing new therapeutics. Typical inhibitors are small molecules targeting various aspects of the HSP70 machinery [121]. These include ATPase and complex inhibitors as well as nucleotide-binding domain (NBD)- or substrate-binding domain (SDB)-targeting inhibitors [121]. Apart from small molecule inhibitors, Minnelide, a Triptolide derivative, has been shown to effectively downregulate HSP70 by inducing the miRNA miR-142–3p [122,123]. Additionally, Minnelide is currently tested in several phase I and II studies, including pancreatic cancer and acute myeloid leukemia [2]. Although targeting iHSP70 and therefore altering its expression or function also affects exHSP70, the linkages and consequences are not well understood.

4.1. HSP70 Antibody

The fact that mHSP70 is uniquely expressed on cancer cells compared to normal tissue makes it an excellent therapeutic target [124,125]. Gabrielle Multhoff's group developed a specific IgG antibody directed against the exposed region of mHSP70 by immunizing mice with a 14-mer peptide (TKDNNLLGRFELSG, known as TKD) that encompasses the amino acids 450–463 of the chaperone [126,127]. This antibody, known as cmHsp70.1 mAb, was shown to bind specifically to mHSP70 in vitro and in vivo [127]. In addition, antibody-dependent cell-mediated cytotoxicity (ADCC), induced by the Fc-region of the antibody, was shown to significantly inhibit tumor growth in a preclinical study in mice [126]. However, IgG antibodies are inefficient in inducing ADCC, primarily via neutrophils, due to the binding of several inhibitory Fc-receptors [128]. Therefore, IgA antibodies should be considered an alternative since they efficiently induce ADCC by binding to the Fcα-receptor (CD89) on neutrophils, monocytes, and macrophages [129]. Furthermore, IgA antibodies do not activate the classical complement pathway, consequently being rather anti-inflammatory than IgG antibodies [130]. The ability to oligomerize enables IgA to bind and crosslink several Fc-receptors, which was reported to release the chemoattractant LTB4, engaging neutrophils for tumor cell killing [131]. Disadvantages, such as the short serum half-life, could be overcome by antibody engineering [132].

The antibody cmHSP70.1 was also reported to be efficiently endocytosed by tumor cells, offering a specific route for targeted therapy [127]. Therefore, superparamagnetic iron oxide nanoparticles (SPIONs), a specific group of magnetic nanoparticles (MNPs), were decorated with the antibody [133,134]. The results showed an increase in retention of the particles in a glioma model in vivo [133]. Generally, MNPs can be used for diagnosis via magnetic resonance imaging or exploited for therapy by applying an alternate magnetic field (AMF) [135]. The AMF results in the spinning of iron particles, generating heat and killing the tumor by hyperthermia [135]. Interestingly, inducing ionizing radiation after injecting the particles significantly increased the retention of the particles in the tumor [134]. This is in line with the findings that radiation increases mHSP70 in tumor cells [136].

An alternative to MNPs are gold nanoparticles (AuNPs), which have a good biocompatibility and are easy to synthesize, although their costs of synthesizing are quite high [136]. They can be used for several applications, including delivering different compounds and photothermal or photodynamic therapy [136]. Recently, AuNPs coated with cmHsp70.1 mAb were shown to enable computed tomography of tumors in a preclinical study [137].

Using extracellular vesicles as specific carriers of drugs is another promising approach. Since EVs are naturally occurring vesicles, side-effects of the carrier should be minimal. Moreover, EVs could be engineered to express HSP70 and specifically target tumor cells while carrying distinct molecules for tumor cell killing. In 2010, Xie and colleagues demonstrated that mHSP70 engineered exosomes efficiently induced DC maturation, which increased T- and NK cell-dependent tumor killing [138]. Other reports showed successful loading of the chemotherapeutic paclitaxel into exosomes, resulting in tumor growth inhibition [139,140].

4.2. HSP70 Peptides

Since exHSP70 was shown to stimulate immune cells for tumor cell killing, the 14mer-peptide, which was used to generate the antibody, was also investigated as a therapeutic option. In the early 2000s, Multhoff and co-workers reported TKD as a recognition structure for NK cells and showed that it stimulated the proliferative and cytolytic activity [113]. Shortly after, TKD was successfully studied in a phase I trial, where NK cells were stimulated ex vivo, together with low doses of Il-2 [141]. This was supported by a recent phase II trial, where the peptide showed promising results in non-small cell lung cancer [142]. Moreover, together with the inhibition of the programmed cell death protein 1 (PD-1), a prolongation in patient survival was reported in several phase II trials of glioblastoma and lung cancer [143,144].

Another group developed a small peptide targeting the SBD of HSP70. The A8 peptide aptamer showed a decrease in MDSC and inhibited tumor progression in vivo [94]. This was due to a higher affinity of mHSP70 to A8 compared to the TLR2, thus blocking HSP70/TLR2 interaction. Therefore, A8 effectively blocked the exosome-mediated activation of MDSCs. Moreover, MDSC proliferation was also hampered since TLR2-dependent production of IL-6 via pSTAT3 was blocked. Interestingly, treating mice with cisplatin or 5-Fluorouracil led to increased HSP70-bearing exosomes, enhancing activation of MDSCs [94].

Recently, Lin and colleagues demonstrated that a 72 bp long peptide (Tx-01) could suppress ovarian cancer migration and growth in vivo [145]. They showed that Tx-01 is internalized by endo- or pinocytosis, where it binds to iHSP70, which consequently blocks the interaction of HSP70 with the Notch1 intracellular domain (NICD). This hampers the nuclear localization of HSP70, decreasing tumor invasion and migration. In addition, Tx-01 can also bind to mHSP70 leading to a rapid internalization, where it subsequently blocks the translocation of HSP70. Furthermore, the binding of Tx-01 to mHSP70 could be used as a prognostic marker in serous ovarian carcinoma [145].

In addition to peptides and antibodies, granzyme B has been reported to bind to mHSP70, leading to internalization and subsequent apoptosis of the cells [44]. Moreover, the enzyme showed significant tumor suppression of HSP70-positive colon carcinoma in vivo [31]. Recently, granzyme B-tagged SPIONs were developed, demonstrating increased survival of mice with glioblastoma or late-stage lung cancer [146]. Magnetic targeting further enhanced the localization of the SPIONs within the tumor. In addition, the nanoparticles could be used to image HSP70-positive tumors by magnetic resonance imaging [146].

4.3. HSP70 Vaccines

Since exHSP70 was shown to trigger T-cell responses, using it as a vaccine is another potential therapy approach. In particular, tumors are stimulated to secrete HSP70 (and antigens), which can subsequently be purified. HSP70-peptide complexes are then used as vaccines, where they mediate cross-presentation of the antigen in DCs, finally activating CD8$^+$ T-cells [147]. This approach was already investigated in the early 2000s, where researchers showed inhibition of tumor growth in a prostate cancer mouse model [147]. Moreover, using HSP70-peptide complexes from the fusion of DCs and tumor cells significantly enhanced DC maturation and T-cell responses [148].

In summary, targeting exHSP70 represents an interesting theranostic approach. Since exHSP70 is uniquely expressed in tumors and not in normal cells, targeting can be used for both imaging and therapeutic approaches. Various strategies are being investigated, including peptides, antibodies, and enzymes, that bind either directly to HSP70 on tumor cells/EVs or target receptors on immune cells. The latter would mimic a vaccination and lead to activation and engagement of immune cells for tumor cell killing. Finally, vehicles such as nanoparticles or EVs can further enhance tumor imaging or treatment by carrying cytotoxic compounds. HSP70 may also be linked as a chaperone to DNA vaccines to enhance immunogenicity in humans [149]. In this study HSP70 was linked to the HPV16 E7 sequence to facilitate uptake by antigen-presenting cells and antigen presentation.

HPVE7-specific T-cell responses were generally of low frequency but increased in subjects in the second and third cohorts. Tumor regression was reported in 3 out of 9 patients; however, significancy of these results needs to be confirmed in further studies.

5. Conclusions

Cellular stress leads to a multitude of molecular responses, guarding cells from damages. One specific response is the upregulation of the inducible HSP70, subsequently protecting cells via chaperone activities or inhibiting apoptosis. However, in cancer, HSP70 is overexpressed, leading to tumor progression and therapy resistance. Moreover, the chaperone is translocated to the extracellular milieu by several pathways, resulting in membrane-bound, EV-associated, or soluble HSP70. Hereby, exHSP70 can interact with immune cells of the tumor microenvironment, where it operates as a double-edged sword. By targeting different receptors on immune cells, it is able to either trigger pro- or anti-tumorigenic responses. Pro-tumorigenic responses include stimulating neutrophils or monocytes to secrete pro-inflammatory cytokines or activating MDSC, which results in reduced T-cell activity. In contrast, anti-tumorigenic responses comprise activation of NK cells, either directly or indirectly via DCs, and activation of T-cell by stimulating monocytes. Moreover, mediating (cross-)presentation of antigens on MHC class I or II molecules also activates the adaptive immune response for tumor cell killing. Since exHSP70 is uniquely expressed in cancer compared to normal tissue, it is a valuable therapeutic target. Therapies can include targeting exHSP70 or its receptors on tumor-, immune cells or EVs. Moreover, HSP70-antigen complexes can be used to prime immune cells for an effective anti-tumor response.

Although it is widely believed that exHSP70 can trigger opposing effects, much is still unknown or critically debated. For instance, cytokine activities of exHSP70 were questioned since this might be a result of endotoxin contamination. Moreover, the actions of HSP70-protein complexes are often overlooked but, depending on the complex, could trigger different responses. Altogether, exHSP70 can emerge as a crucial player in cancer therapy; however, a better understanding of its function is essential. Therefore, more research regarding specific mechanisms as well as novel therapeutic strategies are needed.

Funding: M.L. and E.P.v.S. are funded by the Deutsche Forschungsgemeinschaft (DFG, German Research Foundation)-GRK 2573/1.

Institutional Review Board Statement: Not applicable.

Informed Consent Statement: Not applicable.

Data Availability Statement: Not applicable.

Acknowledgments: We thank Christian Preußer and Viviane Ponath (Phillips University Marburg) for their comments on the manuscript.

Conflicts of Interest: The authors declare no conflict of interest.

References

1. Rosenzweig, R.; Nillegoda, N.B.; Mayer, M.P.; Bukau, B. The Hsp70 Chaperone Network. *Nat. Rev. Mol. Cell Biol.* **2019**, *20*, 665–680. [CrossRef] [PubMed]
2. Albakova, Z.; Armeev, G.A.; Kanevskiy, L.M.; Kovalenko, E.I.; Sapozhnikov, A.M. HSP70 Multi-Functionality in Cancer. *Cells* **2020**, *9*, 587. [CrossRef] [PubMed]
3. Van Niel, G.; D'Angelo, G.; Raposo, G. Shedding Light on the Cell Biology of Extracellular Vesicles. *Nat. Rev. Mol. Cell Biol.* **2018**, *19*, 213–228. [CrossRef] [PubMed]
4. Kalluri, R.; LeBleu, V.S. The Biology, Function, and Biomedical Applications of Exosomes. *Science* **2020**, *367*. [CrossRef]
5. Liu, Y.; Wang, Y.; Lv, Q.; Li, X. Exosomes: From Garbage Bins to Translational Medicine. *Int. J. Pharm.* **2020**, *583*, 119333. [CrossRef] [PubMed]
6. Lancaster, G.I.; Febbraio, M.A. Exosome-Dependent Trafficking of HSP70: A Novel Secretory Pathway for Cellular Stress Proteins. *J. Biol. Chem.* **2005**, *280*, 23349–23355. [CrossRef] [PubMed]

7. Vega, V.L.; Rodríguez-Silva, M.; Frey, T.; Gehrmann, M.; Diaz, J.C.; Steinem, C.; Multhoff, G.; Arispe, N.; Maio, A.D. Hsp70 Translocates into the Plasma Membrane after Stress and Is Released into the Extracellular Environment in a Membrane-Associated Form That Activates Macrophages. *J. Immunol.* **2008**, *180*, 4299–4307. [CrossRef] [PubMed]

8. Multhoff, G.; Botzler, C.; Wiesnet, M.; Müller, E.; Meier, T.; Wilmanns, W.; Issels, R.D. A Stress-Inducible 72-KDa Heat-Shock Protein (HSP72) Is Expressed on the Surface of Human Tumor Cells, but Not on Normal Cells. *Int. J. Cancer* **1995**, *61*, 272–279. [CrossRef]

9. Pockley, A.G.; Shepherd, J.; Corton, J.M. Detection of Heat Shock Protein 70 (Hsp70) and Anti-Hsp70 Antibodies in the Serum of Normal Individuals. *Immunol. Investig.* **1998**, *27*, 367–377. [CrossRef]

10. Owji, H.; Nezafat, N.; Negahdaripour, M.; Hajiebrahimi, A.; Ghasemi, Y. A Comprehensive Review of Signal Peptides: Structure, Roles, and Applications. *Eur. J. Cell Biol.* **2018**, *97*, 422–441. [CrossRef] [PubMed]

11. Mambula, S.S.; Stevenson, M.A.; Ogawa, K.; Calderwood, S.K. Mechanisms for Hsp70 Secretion: Crossing Membranes without a Leader. *Methods* **2007**, *43*, 168–175. [CrossRef]

12. Broquet, A.H.; Thomas, G.; Masliah, J.; Trugnan, G.; Bachelet, M. Expression of the Molecular Chaperone Hsp70 in Detergent-Resistant Microdomains Correlates with Its Membrane Delivery and Release. *J. Biol. Chem.* **2003**, *278*, 21601–21606. [CrossRef]

13. Nylander, S.; Kalies, I. Brefeldin A, but Not Monensin, Completely Blocks CD69 Expression on Mouse Lymphocytes:: Efficacy of Inhibitors of Protein Secretion in Protocols for Intracellular Cytokine Staining by Flow Cytometry. *J. Immunol. Methods* **1999**, *224*, 69–76. [CrossRef]

14. Sezgin, E.; Levental, I.; Mayor, S.; Eggeling, C. The Mystery of Membrane Organization: Composition, Regulation and Roles of Lipid Rafts. *Nat. Rev. Mol. Cell Biol.* **2017**, *18*, 361–374. [CrossRef]

15. Hunter-Lavin, C.; Davies, E.L.; Bacelar, M.M.F.V.G.; Marshall, M.J.; Andrew, S.M.; Williams, J.H.H. Hsp70 Release from Peripheral Blood Mononuclear Cells. *Biochem. Biophys. Res. Commun.* **2004**, *324*, 511–517. [CrossRef]

16. Gehrmann, M.; Liebisch, G.; Schmitz, G.; Anderson, R.; Steinem, C.; De Maio, A.; Pockley, G.; Multhoff, G. Tumor-Specific Hsp70 Plasma Membrane Localization Is Enabled by the Glycosphingolipid Gb3. *PLoS ONE* **2008**, *3*, e1925. [CrossRef]

17. Smulders, L.; Daniels, A.J.; Plescia, C.B.; Berger, D.; Stahelin, R.V.; Nikolaidis, N. Characterization of the Relationship between the Chaperone and Lipid-Binding Functions of the 70-KDa Heat-Shock Protein, HspA1A. *Int. J. Mol. Sci.* **2020**, *21*, 5995. [CrossRef] [PubMed]

18. McCallister, C.; Kdeiss, B.; Nikolaidis, N. Biochemical Characterization of the Interaction between HspA1A and Phospholipids. *Cell Stress Chaperones* **2016**, *21*, 41–53. [CrossRef] [PubMed]

19. Schilling, D.; Gehrmann, M.; Steinem, C.; De Maio, A.; Pockley, A.G.; Abend, M.; Molls, M.; Multhoff, G. Binding of Heat Shock Protein 70 to Extracellular Phosphatidylserine Promotes Killing of Normoxic and Hypoxic Tumor Cells. *FASEB J.* **2009**, *23*, 2467–2477. [CrossRef] [PubMed]

20. Alder, G.M.; Austen, B.M.; Bashford, C.L.; Mehlert, A.; Pasternak, C.A. Heat Shock Proteins Induce Pores in Membranes. *Biosci. Rep.* **1990**, *10*, 509–518. [CrossRef]

21. Bilog, A.D.; Smulders, L.; Oliverio, R.; Labanieh, C.; Zapanta, J.; Stahelin, R.V.; Nikolaidis, N. Membrane Localization of HspA1A, a Stress Inducible 70-KDa Heat-Shock Protein, Depends on Its Interaction with Intracellular Phosphatidylserine. *Biomolecules* **2019**, *9*, 152. [CrossRef]

22. Gastpar, R.; Gehrmann, M.; Bausero, M.A.; Asea, A.; Gross, C.; Schroeder, J.A.; Multhoff, G. Heat Shock Protein 70 Surface-Positive Tumor Exosomes Stimulate Migratory and Cytolytic Activity of Natural Killer Cells. *Cancer Res.* **2005**, *65*, 5238–5247. [CrossRef]

23. Clayton, A.; Turkes, A.; Navabi, H.; Mason, M.D.; Tabi, Z. Induction of Heat Shock Proteins in B-Cell Exosomes. *J. Cell Sci.* **2005**, *118*, 3631–3638. [CrossRef]

24. Takeuchi, T.; Suzuki, M.; Fujikake, N.; Popiel, H.A.; Kikuchi, H.; Futaki, S.; Wada, K.; Nagai, Y. Intercellular Chaperone Transmission via Exosomes Contributes to Maintenance of Protein Homeostasis at the Organismal Level. *Proc. Natl. Acad. Sci. USA* **2015**, *112*, E2497–E2506. [CrossRef] [PubMed]

25. Smith, V.L.; Jackson, L.; Schorey, J.S. Ubiquitination as a Mechanism To Transport Soluble Mycobacterial and Eukaryotic Proteins to Exosomes. *J. Immunol.* **2015**, *195*, 2722–2730. [CrossRef] [PubMed]

26. Moreno-Gonzalo, O.; Villarroya-Beltri, C.; Sánchez-Madrid, F. Post-Translational Modifications of Exosomal Proteins. *Front. Immunol.* **2014**, *5*, 383. [CrossRef] [PubMed]

27. Katzmann, D.J.; Odorizzi, G.; Emr, S.D. Receptor Downregulation and Multivesicular-Body Sorting. *Nat. Rev. Mol. Cell Biol.* **2002**, *3*, 893–905. [CrossRef]

28. Nikita; Porter, C.M.; Truman, A.W.; Truttmann, M.C. Post-translational modifications of Hsp70 family proteins: Expanding the chaperone code. *J. Biol. Chem.* **2020**, *295*, 10689–10708. [CrossRef]

29. Soss, S.E.; Rose, K.L.; Hill, S.; Jouan, S.; Chazin, W.J. Biochemical and Proteomic Analysis of Ubiquitination of Hsc70 and Hsp70 by the E3 Ligase CHIP. *PLoS ONE* **2015**, *10*, e0128240. [CrossRef]

30. Quintana-Gallardo, L.; Martín-Benito, J.; Marcilla, M.; Espadas, G.; Sabidó, E.; Valpuesta, J.M. The Cochaperone CHIP Marks Hsp70- and Hsp90-Bound Substrates for Degradation through a Very Flexible Mechanism. *Sci. Rep.* **2019**, *9*, 5102. [CrossRef]

31. Jiang, J.; Ballinger, C.A.; Wu, Y.; Dai, Q.; Cyr, D.M.; Höhfeld, J.; Patterson, C. CHIP Is a U-Box-Dependent E3 Ubiquitin Ligase: Identification of Hsc70 as a Target for Ubiquitylation. *J. Biol. Chem.* **2001**, *276*, 42938–42944. [CrossRef]

32. Evdonin, A.L.; Guzhova, I.V.; Margulis, B.A.; Medvedeva, N.D. Phospholipse c Inhibitor, U73122, Stimulates Release of Hsp-70 Stress Protein from A431 Human Carcinoma Cells. *Cancer Cell Int.* **2004**, *4*, 2. [CrossRef] [PubMed]

33. Li, Z.; Zhuang, M.; Zhang, L.; Zheng, X.; Yang, P.; Li, Z. Acetylation Modification Regulates GRP78 Secretion in Colon Cancer Cells. *Sci. Rep.* **2016**, *6*, 30406. [CrossRef] [PubMed]
34. Yang, Y.; Fiskus, W.; Yong, B.; Atadja, P.; Takahashi, Y.; Pandita, T.K.; Wang, H.-G.; Bhalla, K.N. Acetylated Hsp70 and KAP1-Mediated Vps34 SUMOylation Is Required for Autophagosome Creation in Autophagy. *Proc. Natl. Acad. Sci. USA* **2013**, *110*, 6841–6846. [CrossRef] [PubMed]
35. Fang, Y.; Wu, N.; Gan, X.; Yan, W.; Morrell, J.C.; Gould, S.J. Higher-Order Oligomerization Targets Plasma Membrane Proteins and HIV Gag to Exosomes. *PLoS Biol.* **2007**, *5*, e158. [CrossRef] [PubMed]
36. Nimmervoll, B.; Chtcheglova, L.A.; Juhasz, K.; Cremades, N.; Aprile, F.A.; Sonnleitner, A.; Hinterdorfer, P.; Vigh, L.; Preiner, J.; Balogi, Z. Cell Surface Localised Hsp70 Is a Cancer Specific Regulator of Clathrin-Independent Endocytosis. *FEBS Lett.* **2015**, *589*, 2747–2753. [CrossRef] [PubMed]
37. Takakuwa, J.E.; Nitika; Knighton, L.E.; Truman, A.W. Oligomerization of Hsp70: Current Perspectives on Regulation and Function. *Front. Mol. Biosci.* **2019**, *6*, 81. [CrossRef] [PubMed]
38. Basu, S.; Binder, R.J.; Suto, R.; Anderson, K.M.; Srivastava, P.K. Necrotic but Not Apoptotic Cell Death Releases Heat Shock Proteins, Which Deliver a Partial Maturation Signal to Dendritic Cells and Activate the NF-KB Pathway. *Int. Immunol.* **2000**, *12*, 1539–1546. [CrossRef]
39. Mambula, S.S.; Calderwood, S.K. Heat Shock Protein 70 Is Secreted from Tumor Cells by a Nonclassical Pathway Involving Lysosomal Endosomes. *J. Immunol.* **2006**, *177*, 7849–7857. [CrossRef]
40. Evdonin, A.L.; Martynova, M.G.; Bystrova, O.A.; Guzhova, I.V.; Margulis, B.A.; Medvedeva, N.D. The Release of Hsp70 from A431 Carcinoma Cells Is Mediated by Secretory-like Granules. *Eur. J. Cell Biol.* **2006**, *85*, 443–455. [CrossRef]
41. Chaplin, D.D. Overview of the Immune Response. *J. Allergy Clin. Immunol.* **2010**, *125*, S3–S23. [CrossRef] [PubMed]
42. Ikwegbue, P.C.; Masamba, P.; Oyinloye, B.E.; Kappo, A.P. Roles of Heat Shock Proteins in Apoptosis, Oxidative Stress, Human Inflammatory Diseases, and Cancer. *Pharmaceuticals* **2017**, *11*, 2. [CrossRef] [PubMed]
43. Joly, A.-L.; Wettstein, G.; Mignot, G.; Ghiringhelli, F.; Garrido, C. Dual Role of Heat Shock Proteins as Regulators of Apoptosis and Innate Immunity. *J. Innate Immun.* **2010**, *2*, 238–247. [CrossRef]
44. Gross, C.; Koelch, W.; DeMaio, A.; Arispe, N.; Multhoff, G. Cell Surface-Bound Heat Shock Protein 70 (Hsp70) Mediates Perforin-Independent Apoptosis by Specific Binding and Uptake of Granzyme B. *J. Biol. Chem.* **2003**, *278*, 41173–41181. [CrossRef] [PubMed]
45. Srivastava, P. Roles of Heat-Shock Proteins in Innate and Adaptive Immunity. *Nat. Rev. Immunol.* **2002**, *2*, 185–194. [CrossRef] [PubMed]
46. Figueiredo, C.; Wittmann, M.; Wang, D.; Dressel, R.; Seltsam, A.; Blasczyk, R.; Eiz-Vesper, B. Heat Shock Protein 70 (HSP70) Induces Cytotoxicity of T-Helper Cells. *Blood* **2009**, *113*, 3008–3016. [CrossRef]
47. Asea, A.; Kraeft, S.-K.; Kurt-Jones, E.A.; Stevenson, M.A.; Chen, L.B.; Finberg, R.W.; Koo, G.C.; Calderwood, S.K. HSP70 Stimulates Cytokine Production through a CD14-Dependant Pathway, Demonstrating Its Dual Role as a Chaperone and Cytokine. *Nat. Med.* **2000**, *6*, 435–442. [CrossRef] [PubMed]
48. Asea, A. Stress Proteins and Initiation of Immune Response: Chaperokine Activity of Hsp72. *Exerc. Immunol. Rev.* **2005**, *11*, 34. [PubMed]
49. Hulina, A.; Grdić Rajković, M.; Jakšić Despot, D.; Jelić, D.; Dojder, A.; Čepelak, I.; Rumora, L. Extracellular Hsp70 Induces Inflammation and Modulates LPS/LTA-Stimulated Inflammatory Response in THP-1 Cells. *Cell Stress Chaperones* **2018**, *23*, 373–384. [CrossRef]
50. Campisi, J.; Fleshner, M. Role of Extracellular HSP72 in Acute Stress-Induced Potentiation of Innate Immunity in Active Rats. *J. Appl. Physiol. (1985)* **2003**, *94*, 43–52. [CrossRef]
51. Asea, A.; Rehli, M.; Kabingu, E.; Boch, J.A.; Bare, O.; Auron, P.E.; Stevenson, M.A.; Calderwood, S.K. Novel Signal Transduction Pathway Utilized by Extracellular HSP70: Role of Toll-like Receptor (TLR) 2 and TLR4. *J. Biol. Chem.* **2002**, *277*, 15028–15034. [CrossRef]
52. Becker, T.; Hartl, F.-U.; Wieland, F. CD40, an Extracellular Receptor for Binding and Uptake of Hsp70-Peptide Complexes. *J. Cell Biol.* **2002**, *158*, 1277–1285. [CrossRef] [PubMed]
53. Gao, B.; Tsan, M.-F. Endotoxin Contamination in Recombinant Human Heat Shock Protein 70 (Hsp70) Preparation Is Responsible for the Induction of Tumor Necrosis Factor Alpha Release by Murine Macrophages. *J. Biol. Chem.* **2003**, *278*, 174–179. [CrossRef]
54. Bausinger, H.; Lipsker, D.; Ziylan, U.; Manié, S.; Briand, J.-P.; Cazenave, J.-P.; Muller, S.; Haeuw, J.-F.; Ravanat, C.; de la Salle, H.; et al. Endotoxin-Free Heat-Shock Protein 70 Fails to Induce APC Activation. *Eur. J. Immunol.* **2002**, *32*, 3708–3713. [CrossRef]
55. Stocki, P.; Wang, X.N.; Dickinson, A.M. Inducible Heat Shock Protein 70 Reduces T Cell Responses and Stimulatory Capacity of Monocyte-Derived Dendritic Cells. *J. Biol. Chem.* **2012**, *287*, 12387–12394. [CrossRef]
56. Stocki, P.; Dickinson, A.M. The Immunosuppressive Activity of Heat Shock Protein 70. *Autoimmune Dis.* **2012**, *2012*, 617213. [CrossRef] [PubMed]
57. Tsan, M.-F.; Gao, B. Pathogen-Associated Molecular Pattern Contamination as Putative Endogenous Ligands of Toll-like Receptors. *J. Endotoxin Res.* **2007**, *13*, 6–14. [CrossRef] [PubMed]
58. Ye, Z.; Gan, Y.-H. Flagellin Contamination of Recombinant Heat Shock Protein 70 Is Responsible for Its Activity on T Cells. *J. Biol. Chem.* **2007**, *282*, 4479–4484. [CrossRef] [PubMed]
59. Fong, J.J.; Sreedhara, K.; Deng, L.; Varki, N.M.; Angata, T.; Liu, Q.; Nizet, V.; Varki, A. Immunomodulatory Activity of Extracellular Hsp70 Mediated via Paired Receptors Siglec-5 and Siglec-14. *EMBO J.* **2015**, *34*, 2775–2788. [CrossRef] [PubMed]

60. Grunwald, M.S.; Ligabue-Braun, R.; Souza, C.S.; Heimfarth, L.; Verli, H.; Gelain, D.P.; Moreira, J.C.F. Putative Model for Heat Shock Protein 70 Complexation with Receptor of Advanced Glycation End Products through Fluorescence Proximity Assays and Normal Mode Analyses. *Cell Stress Chaperones* **2017**, *22*, 99–111. [CrossRef] [PubMed]

61. Somensi, N.; Brum, P.O.; de Miranda Ramos, V.; Gasparotto, J.; Zanotto-Filho, A.; Rostirolla, D.C.; da Silva Morrone, M.; Moreira, J.C.F.; Pens Gelain, D. Extracellular HSP70 Activates ERK1/2, NF-KB and Pro-Inflammatory Gene Transcription Through Binding with RAGE in A549 Human Lung Cancer Cells. *Cell Physiol. Biochem.* **2017**, *42*, 2507–2522. [CrossRef]

62. Broere, F.; van der Zee, R.; van Eden, W. Heat Shock Proteins Are No DAMPs, Rather "DAMPERs". *Nat. Rev. Immunol.* **2011**, *11*, 565. [CrossRef]

63. Ferat-Osorio, E.; Sánchez-Anaya, A.; Gutiérrez-Mendoza, M.; Boscó-Gárate, I.; Wong-Baeza, I.; Pastelin-Palacios, R.; Pedraza-Alva, G.; Bonifaz, L.C.; Cortés-Reynosa, P.; Pérez-Salazar, E.; et al. Heat Shock Protein 70 Down-Regulates the Production of Toll-like Receptor-Induced pro-Inflammatory Cytokines by a Heat Shock Factor-1/Constitutive Heat Shock Element-Binding Factor-Dependent Mechanism. *J. Inflamm.* **2014**, *11*, 19. [CrossRef] [PubMed]

64. Song, M.; Pinsky, M.R.; Kellum, J.A. Heat Shock Factor 1 Inhibits Nuclear Factor-KappaB Nuclear Binding Activity during Endotoxin Tolerance and Heat Shock. *J. Crit. Care* **2008**, *23*, 406–415. [CrossRef] [PubMed]

65. Dokladny, K.; Lobb, R.; Wharton, W.; Ma, T.Y.; Moseley, P.L. LPS-Induced Cytokine Levels Are Repressed by Elevated Expression of HSP70 in Rats: Possible Role of NF-KappaB. *Cell Stress Chaperones* **2010**, *15*, 153–163. [CrossRef] [PubMed]

66. Alberti, G.; Paladino, L.; Vitale, A.M.; Caruso Bavisotto, C.; Conway de Macario, E.; Campanella, C.; Macario, A.J.L.; Marino Gammazza, A. Functions and Therapeutic Potential of Extracellular Hsp60, Hsp70, and Hsp90 in Neuroinflammatory Disorders. *Appl. Sci.* **2021**, *11*, 736. [CrossRef]

67. Noessner, E.; Gastpar, R.; Milani, V.; Brandl, A.; Hutzler, P.J.S.; Kuppner, M.C.; Roos, M.; Kremmer, E.; Asea, A.; Calderwood, S.K.; et al. Tumor-Derived Heat Shock Protein 70 Peptide Complexes Are Cross-Presented by Human Dendritic Cells. *J. Immunol.* **2002**, *169*, 5424–5432. [CrossRef] [PubMed]

68. Thériault, J.R.; Mambula, S.S.; Sawamura, T.; Stevenson, M.A.; Calderwood, S.K. Extracellular HSP70 Binding to Surface Receptors Present on Antigen Presenting Cells and Endothelial/Epithelial Cells. *FEBS Lett.* **2005**, *579*, 1951–1960. [CrossRef]

69. Li, Z.; Menoret, A.; Srivastava, P. Roles of Heat-Shock Proteins in Antigen Presentation and Cross-Presentation. *Curr. Opin. Immunol.* **2002**, *14*, 45–51. [CrossRef]

70. Bendz, H.; Ruhland, S.C.; Pandya, M.J.; Hainzl, O.; Riegelsberger, S.; Braüchle, C.; Mayer, M.P.; Buchner, J.; Issels, R.D.; Noessner, E. Human Heat Shock Protein 70 Enhances Tumor Antigen Presentation through Complex Formation and Intracellular Antigen Delivery without Innate Immune Signaling. *J. Biol. Chem.* **2007**, *282*, 31688–31702. [CrossRef]

71. Haug, M.; Dannecker, L.; Schepp, C.P.; Kwok, W.W.; Wernet, D.; Buckner, J.H.; Kalbacher, H.; Dannecker, G.E.; Holzer, U. The Heat Shock Protein Hsp70 Enhances Antigen-Specific Proliferation of Human CD4+ Memory T Cells. *Eur. J. Immunol.* **2005**, *35*, 3163–3172. [CrossRef]

72. Fischer, N.; Haug, M.; Kwok, W.W.; Kalbacher, H.; Wernet, D.; Dannecker, G.E.; Holzer, U. Involvement of CD91 and Scavenger Receptors in Hsp70-Facilitated Activation of Human Antigen-Specific CD4+ Memory T Cells. *Eur. J. Immunol.* **2010**, *40*, 986–997. [CrossRef] [PubMed]

73. Wang, Y.; Seidl, T.; Whittall, T.; Babaahmady, K.; Lehner, T. Stress-Activated Dendritic Cells Interact with CD4+ T Cells to Elicit Homeostatic Memory. *Eur. J. Immunol.* **2010**, *40*, 1628–1638. [CrossRef]

74. Wang, Y.; Rahman, D.; Lehner, T. A Comparative Study of Stress-Mediated Immunological Functions with the Adjuvanticity of Alum. *J. Biol. Chem.* **2012**, *287*, 17152–17160. [CrossRef] [PubMed]

75. Kane, B.A.; Bryant, K.J.; McNeil, H.P.; Tedla, N.T. Termination of Immune Activation: An Essential Component of Healthy Host Immune Responses. *J. Innate Immun.* **2014**, *6*, 727–738. [CrossRef] [PubMed]

76. Hirsh, M.I.; Hashiguchi, N.; Chen, Y.; Yip, L.; Junger, W.G. Surface Expression of HSP72 by LPS-Stimulated Neutrophils Facilitates GammadeltaT Cell-Mediated Killing. *Eur. J. Immunol.* **2006**, *36*, 712–721. [CrossRef] [PubMed]

77. Zhang, H.; Hu, H.; Jiang, X.; He, H.; Cui, L.; He, W. Membrane HSP70: The Molecule Triggering Gammadelta T Cells in the Early Stage of Tumorigenesis. *Immunol. Investig.* **2005**, *34*, 453–468. [CrossRef] [PubMed]

78. Dar, A.A.; Patil, R.S.; Chiplunkar, S.V. Insights into the Relationship between Toll Like Receptors and Gamma Delta T Cell Responses. *Front. Immunol.* **2014**, *5*, 366. [CrossRef]

79. Kang, H.-J.; Moon, H.-S.; Chung, H.-W. The Expression of FAS-Associated Factor 1 and Heat Shock Protein 70 in Ovarian Cancer. *Obstet. Gynecol. Sci.* **2014**, *57*, 281–290. [CrossRef]

80. Gao, G.; Liu, S.; Yao, Z.; Zhan, Y.; Chen, W.; Liu, Y. The Prognostic Significance of Hsp70 in Patients with Colorectal Cancer Patients: A PRISMA-Compliant Meta-Analysis. *Biomed. Res. Int.* **2021**, *2021*, 5526327. [CrossRef]

81. Thorsteinsdottir, J.; Stangl, S.; Fu, P.; Guo, K.; Albrecht, V.; Eigenbrod, S.; Erl, J.; Gehrmann, M.; Tonn, J.-C.; Multhoff, G.; et al. Overexpression of Cytosolic, Plasma Membrane Bound and Extracellular Heat Shock Protein 70 (Hsp70) in Primary Glioblastomas. *J. Neurooncol.* **2017**, *135*, 443–452. [CrossRef]

82. Chanteloup, G.; Cordonnier, M.; Isambert, N.; Bertaut, A.; Hervieu, A.; Hennequin, A.; Luu, M.; Zanetta, S.; Coudert, B.; Bengrine, L.; et al. Monitoring HSP70 Exosomes in Cancer Patients' Follow up: A Clinical Prospective Pilot Study. *J. Extracell Vesicles* **2020**, *9*, 1766192. [CrossRef]

83. Nowak, M.; Glowacka, E.; Kielbik, M.; Kulig, A.; Sulowska, Z.; Klink, M. Secretion of Cytokines and Heat Shock Protein (HspA1A) by Ovarian Cancer Cells Depending on the Tumor Type and Stage of Disease. *Cytokine* **2017**, *89*, 136–142. [CrossRef] [PubMed]

84. Annunziata, C.M.; Kleinberg, L.; Davidson, B.; Berner, A.; Gius, D.; Tchabo, N.; Steinberg, S.M.; Kohn, E.C. BAG-4/SODD and Associated Antiapoptotic Proteins Are Linked to Aggressiveness of Epithelial Ovarian Cancer. *Clin. Cancer Res.* **2007**, *13*, 6585–6592. [CrossRef] [PubMed]

85. Li, X.-F.; Hua, T.; Li, Y.; Tian, Y.-J.; Huo, Y.; Kang, S. The HSP70 Gene Predicts Prognosis and Response to Chemotherapy in Epithelial Ovarian Cancer. *Ann. Transl. Med.* **2021**, *9*, 806. [CrossRef] [PubMed]

86. Finkernagel, F.; Reinartz, S.; Schuldner, M.; Malz, A.; Jansen, J.M.; Wagner, U.; Worzfeld, T.; Graumann, J.; von Strandmann, E.P.; Müller, R. Dual-Platform Affinity Proteomics Identifies Links between the Recurrence of Ovarian Carcinoma and Proteins Released into the Tumor Microenvironment. *Theranostics* **2019**, *9*, 6601–6617. [CrossRef] [PubMed]

87. Li, Q.-L.; Bu, N.; Yu, Y.-C.; Hua, W.; Xin, X.-Y. Exvivo Experiments of Human Ovarian Cancer Ascites-Derived Exosomes Presented by Dendritic Cells Derived from Umbilical Cord Blood for Immunotherapy Treatment. *Clin. Med. Oncol.* **2008**, *2*, 461–467. [CrossRef] [PubMed]

88. Klink, M.; Nowak, M.; Kielbik, M.; Bednarska, K.; Blus, E.; Szpakowski, M.; Szyllo, K.; Sulowska, Z. The Interaction of HspA1A with TLR2 and TLR4 in the Response of Neutrophils Induced by Ovarian Cancer Cells in Vitro. *Cell Stress Chaperones* **2012**, *17*, 661–674. [CrossRef] [PubMed]

89. Adkins, I.; Sadilkova, L.; Hradilova, N.; Tomala, J.; Kovar, M.; Spisek, R. Severe, but Not Mild Heat-Shock Treatment Induces Immunogenic Cell Death in Cancer Cells. *Oncoimmunology* **2017**, *6*, e1311433. [CrossRef] [PubMed]

90. Park, G.B.; Chung, Y.H.; Kim, D. Induction of Galectin-1 by TLR-Dependent PI3K Activation Enhances Epithelial-Mesenchymal Transition of Metastatic Ovarian Cancer Cells. *Oncol. Rep.* **2017**, *37*, 3137–3145. [CrossRef]

91. Barreca, M.M.; Spinello, W.; Cavalieri, V.; Turturici, G.; Sconzo, G.; Kaur, P.; Tinnirello, R.; Asea, A.A.A.; Geraci, F. Extracellular Hsp70 Enhances Mesoangioblast Migration via an Autocrine Signaling Pathway. *J. Cell Physiol.* **2017**, *232*, 1845–1861. [CrossRef] [PubMed]

92. Lee, K.-J.; Kim, Y.M.; Kim, D.Y.; Jeoung, D.; Han, K.; Lee, S.-T.; Lee, Y.-S.; Park, K.H.; Park, J.H.; Kim, D.J.; et al. Release of Heat Shock Protein 70 (Hsp70) and the Effects of Extracellular Hsp70 on Matric Metalloproteinase-9 Expression in Human Monocytic U937 Cells. *Exp. Mol. Med.* **2006**, *38*, 364–374. [CrossRef] [PubMed]

93. Lee, K.-H.; Jeong, J.; Yoo, C.-G. Positive Feedback Regulation of Heat Shock Protein 70 (Hsp70) Is Mediated through Toll-like Receptor 4-PI3K/Akt-Glycogen Synthase Kinase-3β Pathway. *Exp. Cell Res.* **2013**, *319*, 88–95. [CrossRef] [PubMed]

94. Gobbo, J.; Marcion, G.; Cordonnier, M.; Dias, A.M.M.; Pernet, N.; Hammann, A.; Richaud, S.; Mjahed, H.; Isambert, N.; Clausse, V.; et al. Restoring Anticancer Immune Response by Targeting Tumor-Derived Exosomes With a HSP70 Peptide Aptamer. *J. Natl. Cancer Inst.* **2016**, *108*, djv330. [CrossRef]

95. Diao, J.; Yang, X.; Song, X.; Chen, S.; He, Y.; Wang, Q.; Chen, G.; Luo, C.; Wu, X.; Zhang, Y. Exosomal Hsp70 Mediates Immuno-suppressive Activity of the Myeloid-Derived Suppressor Cells via Phosphorylation of Stat3. *Med. Oncol.* **2015**, *32*, 35. [CrossRef]

96. Chalmin, F.; Ladoire, S.; Mignot, G.; Vincent, J.; Bruchard, M.; Remy-Martin, J.-P.; Boireau, W.; Rouleau, A.; Simon, B.; Lanneau, D.; et al. Membrane-Associated Hsp72 from Tumor-Derived Exosomes Mediates STAT3-Dependent Immunosup-pressive Function of Mouse and Human Myeloid-Derived Suppressor Cells. *J. Clin. Investig.* **2010**, *120*, 457–471. [CrossRef]

97. Rébé, C.; Végran, F.; Berger, H.; Ghiringhelli, F. STAT3 Activation: A Key Factor in Tumor Immunoescape. *JAKSTAT* **2013**, *2*, e23010. [CrossRef]

98. Rodriguez, P.C.; Quiceno, D.G.; Zabaleta, J.; Ortiz, B.; Zea, A.H.; Piazuelo, M.B.; Delgado, A.; Correa, P.; Brayer, J.; Sotomayor, E.M.; et al. Arginase I Production in the Tumor Microenvironment by Mature Myeloid Cells Inhibits T-Cell Receptor Expression and Antigen-Specific T-Cell Responses. *Cancer Res.* **2004**, *64*, 5839–5849. [CrossRef] [PubMed]

99. Bingisser, R.M.; Tilbrook, P.A.; Holt, P.G.; Kees, U.R. Macrophage-Derived Nitric Oxide Regulates T Cell Activation via Reversible Disruption of the Jak3/STAT5 Signaling Pathway. *J. Immunol.* **1998**, *160*, 5729–5734. [PubMed]

100. Wu, F.-H.; Yuan, Y.; Li, D.; Liao, S.-J.; Yan, B.; Wei, J.-J.; Zhou, Y.-H.; Zhu, J.-H.; Zhang, G.-M.; Feng, Z.-H. Extracellular HSPA1A Promotes the Growth of Hepatocarcinoma by Augmenting Tumor Cell Proliferation and Apoptosis-Resistance. *Cancer Lett.* **2012**, *317*, 157–164. [CrossRef] [PubMed]

101. De Larco, J.E.; Wuertz, B.R.K.; Furcht, L.T. The Potential Role of Neutrophils in Promoting the Metastatic Phenotype of Tumors Releasing Interleukin-8. *Clin. Cancer Res.* **2004**, *10*, 4895–4900. [CrossRef] [PubMed]

102. Wheeler, D.S.; Chase, M.A.; Senft, A.P.; Poynter, S.E.; Wong, H.R.; Page, K. Extracellular Hsp72, an Endogenous DAMP, Is Released by Virally Infected Airway Epithelial Cells and Activates Neutrophils via Toll-like Receptor (TLR)-4. *Respir. Res.* **2009**, *10*, 31. [CrossRef]

103. Kelly, M.G.; Alvero, A.B.; Chen, R.; Silasi, D.-A.; Abrahams, V.M.; Chan, S.; Visintin, I.; Rutherford, T.; Mor, G. TLR-4 Signaling Promotes Tumor Growth and Paclitaxel Chemoresistance in Ovarian Cancer. *Cancer Res.* **2006**, *66*, 3859–3868. [CrossRef] [PubMed]

104. Sapi, E.; Alvero, A.B.; Chen, W.; O'Malley, D.; Hao, X.-Y.; Dwipoyono, B.; Garg, M.; Kamsteeg, M.; Rutherford, T.; Mor, G. Resistance of Ovarian Carcinoma Cells to Docetaxel Is XIAP Dependent and Reversible by Phenoxodiol. *Oncol. Res. Featur. Preclin. Clin. Cancer Ther.* **2004**, *14*, 567–578. [CrossRef]

105. Hu, W.; Xu, Z.; Zhu, S.; Sun, W.; Wang, X.; Tan, C.; Zhang, Y.; Zhang, G.; Xu, Y.; Tang, J. Small Extracellular Vesicle-Mediated Hsp70 Intercellular Delivery Enhances Breast Cancer Adriamycin Resistance. *Free Radic. Biol. Med.* **2021**, *164*, 85–95. [CrossRef]

106. Li, H.; Li, Y.; Liu, D.; Sun, H.; Su, D.; Yang, F.; Liu, J. Extracellular HSP70/HSP70-PCs Promote Epithelial-Mesenchymal Transition of Hepatocarcinoma Cells. *PLoS ONE* **2013**, *8*, e84759. [CrossRef]

107. Zhe, Y.; Li, Y.; Liu, D.; Su, D.-M.; Liu, J.-G.; Li, H.-Y. Extracellular HSP70-Peptide Complexes Promote the Proliferation of Hepatocellular Carcinoma Cells via TLR2/4/JNK1/2MAPK Pathway. *Tumour Biol.* **2016**, *37*, 13951–13959. [CrossRef] [PubMed]

108. Rahimi, F.; Karimi, J.; Goodarzi, M.T.; Saidijam, M.; Khodadadi, I.; Razavi, A.N.E.; Nankali, M. Overexpression of Receptor for Advanced Glycation End Products (RAGE) in Ovarian Cancer. *Cancer Biomark.* **2017**, *18*, 61–68. [CrossRef] [PubMed]

109. Wang, D.; Li, T.; Ye, G.; Shen, Z.; Hu, Y.; Mou, T.; Yu, J.; Li, S.; Liu, H.; Li, G. Overexpression of the Receptor for Advanced Glycation Endproducts (RAGE) Is Associated with Poor Prognosis in Gastric Cancer. *PLoS ONE* **2015**, *10*, e0122697. [CrossRef]

110. Prantner, D.; Nallar, S.; Vogel, S.N. The Role of RAGE in Host Pathology and Crosstalk between RAGE and TLR4 in Innate Immune Signal Transduction Pathways. *FASEB J.* **2020**, *34*, 15659–15674. [CrossRef] [PubMed]

111. Sakaguchi, M.; Murata, H.; Yamamoto, K.; Ono, T.; Sakaguchi, Y.; Motoyama, A.; Hibino, T.; Kataoka, K.; Huh, N. TIRAP, an Adaptor Protein for TLR2/4, Transduces a Signal from RAGE Phosphorylated upon Ligand Binding. *PLoS ONE* **2011**, *6*, e23132. [CrossRef] [PubMed]

112. Multhoff, G.; Botzler, C.; Jennen, L.; Schmidt, J.; Ellwart, J.; Issels, R. Heat Shock Protein 72 on Tumor Cells: A Recognition Structure for Natural Killer Cells. *J. Immunol.* **1997**, *158*, 4341–4350.

113. Multhoff, G.; Pfister, K.; Gehrmann, M.; Hantschel, M.; Gross, C.; Hafner, M.; Hiddemann, W. A 14-Mer Hsp70 Peptide Stimulates Natural Killer (NK) Cell Activity. *Cell Stress Chaperones* **2001**, *6*, 337–344. [CrossRef]

114. Specht, H.M.; Ahrens, N.; Blankenstein, C.; Duell, T.; Fietkau, R.; Gaipl, U.S.; Günther, C.; Gunther, S.; Habl, G.; Hautmann, H.; et al. Heat Shock Protein 70 (Hsp70) Peptide Activated Natural Killer (NK) Cells for the Treatment of Patients with Non-Small Cell Lung Cancer (NSCLC) after Radiochemotherapy (RCTx)—From Preclinical Studies to a Clinical Phase II Trial. *Front. Immunol.* **2015**, *6*, 162. [CrossRef]

115. Elsner, L.; Flügge, P.F.; Lozano, J.; Muppala, V.; Eiz-Vesper, B.; Demiroglu, S.Y.; Malzahn, D.; Herrmann, T.; Brunner, E.; Bickeböller, H.; et al. The Endogenous Danger Signals HSP70 and MICA Cooperate in the Activation of Cytotoxic Effector Functions of NK Cells. *J. Cell. Mol. Med.* **2010**, *14*, 992–1002. [CrossRef]

116. Sharapova, T.N.; Romanova, E.A.; Ivanova, O.K.; Yashin, D.V.; Sashchenko, L.P. Hsp70 Interacts with the TREM-1 Receptor Expressed on Monocytes and Thereby Stimulates Generation of Cytotoxic Lymphocytes Active against MHC-Negative Tumor Cells. *Int. J. Mol. Sci.* **2021**, *22*, 6889. [CrossRef]

117. Yashin, D.V.; Ivanova, O.K.; Soshnikova, N.V.; Sheludchenkov, A.A.; Romanova, E.A.; Dukhanina, E.A.; Tonevitsky, A.G.; Gnuchev, N.V.; Gabibov, A.G.; Georgiev, G.P.; et al. Tag7 (PGLYRP1) in Complex with Hsp70 Induces Alternative Cytotoxic Processes in Tumor Cells via TNFR1 Receptor. *J. Biol. Chem.* **2015**, *290*, 21724–21731. [CrossRef] [PubMed]

118. Sashchenko, L.P.; Romanova, E.A.; Ivanova, O.K.; Sharapova, T.N.; Yashin, D.V. FasL and the NKG2D Receptor Are Required for the Secretion of the Tag7/PGRP-S-Hsp70 Complex by the Cytotoxic CD8+ Lymphocytes. *IUBMB Life* **2017**, *69*, 30–36. [CrossRef] [PubMed]

119. Chen, T.; Guo, J.; Han, C.; Yang, M.; Cao, X. Heat Shock Protein 70, Released from Heat-Stressed Tumor Cells, Initiates Antitumor Immunity by Inducing Tumor Cell Chemokine Production and Activating Dendritic Cells via TLR4 Pathway. *J. Immunol.* **2009**, *182*, 1449–1459. [CrossRef]

120. Komarova, E.Y.; Marchenko, L.V.; Zhakhov, A.V.; Nikotina, A.D.; Aksenov, N.D.; Suezov, R.V.; Ischenko, A.M.; Margulis, B.A.; Guzhova, I.V. Extracellular Hsp70 Reduces the Pro-Tumor Capacity of Monocytes/Macrophages Co-Cultivated with Cancer Cells. *Int. J. Mol. Sci.* **2019**, *21*, 59. [CrossRef]

121. Kumar, S.; Stokes, J.; Singh, U.P.; Scissum Gunn, K.; Acharya, A.; Manne, U.; Mishra, M. Targeting Hsp70: A Possible Therapy for Cancer. *Cancer Lett.* **2016**, *374*, 156–166. [CrossRef]

122. Jacobson, B.A.; Chen, E.Z.; Tang, S.; Belgum, H.S.; McCauley, J.A.; Evenson, K.A.; Etchison, R.G.; Jay-Dixon, J.; Patel, M.R.; Raza, A.; et al. Triptolide and Its Prodrug Minnelide Suppress Hsp70 and Inhibit in Vivo Growth in a Xenograft Model of Mesothelioma. *Genes Cancer* **2015**, *6*, 144–152. [CrossRef] [PubMed]

123. MacKenzie, T.N.; Mujumdar, N.; Banerjee, S.; Sangwan, V.; Sarver, A.; Vickers, S.; Subramanian, S.; Saluja, A.K. Triptolide Induces the Expression of MiR-142-3p: A Negative Regulator of Heat Shock Protein 70 and Pancreatic Cancer Cell Proliferation. *Mol. Cancer Ther.* **2013**, *12*, 1266–1275. [CrossRef] [PubMed]

124. Kleinjung, T.; Arndt, O.; Feldmann, H.J.; Bockmühl, U.; Gehrmann, M.; Zilch, T.; Pfister, K.; Schönberger, J.; Marienhagen, J.; Eilles, C.; et al. Heat Shock Protein 70 (Hsp70) Membrane Expression on Head-and-Neck Cancer Biopsy-a Target for Natural Killer (NK) Cells. *Int. J. Radiat. Oncol. Biol. Phys.* **2003**, *57*, 820–826. [CrossRef]

125. Hantschel, M.; Pfister, K.; Jordan, A.; Scholz, R.; Andreesen, R.; Schmitz, G.; Schmetzer, H.; Hiddemann, W.; Multhoff, G. Hsp70 Plasma Membrane Expression on Primary Tumor Biopsy Material and Bone Marrow of Leukemic Patients. *Cell Stress Chaperones* **2000**, *5*, 438–442. [CrossRef]

126. Stangl, S.; Gehrmann, M.; Riegger, J.; Kuhs, K.; Riederer, I.; Sievert, W.; Hube, K.; Mocikat, R.; Dressel, R.; Kremmer, E.; et al. Targeting Membrane Heat-Shock Protein 70 (Hsp70) on Tumors by CmHsp70.1 Antibody. *Proc. Natl. Acad. Sci. USA* **2011**, *108*, 733–738. [CrossRef]

127. Stangl, S.; Gehrmann, M.; Dressel, R.; Alves, F.; Dullin, C.; Themelis, G.; Ntziachristos, V.; Staeblein, E.; Walch, A.; Winkelmann, I.; et al. In Vivo Imaging of CT26 Mouse Tumours by Using CmHsp70.1 Monoclonal Antibody. *J. Cell. Mol. Med.* **2011**, *15*, 874–887. [CrossRef]

128. Dechant, M.; Beyer, T.; Schneider-Merck, T.; Weisner, W.; Peipp, M.; van de Winkel, J.G.J.; Valerius, T. Effector Mechanisms of Recombinant IgA Antibodies against Epidermal Growth Factor Receptor. *J. Immunol.* **2007**, *179*, 2936–2943. [CrossRef]

129. Stockmeyer, B.; Dechant, M.; van Egmond, M.; Tutt, A.L.; Sundarapandiyan, K.; Graziano, R.F.; Repp, R.; Kalden, J.R.; Gramatzki, M.; Glennie, M.J.; et al. Triggering Fc Alpha-Receptor I (CD89) Recruits Neutrophils as Effector Cells for CD20-Directed Antibody Therapy. *J. Immunol.* **2000**, *165*, 5954–5961. [CrossRef]

130. Russell, M.W.; Reinholdt, J.; Kilian, M. Anti-Inflammatory Activity of Human IgA Antibodies and Their Fab Alpha Fragments: Inhibition of IgG-Mediated Complement Activation. *Eur. J. Immunol.* **1989**, *19*, 2243–2249. [CrossRef]

131. Van der Steen, L.; Tuk, C.W.; Bakema, J.E.; Kooij, G.; Reijerkerk, A.; Vidarsson, G.; Bouma, G.; Kraal, G.; de Vries, H.E.; Beelen, R.H.J.; et al. Immunoglobulin A: Fc(Alpha)RI Interactions Induce Neutrophil Migration through Release of Leukotriene B4. *Gastroenterology* **2009**, *137*, 2018–2029. [CrossRef] [PubMed]

132. Lohse, S.; Meyer, S.; Meulenbroek, L.A.P.M.; Jansen, J.H.M.; Nederend, M.; Kretschmer, A.; Klausz, K.; Möginger, U.; Derer, S.; Rösner, T.; et al. An Anti-EGFR IgA That Displays Improved Pharmacokinetics and Myeloid Effector Cell Engagement In Vivo. *Cancer Res.* **2016**, *76*, 403–417. [CrossRef] [PubMed]

133. Shevtsov, M.A.; Yakovleva, L.Y.; Nikolaev, B.P.; Marchenko, Y.Y.; Dobrodumov, A.V.; Onokhin, K.V.; Onokhina, Y.S.; Selkov, S.A.; Mikhrina, A.L.; Guzhova, I.V.; et al. Tumor Targeting Using Magnetic Nanoparticle Hsp70 Conjugate in a Model of C6 Glioma. *Neuro-Oncology* **2014**, *16*, 38–49. [CrossRef]

134. Shevtsov, M.A.; Nikolaev, B.P.; Ryzhov, V.A.; Yakovleva, L.Y.; Marchenko, Y.Y.; Parr, M.A.; Rolich, V.I.; Mikhrina, A.L.; Dobrodumov, A.V.; Pitkin, E.; et al. Ionizing Radiation Improves Glioma-Specific Targeting of Superparamagnetic Iron Oxide Nanoparticles Conjugated with CmHsp70.1 Monoclonal Antibodies (SPION–CmHsp70.1). *Nanoscale* **2015**, *7*, 20652–20664. [CrossRef]

135. Huang, H.S.; Hainfeld, J.F. Intravenous Magnetic Nanoparticle Cancer Hyperthermia. *Int. J. Nanomed.* **2013**, *8*, 2521–2532. [CrossRef]

136. Shevtsov, M.; Huile, G.; Multhoff, G. Membrane Heat Shock Protein 70: A Theranostic Target for Cancer Therapy. *Philos. Trans. R. Soc. B Biol. Sci.* **2018**, *373*, 20160526. [CrossRef]

137. Kimm, M.A.; Shevtsov, M.; Werner, C.; Sievert, W.; Zhiyuan, W.; Schoppe, O.; Menze, B.H.; Rummeny, E.J.; Proksa, R.; Bystrova, O.; et al. Gold Nanoparticle Mediated Multi-Modal CT Imaging of Hsp70 Membrane-Positive Tumors. *Cancers* **2020**, *12*, 1331. [CrossRef]

138. Xie, Y.; Bai, O.; Zhang, H.; Yuan, J.; Zong, S.; Chibbar, R.; Slattery, K.; Qureshi, M.; Wei, Y.; Deng, Y.; et al. Membrane-Bound HSP70-Engineered Myeloma Cell-Derived Exosomes Stimulate More Efficient CD8+ CTL- and NK-Mediated Antitumour Immunity than Exosomes Released from Heat-Shocked Tumour Cells Expressing Cytoplasmic HSP70. *J. Cell. Mol. Med.* **2010**, *14*, 2655–2666. [CrossRef] [PubMed]

139. Kim, M.S.; Haney, M.J.; Zhao, Y.; Mahajan, V.; Deygen, I.; Klyachko, N.L.; Inskoe, E.; Piroyan, A.; Sokolsky, M.; Okolie, O.; et al. Development of Exosome-Encapsulated Paclitaxel to Overcome MDR in Cancer Cells. *Nanomedicine* **2016**, *12*, 655–664. [CrossRef] [PubMed]

140. Pascucci, L.; Coccè, V.; Bonomi, A.; Ami, D.; Ceccarelli, P.; Ciusani, E.; Viganò, L.; Locatelli, A.; Sisto, F.; Doglia, S.M.; et al. Paclitaxel Is Incorporated by Mesenchymal Stromal Cells and Released in Exosomes That Inhibit in Vitro Tumor Growth: A New Approach for Drug Delivery. *J. Control. Release* **2014**, *192*, 262–270. [CrossRef] [PubMed]

141. Krause, S.W.; Gastpar, R.; Andreesen, R.; Gross, C.; Ullrich, H.; Thonigs, G.; Pfister, K.; Multhoff, G. Treatment of Colon and Lung Cancer Patients with Ex Vivo Heat Shock Protein 70-Peptide-Activated, Autologous Natural Killer Cells: A Clinical Phase I Trial. *Clin. Cancer Res.* **2004**, *10*, 3699–3707. [CrossRef] [PubMed]

142. Multhoff, G.; Seier, S.; Stangl, S.; Sievert, W.; Shevtsov, M.; Werner, C.; Pockley, A.G.; Blankenstein, C.; Hildebrandt, M.; Offner, R.; et al. Targeted Natural Killer Cell–Based Adoptive Immunotherapy for the Treatment of Patients with NSCLC after Radiochemotherapy: A Randomized Phase II Clinical Trial. *Clin. Cancer Res.* **2020**, *26*, 5368–5379. [CrossRef]

143. Kokowski, K.; Stangl, S.; Seier, S.; Hildebrandt, M.; Vaupel, P.; Multhoff, G. Radiochemotherapy Combined with NK Cell Transfer Followed by Second-Line PD-1 Inhibition in a Patient with NSCLC Stage IIIb Inducing Long-Term Tumor Control: A Case Study. *Strahlenther. Onkol.* **2019**, *195*, 352–361. [CrossRef] [PubMed]

144. Shevtsov, M.; Pitkin, E.; Ischenko, A.; Stangl, S.; Khachatryan, W.; Galibin, O.; Edmond, S.; Lobinger, D.; Multhoff, G. Ex Vivo Hsp70-Activated NK Cells in Combination With PD-1 Inhibition Significantly Increase Overall Survival in Preclinical Models of Glioblastoma and Lung Cancer. *Front. Immunol.* **2019**, *10*, 454. [CrossRef]

145. Lin, C.-N.; Tsai, Y.-C.; Hsu, C.-C.; Liang, Y.-L.; Wu, Y.-Y.; Kang, C.-Y.; Lin, C.-H.; Hsu, P.-H.; Lee, G.-B.; Hsu, K.-F. An Aptamer Interacting with Heat Shock Protein 70 Shows Therapeutic Effects and Prognostic Ability in Serous Ovarian Cancer. *Mol. Ther. Nucleic Acids* **2021**, *23*, 757–768. [CrossRef]

146. Shevtsov, M.; Stangl, S.; Nikolaev, B.; Yakovleva, L.; Marchenko, Y.; Tagaeva, R.; Sievert, W.; Pitkin, E.; Mazur, A.; Tolstoy, P.; et al. Granzyme B Functionalized Nanoparticles Targeting Membrane Hsp70-Positive Tumors for Multimodal Cancer Theranostics. *Small* **2019**, *15*, 1900205. [CrossRef]

147. Vanaja, D.K.; Grossmann, M.E.; Celis, E.; Young, C.Y. Tumor Prevention and Antitumor Immunity with Heat Shock Protein 70 Induced by 15-Deoxy-Delta12,14-Prostaglandin J2 in Transgenic Adenocarcinoma of Mouse Prostate Cells. *Cancer Res.* **2000**, *60*, 4714–4718.

148. Gong, J.; Zhang, Y.; Durfee, J.; Weng, D.; Liu, C.; Koido, S.; Song, B.; Apostolopoulos, V.; Calderwood, S.K. A Heat Shock Protein 70-Based Vaccine with Enhanced Immunogenicity for Clinical Use. *J. Immunol.* **2010**, *184*, 488–496. [CrossRef] [PubMed]

149. Trimble, C.L.; Peng, S.; Kos, F.; Gravitt, P.; Viscidi, R.; Sugar, E.; Pardoll, D.; Wu, T.C. A phase I trial of a human papillomavirus DNA vaccine for HPV16+ cervical intraepithelial neoplasia 2/3. *Clin. Cancer Res.* **2009**, *15*, 361–367. [CrossRef]

Review

Exosomes: A Forthcoming Era of Breast Cancer Therapeutics

Banashree Bondhopadhyay [1,†], Sandeep Sisodiya [1,2,†], Faisal Abdulrahman Alzahrani [3], Muhammed A. Bakhrebah [4], Atul Chikara [1,2], Vishakha Kasherwal [1,5], Asiya Khan [6,7], Jyoti Rani [1], Sajad Ahmad Dar [8], Naseem Akhter [9], Pranay Tanwar [7], Usha Agrawal [10] and Showket Hussain [1,*]

[1] ICMR-National Institute of Cancer Prevention and Research, Noida 201301, India; bbanerjee218@gmail.com (B.B.); sandeepsisodiya99@gmail.com (S.S.); atul.chikara5@gmail.com (A.C.); vishakha.kasherwal@s.amity.edu (V.K.); jyotiranibio2018@gmail.com (J.R.)

[2] Symbiosis School of Biological Sciences, Symbiosis International (Deemed University), Pune 411004, India

[3] Department of Biochemistry, Faculty of Science, Embryonic Stem Cells Unit, King Fahd Medical Research Center, King Abdulaziz University, Jeddah 21589, Saudi Arabia; faahalzahrani@kau.edu.sa

[4] Life Science and Environment Research Institute, King Abdulaziz City for Science and Technology (KACST), Riyadh 11442, Saudi Arabia; mbakhrbh@kacst.edu.sa

[5] Amity Institute of Molecular Medicine and Stem Cell Research, Amity University, Noida 201313, India

[6] Centre for Medical Biotechnology, Amity Institute of Biotechnology, Amity University, Noida 201313, India; asiya.khan@student.amity.edu

[7] Laboratory Oncology Unit, Dr. Bheem Rao Ambedkar Institute Rotary Cancer Hospital (Dr. BRA-IRCH), All India Institute of Medical Sciences, Ansari Nagar, New Delhi 110023, India; pranaytanwar@aiims.edu

[8] Research and Scientific Studies Unit, College of Nursing, Jazan University, Jazan 45142, Saudi Arabia; sdar@jazanu.edu.sa

[9] Department of Laboratory Medicine, Faculty of Applied Medical Sciences, Albaha University, Albaha 65411, Saudi Arabia; nakhter@bu.edu.sa

[10] ICMR-National Institute of Pathology, New Delhi 110029, India; director-nip@icmr.gov.in

[*] Correspondence: showket.hussain@gov.in

[†] These authors contributed equally.

Citation: Bondhopadhyay, B.; Sisodiya, S.; Alzahrani, F.A.; Bakhrebah, M.A.; Chikara, A.; Kasherwal, V.; Khan, A.; Rani, J.; Dar, S.A.; Akhter, N.; et al. Exosomes: A Forthcoming Era of Breast Cancer Therapeutics. *Cancers* 2021, 13, 4672. https://doi.org/10.3390/cancers13184672

Academic Editors: Farrukh Aqil, Ramesh Chandra Gupta and Álvaro González

Received: 21 May 2021
Accepted: 29 June 2021
Published: 17 September 2021

Publisher's Note: MDPI stays neutral with regard to jurisdictional claims in published maps and institutional affiliations.

Simple Summary: Breast cancer prevalence is a major challenge worldwide due to the lack of early diagnostics and treatment modalities. In this era of technological advancements, researchers are exploring several grey areas in breast cancer research, which may lead to the appropriate point of care, non-invasive and diagnostic aid for early breast cancer detection and management. Exosome-based research, an emerging area, endeavors to locate and elucidate the role of exosomes in breast cancer diagnostics, immune response and clinical outcomes. This review may provide insights on small extracellular vesicles research and their role in breast cancer. Future extensive studies on exosome biology in conjunction with cancer genetics shall undoubtedly open up new vistas in exosome-based diagnostics for early cancer detection and therapeutics.

Abstract: Despite the recent advancements in therapeutics and personalized medicine, breast cancer remains one of the most lethal cancers among women. The prognostic and diagnostic aids mainly include assessment of tumor tissues with conventional methods towards better therapeutic strategies. However, current era of gene-based research may influence the treatment outcome particularly as an adjunct to diagnostics by exploring the role of non-invasive liquid biopsies or circulating markers. The characterization of tumor milieu for physiological fluids has been central to identifying the role of exosomes or small extracellular vesicles (sEVs). These exosomes provide necessary communication between tumor cells in the tumor microenvironment (TME). The manipulation of exosomes in TME may provide promising diagnostic/therapeutic strategies, particularly in triple-negative breast cancer patients. This review has described and highlighted the role of exosomes in breast carcinogenesis and how they could be used or targeted by recent immunotherapeutics to achieve promising intervention strategies.

Keywords: exosomes; small extracellular vesicles; breast cancer; cancer aggressiveness; multi-drug resistance; diagnosis; immune response; immunotherapy

1. Introduction

Breast cancer, a heterogeneous disease, is a common cause of death in females worldwide [1–3]. The current treatment strategies are based on the expression pattern of the estrogen receptor (ER), the progesterone receptor (PR) and the ERBB2 receptor (Her2) profile [4,5]. Recently, breast cancer survival rate has improved due to outcomes in the primary molecular sub-classification when administered with targeted therapies such as hormone therapy and HER2-targeted therapy (e.g., trastuzumab) [6]. As per the gene expression pattern of breast cancer patients, clustering leads to five different molecular subtypes of breast cancer, i.e., normal type, basal type, Her2-rich, luminal A and luminal B [7], and classifies ER- breast cancer into four different subtypes and triple-negative breast cancers (TNBCs) into six subtypes [8]. The current understanding of breast cancer biology has led to significant improvements in diagnostic and prognostic methods and enhanced novel targeted therapies. However, the limited knowledge about the molecular processes or mechanisms involved in breast cancer pathogenesis has led to restricted therapeutic approaches and poor prognosis of breast cancer patients. Studies have recently elucidated the role of a typical vesicular structure of 30–150 nm diameter called exosomes and/or small extracellular vesicles (sEVs), secreted by various immune cells such as dendritic and Chimeric Antigen Receptor T cells (CAR-T) cells to provide robust diagnostics and therapeutic interventions [9,10]. In the year 1985, exosomes were initially described as a budding membrane of intracellular vesicles [11]. However, recently, stem cells, endothelial cells, dendritic cells, B cells, T cells and especially cancer cells were found to secrete exosomes [12], that can play a crucial role in cell signaling communication, in both paracrine and autocrine manner [13]. Exosomes also assist in transporting various molecules, including proteins, lipids, DNA, mRNA, micro RNAs (miRNA) and lncRNA (Long noncoding RNA) [14,15]. Moreover, exosomes are found amply in pathological and/or physiological fluids, such as breast milk, cerebrospinal fluid, serum, saliva, urine, plasma and ascites [16], making them promising target molecules as cancer cells release more exosomes than non-cancer cells.

2. Exosomes: Structure and Functions

Exosomes, first identified by Johnstone et al., are nanovesicles derived from cultured monolayer cells [17], made of growing intracellular endosomes that produce multicellular bodies (MVBs) fused with plasma membranes to secrete exosomes out of the cells [11,18]. Exosomes are lipid vesicles with a bilayer structure and a diameter of 30 to 150 nm [10,19,20], and a buoyant density of 1.13 g/mL to 1.19 g/mL [21]; formed during the process of endosomal maturation by dependent and independent endosomal sorting complexes required for transport (ESCRT) processes [22]. They express several proteins including protein/tetraspanin markers such as TSG101, ALIX, CD63, HSP70, tetraspanin 1–19, Putative tetraspanin-19, Uroplakin-1a,1b, Peripherin-2, CD Antigen 9, 63, 81, 82, 151 and Leucocyte surface antigen CD53, CD37 which play a key role in vesicle detection [23–25].

Exosomes are a crucial element in the metastasis, development and treatment efficacy of cancer. They also play a key role in tumor development owing to their ten times higher secretion efficiency in cancer cells than in normal cells, resulting in cellular contact in the tumor niche through nucleic acid and oncogenic protein transmission [26–34]. The absorption of exosomes induces upregulation of genes related to angiogenesis, leading to proliferation, migration and germination of endothelial cells [35]. In the premetastatic niche, exosomes help in epithelial to mesenchymal transition (EMT) through distant metastasis [36–38], and also contribute to cancer-associated fibroblasts (CAFs) for the enhancement of cancer aggressiveness. Exosomes are also involved in neutrophil deployment, growth and stimulation of myeloid-derived suppressor cells (MDSC), inhibition of dendritic cell (DC) differentiation, inhibition of natural killer cells (NK) cytotoxicity, induction of M2 polarization of macrophages, development of regulatory T cells (Treg) and induction of apoptosis of cytotoxic T (Tc) cells [39–42]. Exosomes not only contribute to the growth of cancer cells but also provide chemoresistance to the neighboring cells in the tumor microenvironment against various chemotherapeutic agents, displaying the role as a safeguard for other cancer cells [43,44]. Various in-vitro studies and clinical studies on breast cancer have demonstrated that exosomes might contribute to miRNA processing delivery and result in induction of tumor formation and/or transformation in non-tumorigenic breast cells [45]. In addition, autocrine signaling has been found to trigger further cancer progression via exosomes derived from the cancer cells. For example, exosomes extracted from in vitro gastric cancer cells encourage growth via Akt/PI3K (Phosphoinositide 3-Kinase), MAPK (Mitogen Activated Protein Kinase) and Notch-1 dependent signaling pathways [46,47]. Overall, cancer cells can customize isomorphic exosomes to guide cancer progression by targeting the different molecules and processes related to breast carcinogenesis (Figure 1).

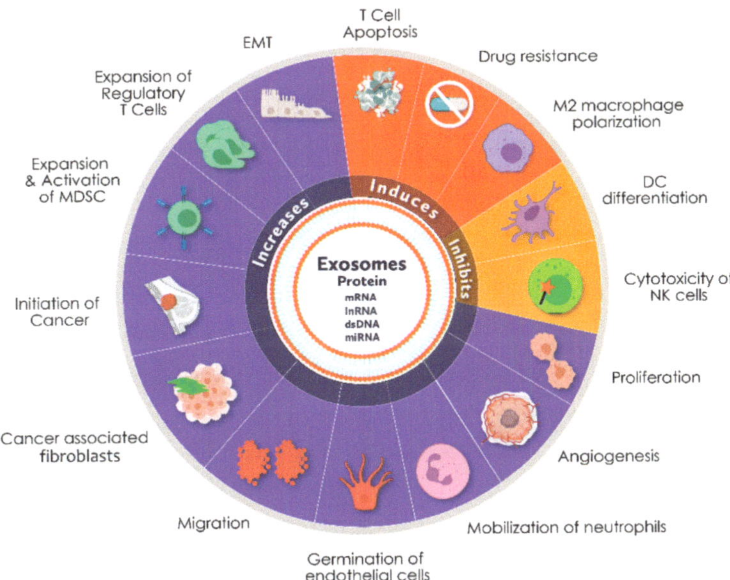

Figure 1. The schematic figure represents the functional abilities of exosomes that may be involved in various cellular processes during breast carcinogenesis (Icons are created with biorender.com (accessed on 20 June 2021)).

3. Origin-Based Types of Exosomes

There are several types of exosomes depending on their site of origin: DCs-derived exosomes, Tumor-derived exosomes, Ascites-derived exosomes (Aexs), CTL derived exosomes, CAR-T (Cytotoxic T Lymphocyte) cells-derived exosomes, Mesenchymal stem cell-derived exosomes (MSCs) and natural source derived exosomes which are discussed below in detail and represented in Figure 2.

3.1. DCs-Derived Exosomes (Dexs)

Dendritic cells (DCs), involved in the first stage of cancer immunity, aims to activate tumor-specific cytotoxic lymphocytes, leading to the destruction of tumor cells [48]. The first FDA-approved DC vaccine to be used as immunotherapy for castration-resistant prostate cancer showed an average survival of 4.1 months (25.8 months in the ciprofloxacin-T group and 21.7 months in the placebo group) [49]. However, the DC vaccine consists of living cells, making it really expensive in terms of storage and stability over a longer period of time. Dexs carry numerous DC molecules associated with immune function including peptide/Major Histocompatibility Complex (MHC) complexes that trigger the response of antigen-specific T lymphocytes [50,51], and co-stimulatory molecules, in particular CD80, CD83, CD86, which further aid in the enhancement and initiation of T lymphocyte cells. Exposing adenocarcinoma cells to Dex treatment also causes an increase in the induction of interferon secretion [52,53]. These studies suggest that Dex maintains an essential immunostimulatory power of DCs, which could become a promising tool for cancer immunotherapy in future.

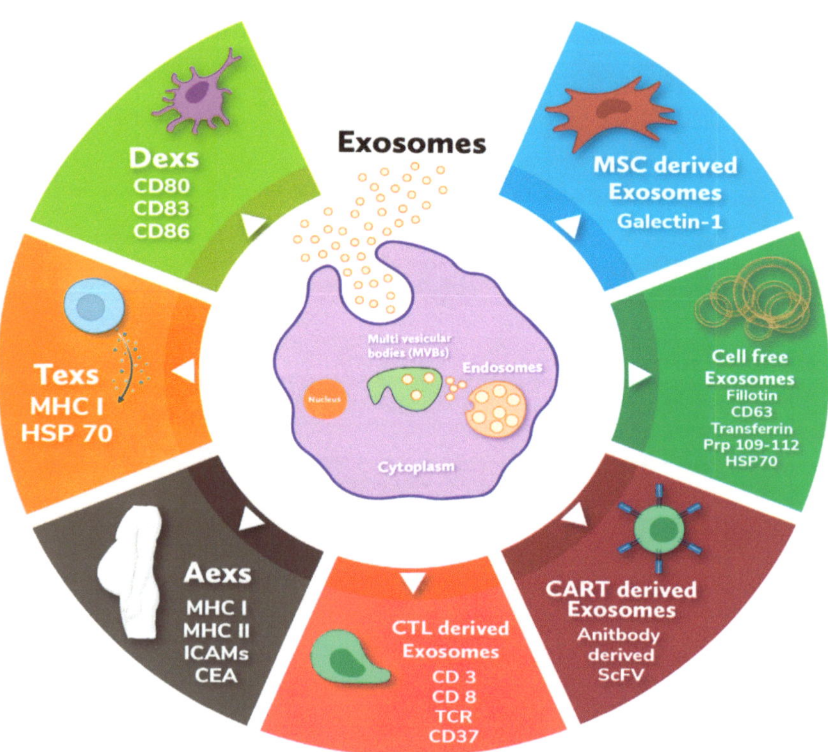

Figure 2. The diagrammatic representation depicts the release of different types of exosomes from the cells and their molecular markers (Icons are created with biorender.com (accessed on 20 June 2021)).

3.2. Tumor-Derived Exosomes (Texs)

Texs carry MHC-I, HSP70 and antigens speculated to be the source of specific stimuli against immune response exerted by cancer cells. Texs elicit an enhanced anti-tumor reaction more efficiently than the cancer cell debris, apoptotic materials and irradiated tumor cells [54]. HSP70, a stress-inducible exosomal heat shock protein that promotes NK cell activation and cancer cell lysis via granzyme B, acts as an endogenous danger signal to increase the immunogenicity of tumors by induction of CTL response [55]. Texs can also effectively release a variety of tumor antigens to DC; thus, they can be exploited as antigen carriers for cancer immunotherapeutics [56]. Texs are known to play a key role in cancer growth and progression, such as inducing apoptosis in activated CD8+ T cells, inhibiting immune cell proliferation, interfering with the monocyte differentiation, suppressing NK cell activity and encouraging Treg and MDSC expansion [57]. These effects come by directly suppressing the proliferation and inhibiting the cytotoxicity of NK cells or binding directly to the T-cells associated with HER2 receptors leading to the activation of multiple cells to inhibit tumor growth. In addition, the removal of PD-L1 leads to the anti-tumor properties, hence becoming as one of the potential therapeutic target [58]. Similar to the Dexs, Texs might also become a potential and immunogenic acellular vaccine [59].

3.3. Ascites-Derived Exosomes (Aexs)

Aexes are another form of exosome shown to play an important role in carcinogenesis. Aexs contain MHC-I and –II molecules, co-stimulatory molecules, ICAMs and the immunogenic carcinoembryonic antigen (CEA) which APCs may recognize. Initial clinical trials in advanced CRC patients have shown promising anti -tumor response of Aexs along with GM-CSF (Granulocyte-Macrophage Cell Simulating Factor) and may serve as alternative to immunotherapy [60].

3.4. CTLs Derived Exosomes

In the year 1989, Peters et al. suggested that exosomes derived from human T cells participate in the interaction of CTLs and the target cells [61]. However, in specificity towards CTLs, the presence of CD3, CD8 and TCR on CTLs derived exosomes could provide cytotoxicity to the targeted cells through TCR (T- Cell Receptor) interaction with the antigen/MHC-I complex. Such interaction may result in the target cell death [62], due to the presence of cytotoxic compounds in exosomes, including perforin, granzymes and lysosomal enzymes [63]. Early studies have emphasized that the accelerated secretion of exosomes by CTLs through TCR activation and TCR/CD3ζ complex has existed on the surface membrane of exosomes derived from human CTL [64], resulting in the rapid elimination of the target cell and thus serving and contributing to the adaptive immunity.

3.5. Exosomes Derived from CAR-T Cells

CAR-T cell-derived exosomes may possess antibody-derived single-chain variable fragment (scFv), a promising alternative to cell therapy. Cellular communications between CAR-T or CTL and cancer cells are required for the anti-tumor effect of CAR-T cells and CTLs especially in an aggressive tumor. Both CAR-T cells and CTLs interaction with the cancer cells require the penetration of the CAR-T or CTLs cells in the tumor. However, the tumor milieu can limit the mode of action of CAR-T cells and CTLs as the scFv may influence the CAR-T cell function [65]. Consequently, this may limit clinical application of CAR-T-based cell therapy particularly in many solid tumors [66]. However, the adoptive transfer of CAR-T cells proposes an innovative method in cancer immunotherapy by provoking prompt and long-lasting clinical responses albeit with acute toxicities [67]. The exosomes released by CAR-T cells carry CAR on their surface, which helps in releasing highly cytotoxic molecules, thus inhibiting tumor growth. CAR exosomes do not express programmed cell death protein 1 (PD1) and, in contrast with CAR-T cells, their anti-tumor effect is uninfluenced by recombinant PD-L1 treatment. In addition, CAR exosomes have less toxicity and thus safer than CAR-T based cell therapy [63]. Having said that, CAR-T cell administered in vivo have shown significant tumor suppression and thus the use of CAR-T cell exosomes against triple negative breast cancer (TBNC) expressing MSLN may provide significant therapeutic benefit [68].

3.6. Mesenchymal Stem Cell-Derived Exosomes (MSCs)

MSCs are the important components in tissue repair/wound healing and can also produce exosomes at a very large scale [67]. MSC-exosomes also play a role in apoptosis of the activated T cells as they express galactin-1, a carbohydrate-binding protein that binds to the distinct set of glycoprotein receptors and acts extracellularly to induce cell death. MSCs can also pack mRNA into exosomes, preventing tumor migration and infiltration to distant areas. MSC-derived exosomes can also transmit extracellular miR-143 to osteosarcoma cells, which significantly decreases the migration of osteosarcoma cells. In addition, the discharge of MSC-derived exosomes miR-23B causes cell cycle suppression and dysfunction of breast cancer cells, thus preventing cancer cell migration and infiltration [69,70]. These exosomes stimulate the secretion of Interleukin-6 (IL-6), Interferon-γ (IFN-γ), Tumor necrosis factor-α (TNF-α) along with Activated B cells, T cells and Antigen presenting Cells (APCs) containing HoxB4. This affects the DC maturation and promotes T cell proliferation, differentiation and activation through the WNT signaling pathway [67]. These findings need to be further explored extensively for better therapeutics.

3.7. Exosomes Derived from Natural Sources

Interestingly, exosomes are also derived from plant sources and food/edible materials. Food derived exosomes (FDEs) are involved in the transport of biomolecules for cell-to-cell communication. These small vesicles (50–300 nm) are surrounded by a phospholipid bilayer and form intraluminal vesicles (ILVs) in multi vesicular bodies. These bodies fuse with the plasma membrane to produce ILVs in the extracellular environment and are referred to as exosomes [71,72]. Plant-derived exosome-like particles have gained much attention because of their source of origin and are known as Plant-derived edible nanoparticles (PDENs). They are found in the paramural space of plants and are identical in structure and function to their mammalian counterparts [73]. PDENs respond differently in different biological conditions; variation in the size and surface charge of exosomes depends on the plant source and environment. In the stomach and intestinal environment, grape-derived exosome-like vesicles reduced in size compared to vesicles suspended in water, while a fraction of ginger-derived vesicles expanded in stomach and intestine [74]. In addition, large number of exosome-like vesicles has been identified from ginger (Aprox. 50 mg per 1 kg of ginger) which are rich in proteins, lipids and other nuclear components [26]. The epidemiological studies suggest that continuous human exposure to exosomes of pasteurized milk may confer substantial risk for the development of chronic diseases including obesity, type 2 diabetes mellitus, osteoporosis, common cancers such as prostate, breast, liver, B-cells and Parkinson's disease [26].

4. Exosomal Biomarkers in Breast Cancer

As the basic principles of exosome biology and their relationship with cancer and drug resistance are better understood, exosomes and the tumor microenvironment are increasingly becoming attractive targets for clinical applications; primarily due to their versatile role in carcinogenesis in terms of cancer diagnostic and treatment response [75]. Subsequently, exosomes can portray the entire tumor milieu because of their ability to be secreted from any cancer cell type [76]. It has been observed that exosome secretion has a direct relationship with the severity of cancer lesions, which may not only detect the disease but also the type of disease [76,77]. Circulating exosome-encapsulated miRNAs have been observed as ideal biomarkers for breast cancer for its good correlation with disease progression. For example, significantly high amount of exosomal miRNAs such as ci-miRNA-27 and ci-miRNA-365 are found in triple negative breast cancer patients compared to hormone receptor positive breast cancer patients [77–80]. Exosomes are shown to preserve miRNAs as cell-free miRNAs, as they are found in purified human peripheral blood micro-vessels. Subsequently, various studies show exosomal miRNAs in the blood as novel biomarkers for the diagnostic and prognostic evaluation of various human cancers including breast cancer [81–87]. In situ detection of miRNAs has highlighted that miR-21 could be a potential biomarker for both MCF-7 cells derived and normal cell-derived exosomes. In addition, miR-16 was also found to be transferred from murine breast cancer-derived TAMs via tumor-derived exosomes that prevent infiltration and polarization of macrophages in the tumor niche [88]. Exosomes derived from TAMs, containing miR-223 promote the invasive potential of breast cancer cells, thus promoting tumorigenesis [89]. Studies have also shown that the elevated level of TAMs resulting in a poor prognosis of breast cancer. However, TAM-derived exosomes might play a significant role in controlling disease progression and treatment via miRNA secretion. Consequently, exosomal miRNAs may critically impact breast cancer proliferation: metastasis, drug resistance, microenvironment formation and immune response. Some significant miRNAs are discussed in Table 1. Moreover, isolation of tumor markers in liquid biopsies is easy and cost-effective than solid tissue biopsies [90]. However, the physiognomies of circulating tumor cells (CTC) and cell-free DNA (cf-DNA) related to cancer are still unclear as compared to the exosomes of solid tumor biopsies. Furthermore, cf-DNAs carry mutations distinctively of the consistent primary tumors. In contrast, more circulating tumor DNA clearance is usually observed in the liver or kidneys, indicating steadiness and pathogenicity of circulating tumor DNA [91]. Exosomes containing different markers are represented in Table 2 and Figure 2.

Table 1. List of some important exosomal miRNAs related to breast cancer.

S.No.	Description/Function	miRNAs Involved	Refs.
1	Exosomal miRNAs in breast cancer cell proliferation and apoptosis	miR-10a, miR-10b, miR-21, miR-27a, miR-155 and miR-373	[45]
		miR-21 and miR-10b	[92,93]
		miR-128	[94]
2	Exosomal miRNAs in breast cancer metastasis	miR-200a, miR-200b, miR-200c, miR-429 and miR-141	[95]
		miR-200c and miR-141	[95]
3	Exosomal miRNAs in drug sensitivity and resistance in breast cancer	miR-100, miR-17, miR-222, miR-342–3p and miR-451	[44]
		miR-4443, miR-574–3p, miR-7847–3p, miR-423–5p, miR-4298, miR-3178, miR-6780b-3p, miR-7107–5p, miR-744–5p, miR-4258, miR-138–5p and miR-210–3p	[96]
		miR-221/222	[97]
		miR-9	[98]
		miR-939	[99]
		miRNA-122	[100]
		miR-23b and miR-320b	[101]
4	Exosomal miRNAs in breast cancer tumor microenvironment	miR-21, miR-378e and miR 143	[102]
		miR-127, miR-197, miR-222 and miR-223	[103]
		MiR-503	[104]
		Exosomal miR-198, miR-26a, miR34a and miR-494	[19]
		miR-134	[105]
		miR-182	[106]
		miR-101 and miR-372	[107]
		miR-21 and miR-1246	[80]
		exosomal miR-1246	[108]
		miR-105	[98]
		miRNA-10b	[109]
		miR19a	[110]
		miR-338-3p, miR-340-5p and miR124-3p	[111]
		miR-29b-3p, miR-20b-5p, miR17-5p, miR-130a-3p, miR-18a-5p, miR-195-5p, miR-486-5p and miR-93-5p	[111]
		miR-221/222	[112]
		miRNA-451	[113]

Table 2. List of exosomal protein markers involved in breast cancer.

S.No.	Expression Site	Protein Markers	Refs.
1	Serum/pleural effusion-derived exosomes from breast cancer patients or cell lines	ADAM10, HSP70, CD9, Annexin1,	[114]
		TrpC5	[115]
		Glucose transporter 1 (GLUT-1), glypican 1 (GPC-1),	[116]
		Glutathione S-transferase P1(GSTP-1)	[117]
		HER-2	[118]
		Survivin (Survivn 2B)	[119]
		P-glycoprotein/TrpC5/ABCG2	[120]
		Ubiquitin carboxyl terminal hydrolase-L1 (UCH-L1)	[121]
		CD24, tetraspanins and epithelial cell adhesion molecule (EpCam)	[122]
2	Plasma	Developmental endothelial Locus-1 (Del-1) and fibronectin	[123,124]
		Fibronectin	[124]
4	Total blood	SERPINA1, KRT6B and SOCS3, IGF2R	[125]

5. Exosomes in Breast Cancer Aggressiveness

Communication of cancer cells with neighboring cells is crucial for tumor development, and it may happen through direct cell to cell or intracellularly with the help of some secretory molecules [126,127]. Exosomes produced from tumors are capable of promoting tumor cell proliferation and metastasis. Apart from their pro-tumorigenic activities, exosomes also contribute to tumor-tumor communication via chemoresistance transmission. Corcoran and colleagues first demonstrated that exosomes could convey Docetaxel resistance in prostate cancer [128], similar events have been observed in a variety of tumors such as hepatocellular, lung, liver including breast cancers [129–131]. Exosomes derived from tumors also interact with non-transformed differentiated cells, triggering the development of malignant characteristics in these target cells. For example, exosomes mediate intercellular communication between neoplastic and normal cells, resulting in the latter developing a pro-inflammatory phenotype. Exosomes from arsenite-treated liver cells were demonstrated to activate the IL6, IL8/STAT3 pathway, thereby increasing miR155 expression and inflammatory characteristics in normal liver cells [132].

In addition, tumor-derived exosomes play a critical role in tumor invasion by promoting tumor cell viability along with extracellular matrix degradation through matrix metalloproteinases (MMPs). They also exclude apoptosis-inducing proteins, specifically leading to the escape of tumor cells from immune surveillance [133,134]. HSP90+ exosomes derived from metastatic breast cancer cells and released exotically with the help of rab27b, can promote tumor invasion via degradation of extracellular matrix and activation of MMP2 [135]. Studies have also highlighted that exosome derived from linoleic acid-induced MDA-MB-231 can reduce E-cadherin expression while enhancing the expression of Snail 1/Snail 2, Twist 1/Twist 2, Vimentin, N-cadherin and Sip1 [136]. It has also been observed that exosomes derived from breast cancer cells contain miR-105, which regulates the tight junction protein ZO1 in recipient endothelial cells, may lead to augmented vascular permeability by downregulating the levels of ZOI [98]. Furthermore, recent research suggest that breast cancer-derived exosomes play a compelling role in organ-specific metastasis and angiogenesis as they contain annexin A2, which mediates brain and lung metastasis in particular [137]. An improved understanding of their mechanism may allow important therapeutic implications.

6. Exosomes as Drug Carriers

Exosomes have a low immune prototype, and thus have minor adverse effects [138]. Furthermore, exosomes can easily enter cells due to interactions between exosome membrane proteins and recipient cells [139], which makes them the most effective natural carrier for drug delivery. However, the origin of exosomes, techniques of purification, forms of drug loading and the final drug delivery system needs to be elucidated [17]. Tumor derived exosomes can deliver drugs more precisely to tumor cells and suppress tumor progression as seen in case of paclitaxel delivery to prostate cancer [140]. Similarly, exosomes from pancreatic cancer cells could effectively transfer curcumin to pancreatic cancer cells and cause considerable cell death [141]. In general, drug-loaded exosomes show better efficacy than chemical drugs alone. Furthermore, Kim et al. discovered that paclitaxel-loaded macrophage-derived exosomes had higher stability and loading efficiency than other drug-loading approaches, inhibiting Lewis Lung Carcinoma cell proliferation more effectively and showing anti-tumor activity in a murine Lewis Lung Carcinoma model [142]. In addition, Yong T et al. developed biocompatible tumor cell-exocytosed exosome-sheathed PSiNPs (E-PSiNPs) as a drug carrier for targeted cancer chemotherapy, which resulted in greater in vivo enrichment in total tumor cells and side population cells with CSC-like characteristics. The treatment also showed remarkable anticancer and CSC-killing activity in subcutaneous, orthotopic and metastatic tumors [143]. The administration of doxorubicin-loaded exosomes generated from DCs can significantly decrease breast tumor cell proliferation with no toxicity in mice. When DC-derived exosomes are combined with specific IRGD peptides, the exosomes have the ability to target breast cancer more effectively than a chemical formulation alone [144]. Exosomes containing cisplatin can prolong the life of ovarian cancer mice without generating liver or kidney side effects, which is an advantage over cisplatin alone. Additionally, exosomes containing cisplatin have an anti-tumor impact, in vivo and in vitro [145]. Curcumin loaded exosome of a murine lymphoma cell line may be successfully transferred to brain tissue, causing microglia death in the brain. These findings suggest that the strategy could provide a noninvasive and innovative therapeutic approach for treating brain inflammatory illnesses [146]. Mesenchymal stem cell-derived exosomes have been used to load miR-146b, resulting in effective inhibition of tumor growth [147,148]. These findings suggest that exosomes may be used as effective drug delivery vehicle with minimal side effects, however, more evidence are needed to use exosomes as drug delivery system.

7. Exosomes in Multidrug Resistance

Breast cancer exosomes can bind to selective therapeutic antibodies that can lead to treatment failure due to drug adsorption. Exosomes isolated from Her2+ breast cancer cell supernatants or serum can bind to trastuzumab, inhibiting its activity. The finding suggests that Her2+ exosomes may be used as a biomarker in trastuzumab-resistant tumor aggressiveness [118,149,150]. Various molecule such as transient receptor channel 5 (TrpC5), P-glycoprotein (P-gp), Survivin, DOX, mtDNA, Glutathione S-transferase P1 (GSTP1), ubiquitin carboxy-terminal hydrolase L1 (UCH-L1) etc. are linked with exosome mediated drug resistance [115,117,120,121,150–152]. Therefore, tumor derived exosome may not only serve as non-invasive biomarkers to explore the mechanism of drug resistance in breast cancer cases but also lead to personalized medicine or therapeutic interventions.

8. Exosomes in Breast Cancer Diagnosis

Recent research shows the presence of exosomes in nearly all body fluids, including blood, urine, saliva, breast milk, cerebrospinal fluid, semen, amniotic fluid and ascites [153]. Few studies have also proposed the utility of exosomes in the diagnosis and prognosis of different types of cancers. Particularly, in breast cancer, differential secretion of exosomes displaying an array of proteins such as Tetraspanin CD9, HSP70, Annexin-1 and metalloprotease ADAM10 at various stages of breast cancer may contribute to an accurate diagnosis and prognosis [114,150]. For example, tetraspanin CD63, an integrin-binding partner exclusively present on exosomes, expression correlates inversely with the cancer metastasis [154–156]. Del-1 and exosomal survival-2B (pro-apoptotic protein) can be used for differentiating benign/non-cancerous breast tumor [123] and a diagnostic and/or prognostic marker in patients with early breast cancer, respectively [119]. Along with several proteins, tumor-derived exosomal miRNAs such as miR16 also contribute to tumor evasion, leading to tumor progression. Mechanistically, exosomes derived from cancerous cells modifies the tumor microenvironment, which can eventually trigger immune cells to release epigallocatechin gallate (EGCG) [157]. Further mechanistic elucidation of proteins and miRNAs derived from circulating plasma exosomes can act as an early diagnostic, prognostic as well as therapeutic tool in cases of breast cancer.

9. Exosomes in Immune Response and Immunotherapy

Recent research findings indicate a distinct advantage of immunotherapy over existing conventional therapies [158]. Exosomes, derived from the cancer cells including breast cancer can modify the immune response by interacting with various immune cells, e.g., macrophages, regulatory T cells (Tregs), dendritic cells (DCs) and T cells [159]. Studies in breast cancer have also demonstrated that exosomal miRNAs transport stimulate the macrophages and contribute to angiogenesis [160]. Exosomes derived from murine breast cancer 4T1 cells took up fibronectin leading to an active interaction with immune cells when co-cultured with tumor infiltrating leukocytes [161]. The release of protein-coated exosomes called PD-L1, part of immune checkpoint protein family actively involved in immune surveillance, in melanoma skin cancer models and in blood samples of the people treated for breast and lung cancer [162] suggest a novel method to increase the efficacy of exosomes dependent tumor vaccines.

In addition, immunocyte exosomes include cytokines that govern inflammatory responses, innate immunity and lymphocyte production, among other processes. The research team of Gao, et al. found that Dex contains TNF-α, which could activate NF-KB by releasing membrane-bound TNF-α suggesting an involvement in endothelial inflammation and atherosclerosis [163]. Exosomes released by DCs, and macrophages include membrane-bound IL-1, which could be involved in inflammation [164,165]. Wang et al. found that the TGF-β-containing thymic cell-based exosomes boost T-cell development to Foxp3+ Tregs, the differentiation of CD4+CD25 T-cells from Tregs into the effector and their in vitro and in vivo proliferation [165]. Findings mentioned above suggest that exosomes may control key immunologic processes, release cytokines, regulate inflammatory response and innate immunity and also imply that immune cell exosomes may govern stem cell mobilization, tissue remodeling and immunological regulation.

10. Clinical Application of Exosomes

Recently, cancer cells secreted exosomes have become one of the emerging research areas in understanding cancer, especially breast carcinogenesis. Additionally, it also provides us an opportunity to explore biomarkers for better diagnosis and prognosis at an early stage [166,167]. In 2016, two test kits on fluid biopsy-based approaches were available to detect prostate and lung diseases (ExoDx® prostate and ExoDx® Lung, Exosome Diagnostics Inc., Waltham, MA, U.S.A) [87,168,169]. Breast cancer-derived exosomes have also been considered as a potential indicator of cancer progression [114]. However, further investigation is needed. Several proteins, including epidermal growth factor receptor (EGFR), survival apoptosis inhibitor, carcinogenic marker CD24, localized adhesive kinase (FAK) and surface cell proteoglycan glycan-1, are significantly overexpressed in the breast cancer patient's serum-derived exosomes as compared to the healthy donors [119,170]. Researchers found a higher level of exosomes derived 27-hydroxycholesterol exosomes in MCF-7 when compared to MDA-MB-231 cells [171,172]. Exosomes derived epigallocatechin gallate (EGCG)-treated breast cancer cells when incubated with TAM in vivo, were found to repress M2 polarization and NF-κB signaling led to anti-tumor immune response [88]. Such studies indicate the potential use of exosomes as a promising agent for drug delivery vehicles in anti-tumor therapy. Furthermore, exosomes can both spread and curb the infections, and thus are considered as suitable candidates for developing vaccines for prevention and treatment [173]. The vaccine developed from exosomes was effective in anti-tumor immunity, however, further research is warranted to exploit its potency as a therapeutic candidate.

11. Future Prospects

Availability of limited therapies against breast cancer particularly TNBCs cause higher mortality than other subtypes among breast cancer patients. Ample evidence indicates the role of exosome and/or sEVs in carcinogenesis and thus can be used for diagnosis. Moreover, exosomes may act as a bridge for cellular communication in the tumor microenvironment resulting in tumor development, invasion, metastasis and drug resistance. Apart from their role in cancer progression, these could serve as a potential vehicle for inhibiting tumor growth and development by manipulating them for drug development and immune-surveillance.

Recent path-breaking research tools such as immunotherapeutics (PDL-1, CAR-T, etc.) have immensely benefitted patients. Importantly, exosome-based immunotherapeutics exoPDL1, type of exosome-related immunotherapy, can be used to design drugs with minimum toxicity and greater clinical benefits. The small size of the exosomes makes them useful natural carriers for drug delivery into the cancer cell and may significantly contribute to therapeutic use.

12. Conclusions

It is desirable that the ongoing efforts in cancer research should not only focus on the role of exosomes or sEVs in vitro but also on their significance in liquid biopsies, immunotherapy, drug designing and drug delivery systems to benefit patients greatly particularly in triple- negative breast cancer patients.

Author Contributions: Conceptualization, S.H., investigation, S.H. and S.S., data curation, B.B. and S.S.; writing—original draft preparation, B.B., S.S., F.A.A., M.A.B., A.C., V.K., J.R. and A.K., writing and editing, S.H., B.B., S.S., F.A.A., M.A.B., A.C., V.K., J.R., A.K., S.A.D., N.A., P.T. and U.A.; supervision, S.H. and S.A.D. All authors have read and agreed to the published version of the manuscript.

Funding: This study is supported by the Indian Council of Medical Research, New Delhi, India under grant number 5/13/1/TF/NICPR/16/NCD-III.

Acknowledgments: The authors acknowledge the support of Indian Council of Medical Research, New Delhi, India to S.H., King Abdulaziz University, Jeddah to F.A.A. and King Abdulaziz City for Science and Technology (KACST) to M.A.B.

Conflicts of Interest: Authors declare no conflict of interest.

References

1. Bray, F.; Ferlay, J.; Soerjomataram, I.; Siegel, R.L.; Torre, L.A.; Jemal, A. Global cancer statistics 2018: GLOBOCAN estimates of incidence and mortality worldwide for 36 cancers in 185 countries. *CA Cancer J. Clin.* **2018**, *68*, 394–424. [CrossRef]
2. Nazir, S.U.; Kumar, R.; Dil, A.; Rasool, I.; Bondhopadhyay, B.; Singh, A.; Tripathi, R.; Singh, N.; Khan, A.; Tanwar, P.; et al. Differential expression of Ets-1 in breast cancer among North Indian population. *J. Cell. Biochem.* **2019**, *120*, 14552–14561. [CrossRef] [PubMed]
3. Nazir, S.U.; Kumar, R.; Singh, A.; Khan, A.; Tanwar, P.; Tripathi, R.; Mehrotra, R.; Hussain, S. Breast cancer invasion and progression by MMP-9 through Ets-1 transcription factor. *Gene* **2019**, *711*, 143952. [CrossRef]
4. Perou, C.M.; Sørlie, T.; Eisen, M.B.; van de Rijn, M.; Jeffrey, S.S.; Rees, C.A.; Pollack, J.R.; Ross, D.T.; Johnsen, H.; Akslen, L.A.; et al. Molecular portraits of human breast tumours. *Nature* **2000**, *406*, 747–752. [CrossRef] [PubMed]
5. Gupta, G.K.; Collier, A.L.; Lee, D.; Hoefer, R.A.; Zheleva, V.; Siewertsz van Reesema, L.L.; Tang-Tan, A.M.; Guye, M.L.; Chang, D.Z.; Winston, J.S.; et al. Perspectives on Triple-Negative Breast Cancer: Current Treatment Strategies, Unmet Needs, and Potential Targets for Future Therapies. *Cancers* **2020**, *12*, 2392. [CrossRef] [PubMed]

6. Curtis, C.; Shah, S.P.; Chin, S.-F.; Turashvili, G.; Rueda, O.M.; Dunning, M.J.; Speed, D.; Lynch, A.G.; Samarajiwa, S.; Yuan, Y.; et al. The genomic and transcriptomic architecture of 2,000 breast tumours reveals novel subgroups. *Nature* **2012**, *486*, 346–352. [CrossRef]

7. Koboldt, D.C.; Fulton, R.S.; McLellan, M.D.; Schmidt, H.; Kalicki-Veizer, J.; McMichael, J.F.; Fulton, L.L.; Dooling, D.J.; Ding, L.; Mardis, E.R.; et al. Comprehensive molecular portraits of human breast tumours. *Nature* **2012**, *490*, 61–70. [CrossRef]

8. Marusyk, A.; Almendro, V.; Polyak, K. Intra-tumour heterogeneity: A looking glass for cancer? *Nat. Rev. Cancer* **2012**, *12*, 323–334. [CrossRef]

9. Théry, C.; Witwer, K.W.; Aikawa, E.; Alcaraz, M.J.; Anderson, J.D.; Andriantsitohaina, R.; Antoniou, A.; Arab, T.; Archer, F.; Atkin-Smith, G.K.; et al. Minimal information for studies of extracellular vesicles 2018 (MISEV2018): A position statement of the International Society for Extracellular Vesicles and update of the MISEV2014 guidelines. *J. Extracell. Vesicles* **2018**, *7*, 1535750. [CrossRef]

10. Giordano, C.; La Camera, G.; Gelsomino, L.; Barone, I.; Bonofiglio, D.; Andò, S.; Catalano, S. The Biology of Exosomes in Breast Cancer Progression: Dissemination, Immune Evasion and Metastatic Colonization. *Cancers* **2020**, *12*, 2179. [CrossRef]

11. Pan, B.T.; Teng, K.; Wu, C.; Adam, M.; Johnstone, R.M. Electron microscopic evidence for externalization of the transferrin receptor in vesicular form in sheep reticulocytes. *J. Cell Biol.* **1985**, *101*, 942–948. [CrossRef] [PubMed]

12. Yu, B.; Zhang, X.; Li, X. Exosomes derived from mesenchymal stem cells. *Int. J. Mol. Sci.* **2014**, *15*, 4142–4157. [CrossRef]

13. Stahl, P.D.; Barbieri, M.A. Multivesicular Bodies and Multivesicular Endosomes: The "Ins and Outs" of Endosomal Traffic. *Sci. Signal.* **2002**, *2002*, pe32. [CrossRef] [PubMed]

14. Javeed, N.; Mukhopadhyay, D. Exosomes and their role in the micro-/macro-environment: A comprehensive review. *J. Biomed. Res.* **2017**, *31*, 386–394. [CrossRef] [PubMed]

15. Lee, Y.; El Andaloussi, S.; Wood, M.J. Exosomes and microvesicles: Extracellular vesicles for genetic information transfer and gene therapy. *Hum. Mol. Genet.* **2012**, *21*, R125–R134. [CrossRef]

16. Aursulesei, V.; Vasincu, D.; Timofte, D.; Vrajitoriu, L.; Gatu, I.; Iacob, D.D.; Ghizdovat, V.; Buzea, C.; Agop, M. New mechanisms of vesicles migration. *Gen. Physiol. Biophys.* **2016**, *35*, 287–298. [CrossRef] [PubMed]

17. Johnstone, R.M.; Adam, M.; Hammond, J.R.; Orr, L.; Turbide, C. Vesicle formation during reticulocyte maturation. Association of plasma membrane activities with released vesicles (exosomes). *J. Biol. Chem.* **1987**, *262*, 9412–9420. [CrossRef]

18. Harding, C.; Heuser, J.; Stahl, P. Receptor-mediated endocytosis of transferrin and recycling of the transferrin receptor in rat reticulocytes. *J. Cell Biol.* **1983**, *97*, 329–339. [CrossRef]

19. Kruger, S.; Abd Elmageed, Z.Y.; Hawke, D.H.; Wörner, P.M.; Jansen, D.A.; Abdel-Mageed, A.B.; Alt, E.U.; Izadpanah, R. Molecular characterization of exosome-like vesicles from breast cancer cells. *BMC Cancer* **2014**, *14*, 44. [CrossRef]

20. Harris, D.A.; Patel, S.H.; Gucek, M.; Hendrix, A.; Westbroek, W.; Taraska, J.W. Exosomes released from breast cancer carcinomas stimulate cell movement. *PLoS ONE* **2015**, *10*, e0117495. [CrossRef]

21. Théry, C.; Ostrowski, M.; Segura, E. Membrane vesicles as conveyors of immune responses. *Nat. Rev. Immunol.* **2009**, *9*, 581–593. [CrossRef]

22. Raposo, G.; Stoorvogel, W. Extracellular vesicles: Exosomes, microvesicles, and friends. *J. Cell Biol.* **2013**, *200*, 373–383. [CrossRef] [PubMed]

23. Théry, C.; Boussac, M.; Véron, P.; Ricciardi-Castagnoli, P.; Raposo, G.; Garin, J.; Amigorena, S. Proteomic analysis of dendritic cell-derived exosomes: A secreted subcellular compartment distinct from apoptotic vesicles. *J. Immunol.* **2001**, *166*, 7309–7318. [CrossRef]

24. Colombo, M.; Raposo, G.; Théry, C. Biogenesis, secretion, and intercellular interactions of exosomes and other extracellular vesicles. *Annu. Rev. Cell Dev. Biol.* **2014**, *30*, 255–289. [CrossRef] [PubMed]

25. Jankovičová, J.; Sečová, P.; Michalková, K.; Antalíková, J. Tetraspanins, More than Markers of Extracellular Vesicles in Reproduction. *Int. J. Mol. Sci.* **2020**, *21*, 7568. [CrossRef] [PubMed]

26. Zhang, M.; Viennois, E.; Prasad, M.; Zhang, Y.; Wang, L.; Zhang, Z.; Han, M.K.; Xiao, B.; Xu, C.; Srinivasan, S.; et al. Edible ginger-derived nanoparticles: A novel therapeutic approach for the prevention and treatment of inflammatory bowel disease and colitis-associated cancer. *Biomaterials* **2016**, *101*, 321–340. [CrossRef]

27. Lee, J.U.; Kim, S.; Sim, S.J. SERS-based Nanoplasmonic Exosome Analysis: Enabling Liquid Biopsy for Cancer Diagnosis and Monitoring Progression. *BioChip J.* **2020**, *14*, 231–241. [CrossRef]

28. Gercel-Taylor, C.; Atay, S.; Tullis, R.H.; Kesimer, M.; Taylor, D.D. Nanoparticle analysis of circulating cell-derived vesicles in ovarian cancer patients. *Anal. Biochem.* **2012**, *428*, 44–53. [CrossRef]

29. Mao, L.; Li, X.; Gong, S.; Yuan, H.; Jiang, Y.; Huang, W.; Sun, X.; Dang, X. Serum exosomes contain ECRG4 mRNA that suppresses tumor growth via inhibition of genes involved in inflammation, cell proliferation, and angiogenesis. *Cancer Gene Ther.* **2018**, *25*, 248–259. [CrossRef]

30. Zhang, X.; Yuan, X.; Shi, H.; Wu, L.; Qian, H.; Xu, W. Exosomes in cancer: Small particle, big player. *J. Hematol. Oncol.* **2015**, *8*, 83. [CrossRef]

31. Kosaka, N. Decoding the Secret of Cancer by Means of Extracellular Vesicles. *J. Clin. Med.* **2016**, *5*, 22. [CrossRef]

32. Roma-Rodrigues, C.; Fernandes, A.R.; Baptista, P.V. Exosome in tumour microenvironment: Overview of the crosstalk between normal and cancer cells. *BioMed Res. Int.* **2014**, *2014*, 179486. [CrossRef] [PubMed]

33. Minciacchi, V.R.; Freeman, M.R.; Di Vizio, D. Extracellular vesicles in cancer: Exosomes, microvesicles and the emerging role of large oncosomes. *Semin. Cell Dev. Biol.* **2015**, *40*, 41–51. [CrossRef] [PubMed]
34. Sinha, D.; Roy, S.; Saha, P.; Chatterjee, N.; Bishayee, A. Trends in Research on Exosomes in Cancer Progression and Anticancer Therapy. *Cancers* **2021**, *13*, 326. [CrossRef] [PubMed]
35. Liu, Y.; Shi, K.; Chen, Y.; Wu, X.; Chen, Z.; Cao, K.; Tao, Y.; Chen, X.; Liao, J.; Zhou, J. Exosomes and Their Role in Cancer Progression. *Front. Oncol.* **2021**. [CrossRef]
36. Endres, M.; Kneitz, S.; Orth, M.F.; Perera, R.K.; Zernecke, A.; Butt, E. Regulation of matrix metalloproteinases (MMPs) expression and secretion in MDA-MB-231 breast cancer cells by LIM and SH3 protein 1 (LASP1). *Oncotarget* **2016**, *7*, 64244–64259. [CrossRef]
37. Sánchez, C.A.; Andahur, E.I.; Valenzuela, R.; Castellón, E.A.; Fullá, J.A.; Ramos, C.G.; Triviño, J.C. Exosomes from bulk and stem cells from human prostate cancer have a differential microRNA content that contributes cooperatively over local and pre-metastatic niche. *Oncotarget* **2016**, *7*, 3993–4008. [CrossRef]
38. Sceneay, J.; Smyth, M.J.; Möller, A. The pre-metastatic niche: Finding common ground. *Cancer Metastasis Rev.* **2013**, *32*, 449–464. [CrossRef]
39. Yang, C.; Ruffner, M.A.; Kim, S.H.; Robbins, P.D. Plasma-derived MHC class II+ exosomes from tumor-bearing mice suppress tumor antigen-specific immune responses. *Eur. J. Immunol.* **2012**, *42*, 1778–1784. [CrossRef]
40. Chornoguz, O.; Grmai, L.; Sinha, P.; Artemenko, K.A.; Zubarev, R.A.; Ostrand-Rosenberg, S. Proteomic pathway analysis reveals inflammation increases myeloid-derived suppressor cell resistance to apoptosis. *Mol. Cell. Proteom.* **2011**, *10*. [CrossRef]
41. Taylor, D.D.; Gercel-Taylor, C. Exosomes/microvesicles: Mediators of cancer-associated immunosuppressive microenvironments. *Semin. Immunopathol.* **2011**, *33*, 441–454. [CrossRef]
42. Condamine, T.; Gabrilovich, D.I. Molecular mechanisms regulating myeloid-derived suppressor cell differentiation and function. *Trends Immunol.* **2011**, *32*, 19–25. [CrossRef]
43. Wang, T.; Diaz, A.J.; Yen, Y. The role of peroxiredoxin II in chemoresistance of breast cancer cells. *Breast Cancer* **2014**, *6*, 73–80. [CrossRef]
44. Chen, W.X.; Liu, X.M.; Lv, M.M.; Chen, L.; Zhao, J.H.; Zhong, S.L.; Ji, M.H.; Hu, Q.; Luo, Z.; Wu, J.Z.; et al. Exosomes from drug-resistant breast cancer cells transmit chemoresistance by a horizontal transfer of microRNAs. *PLoS ONE* **2014**, *9*, e95240. [CrossRef]
45. Melo, S.A.; Sugimoto, H.; O'Connell, J.T.; Kato, N.; Villanueva, A.; Vidal, A.; Qiu, L.; Vitkin, E.; Perelman, L.T.; Melo, C.A.; et al. Cancer exosomes perform cell-independent microRNA biogenesis and promote tumorigenesis. *Cancer Cell* **2014**, *26*, 707–721. [CrossRef] [PubMed]
46. Qu, J.L.; Qu, X.J.; Zhao, M.F.; Teng, Y.E.; Zhang, Y.; Hou, K.Z.; Jiang, Y.H.; Yang, X.H.; Liu, Y.P. Gastric cancer exosomes promote tumour cell proliferation through PI3K/Akt and MAPK/ERK activation. *Dig. Liver Dis.* **2009**, *41*, 875–880. [CrossRef] [PubMed]
47. Ristorcelli, E.; Beraud, E.; Mathieu, S.; Lombardo, D.; Verine, A. Essential role of Notch signaling in apoptosis of human pancreatic tumoral cells mediated by exosomal nanoparticles. *Int. J. Cancer* **2009**, *125*, 1016–1026. [CrossRef]
48. Chen, D.S.; Mellman, I. Oncology meets immunology: The cancer-immunity cycle. *Immunity* **2013**, *39*, 1–10. [CrossRef] [PubMed]
49. Dubensky, T.W., Jr.; Skoble, J.; Lauer, P.; Brockstedt, D.G. Killed but metabolically active vaccines. *Curr. Opin. Biotechnol.* **2012**, *23*, 917–923. [CrossRef] [PubMed]
50. Admyre, C.; Johansson, S.M.; Paulie, S.; Gabrielsson, S. Direct exosome stimulation of peripheral human T cells detected by ELISPOT. *Eur. J. Immunol.* **2006**, *36*, 1772–1781. [CrossRef]
51. Utsugi-Kobukai, S.; Fujimaki, H.; Hotta, C.; Nakazawa, M.; Minami, M. MHC class I-mediated exogenous antigen presentation by exosomes secreted from immature and mature bone marrow derived dendritic cells. *Immunol. Lett.* **2003**, *89*, 125–131. [CrossRef]
52. Zitvogel, L.; Regnault, A.; Lozier, A.; Wolfers, J.; Flament, C.; Tenza, D.; Ricciardi-Castagnoli, P.; Raposo, G.; Amigorena, S. Eradication of established murine tumors using a novel cell-free vaccine: Dendritic cell-derived exosomes. *Nat. Med.* **1998**, *4*, 594–600. [CrossRef] [PubMed]
53. Andre, F.; Escudier, B.; Angevin, E.; Tursz, T.; Zitvogel, L. Exosomes for cancer immunotherapy. *Ann. Oncol.* **2004**, *15*, iv141–iv144. [CrossRef]
54. Wolfers, J.; Lozier, A.; Raposo, G.; Regnault, A.; Théry, C.; Masurier, C.; Flament, C.; Pouzieux, S.; Faure, F.; Tursz, T.; et al. Tumor-derived exosomes are a source of shared tumor rejection antigens for CTL cross-priming. *Nat. Med.* **2001**, *7*, 297–303. [CrossRef]
55. Elsner, L.; Muppala, V.; Gehrmann, M.; Lozano, J.; Malzahn, D.; Bickeböller, H.; Brunner, E.; Zientkowska, M.; Herrmann, T.; Walter, L.; et al. The heat shock protein HSP70 promotes mouse NK cell activity against tumors that express inducible NKG2D ligands. *J. Immunol.* **2007**, *179*, 5523–5533. [CrossRef]
56. Pitt, J.M.; Charrier, M.; Viaud, S.; André, F.; Besse, B.; Chaput, N.; Zitvogel, L. Dendritic cell-derived exosomes as immunotherapies in the fight against cancer. *J. Immunol.* **2014**, *193*, 1006–1011. [CrossRef] [PubMed]
57. Olejarz, W.; Dominiak, A.; Żołnierzak, A.; Kubiak-Tomaszewska, G.; Lorenc, T. Tumor-Derived Exosomes in Immunosuppression and Immunotherapy. *J. Immunol. Res.* **2020**, *2020*, 6272498. [CrossRef]
58. Poggio, M.; Hu, T.; Pai, C.-C.; Chu, B.; Belair, C.D.; Chang, A.; Montabana, E.; Lang, U.E.; Fu, Q.; Fong, L.; et al. Suppression of Exosomal PD-L1 Induces Systemic Anti-tumor Immunity and Memory. *Cell* **2019**, *177*, 414–427.e413. [CrossRef]
59. Lee, E.-Y.; Park, K.-S.; Yoon, Y.J.; Lee, J.; Moon, H.-G.; Jang, S.C.; Choi, K.-H.; Kim, Y.-K.; Gho, Y.S. Therapeutic Effects of Autologous Tumor-Derived Nanovesicles on Melanoma Growth and Metastasis. *PLoS ONE* **2012**, *7*, e33330. [CrossRef]

60. Dai, S.; Wei, D.; Wu, Z.; Zhou, X.; Wei, X.; Huang, H.; Li, G. Phase I clinical trial of autologous ascites-derived exosomes combined with GM-CSF for colorectal cancer. *Mol. Ther.* **2008**, *16*, 782–790. [CrossRef] [PubMed]
61. Peters, P.J.; Geuze, H.J.; Van der Donk, H.A.; Slot, J.W.; Griffith, J.M.; Stam, N.J.; Clevers, H.C.; Borst, J. Molecules relevant for T cell-target cell interaction are present in cytolytic granules of human T lymphocytes. *Eur. J. Immunol.* **1989**, *19*, 1469–1475. [CrossRef] [PubMed]
62. Gao, D.; Jiang, L. Exosomes in cancer therapy: A novel experimental strategy. *Am. J. Cancer Res.* **2018**, *8*, 2165–2175. [PubMed]
63. Fu, W.; Lei, C.; Liu, S.; Cui, Y.; Wang, C.; Qian, K.; Li, T.; Shen, Y.; Fan, X.; Lin, F.; et al. CAR exosomes derived from effector CAR-T cells have potent antitumour effects and low toxicity. *Nat. Commun.* **2019**, *10*, 4355. [CrossRef] [PubMed]
64. Blanchard, N.; Lankar, D.; Faure, F.; Regnault, A.; Dumont, C.; Raposo, G.; Hivroz, C. TCR activation of human T cells induces the production of exosomes bearing the TCR/CD3/zeta complex. *J. Immunol.* **2002**, *168*, 3235–3241. [CrossRef] [PubMed]
65. Dotti, G.; Gottschalk, S.; Savoldo, B.; Brenner, M.K. Design and development of therapies using chimeric antigen receptor-expressing T cells. *Immunol. Rev.* **2014**, *257*, 107–126. [CrossRef]
66. Srivastava, S.; Riddell, S.R. Chimeric Antigen Receptor T Cell Therapy: Challenges to Bench-to-Bedside Efficacy. *J. Immunol.* **2018**, *200*, 459–468. [CrossRef]
67. Tang, X.J.; Sun, X.Y.; Huang, K.M.; Zhang, L.; Yang, Z.S.; Zou, D.D.; Wang, B.; Warnock, G.L.; Dai, L.J.; Luo, J. Therapeutic potential of CAR-T cell-derived exosomes: A cell-free modality for targeted cancer therapy. *Oncotarget* **2015**, *6*, 44179–44190. [CrossRef] [PubMed]
68. Yang, P.; Cao, X.; Cai, H.; Feng, P.; Chen, X.; Zhu, Y.; Yang, Y.; An, W.; Yang, Y.; Jie, J. The exosomes derived from CAR-T cell efficiently target mesothelin and reduce triple-negative breast cancer growth. *Cell. Immunol.* **2021**, *360*, 104262. [CrossRef]
69. Zhou, J.; Tan, X.; Tan, Y.; Li, Q.; Ma, J.; Wang, G. Mesenchymal Stem Cell Derived Exosomes in Cancer Progression, Metastasis and Drug Delivery: A Comprehensive Review. *J. Cancer* **2018**, *9*, 3129–3137. [CrossRef]
70. Kordelas, L.; Rebmann, V.; Ludwig, A.K.; Radtke, S.; Ruesing, J.; Doeppner, T.R.; Epple, M.; Horn, P.A.; Beelen, D.W.; Giebel, B. MSC-derived exosomes: A novel tool to treat therapy-refractory graft-versus-host disease. *Leukemia* **2014**, *28*, 970–973. [CrossRef]
71. An, Q.; van Bel, A.J.; Hückelhoven, R. Do plant cells secrete exosomes derived from multivesicular bodies? *Plant Signal. Behav.* **2007**, *2*, 4–7. [CrossRef]
72. Tanchak, M.A.; Fowke, L.C. The morphology of multivesicular bodies in soybean protoplasts and their role in endocytosis. *Protoplasma* **1987**, *138*, 173–182. [CrossRef]
73. Akuma, P.; Okagu, O.D.; Udenigwe, C.C. Naturally Occurring Exosome Vesicles as Potential Delivery Vehicle for Bioactive Compounds. *Front. Sustain. Food Syst.* **2019**, *3*, 23. [CrossRef]
74. Mu, J.; Zhuang, X.; Wang, Q.; Jiang, H.; Deng, Z.-B.; Wang, B.; Zhang, L.; Kakar, S.; Jun, Y.; Miller, D.; et al. Interspecies communication between plant and mouse gut host cells through edible plant derived exosome-like nanoparticles. *Mol. Nutr. Food Res.* **2014**, *58*, 1561–1573. [CrossRef]
75. György, B.; Hung, M.E.; Breakefield, X.O.; Leonard, J.N. Therapeutic applications of extracellular vesicles: Clinical promise and open questions. *Annu. Rev. Pharmacol. Toxicol.* **2015**, *55*, 439–464. [CrossRef]
76. Peinado, H.; Alečković, M.; Lavotshkin, S.; Matei, I.; Costa-Silva, B.; Moreno-Bueno, G.; Hergueta-Redondo, M.; Williams, C.; García-Santos, G.; Ghajar, C.; et al. Melanoma exosomes educate bone marrow progenitor cells toward a pro-metastatic phenotype through MET. *Nat. Med.* **2012**, *18*, 883–891. [CrossRef] [PubMed]
77. Joyce, D.P.; Kerin, M.J.; Dwyer, R.M. Exosome-encapsulated microRNAs as circulating biomarkers for breast cancer. *Int. J. Cancer* **2016**, *139*, 1443–1448. [CrossRef]
78. Nabet, B.Y.; Qiu, Y.; Shabason, J.E.; Wu, T.J.; Yoon, T.; Kim, B.C.; Benci, J.L.; DeMichele, A.M.; Tchou, J.; Marcotrigiano, J.; et al. Exosome RNA Unshielding Couples Stromal Activation to Pattern Recognition Receptor Signaling in Cancer. *Cell* **2017**, *170*, 352–366.e313. [CrossRef] [PubMed]
79. Stevic, I.; Müller, V.; Weber, K.; Fasching, P.A.; Karn, T.; Marmé, F.; Schem, C.; Stickeler, E.; Denkert, C.; van Mackelenbergh, M.; et al. Specific microRNA signatures in exosomes of triple-negative and HER2-positive breast cancer patients undergoing neoadjuvant therapy within the GeparSixto trial. *BMC Med.* **2018**, *16*, 179. [CrossRef] [PubMed]
80. Hannafon, B.N.; Trigoso, Y.D.; Calloway, C.L.; Zhao, Y.D.; Lum, D.H.; Welm, A.L.; Zhao, Z.J.; Blick, K.E.; Dooley, W.C.; Ding, W.Q. Plasma exosome microRNAs are indicative of breast cancer. *Breast Cancer Res.* **2016**, *18*, 90. [CrossRef] [PubMed]
81. Kinoshita, T.; Yip, K.W.; Spence, T.; Liu, F.F. MicroRNAs in extracellular vesicles: Potential cancer biomarkers. *J. Hum. Genet.* **2017**, *62*, 67–74. [CrossRef]
82. Tetta, C.; Ghigo, E.; Silengo, L.; Deregibus, M.C.; Camussi, G. Extracellular vesicles as an emerging mechanism of cell-to-cell communication. *Endocrine* **2013**, *44*, 11–19. [CrossRef] [PubMed]
83. Yan, S.; Han, B.; Gao, S.; Wang, X.; Wang, Z.; Wang, F.; Zhang, J.; Xu, D.; Sun, B. Exosome-encapsulated microRNAs as circulating biomarkers for colorectal cancer. *Oncotarget* **2017**, *8*, 60149–60158. [CrossRef] [PubMed]
84. Weidle, U.H.; Dickopf, S.; Hintermair, C.; Kollmorgen, G.; Birzele, F.; Brinkmann, U. The Role of micro RNAs in Breast Cancer Metastasis: Preclinical Validation and Potential Therapeutic Targets. *Cancer Genom. Proteom.* **2018**, *15*, 17–39. [CrossRef]
85. He, Y.; Deng, F.; Yang, S.; Wang, D.; Chen, X.; Zhong, S.; Zhao, J.; Tang, J. Exosomal microRNA: A novel biomarker for breast cancer. *Biomark. Med.* **2018**, *12*, 177–188. [CrossRef] [PubMed]
86. Sempere, L.F.; Keto, J.; Fabbri, M. Exosomal MicroRNAs in Breast Cancer towards Diagnostic and Therapeutic Applications. *Cancers* **2017**, *9*, 71. [CrossRef] [PubMed]

87. Jayaseelan, V.P. Emerging role of exosomes as promising diagnostic tool for cancer. *Cancer Gene Ther.* **2020**, *27*, 395–398. [CrossRef]
88. Jang, J.Y.; Lee, J.K.; Jeon, Y.K.; Kim, C.W. Exosome derived from epigallocatechin gallate treated breast cancer cells suppresses tumor growth by inhibiting tumor-associated macrophage infiltration and M2 polarization. *BMC Cancer* **2013**, *13*, 421. [CrossRef] [PubMed]
89. Yang, M.; Chen, J.; Su, F.; Yu, B.; Su, F.; Lin, L.; Liu, Y.; Huang, J.D.; Song, E. Microvesicles secreted by macrophages shuttle invasion-potentiating microRNAs into breast cancer cells. *Mol. Cancer* **2011**, *10*, 117. [CrossRef]
90. Kalra, H.; Adda, C.G.; Liem, M.; Ang, C.S.; Mechler, A.; Simpson, R.J.; Hulett, M.D.; Mathivanan, S. Comparative proteomics evaluation of plasma exosome isolation techniques and assessment of the stability of exosomes in normal human blood plasma. *Proteomics* **2013**, *13*, 3354–3364. [CrossRef]
91. Qin, Z.; Ljubimov, V.A.; Zhou, C.; Tong, Y.; Liang, J. Cell-free circulating tumor DNA in cancer. *Chin. J. Cancer* **2016**, *35*, 36. [CrossRef] [PubMed]
92. Ma, L.; Teruya-Feldstein, J.; Weinberg, R.A. Tumour invasion and metastasis initiated by microRNA-10b in breast cancer. *Nature* **2007**, *449*, 682–688. [CrossRef] [PubMed]
93. Yan, L.X.; Wu, Q.N.; Zhang, Y.; Li, Y.Y.; Liao, D.Z.; Hou, J.H.; Fu, J.; Zeng, M.S.; Yun, J.P.; Wu, Q.L.; et al. Knockdown of miR-21 in human breast cancer cell lines inhibits proliferation, in vitro migration and in vivo tumor growth. *Breast Cancer Res.* **2011**, *13*, R2. [CrossRef] [PubMed]
94. Wei, Y.; Li, M.; Cui, S.; Wang, D.; Zhang, C.-Y.; Zen, K.; Li, L. Shikonin Inhibits the Proliferation of Human Breast Cancer Cells by Reducing Tumor-Derived Exosomes. *Molecules* **2016**, *21*, 777. [CrossRef] [PubMed]
95. Le, M.T.N.; Hamar, P.; Guo, C.; Basar, E.; Perdigão-Henriques, R.; Balaj, L.; Lieberman, J. miR-200-containing extracellular vesicles promote breast cancer cell metastasis. *J. Clin. Investig.* **2014**, *124*, 5109–5128. [CrossRef]
96. Zhong, S.; Chen, X.; Wang, D.; Zhang, X.; Shen, H.; Yang, S.; Lv, M.; Tang, J.; Zhao, J. MicroRNA expression profiles of drug-resistance breast cancer cells and their exosomes. *Oncotarget* **2016**, *7*, 19601–19609. [CrossRef]
97. Wei, Y.; Lai, X.; Yu, S.; Chen, S.; Ma, Y.; Zhang, Y.; Li, H.; Zhu, X.; Yao, L.; Zhang, J. Exosomal miR-221/222 enhances tamoxifen resistance in recipient ER-positive breast cancer cells. *Breast Cancer Res. Treat.* **2014**, *147*, 423–431. [CrossRef]
98. Zhou, W.; Fong, M.Y.; Min, Y.; Somlo, G.; Liu, L.; Palomares, M.R.; Yu, Y.; Chow, A.; O'Connor, S.T.F.; Chin, A.R.; et al. Cancer-secreted miR-105 destroys vascular endothelial barriers to promote metastasis. *Cancer Cell* **2014**, *25*, 501–515. [CrossRef] [PubMed]
99. Di Modica, M.; Regondi, V.; Sandri, M.; Iorio, M.V.; Zanetti, A.; Tagliabue, E.; Casalini, P.; Triulzi, T. Breast cancer-secreted miR-939 downregulates VE-cadherin and destroys the barrier function of endothelial monolayers. *Cancer Lett.* **2017**, *384*, 94–100. [CrossRef]
100. Fong, M.Y.; Zhou, W.; Liu, L.; Alontaga, A.Y.; Chandra, M.; Ashby, J.; Chow, A.; O'Connor, S.T.F.; Li, S.; Chin, A.R.; et al. Breast-cancer-secreted miR-122 reprograms glucose metabolism in premetastatic niche to promote metastasis. *Nat. Cell Biol.* **2015**, *17*, 183–194. [CrossRef]
101. Hannafon, B.N.; Carpenter, K.J.; Berry, W.L.; Janknecht, R.; Dooley, W.C.; Ding, W.-Q. Exosome-mediated microRNA signaling from breast cancer cells is altered by the anti-angiogenesis agent docosahexaenoic acid (DHA). *Mol. Cancer* **2015**, *14*, 133. [CrossRef] [PubMed]
102. Donnarumma, E.; Fiore, D.; Nappa, M.; Roscigno, G.; Adamo, A.; Iaboni, M.; Russo, V.; Affinito, A.; Puoti, I.; Quintavalle, C.; et al. Cancer-associated fibroblasts release exosomal microRNAs that dictate an aggressive phenotype in breast cancer. *Oncotarget* **2017**, *8*, 19592–19608. [CrossRef] [PubMed]
103. Lim, P.K.; Bliss, S.A.; Patel, S.A.; Taborga, M.; Dave, M.A.; Gregory, L.A.; Greco, S.J.; Bryan, M.; Patel, P.S.; Rameshwar, P. Gap Junction–Mediated Import of MicroRNA from Bone Marrow Stromal Cells Can Elicit Cell Cycle Quiescence in Breast Cancer Cells. *Cancer Res.* **2011**, *71*, 1550. [CrossRef]
104. Bovy, N.; Blomme, B.; Frères, P.; Dederen, S.; Nivelles, O.; Lion, M.; Carnet, O.; Martial, J.A.; Noël, A.; Thiry, M.; et al. Endothelial exosomes contribute to the antitumor response during breast cancer neoadjuvant chemotherapy via microRNA transfer. *Oncotarget* **2015**, *6*, 10253–10266. [CrossRef]
105. O'Brien, K.; Lowry, M.C.; Corcoran, C.; Martinez, V.G.; Daly, M.; Rani, S.; Gallagher, W.M.; Radomski, M.W.; MacLeod, R.A.F.; O'Driscoll, L. miR-134 in extracellular vesicles reduces triple-negative breast cancer aggression and increases drug sensitivity. *Oncotarget* **2015**, *6*, 32774–32789. [CrossRef]
106. Mihelich, B.L.; Dambal, S.; Lin, S.; Nonn, L. miR-182, of the miR-183 cluster family, is packaged in exosomes and is detected in human exosomes from serum, breast cells and prostate cells. *Oncol. Lett.* **2016**, *12*, 1197–1203. [CrossRef]
107. Eichelser, C.; Stückrath, I.; Müller, V.; Milde-Langosch, K.; Wikman, H.; Pantel, K.; Schwarzenbach, H. Increased serum levels of circulating exosomal microRNA-373 in receptor-negative breast cancer patients. *Oncotarget* **2014**, *5*, 9650–9663. [CrossRef] [PubMed]
108. Li, X.J.; Ren, Z.J.; Tang, J.H.; Yu, Q. Exosomal MicroRNA MiR-1246 Promotes Cell Proliferation, Invasion and Drug Resistance by Targeting CCNG2 in Breast Cancer. *Cell. Physiol. Biochem.* **2017**, *44*, 1741–1748. [CrossRef] [PubMed]
109. Singh, R.; Pochampally, R.; Watabe, K.; Lu, Z.; Mo, Y.-Y. Exosome-mediated transfer of miR-10b promotes cell invasion in breast cancer. *Mol. Cancer* **2014**, *13*, 256. [CrossRef]
110. Zhang, L.; Zhang, S.; Yao, J.; Lowery, F.J.; Zhang, Q.; Huang, W.-C.; Li, P.; Li, M.; Wang, X.; Zhang, C.; et al. Microenvironment-induced PTEN loss by exosomal microRNA primes brain metastasis outgrowth. *Nature* **2015**, *527*, 100–104. [CrossRef] [PubMed]

111. Sueta, A.; Yamamoto, Y.; Tomiguchi, M.; Takeshita, T.; Yamamoto-Ibusuki, M.; Iwase, H. Differential expression of exosomal miRNAs between breast cancer patients with and without recurrence. *Oncotarget* **2017**, *8*, 69934–69944. [CrossRef]
112. Yu, D.-D.; Wu, Y.; Zhang, X.-H.; Lv, M.-M.; Chen, W.-X.; Chen, X.; Yang, S.-J.; Shen, H.; Zhong, S.-L.; Tang, J.-H.; et al. Exosomes from adriamycin-resistant breast cancer cells transmit drug resistance partly by delivering miR-222. *Tumor Biol.* **2016**, *37*, 3227–3235. [CrossRef] [PubMed]
113. Chen, W.-X.; Zhong, S.-L.; Ji, M.-H.; Pan, M.; Hu, Q.; Lv, M.-M.; Luo, Z.; Zhao, J.-H.; Tang, J.-H. MicroRNAs delivered by extracellular vesicles: An emerging resistance mechanism for breast cancer. *Tumor Biol.* **2014**, *35*, 2883–2892. [CrossRef]
114. Galindo-Hernandez, O.; Villegas-Comonfort, S.; Candanedo, F.; González-Vázquez, M.C.; Chavez-Ocaña, S.; Jimenez-Villanueva, X.; Sierra-Martinez, M.; Salazar, E.P. Elevated Concentration of Microvesicles Isolated from Peripheral Blood in Breast Cancer Patients. *Arch. Med. Res.* **2013**, *44*, 208–214. [CrossRef] [PubMed]
115. Ma, X.; Chen, Z.; Hua, D.; He, D.; Wang, L.; Zhang, P.; Wang, J.; Cai, Y.; Gao, C.; Zhang, X.; et al. Essential role for TrpC5-containing extracellular vesicles in breast cancer with chemotherapeutic resistance. *Proc. Natl. Acad. Sci. USA* **2014**, *111*, 6389–6394. [CrossRef] [PubMed]
116. Risha, Y.; Minic, Z.; Ghobadloo, S.M.; Berezovski, M.V. The proteomic analysis of breast cell line exosomes reveals disease patterns and potential biomarkers. *Sci. Rep.* **2020**, *10*, 13572. [CrossRef]
117. Yang, S.J.; Wang, D.D.; Li, J.; Xu, H.Z.; Shen, H.Y.; Chen, X.; Zhou, S.Y.; Zhong, S.L.; Zhao, J.H.; Tang, J.H. Predictive role of GSTP1-containing exosomes in chemotherapy-resistant breast cancer. *Gene* **2017**, *623*, 5–14. [CrossRef] [PubMed]
118. Ciravolo, V.; Huber, V.; Ghedini, G.C.; Venturelli, E.; Bianchi, F.; Campiglio, M.; Morelli, D.; Villa, A.; Mina, P.D.; Menard, S.; et al. Potential role of HER2-overexpressing exosomes in countering trastuzumab-based therapy. *J. Cell. Physiol.* **2012**, *227*, 658–667. [CrossRef]
119. Khan, S.; Bennit, H.F.; Turay, D.; Perez, M.; Mirshahidi, S.; Yuan, Y.; Wall, N.R. Early diagnostic value of survivin and its alternative splice variants in breast cancer. *BMC Cancer* **2014**, *14*, 176. [CrossRef]
120. Jaiswal, R.; Luk, F.; Dalla, P.V.; Grau, G.E.R.; Bebawy, M. Breast Cancer-Derived Microparticles Display Tissue Selectivity in the Transfer of Resistance Proteins to Cells. *PLoS ONE* **2013**, *8*, e61515. [CrossRef]
121. Ning, K.; Wang, T.; Sun, X.; Zhang, P.; Chen, Y.; Jin, J.; Hua, D. UCH-L1-containing exosomes mediate chemotherapeutic resistance transfer in breast cancer. *J. Surg. Oncol.* **2017**, *115*, 932–940. [CrossRef] [PubMed]
122. Rupp, A.-K.; Rupp, C.; Keller, S.; Brase, J.C.; Ehehalt, R.; Fogel, M.; Moldenhauer, G.; Marmé, F.; Sültmann, H.; Altevogt, P. Loss of EpCAM expression in breast cancer derived serum exosomes: Role of proteolytic cleavage. *Gynecol. Oncol.* **2011**, *122*, 437–446. [CrossRef] [PubMed]
123. Moon, P.G.; Lee, J.E.; Cho, Y.E.; Lee, S.J.; Jung, J.H.; Chae, Y.S.; Bae, H.I.; Kim, Y.B.; Kim, I.S.; Park, H.Y.; et al. Identification of Developmental Endothelial Locus-1 on Circulating Extracellular Vesicles as a Novel Biomarker for Early Breast Cancer Detection. *Clin. Cancer Res.* **2016**, *22*, 1757–1766. [CrossRef] [PubMed]
124. Moon, P.-G.; Lee, J.-E.; Cho, Y.-E.; Lee, S.J.; Chae, Y.S.; Jung, J.H.; Kim, I.-S.; Park, H.Y.; Baek, M.-C. Fibronectin on circulating extracellular vesicles as a liquid biopsy to detect breast cancer. *Oncotarget* **2016**, *7*, 40189–40199. [CrossRef] [PubMed]
125. Tutanov, O.; Proskura, K.; Kamyshinsky, R.; Shtam, T.; Tsentalovich, Y.; Tamkovich, S. Proteomic Profiling of Plasma and Total Blood Exosomes in Breast Cancer: A Potential Role in Tumor Progression, Diagnosis, and Prognosis. *Front. Oncol.* **2020**, *10*, 2173. [CrossRef]
126. Berchem, G.; Noman, M.Z.; Bosseler, M.; Paggetti, J.; Baconnais, S.; Le Cam, E.; Nanbakhsh, A.; Moussay, E.; Mami-Chouaib, F.; Janji, B.; et al. Hypoxic tumor-derived microvesicles negatively regulate NK cell function by a mechanism involving TGF-β and miR23a transfer. *Oncoimmunology* **2015**, *5*, e1062968. [CrossRef]
127. Besse, B.; Charrier, M.; Lapierre, V.; Dansin, E.; Lantz, O.; Planchard, D.; Le Chevalier, T.; Livartoski, A.; Barlesi, F.; Laplanche, A.; et al. Dendritic cell-derived exosomes as maintenance immunotherapy after first line chemotherapy in NSCLC. *Oncoimmunology* **2015**, *5*, e1071008. [CrossRef] [PubMed]
128. Corcoran, C.; Rani, S.; O'Brien, K.; O'Neill, A.; Prencipe, M.; Sheikh, R.; Webb, G.; McDermott, R.; Watson, W.; Crown, J.; et al. Docetaxel-resistance in prostate cancer: Evaluating associated phenotypic changes and potential for resistance transfer via exosomes. *PLoS ONE* **2012**, *7*, e50999. [CrossRef] [PubMed]
129. Takahashi, K.; Yan, I.K.; Kogure, T.; Haga, H.; Patel, T. Extracellular vesicle-mediated transfer of long non-coding RNA ROR modulates chemosensitivity in human hepatocellular cancer. *FEBS Open Bio* **2014**, *4*, 458–467. [CrossRef]
130. Xiao, X.; Yu, S.; Li, S.; Wu, J.; Ma, R.; Cao, H.; Zhu, Y.; Feng, J. Exosomes: Decreased sensitivity of lung cancer A549 cells to cisplatin. *PLoS ONE* **2014**, *9*, e89534. [CrossRef]
131. Kong, J.N.; He, Q.; Wang, G.; Dasgupta, S.; Dinkins, M.B.; Zhu, G.; Kim, A.; Spassieva, S.; Bieberich, E. Guggulsterone and bexarotene induce secretion of exosome-associated breast cancer resistance protein and reduce doxorubicin resistance in MDA-MB-231 cells. *Int. J. Cancer* **2015**, *137*, 1610–1620. [CrossRef]
132. Chen, C.; Luo, F.; Liu, X.; Lu, L.; Xu, H.; Yang, Q.; Xue, J.; Shi, L.; Li, J.; Zhang, A.; et al. NF-kB-regulated exosomal miR-155 promotes the inflammation associated with arsenite carcinogenesis. *Cancer Lett.* **2017**, *388*, 21–33. [CrossRef]
133. Nawaz, M.; Shah, N. Extracellular Vesicles and Matrix Remodeling Enzymes: The Emerging Roles in Extracellular Matrix Remodeling, Progression of Diseases and Tissue Repair. *Cells* **2018**, *7*, 167. [CrossRef] [PubMed]
134. Sung, B.H.; Ketova, T.; Hoshino, D.; Zijlstra, A. Directional cell movement through tissues is controlled by exosome secretion. *Nat. Commun.* **2015**, *6*, 7164. [CrossRef] [PubMed]

135. Hendrix, A.; Maynard, D.; Pauwels, P.; Braems, G.; Denys, H.; Van den Broecke, R.; Lambert, J.; Van Belle, S.; Cocquyt, V.; Gespach, C.; et al. Effect of the secretory small GTPase Rab27B on breast cancer growth, invasion, and metastasis. *J. Natl. Cancer Inst.* **2010**, *102*, 866–880. [CrossRef]

136. Galindo-Hernandez, O.; Serna-Marquez, N.; Castillo-Sanchez, R.; Salazar, E.P. Extracellular vesicles from MDA-MB-231 breast cancer cells stimulated with linoleic acid promote an EMT-like process in MCF10A cells. *Prostaglandins Leukot. Essent. Fat. Acids* **2014**, *91*, 299–310. [CrossRef]

137. Maji, S.; Chaudhary, P.; Akopova, I.; Nguyen, P.M.; Hare, R.J.; Gryczynski, I.; Vishwanatha, J.K. Exosomal Annexin II Promotes Angiogenesis and Breast Cancer Metastasis. *Mol. Cancer Res.* **2017**, *15*, 93–105. [CrossRef]

138. Kim, M.S.; Haney, M.J.; Zhao, Y.; Yuan, D.; Deygen, I.; Klyachko, N.L.; Kabanov, A.V.; Batrakova, E.V. Engineering macrophage-derived exosomes for targeted paclitaxel delivery to pulmonary metastases: In vitro and in vivo evaluations. *Nanomed. Nanotechnol. Biol. Med.* **2018**, *14*, 195–204. [CrossRef]

139. Mulcahy, L.A.; Pink, R.C.; Carter, D.R.F. Routes and mechanisms of extracellular vesicle uptake. *J. Extracell. Vesicles* **2014**, *3*. [CrossRef]

140. Saari, H.; Lázaro-Ibáñez, E.; Viitala, T.; Vuorimaa-Laukkanen, E.; Siljander, P.; Yliperttula, M. Microvesicle- and exosome-mediated drug delivery enhances the cytotoxicity of Paclitaxel in autologous prostate cancer cells. *J. Control. Release* **2015**, *220*, 727–737. [CrossRef] [PubMed]

141. Osterman, C.J.D.; Lynch, J.C.; Leaf, P.; Gonda, A.; Ferguson Bennit, H.R.; Griffiths, D.; Wall, N.R. Curcumin Modulates Pancreatic Adenocarcinoma Cell-Derived Exosomal Function. *PLoS ONE* **2015**, *10*, e0132845. [CrossRef] [PubMed]

142. Kim, M.S.; Haney, M.J.; Zhao, Y.; Mahajan, V.; Deygen, I.; Klyachko, N.L.; Inskoe, E.; Piroyan, A.; Sokolsky, M.; Okolie, O.; et al. Development of exosome-encapsulated paclitaxel to overcome MDR in cancer cells. *Nanomed. Nanotechnol. Biol. Med.* **2016**, *12*, 655–664. [CrossRef]

143. Yong, T.; Zhang, X.; Bie, N.; Zhang, H.; Zhang, X.; Li, F.; Hakeem, A.; Hu, J.; Gan, L.; Santos, H.A.; et al. Tumor exosome-based nanoparticles are efficient drug carriers for chemotherapy. *Nat. Commun.* **2019**, *10*, 3838. [CrossRef]

144. Tian, Y.; Li, S.; Song, J.; Ji, T.; Zhu, M.; Anderson, G.J.; Wei, J.; Nie, G. A doxorubicin delivery platform using engineered natural membrane vesicle exosomes for targeted tumor therapy. *Biomaterials* **2014**, *35*, 2383–2390. [CrossRef] [PubMed]

145. Tang, K.; Zhang, Y.; Zhang, H.; Xu, P.; Liu, J.; Ma, J.; Lv, M.; Li, D.; Katirai, F.; Shen, G.-X.; et al. Delivery of chemotherapeutic drugs in tumour cell-derived microparticles. *Nat. Commun.* **2012**, *3*, 1282. [CrossRef]

146. Munoz, J.L.; Bliss, S.A.; Greco, S.J.; Ramkissoon, S.H.; Ligon, K.L.; Rameshwar, P. Delivery of Functional Anti-miR-9 by Mesenchymal Stem Cell-derived Exosomes to Glioblastoma Multiforme Cells Conferred Chemosensitivity. *Mol. Ther. Nucleic Acids* **2013**, *2*, e126. [CrossRef]

147. Shtam, T.A.; Kovalev, R.A.; Varfolomeeva, E.Y.; Makarov, E.M.; Kil, Y.V.; Filatov, M.V. Exosomes are natural carriers of exogenous siRNA to human cells in vitro. *Cell Commun. Signal.* **2013**, *11*, 88. [CrossRef]

148. Pan, Q.; Ramakrishnaiah, V.; Henry, S.; Fouraschen, S.; de Ruiter, P.E.; Kwekkeboom, J.; Tilanus, H.W.; Janssen, H.L.A.; van der Laan, L.J.W. Hepatic cell-to-cell transmission of small silencing RNA can extend the therapeutic reach of RNA interference (RNAi). *Gut* **2012**, *61*, 1330. [CrossRef]

149. Dong, X.; Bai, X.; Ni, J.; Zhang, H.; Duan, W.; Graham, P.; Li, Y. Exosomes and breast cancer drug resistance. *Cell Death Dis.* **2020**, *11*, 987. [CrossRef]

150. Jabbari, N.; Akbariazar, E.; Feqhhi, M.; Rahbarghazi, R.; Rezaie, J. Breast cancer-derived exosomes: Tumor progression and therapeutic agents. *J. Cell. Physiol.* **2020**, *235*, 6345–6356. [CrossRef] [PubMed]

151. Ma, X.; Cai, Y.; He, D.; Zou, C.; Zhang, P.; Lo, C.Y.; Xu, Z.; Chan, F.L.; Yu, S.; Chen, Y.; et al. Transient receptor potential channel TRPC5 is essential for P-glycoprotein induction in drug-resistant cancer cells. *Proc. Natl. Acad. Sci. USA* **2012**, *109*, 16282–16287. [CrossRef] [PubMed]

152. Dong, Y.; Pan, Q.; Jiang, L.; Chen, Z.; Zhang, F.; Liu, Y.; Xing, H.; Shi, M.; Li, J.; Li, X.; et al. Tumor endothelial expression of P-glycoprotein upon microvesicular transfer of TrpC5 derived from adriamycin-resistant breast cancer cells. *Biochem. Biophys. Res. Commun.* **2014**, *446*, 85–90. [CrossRef]

153. Zhang, Y.; Liu, Y.; Liu, H.; Tang, W.H. Exosomes: Biogenesis, biologic function and clinical potential. *Cell Biosci.* **2019**, *9*, 19. [CrossRef] [PubMed]

154. Yang, X.; Kovalenko, O.V.; Tang, W.; Claas, C.; Stipp, C.S.; Hemler, M.E. Palmitoylation supports assembly and function of integrin-tetraspanin complexes. *J. Cell Biol.* **2004**, *167*, 1231–1240. [CrossRef]

155. Chirco, R.; Liu, X.W.; Jung, K.K.; Kim, H.R. Novel functions of TIMPs in cell signaling. *Cancer Metastasis Rev.* **2006**, *25*, 99–113. [CrossRef]

156. Menck, K.; Klemm, F.; Gross, J.C.; Pukrop, T.; Wenzel, D.; Binder, C. Induction and transport of Wnt 5a during macrophage-induced malignant invasion is mediated by two types of extracellular vesicles. *Oncotarget* **2013**, *4*, 2057–2066. [CrossRef]

157. Graner, M.W.; Schnell, S.; Olin, M.R. Tumor-derived exosomes, microRNAs, and cancer immune suppression. *Semin. Immunopathol.* **2018**, *40*, 505–515. [CrossRef]

158. Nam, G.-H.; Choi, Y.; Kim, G.B.; Kim, S.; Kim, S.A.; Kim, I.-S. Emerging Prospects of Exosomes for Cancer Treatment: From Conventional Therapy to Immunotherapy. *Adv. Mater.* **2020**, *32*, 2002440. [CrossRef] [PubMed]

159. Bondhopadhyay, B.; Sisodiya, S.; Chikara, A.; Khan, A.; Tanwar, P.; Afroze, D.; Singh, N.; Agrawal, U.; Mehrotra, R.; Hussain, S. Cancer immunotherapy: A promising dawn in cancer research. *Am. J. Blood Res.* **2020**, *10*, 375–385. [PubMed]

160. Tsutsui, S.; Yasuda, K.; Suzuki, K.; Tahara, K.; Higashi, H.; Era, S. Macrophage infiltration and its prognostic implications in breast cancer: The relationship with VEGF expression and microvessel density. *Oncol. Rep.* **2005**, *14*, 425–431. [CrossRef]
161. Deng, Z.; Cheng, Z.; Xiang, X.; Yan, J.; Zhuang, X.; Liu, C.; Jiang, H.; Ju, S.; Zhang, L.; Grizzle, W.; et al. Tumor cell cross talk with tumor-associated leukocytes leads to induction of tumor exosomal fibronectin and promotes tumor progression. *Am. J. Pathol.* **2012**, *180*, 390–398. [CrossRef]
162. NCI Staff. *Exosomes May Help Tumors Evade Immune System*; National Cancer Institute: Bethesda, MD, USA, 2018.
163. Gao, W.; Liu, H.; Yuan, J.; Wu, C.; Huang, D.; Ma, Y.; Zhu, J.; Ma, L.; Guo, J.; Shi, H.; et al. Exosomes derived from mature dendritic cells increase endothelial inflammation and atherosclerosis via membrane TNF-α mediated NF-κB pathway. *J. Cell. Mol. Med.* **2016**, *20*, 2318–2327. [CrossRef] [PubMed]
164. Pizzirani, C.; Ferrari, D.; Chiozzi, P.; Adinolfi, E.; Sandonà, D.; Savaglio, E.; Di Virgilio, F. Stimulation of P2 receptors causes release of IL-1β–loaded microvesicles from human dendritic cells. *Blood* **2006**, *109*, 3856–3864. [CrossRef] [PubMed]
165. Qu, Y.; Franchi, L.; Nunez, G.; Dubyak, G.R. Nonclassical IL-1β Secretion Stimulated by P2X7 Receptors Is Dependent on Inflammasome Activation and Correlated with Exosome Release in Murine Macrophages. *J. Immunol.* **2007**, *179*, 1913. [CrossRef]
166. Logozzi, M.; De Milito, A.; Lugini, L.; Borghi, M.; Calabrò, L.; Spada, M.; Perdicchio, M.; Marino, M.L.; Federici, C.; Iessi, E.; et al. High levels of exosomes expressing CD63 and caveolin-1 in plasma of melanoma patients. *PLoS ONE* **2009**, *4*, e5219. [CrossRef] [PubMed]
167. Silva, J.; Garcia, V.; Rodriguez, M.; Compte, M.; Cisneros, E.; Veguillas, P.; Garcia, J.M.; Dominguez, G.; Campos-Martin, Y.; Cuevas, J.; et al. Analysis of exosome release and its prognostic value in human colorectal cancer. *Genes Chromosom. Cancer* **2012**, *51*, 409–418. [CrossRef]
168. Piombino, C.; Mastrolia, I.; Omarini, C.; Candini, O.; Dominici, M.; Piacentini, F.; Toss, A. The Role of Exosomes in Breast Cancer Diagnosis. *Biomedicines* **2021**, *9*, 312. [CrossRef]
169. Tutrone, R.; Donovan, M.J.; Torkler, P.; Tadigotla, V.; McLain, T.; Noerholm, M.; Skog, J.; McKiernan, J. Clinical utility of the exosome based ExoDx Prostate(IntelliScore) EPI test in men presenting for initial Biopsy with a PSA 2–10 ng/mL. *Prostate Cancer Prostatic Dis.* **2020**, *23*, 607–614. [CrossRef]
170. Golubovskaya, V.M. Focal adhesion kinase and cross-linked signaling in cancer. *Anti Cancer Agents Med. Chem.* **2014**, *14*, 2. [CrossRef]
171. Roberg-Larsen, H.; Lund, K.; Seterdal, K.E.; Solheim, S.; Vehus, T.; Solberg, N.; Krauss, S.; Lundanes, E.; Wilson, S.R. Mass spectrometric detection of 27-hydroxycholesterol in breast cancer exosomes. *J. Steroid Biochem. Mol. Biol.* **2017**, *169*, 22–28. [CrossRef]
172. Wu, C.Y.; Du, S.L.; Zhang, J.; Liang, A.L.; Liu, Y.J. Exosomes and breast cancer: A comprehensive review of novel therapeutic strategies from diagnosis to treatment. *Cancer Gene Ther.* **2017**, *24*, 6–12. [CrossRef]
173. Liu, Y.; Gu, Y.; Cao, X. The exosomes in tumor immunity. *Oncoimmunology* **2015**, *4*, e1027472. [CrossRef] [PubMed]

cancers

Article

Platelet Microparticles Protect Acute Myelogenous Leukemia Cells against Daunorubicin-Induced Apoptosis

Daniel Cacic [1,*], Håkon Reikvam [2,3], Oddmund Nordgård [1,4], Peter Meyer [1] and Tor Hervig [2,5]

1. Department of Hematology and Oncology, Stavanger University Hospital, 4068 Stavanger, Norway; oddmund.nordgard@sus.no (O.N.); peter.albert.meyer@sus.no (P.M.)
2. Department of Clinical Science, University of Bergen, 5021 Bergen, Norway; hakon.reikvam@uib.no (H.R.); tor.audun.hervig@helse-fonna.no (T.H.)
3. Department of Medicine, Haukeland University Hospital, 5021 Bergen, Norway
4. Department of Chemistry, Bioscience and Environmental Engineering, University of Stavanger, 4036 Stavanger, Norway
5. Laboratory of Immunology and Transfusion Medicine, Haugesund Hospital, 5528 Haugesund, Norway
* Correspondence: daniel.limi.cacic@sus.no

Simple Summary: Activated or apoptotic platelets both shed platelet microparticles that are proven to be internalized by many different cell types, including cancer cells. Here, we have investigated whether platelet microparticles can transfer their contents to the monocytic leukemia cell line THP-1 and if this could change cell activity and resistance to chemotherapy. We show that platelet microparticles were internalized by THP-1 cells and that platelet-associated microRNAs were elevated after a brief co-incubation. Furthermore, differentiation toward macrophages was induced and cell cycle progression, proliferation, and mitochondrial activity were decreased. Co-incubation with platelet microparticles increased chemotherapy resistance, which also was evident in acute myelogenous leukemia cells from patient samples, and it could be explained by the decrease in cell activity. Thus, platelet microparticles may have a role in the evolution of acute myelogenous leukemia and contribute to development of chemotherapy resistance, making them an interesting target for treatment.

Abstract: The role of platelets in cancer development and progression is increasingly evident, and several platelet–cancer interactions have been discovered, including the uptake of platelet micropar-ticles (PMPs) by cancer cells. PMPs inherit a myriad of proteins and small RNAs from the parental platelets, which in turn can be transferred to cancer cells following internalization. However, the exact effect this may have in acute myelogenous leukemia (AML) is unknown. In this study, we sought to investigate whether PMPs could transfer their contents to the THP-1 cell line and if this could change the biological behavior of the recipient cells. Using acridine orange stained PMPs, we demonstrated that PMPs were internalized by THP-1 cells, which resulted in increased levels of miR-125a, miR-125b, and miR-199. In addition, co-incubation with PMPs protected THP-1 and primary AML cells against daunorubicin-induced cell death. We also showed that PMPs impaired cell growth, partially inhibited cell cycle progression, decreased mitochondrial membrane potential, and induced differentiation toward macrophages in THP-1 cells. Our results suggest that this altering of cell phenotype, in combination with decrease in cell activity may offer resistance to daunorubicin-induced apoptosis, as serum starvation also yielded a lower frequency of dead and apoptotic cells when treated with daunorubicin.

Keywords: acute myelogenous leukemia; platelets; microparticles; apoptosis

Citation: Cacic, D.; Reikvam, H.; Nordgård, O.; Meyer, P.; Hervig, T. Platelet Microparticles Protect Acute Myelogenous Leukemia Cells against Daunorubicin-Induced Apoptosis. *Cancers* **2021**, *13*, 1870. https://doi.org/10.3390/cancers13081870

Academic Editors: Farrukh Aqil and Ramesh Chandra Gupta

Received: 28 February 2021
Accepted: 11 April 2021
Published: 14 April 2021

Publisher's Note: MDPI stays neutral with regard to jurisdictional claims in published maps and institutional affiliations.

1. Introduction

Acute myelogenous leukemia (AML) is a bone marrow malignancy originating in hematopoietic stem and progenitor cells [1–3]. The average 5-year survival rate for de novo disease is approximately 50% in younger patients [4], but this may vary widely depending on the occurrence of a selection of genetic aberrances. According to the 2017 European LeukemiaNet genetic risk stratification of AML, survival varies from 20% to over 60% [5]. Curative treatment involves intensive chemotherapy and, for select high-risk patient groups, the addition of consolidating treatment with allogenic stem cell transplantation, which carries the risk of a fatal outcome [6]. Thus, there is a need for a better understanding of tumorigenesis and evolution of the disease to improve treatment strategies.

Platelet–cancer interactions are becoming increasingly evident, and there is proof of cancer disease fundamentally altering the platelet transcriptome [7]. In aggregates with cancer cells, platelet function is hijacked to evade the NK cell response [8,9] and induce cancer cell epithelial–mesenchymal transition to facilitate metastasis [10,11]. Platelets are also important mediators for the development and maintenance of the cancer cell microenvironment [12,13].

Platelet microparticles (PMPs) are small membranous platelet particles (<1000 nm), which either bud off as a result of platelet activation [14] or as apoptotic bodies [15,16]. These microparticles are internalized by a variety of cell types, transferring their contents during this process [16–19]. PMPs contain a selection of the myriad of parent platelet alpha granule proteins [20,21] and platelet-associated microRNAs [22,23], which may potentially affect the biological behavior of the cells that have internalized them. This transfer of microRNAs has been demonstrated in a number of cancer models [24–26], where, although the effects are dependent on the cancer type and model, the PMPs appear to have both pro and anti-tumoral properties.

Targeting anti-apoptotic proteins is a novel strategy in the treatment of AML [27]. BCL2 is an important regulator of the intrinsic or mitochondrial apoptosis pathway, inhibiting BCL2 Antagonist/Killer (BAK) and BCL2 Associated X, Apoptosis Regulator (BAX) oligomerization, thus preventing pore formation in the outer mitochondrial membrane and subsequently leading to the leakage of cytochrome c and activation of caspase-9 [28]. Both platelet releasate and lysate seem to counter the effects of agents that specifically target this pathway, revealing an anti-apoptotic potential of platelets in AML [29]. There is also evidence that intrinsic apoptosis can be affected by the transfection of certain microRNAs, which are also found to be overexpressed in AML and present in platelets, indicating the potential relevance of these regulatory RNA molecules in an interaction between AML cells and platelets [30,31].

The role of microRNAs in AML is further supported by several studies showing an association of microRNA expression with mortality and chemotherapy resistance in patients, in whom several of the microRNAs are known to be present in high concentrations in platelets and platelet microparticles [32–34]. In this study, we aimed to assess whether PMPs could be taken up by AML cells and if this would change the AML cells' microRNA levels and in vitro chemotherapy resistance.

2. Results

2.1. Platelet-Associated microRNAs Are Increased in THP-1 Cells after PMP Co-Incubation

To examine whether platelet microparticles could be internalized by AML cells, we cultured cells from the monocytic AML cell line THP-1, with acridine orange (AO)-stained PMPs for 18 h. There was a PMP concentration-dependent increase of fluorescence in the co-incubated cells, and fluorescence microscopy revealed that that the stain was indeed dispersed within the cell nucleus and not located to bound microparticles (Figure 1A,B). To further investigate whether this PMP internalization could increase microRNA levels, we analyzed a selection of platelet-associated microRNAs in THP-1 cells after 18 h of co-incubation with PMPs. miR-125a-5p, miR-125b-5p, and miR-199-5p levels were all markedly increased (Figure 1C). This was particularly true for miR-199-5p, where levels were undetectable without PMP co-incubation in 2/3 samples (range 37.55–not detected), versus an average Cq value of 33.20 (range 33.13–33.24) with PMP co-incubation. These findings give indirect proof that microRNAs can be transferred from platelets to THP-1 cells through PMP internalization.

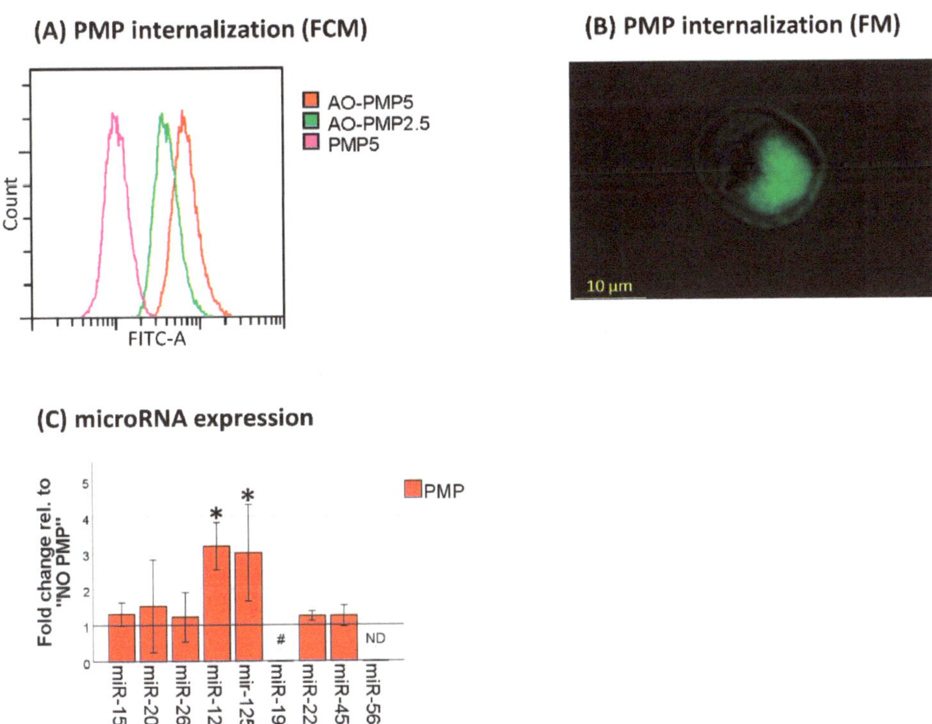

Figure 1. Internalization of platelet microparticles (PMPs) in THP-1 cells after 18 h of co-incubation. (**A**) Transfer of acridine orange from stained PMPs analyzed by flow cytometry (FCM). Histogram plot from a representative experiment ($n = 2$). Number following different PMP groups denotes the final concentration in million PMPs per mL medium. AO, acridine orange. (**B**) Transfer of acridine orange from stained PMPs analyzed by fluorescence microscopy (FM) at $400\times$ magnification. (**C**) Changes in levels of microRNAs ($n = 3$). microRNA data were calculated as fold change from THP-1 without PMP co-incubation, normalized for *BCR*. *p* values were calculated using the one-sample *t* test. * $p < 0.05$. #, fold change was not calculated as levels were undetectable in 2/3 replicates for THP-1 without PMP co-incubation. ND, not detected.

2.2. PMPs Lead to Increased Resistance of THP-1 Cells to DNR

Both miR-125a-5p and miR-125b-5p have been associated with resistance to chemotherapy in retroviral transduction studies [30,31]. Therefore, we examined whether the cytotoxic effect of daunorubicin (DNR), a common front-line chemotherapeutic in AML, could be influenced by PMP internalization. Co-incubation of THP-1 cells with PMPs decreased the relative frequency of dead and apoptotic cells in a concentration-dependent manner following treatment with DNR (Figure 2A). Thus, for all other analyses, PMPs were co-incubated at a concentration of 1.5×10^7 per mL medium, unless otherwise specified, as this generated the highest chemoprotective effect. Vector control experiments, where the supernatant of isolated PMPs was added to Iscove's Modified Dulbecco's Medium (IMDM) + 10% FBS medium at a concentration of 5%, only had a small and non-significant effect on resistance to DNR ($p = 0.109$).

(A) Apoptosis and cell death – THP1

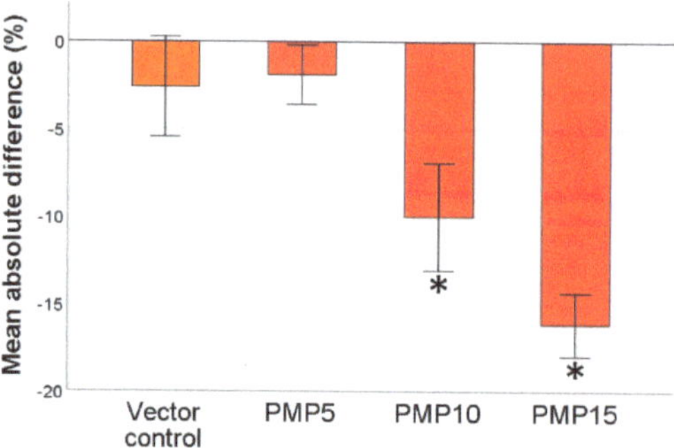

(B) Apoptosis and cell death – primary AML

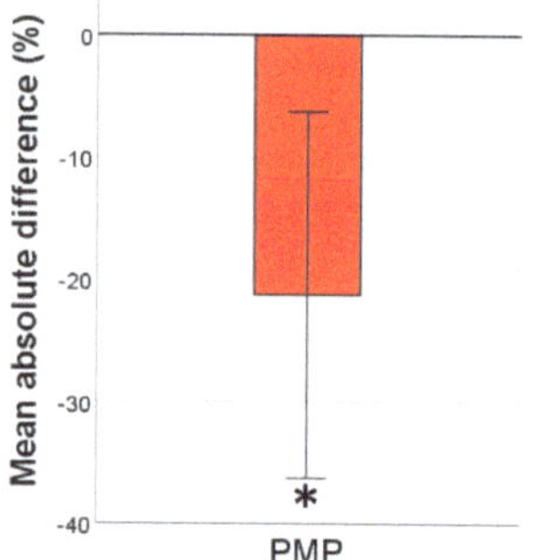

Figure 2. Apoptosis and cell death in THP-1 and primary acute myelogenous leukemia (AML) cells treated with 0.5 μM daunorubicin. (**A**) Difference in frequency of dead and apoptotic cells in THP-1 cells with or without co-incubation with platelet microparticles (PMPs) (*n* = 3). Number following different PMP groups denotes final concentration in million PMPs per mL medium. (**B**) Evaluation of the effect of PMPs on the frequency of dead and apoptotic cells in primary AML cells (nine patient samples, *n* = 4). *p* values were calculated using the one-sample *t* test (test value = 0). * *p* < 0.05.

2.3. Co-Incubation of Primary AML Cells with PMPs Also Increased Resistance to DNR

As there are limitations for the clinical relevance of cell line AML models [35], we examined whether the chemoprotective effect of PMPs could be observed in cells derived from AML patients. Using a similar approach, albeit with serum-free conditions, our results showed an identical effect on primary AML cells where we identified a significantly lower frequency of dead and apoptotic cells when PMPs were added (Figure 2B). The average absolute reduction in dead and apoptotic cells in individual patient samples ranged from 0.2 to 55.9% and was significant in 8/9 patients (Figure S1). One-way ANOVA analysis of the effects on THP-1 cells of PMPs from different releasates (when used in the primary AML experiments as quality controls) revealed no significant inter-releasate batch difference (*p* = 0.823).

2.4. THP-1 Cells Co-Incubated with PMPs Had Lower Caspase-9 Activity Following DNR-Treatment

Using microRNA databases (miRDB [36] and TargetScanHuman [37]), we found several predicted target mRNAs for miR-125a-5p and miR-125b-5p with important roles in the intrinsic apoptotic pathway, such as the pro-apoptotic BCL2 family proteins, BCL2 Modifying Factor (BMF) and BAK1 [38]. To investigate whether PMP internalization could influence the intrinsic apoptotic pathway, THP-1 cells were co-incubated with PMPs and treated with DNR using the established approach. Then, the cells were analyzed for caspase-9 activation, which is a downstream effect of mitochondrial outer membrane permeabilization (MOMP). There was a lower frequency of caspase-9 positive cells in the THP-1 cell cultures co-incubated with PMPs (Figure 3), suggesting that the chemoprotective effect of PMPs could be the result of the effects on the intrinsic apoptotic pathway upstream of caspase-9 activation.

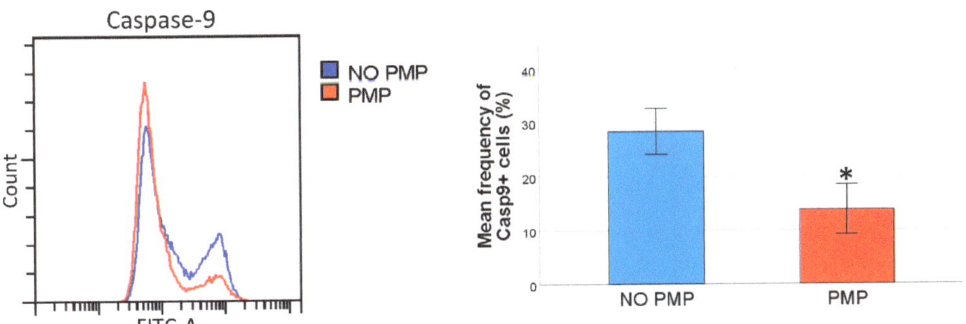

Figure 3. Caspase-9 activation in daunorubicin (DNR)-induced cell death. DNR-treated THP-1 cells were analyzed for caspase-9 activation ($n = 3$). Caspase-9 positive cells were identified as the distinct second peak in the flow histogram. p values were calculated using the paired-sample t test. * $p < 0.05$.

2.5. Decreased Cell Activity Protected THP-1 Cells against DNR

Cytotoxic chemotherapy is believed to be most potent in highly proliferating cancer cells. Inducing cellular dormancy to decrease DNA replication should theoretically offer a chemoprotective effect, as it would prevent DNR-triggered DNA damage. Thus, we investigated whether serum starvation would decrease cell activity and subsequently protect THP-1 cells against DNR. We analyzed proliferation, cell cycle distribution, and mitochondrial membrane potential, and we observed that 48 h of serum starvation in THP-1 cells induced a significant growth arrest (Figure 4A–C). To evaluate whether serum starvation affected DNR-resistance, we compared apoptosis and cell death in THP-1 cells, with or without serum starvation, 24 h after treatment with 0.5 µM DNR. We showed a marked reduction in the frequency of dead and apoptotic cells (Figure 4D).

To investigate whether the apparent chemoprotective effect of PMP co-incubation may be the result of a similar decrease in cell activity, we measured the effects of PMPs on cell proliferation, cell cycle distribution, and mitochondrial membrane potential. Our results showed that co-incubation with PMPs increased the frequency of cells in the G0/G1 cell phase, reduced mitochondrial membrane potential, and inhibited cell proliferation (Figure 5A–C). These findings lead us to believe that PMPs may protect THP-1 cells from DNR-induced cell death by partially inhibiting cell cycle progression and proliferation. Co-incubation with PMPs did not alter mRNA or protein levels of CDK4 (Figure 5D,E), which is fundamental for THP-1 viability and normal cell cycle progression [39,40].

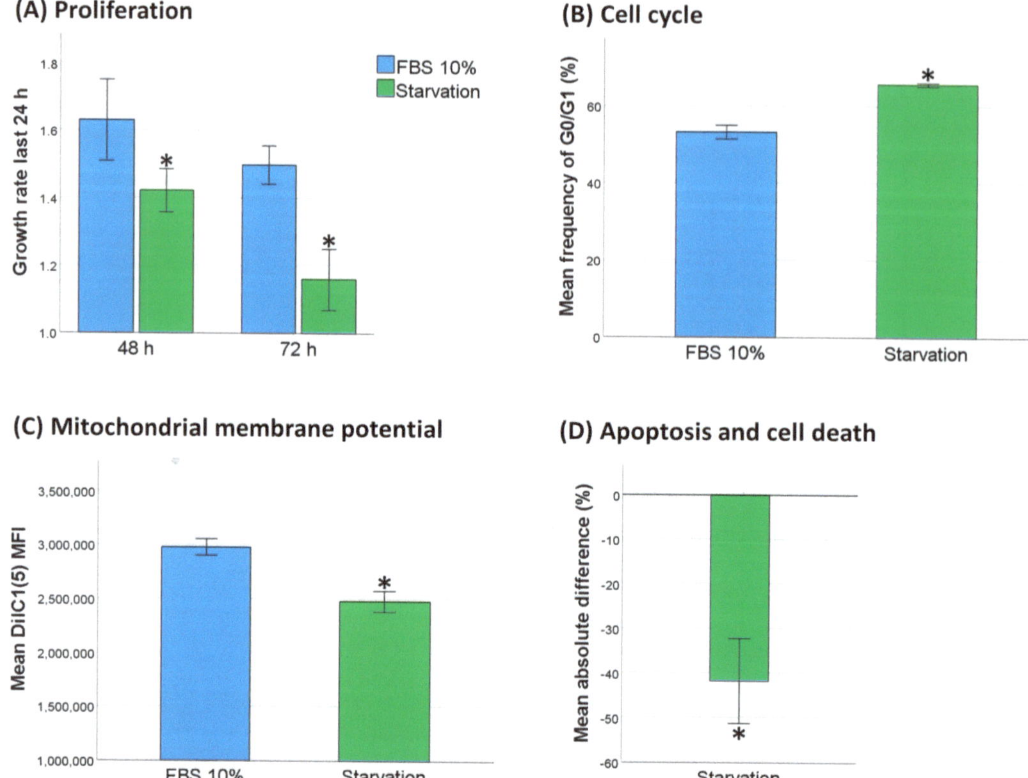

Figure 4. Effects of serum starvation on cell activity and daunorubicin (DNR)-resistance in THP-1 cells. (**A**) Daily proliferation rate analyzed by flow cytometric counting ($n = 4$). (**B**) Cell cycle analysis after 48 h of serum starvation ($n = 3$). Cells in the G0/G1 cell phase were gated. (**C**) Mitochondrial membrane potential after 48 h of serum starvation ($n = 3$), mean fluorescence intensity (MFI) data. (**D**) Difference in DNR-induced cell death and apoptosis after 48 h of serum starvation compared to standard conditions. (**A–C**) were compared using the paired-sample t test. (**D**) was compared using the one-sample t test (test value = 0). * $p < 0.05$.

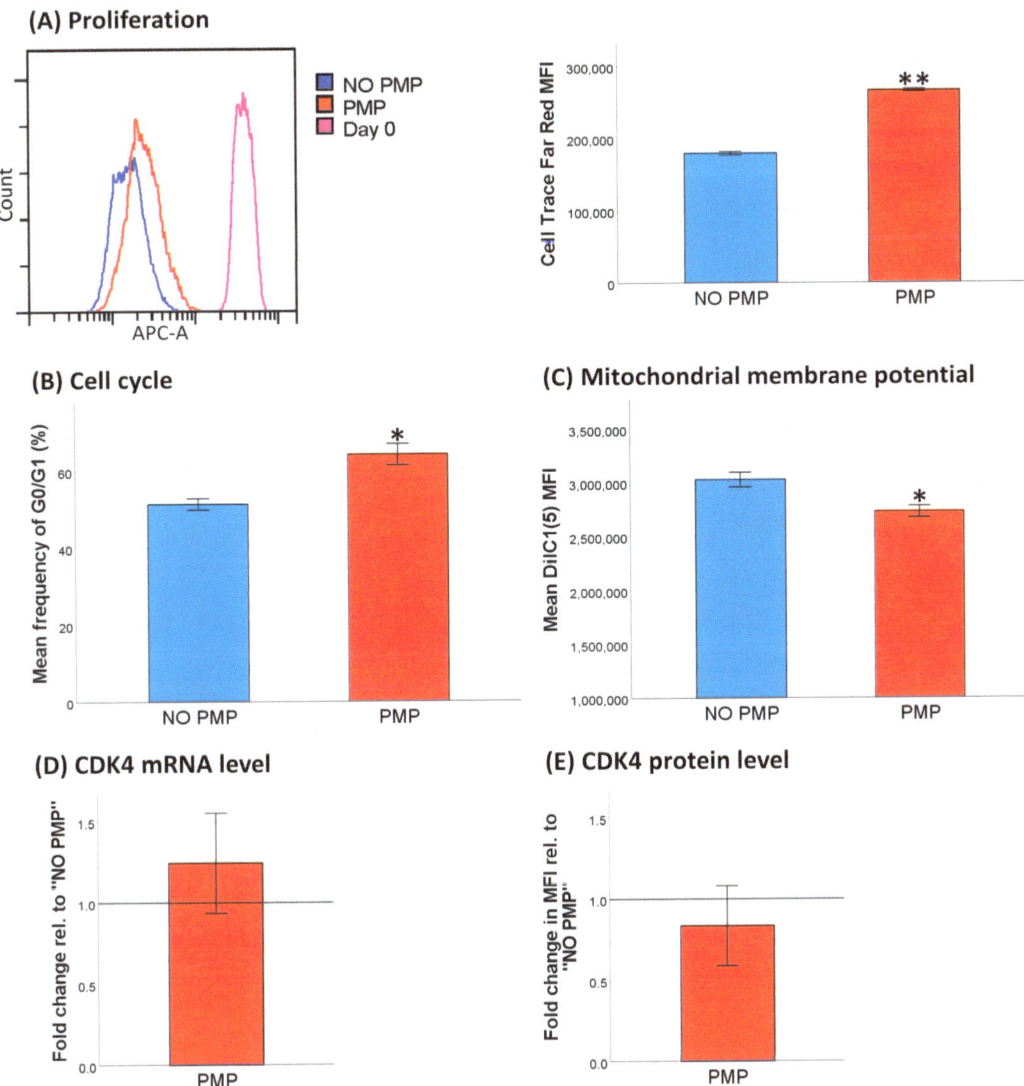

Figure 5. Effects of platelet microparticles (PMPs) on cell activity in THP-1 cells. (**A**) Proliferation analysis by Cell Trace Figure 5. (*n* = 5). (**B**) Cell cycle analysis (*n* = 3). Cells in G0/G1 cell phase were gated. (**C**) Mitochondrial membrane potential (*n* = 4), mean fluorescence intensity (MFI) data. (**D**) CDK4 mRNA levels (*n* = 3). (**E**) CDK4 protein levels (*n* = 5). (**B–E**) were analyzed after 24 h of PMP co-incubation. mRNA data are calculated as fold change from THP-1 without PMP co-incubation, normalized for *ACTB*. Protein data are calculated as fold change in MFI from THP-1 without PMP co-incubation. Data were compared using the paired-sample *t* test for data pairs or the one-sample Wilcoxon signed-rank test for ratios. * $p < 0.05$. ** $p < 0.001$.

2.6. PMP Co-Incubation Increased Differentiation of THP-1 Cells toward Macrophages

THP-1 cells are capable of macrophage differentiation, leading to cell growth arrest. Thus, we wanted to examine if increased differentiation of the cells could contribute to the observed decrease in cell cycle progression. PMP co-incubation increased both forward scatter and side scatter (Figure 6A), indicating increased cell size and granularity, which are two hallmarks of macrophage differentiation [41]. Surprisingly, we could not corroborate the forward scatter findings with measurement of cell cross-sectional area using the particle analysis function in the ImageJ software with pictures taken under an inverted phase-contrast microscope (Figure 6B). However, co-incubation with PMPs led to a significant increase in CD14 antigen expression (Figure 6C). Thus, the decrease in cell cycle progression, and therefore part of the chemoprotective effect, could stem from differentiation of the cells toward macrophages.

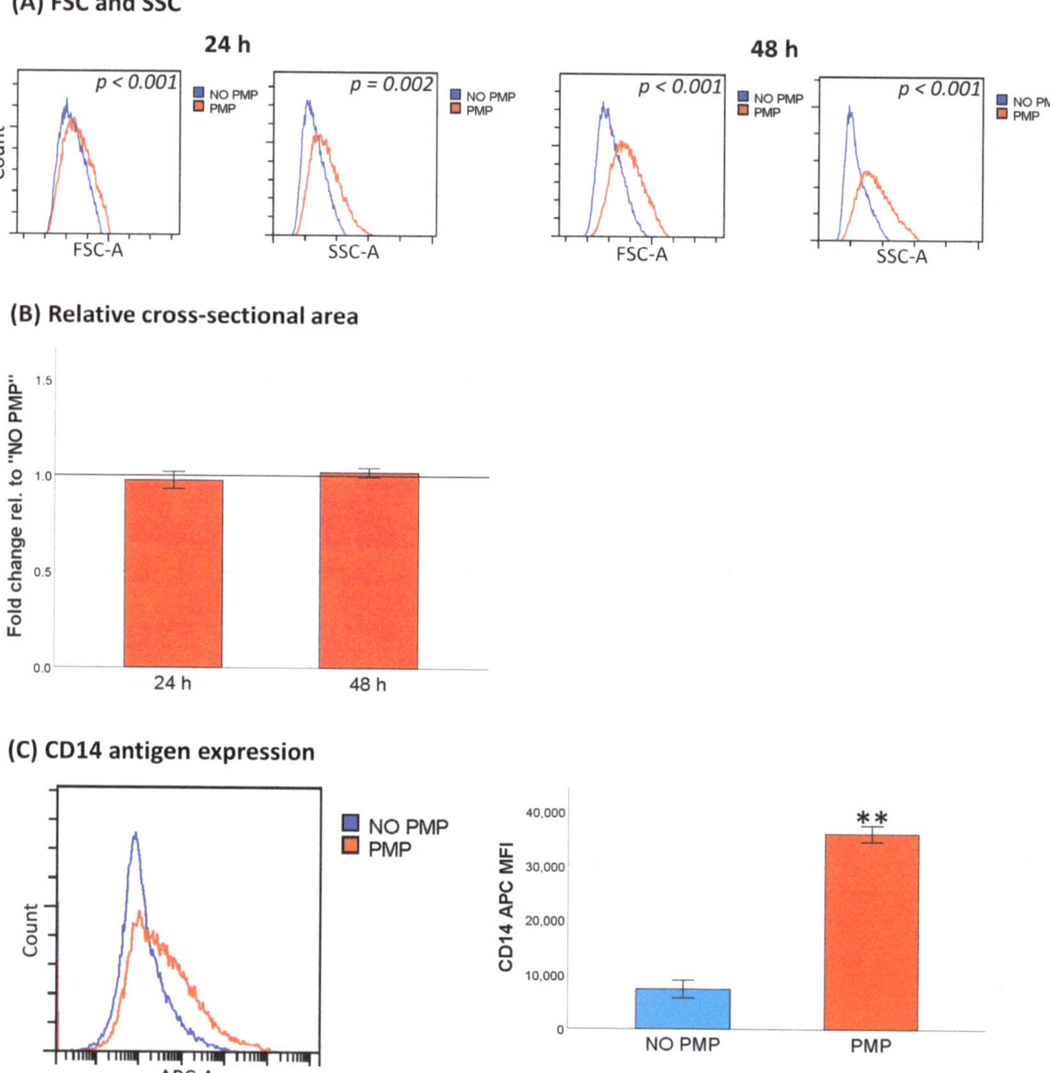

Figure 6. Phenotypical changes induced by platelet microparticles (PMPs) co-incubation. (**A**) Forward scatter (FSC) and side scatter (SSC; $n = 4$). (**B**) Ratio of cell cross-sectional area ($n = 4$). (**C**) CD14 antigen expression after 48 h of culture ($n = 5$). Data were compared using the paired-sample *t* test for data pairs or the one-sample *t* test for ratios. ** $p < 0.001$.

3. Discussion

The important role of platelets in cancer development, progression, and metastasis is becoming increasingly clear, and there are several known mechanisms in this interplay, including the transfer and uptake of platelet microparticles. We showed that co-incubating primary AML cells and THP-1 cells with PMPs increased their DNR resistance. In addition we demonstrated the inhibition of THP-1 cell proliferation, cell cycle progression, mitochondrial membrane potential, and induction of differentiation toward macrophages.

We observed internalization of PMPs by THP-1 cells and a subsequent increase of platelet-associated microRNAs. The internalization was observed using AO stained PMPs. AO will stain DNA, RNA, and acidic vesicles. Platelets do not contain DNA, other than a small amount of mtDNA. However, platelets and PMPs both contain RNA and lysozomes, making AO staining suitable for our purpose.

Co-incubation of THP-1 or primary AML cells with PMPs increased their resistance to DNR-treatment at a concentration of 0.5 µM, which is close to the peak plasma concentration measured in patients receiving 60 mg/m^2 DNR [42]. In AML, DNR is routinely administrated at dosages of 45–90 mg/m^2 [43]. This observed chemoprotective effect could be the result of microRNA transfer from PMPs. The overexpression of both miR-125a and miR-125b have been associated with DNR resistance in AML cell lines, including THP-1 [30,31]. Using the transduction of THP-1 cells with murine stem cell virus (MSCV), these studies obtained a 4 and 4.9-fold increase of miR-125a and miR-125b. In our study, the respective levels were increased 3.0 and 3.2-fold. The transduction of miR-125a and miR-125b resulted in downregulation of the apoptotic proteins Grk2 and Puma [30,31]. The latter is a member of the pro-apoptotic BH3-only proteins of the intrinsic apoptosis pathway [38], where we show a decrease in activation with PMP co-incubation. These proteins are known to be crucial participants in apoptosis, as genetic knockout models are protected against several apoptotic stimuli [44]. However, the cited studies with the transduction of miR-125a and miR-125b did not include an analysis of cell activity in the THP-1 cells associated with microRNA overexpression, which may be decisive for the actual DNA-damage induced by DNR. Other studies have linked the ectopic expression of miR-125a with proliferation inhibition, although in solid tumor cell lines [45,46].

PMPs contain several hundred different proteins and small RNAs, meaning the underlying mechanism for chemoprotection are likely more complex than that reflected in the single microRNA transduction studies. The anti-cancer effect of DNR and other anthracyclines are believed to mainly be a result of interference with topoisomerase II (Top2) enzyme activity [47]; however, other mechanisms have been identified [48]. Top2 introduces double-strand DNA breaks during replication [47]; thus, an inhibition of proliferation should decrease the efficiency of Top2 poisons. We showed that decrease of cell activity through serum starvation protects THP-1 cells against DNR-triggered apoptosis and cell death, and we suggest that PMPs could offer chemoprotection through this mechanism.

Vasina and colleagues have previously shown that microparticles from apoptotic platelets can induce macrophage differentiation in THP-1 cells after 7 days of co-incubation [16]. Here, we show prominent upregulation of CD14 antigen already after 48 h using platelet microparticles from activated platelets from platelet concentrates containing a mixture of PMPs generated by activation and apoptosis, better resembling the in vivo milieu. Seemingly, there are conflicting results regarding cell size analyses, as forward scatter and side scatter were increased, but the measured cross-sectional cell area was unchanged. However, we believe the increase in light scatter was affected by morphological changes with more vacuolization in the cells treated with PMPs.

The observed differentiation effect can at least partially explain the decrease in cell cycle progression, as THP-1 cells treated with phorbol myristate acetate for macrophage differentiation only exit G1 phase to a little extent [49]. CDK4 mRNA and protein levels were unchanged, and the decrease in cell cycle progression would appear to be the result of a downstream target. However, notable downregulation of the CDK4 gene is known to be a later event in macrophage differentiation of THP-1 cells [49]. We have not identified the exact substances that initiate differentiation or lead to the inhibition of cell cycle progression. The latter could partially be an independent process, because contrary to our findings, mitochondrial activity is increased with macrophage differentiation [41]. Transforming growth factor beta (TGF-β) is a potent cell cycle regulator known to be present in platelets. TGF-β induces dormancy or quiescence through several mechanisms, but it cannot be entirely responsible for the observed chemoprotective effect of PMPs, as it is known to be abundant in the platelet secretome [50,51], and we did not observe any significant effects on the frequency of dead and apoptotic cells in our vector control experiments.

Several research groups have reported that PMPs affect cancer development. Michael and colleagues showed that PMPs could infiltrate solid tumors and inhibit the growth of lung and colon cancer [26]. Others have linked PMPs to increased epithelial–mesenchymal transition and metastatic capacity in ovarian cancer [25] and lung cancer invasion [24]. Recent evidence has shown that platelets can have a bimodal effect in colorectal cancer where they inhibit growth but promote metastasis [52]. An extensive review of the role of PMPs in cancer progression can be found elsewhere [53].

Our protocol for the quantitation of microparticles has some limitations, as the PMP number per mL releasate varied on average by 10.0%, but it ranged from 0.4 to 34.9% between technical replicates. Thus, the final concentration of PMPs in the culture media may have varied extensively in some experiments. Accordingly, the concentration of microRNAs and proteins will vary from batch to batch of platelet concentrates. We accounted for the latter when we chose to use pooled platelet concentrates derived from four different donors. Furthermore, we found no inter-batch differences with respect to chemoprotective effect.

The induction of resistance to DNR by PMPs could have significant clinical relevance. Inhibiting the production of PMPs may present a potential therapeutic approach in AML to increase chemosensitivity. This can easily be achieved with platelet inhibition [54]. Platelet inhibition has also previously been linked to both lower cancer incidence and improved cancer-specific survival [55–58], although the exact mechanism is unknown. On the flip side, the differentiation of AML cells by PMPs might be beneficial to inhibit evolution of the disease.

4. Materials and Methods

4.1. Cell Line

The THP-1 cell line was purchased from ATCC (American Type Culture Collection; Manassas, VA, USA) and maintained in Iscove's Modified Dulbecco's Medium (IMDM; Thermo Fisher Scientific, Waltham, MA, USA) + 10% FBS (Sigma Aldrich, St. Louis, MO, USA). Only cells in the exponential growth phase were used, and cultures were kept for less than three months.

4.2. Primary AML Cells

Primary AML cells were isolated by density gradient separation of peripheral blood from consenting patients at the Department of Medicine, Section of Hematology, Haukeland University Hospital (Bergen, Norway). The cells were cryopreserved in liquid nitrogen until use. The cryopreservation solution consisted of insulin-free RPMI 1640 (Sigma Aldrich), supplemented with 10% dimethylsulfoxide and 20% FBS. Primary AML cells were cultured in StemSpan Serum-Free Expansion Medium (Stem Cell Technologies, Vancouver, BC, Canada) with the addition of the following recombinant cytokines in a final concentration of 20 ng/mL: stem cell factor (Peprotech EC, London, UK), G-CSF (Peprotech), and FMS-like tyrosine kinase 3 ligand (Peprotech). Charateristics of the AML patients can be found in Table 1.

Table 1. Characteristics of primary AML patients.

#	Sex	Age	Prev. Myeloid Disease	FAB	Cytogenetics	FLT3	NPM1	CEBPA	CD11b	CD14	CD33	CD34	CD45	CD64	CD117	HLA-DR	L-MPO
1	M	22	No	M5	del(9)	ITD (low ratio)	wt		neg	neg	pos	pos	dim		pos	pos	dim
2	F	60	PV		del(7)	wt	wt	wt	neg	neg		pos	dim		pos	pos	
3	F	56	No	M4	inv(16)	wt	wt		neg	neg	pos	pos	dim	neg	pos	pos	pos
4	M	44	No	M4	inv(16)	wt	wt		neg	neg	neg	neg	dim	dim	neg	pos	
5	F	92	No	M1					neg	neg	dim	neg	dim	dim	pos	pos	pos
6	M	49	No	M4	45, XY	wt	ins	wt	neg	neg		neg	dim	dim	pos	pos	
7	M	76	No	M5	Normal	wt	ins	wt	hetero	hetero		neg	hetero	pos	hetero	pos	
8	F	95	No	M4	Normal	wt	wt	wt				dim	dim		dim		
9	M	29	No	M4	Normal	ITD (high ratio)	wt	wt	neg	neg	neg	pos	dim	neg	neg	pos	pos

patient number.

4.3. Platelet Concentrate

Routinely prepared platelet concentrates pooled from four donors (Tacsi system; Terumo BCT, Lakewood, CO, USA) were provided by the Department of Immunology and Transfusion Medicine, Stavanger University Hospital (Stavanger, Norway), after written consent from the donors. The platelet concentrations were $0.88–1.08 \times 10^9$ per mL. Leukocytes were removed by filtration to a residual level of $<1.00 \times 10^6$. In the final concentrate, the storage medium contained approximately 65% additive solution (PAS-III, Baxter, Lake Zurich, IL, USA) and 35% plasma.

4.4. Platelet Releasate

The platelet concentrate was transferred from the blood bag to separate 50 mL tubes and incubated with a final concentration of 1 U/mL human thrombin (Sigma Aldrich) for one hour in a 37 °C water bath. The tubes were gently agitated every 5 min. The platelet releasate was centrifuged for 10 min at $900\times g$, and the supernatant was transferred to new 50 mL tubes. The samples were stored at -80 °C. Fibrin clots that appeared after thawing were plucked using a 10 mL serological pipette.

4.5. Platelet Microparticles Isolation, Co-Culture, and Measurement

Platelet releasate was centrifuged at $15,000\times g$ for 90 min at room temperature and the supernatant was carefully poured off. To examine the biological effects of platelet microparticles, a mastermix of StemSpan + cytokines (for primary cells), or IMDM + 10% FBS (for THP-1 cells), was used to resuspend PMPs before transfer to cell culture and thoroughly mixed with the cells by pipetting. Two hours after the PMPs were added to the cell cultures, the wells were mixed again by pipetting. For quantitation, the microparticles were resuspended in 400 µL of 0.22 µm filtered Annexin V Binding Buffer (Miltenyi Biotec, Bergisch Gladbach, Germany), before 200 µL of the solution was transferred to a second tube. Twenty µL of Annexin V FITC (Miltenyi Biotec), and 2 µL of anti CD61 APC (clone Y2/51; Miltenyi Biotec), or 22 µL of 0.22 µm filtered Annexin V Binding Buffer for an unstained control, were added and incubated for 15 min at room temperature. Finally, 278 µL of 0.22 µm filtered Annexin V Binding Buffer and 50 µL CountBright beads (Thermo Fisher Scientific) were added before analysis. Microparticle gates were set with Megamix-PLUS FSC beads (size range of beads: 0.3 to 0.9 µm; BioCytex, Marseille, France) using the side scatter channel, according to Poncelet and colleagues [59]. At least 2500 bead events were collected. This as well as all other flow cytometric analyses were performed on a CytoFLEX flow cytometer (Beckman Coulter, Brea, CA, USA) using CytExpert ver. 2.4 acquisition and analysis software (Beckman Coulter).

4.6. Acridine Orange Staining of Platelet Microparticles

Platelet releasate was stained with 100 µg/mL of acridine orange (Thermo Fisher Scientific) and incubated for 30 min at room temperature. Then, the solution was washed and centrifuged two times at $15,000\times g$ for 90 min. Tubes were changed after the first wash step to avoid any contamination of acridine orange that may have adhered to the plastic. Finally, the PMPs were resuspended in IMDM + 10% FBS and co-cultured with THP-1 cells for 18 h at a concentration of 5×10^6 PMPs per mL. The cells were harvested and washed twice in Dulbecco's phosphate-buffered saline (DPBS; Sigma Aldrich) before analysis with flow cytometry using the FITC channel, and with a Zeiss Axioplan 2ie MOT fluorescence microscope (Carl Zeiss, Göttingen, Germany) using an SpGreen filter. At least 25,000 gated cells were collected for flow cytometric analysis.

4.7. mRNA and microRNA Analysis

Total RNA was isolated using the miRNeasy kit (QIAGEN GmbH, Hilden, Germany), and RNA concentration was measured on a NanoDrop 2000 spectrophotometer (Thermo Fisher Scientific). Reverse transcription was performed with the TaqMan MicroRNA Reverse Transcription Kit (Thermo Fisher Scientific), and the High-Capacity cDNA Reverse Transcription Kit (Thermo Fisher Scientific). Real-Time PCR was done on the Mx3005P qPCR system (Agilent Technologies, Palo Alto, CA, USA) using TaqMan MicroRNA Assays (Thermo Fisher Scientific) with the TaqMan Universal Master Mix for microRNA analyses (Thermo Fisher Scientific), and the TaqMan gene expression assays (Thermo Fisher Scientific) with the TaqMan Gene Expression Master Mix (Thermo Fisher Scientific) for mRNA analyses, following the manufacturer's instructions. *BCR* or *ACTB* were used as reference genes, and the relative expression was calculated using the $2^{-\Delta\Delta Cq}$ method. $\Delta\Delta Cq$ was calculated as ΔCq value (target gene minus reference gene) for cells without PMP co-incubation minus ΔCq value for PMP co-incubated cells. For a comprehensive list of the microRNA and mRNA assays used in this study, see Table S1.

4.8. Daunorubicin Apoptosis Assay

Approximately 5×10^5 cells per mL of resuscitated primary AML cells, or THP-1 cells in exponential growth phase, were cultured under aforementioned conditions with or without PMPs for 24 h. The cells were then treated with 0.5 µM daunorubicin hydrochloride (Sigma Aldrich) for another 24 h before further analysis. THP-1 cells were also used as a quality control for the experiments with primary AML cells. Then they were kept in the same batch of StemSpan + cytokines to detect any false negative results in case of issues with the PMP isolation. Cell viability was analyzed with the Annexin V-FITC kit (Miltenyi Biotec), strictly following the manufacturer's instructions. Dead and apoptotic cells were analyzed using flow cytometry and gated out in a single gate using a pseudo color plot of FITC-A versus PerCP Cy 5.5-A after doublet discrimination. For analysis with primary AML cells, contaminating cells were gated out based on light scatter properties. See Figure S2 for gating strategies. At least 25,000 gated cells were collected.

4.9. Caspase-9 Activity

Caspase-9 activity in daunorubicin-treated THP-1 cells was measured using the Casp-GLOW Fluorescein Active Caspase-9 Staining Kit (Thermo Fisher Scientific). Approximately 5×10^5 cells in 0.3 mL IMDM + 10% FBS were stained with 1 µL FITC-LEHD-FMK and incubated for 30 min in a CO_2 incubator before washing twice with the supplied wash medium and analysis with flow cytometry. Both untreated and treated, but not stained THP-1 cells, were used as negative controls. At least 25,000 gated cells were collected.

4.10. Mitochondrial Membrane Potential

Mitochondrial membrane potential was assessed using the MitoProbe DiIC1(5) Assay Kit (Thermo Fisher Scientific). THP-1 cells were cultured with or without PMPs for 24 h before 5×10^5 cells in 1 mL IMDM + 10% FBS were stained with DiIC1(5) using carbonyl cyanide 3-chlorophenylhydrazone-treated cells as a correction for background signal and incubated for 30 min following the manufacturer's instructions. After doublet discrimination, MFI (mean fluorescence intensity) values of gated cells were compared using the APC channel on the flow cytometer. See Figure S3A and B for gating strategy. At least 30,000 gated cells were collected.

4.11. Cell Cycle Analysis

THP-1 cells were incubated for 24 h in IMDM + 10% FBS with or without PMPs. Cells were washed and 5×10^5 cells were stained with 10 µM Vybrant Dye Cycle Green Stain (Thermo Fisher Scientific) and incubated for 30 min in a 37 °C water bath. Immediately after incubation, the cells were analyzed using the FITC channel on the flow cytometer. 2N cells, representing G0/G1 cell phase, were gated out after doublet discrimination. See Figure S3A and C for gating strategy. At least 10,000 gated cells were collected.

4.12. Flow Cytometry Proliferation Analysis

Proliferation analysis was performed with the Cell Trace Far Red Proliferation Kit (Thermo Fisher Scientific) and analyzed with the APC channel on the flow cytometer. On day 0, THP-1 cells at a concentration of 1×10^6 per mL were stained with 5 µM Far Red reagent in DPBS and incubated briefly for 5 min in a 37 °C water bath to avoid excessive cell toxicity. The stained cells were washed with IMDM + 20% FBS and cultured as previously described. Medium with or without PMPs was added on days 2 and 4 to keep concentration of cells below 8×10^5 per mL. A sample of the cells was analyzed on day 0 to identify baseline MFI. At least 25,000 gated cells were collected.

4.13. Flow Cytometry Immunophenotyping

THP-1 cells were cultured under aforementioned conditions with or without PMP co-incubation and harvested after 48 h. Approximately 1×10^6 cells were washed in DPBS, resuspended in 98 µL of DPBS containing 0.5% BSA, and labeled with 2 µL of anti-CD14 APC (clone REA599; Miltenyi Biotec). The cells were incubated for 10 min at 4 °C and washed before analysis. An unstained sample was used to determine background signal. At least 30,000 gated cells were collected.

4.14. Measurement of CDK4 by Indirect Intracellular Flow Cytomtery

THP-1 cells were cultured for 24 h before harvest and analyzed for intracellular protein using the published protocol by Ludwig and colleagues [60]. Briefly, cells were fixed and permeabilized using the eBioscience Foxp3/Transcription Factor Staining Buffer Set (Thermo Fisher Scientific). Then, cells were incubated with unconjugated anti-CDK4 (clone DCS-31; Thermo Fisher Scientific) and labeled with the proper conjugated secondary antibody. Dilution and incubation time can be found in Table S2. A "no primary antibody" sample was used to subtract background signal. At least 25,000 gated cells were collected.

4.15. Measurement of Cell Cross-Sectional Area

For measurement of the cross-sectional area, cultured cells were transferred to a Bürker chamber to minimize the physical cell membrane manipulation and assessed under an inverted phase-contrast microscope. Four representative fields per technical replicate at 100× magnification were captured using an Olympus Pen Lite E-PL5 camera (Olympus, Tokyo, Japan). Pictures were analyzed using the particle analysis function of the ImageJ software ver. 1.52k [61]. Image optimization and thresholding was performed as described in the Supplementary Methods.

4.16. Serum Starvation

In separate experiments, analysis of daunorubicin-induced apoptosis and cell death, cell cycle, and mitochondrial membrane potential were performed in serum-starved THP-1 cells without PMP co-incubation. Cells in the exponential growth phase were washed, resuspended, and kept for 48 h in IMDM before further analysis, as described in the separate sections. For measurement of proliferation rate, cells were resuspended at a concentration of 4×10^5 per mL IMDM, with or without 10% FBS, and counted using the flow cytometer after 24, 48, and 72 h.

4.17. Statistical Analysis

Statistical analyses were performed using the IBM SPSS 26 software (IBM Corp., Armonk, NY, USA). Comparison between experimental groups was performed using tests for paired or independent data when appropriate. The data were checked for normality using P-P plots, Shapiro–Wilks test, and Kolmogorov–Smirnov test. A p value < 0.05 was considered significant. Mean values are reported with a 95% confidence interval unless otherwise specified. "n" denotes technical replicates.

5. Conclusions

We show that PMP co-incubation decreases mitochondrial membrane potential, inhibits cell cycle progression, decreases proliferation, and induces differentiation toward macrophages in THP-1 cells. This differentiation effect, combined with decrease in cell activity, may explain the observed protection against daunorubicin-induced cell death, which is also evident in primary AML cells.

Our results warrant further research to explore the in vivo effects of platelet microparticles in AML, both as anti-apoptotic agents, and as modulators of the disease, as they represent possible therapeutic targets through the use of platelet inhibitors.

Supplementary Materials: The following are available online at https://www.mdpi.com/article/10.3390/cancers13081870/s1, Figure S1: Resistance to daunorubicin (DNR)-induced cell death in primary acute myelogenous leukemia (AML) cells, individual data, Figure S2: Gating strategy for flow cytometric apoptosis assay, Figure S3: Gating strategy for flow cytometric analyses, Table S1: TaqMan assays used in this study, Table S2: Description of antibodies used in this study.

Author Contributions: Conceptualization, D.C.; Methodology, D.C.; Software, D.C.; Validation, D.C.; Formal Analysis, D.C.; Investigation, D.C.; Resources, D.C., H.R. and T.H.; Writing—Original Draft Preparation, D.C.; Writing—Review and Editing, D.C., H.R., O.N., P.M. and T.H.; Visualization, D.C.; Supervision, T.H.; Project Administration, D.C. and T.H.; Funding Acquisition, D.C. and T.H. All authors have read and agreed to the published version of the manuscript.

Funding: This work was supported by Western Norway Regional Health Authorities and Bergen Stem Cell consortium.

Institutional Review Board Statement: The study was conducted according to the guidelines of the Declaration of Helsinki, and approved by the regional ethics committee, *Regional Etisk Komite Vest* (ref 8144) on 16 May 2017.

Informed Consent Statement: Informed consent was obtained from all subjects involved in the study.

Data Availability Statement: The data presented in this study are available on request from the corresponding author.

Acknowledgments: We would like to thank the Department of Immunology and Transfusion Medicine, Stavanger University Hospital and the Department of Medicine, Section of Hematology, Haukeland University Hospital for providing the biological materials used in this study. Graphical abstracted is created with BiroRender.com.

Conflicts of Interest: The authors declared no potential conflict of interest.

References

1. Sarry, J.-E.; Murphy, K.; Perry, R.; Sanchez, P.V.; Secreto, A.; Keefer, C.; Swider, C.R.; Strzelecki, A.-C.; Cavelier, C.; Récher, C.; et al. Human acute myelogenous leukemia stem cells are rare and heterogeneous when assayed in nod/scid/il2rγc-deficient mice. *J. Clin. Investig.* **2011**, *121*, 384–395. [CrossRef] [PubMed]
2. Ng, S.W.K.; Mitchell, A.; Kennedy, J.A.; Chen, W.C.; McLeod, J.; Ibrahimova, N.; Arruda, A.; Popescu, A.; Gupta, V.; Schimmer, A.D.; et al. A 17-gene stemness score for rapid determination of risk in acute leukaemia. *Nature* **2016**, *540*, 433–437. [CrossRef]

3. Krivtsov, A.V.; Twomey, D.; Feng, Z.; Stubbs, M.C.; Wang, Y.; Faber, J.; Levine, J.E.; Wang, J.; Hahn, W.C.; Gilliland, D.G.; et al. Transformation from committed progenitor to leukaemia stem cell initiated by mll-af9. *Nature* **2006**, *442*, 818–822. [CrossRef] [PubMed]
4. Kantarjian, H. Acute myeloid leukemia–major progress over four decades and glimpses into the future. *Am. J. Hematol.* **2016**, *91*, 131–145. [CrossRef] [PubMed]
5. Herold, T.; Rothenberg-Thurley, M.; Grunwald, V.V.; Janke, H.; Goerlich, D.; Sauerland, M.C.; Konstandin, N.P.; Dufour, A.; Schneider, S.; Neusser, M.; et al. Validation and refinement of the revised 2017 european leukemianet genetic risk stratification of acute myeloid leukemia. *Leukemia* **2020**, *34*, 3161–3172. [CrossRef] [PubMed]
6. Döhner, H.; Estey, E.; Grimwade, D.; Amadori, S.; Appelbaum, F.R.; Büchner, T.; Dombret, H.; Ebert, B.L.; Fenaux, P.; Larson, R.A.; et al. Diagnosis and management of aml in adults: 2017 eln recommendations from an international expert panel. *Blood* **2017**, *129*, 424–447. [CrossRef]
7. Zhang, Q.; Hu, H.; Liu, H.; Jin, J.; Zhu, P.; Wang, S.; Shen, K.; Hu, Y.; Li, Z.; Zhan, P.; et al. Rna sequencing enables systematic identification of platelet transcriptomic alterations in nsclc patients. *Biomed. Pharmacother.* **2018**, *105*, 204–214. [CrossRef]
8. Nieswandt, B.; Hafner, M.; Echtenacher, B.; Männel, D.N. Lysis of tumor cells by natural killer cells in mice is impeded by platelets. *Cancer Res.* **1999**, *59*, 1295–1300.
9. Kopp, H.G.; Placke, T.; Salih, H.R. Platelet-derived transforming growth factor-beta down-regulates nkg2d thereby inhibiting natural killer cell antitumor reactivity. *Cancer Res.* **2009**, *69*, 7775–7783. [CrossRef]
10. Yu, M.; Bardia, A.; Wittner, B.S.; Stott, S.L.; Smas, M.E.; Ting, D.T.; Isakoff, S.J.; Ciciliano, J.C.; Wells, M.N.; Shah, A.M.; et al. Circulating breast tumor cells exhibit dynamic changes in epithelial and mesenchymal composition. *Science* **2013**, *339*, 580–584. [CrossRef]
11. Takemoto, A.; Okitaka, M.; Takagi, S.; Takami, M.; Sato, S.; Nishio, M.; Okumura, S.; Fujita, N. A critical role of platelet tgf-β release in podoplanin-mediated tumour invasion and metastasis. *Sci. Rep.* **2017**, *7*, 42186. [CrossRef] [PubMed]
12. Palacios-Acedo, A.L.; Mège, D.; Crescence, L.; Dignat-George, F.; Dubois, C.; Panicot-Dubois, L. Platelets, thrombo-inflammation, and cancer: Collaborating with the enemy. *Front. Immunol.* **2019**, *10*, 1805. [CrossRef] [PubMed]
13. Yan, M.; Jurasz, P. The role of platelets in the tumor microenvironment: From solid tumors to leukemia. *Biochim. Biophys. Acta* **2016**, *1863*, 392–400. [CrossRef] [PubMed]
14. De Paoli, S.H.; Tegegn, T.Z.; Elhelu, O.K.; Strader, M.B.; Patel, M.; Diduch, L.L.; Tarandovskiy, I.D.; Wu, Y.; Zheng, J.; Ovanesov, M.V.; et al. Dissecting the biochemical architecture and morphological release pathways of the human platelet extracellular vesiculome. *Cell. Mol. Life Sci. Cmls* **2018**, *75*, 3781–3801. [CrossRef]
15. Black, A.; Orsó, E.; Kelsch, R.; Pereira, M.; Kamhieh-Milz, J.; Salama, A.; Fischer, M.B.; Meyer, E.; Frey, B.M.; Schmitz, G. Analysis of platelet-derived extracellular vesicles in plateletpheresis concentrates: A multicenter study. *Transfusion* **2017**, *57*, 1459–1469. [CrossRef]
16. Vasina, E.M.; Cauwenberghs, S.; Staudt, M.; Feijge, M.A.H.; Weber, C.; Koenen, R.R.; Heemskerk, J.W.M. Aging- and activation-induced platelet microparticles suppress apoptosis in monocytic cells and differentially signal to proinflammatory mediator release. *Am. J. Blood Res.* **2013**, *3*, 107–123.
17. Laffont, B.; Corduan, A.; Ple, H.; Duchez, A.C.; Cloutier, N.; Boilard, E.; Provost, P. Activated platelets can deliver mrna regulatory ago2*microrna complexes to endothelial cells via microparticles. *Blood* **2013**, *122*, 253–261. [CrossRef]
18. Risitano, A.; Beaulieu, L.M.; Vitseva, O.; Freedman, J.E. Platelets and platelet-like particles mediate intercellular rna transfer. *Blood* **2012**, *119*, 6288–6295. [CrossRef]
19. Anene, C.; Graham, A.M.; Boyne, J.; Roberts, W. Platelet microparticle delivered microrna-let-7a promotes the angiogenic switch. *Biochim. Biophys. Acta Mol. Basis Dis.* **2018**, *1864*, 2633–2643. [CrossRef]
20. Maynard, D.M.; Heijnen, H.F.; Horne, M.K.; White, J.G.; Gahl, W.A. Proteomic analysis of platelet alpha-granules using mass spectrometry. *J. Thromb. Haemost.* **2007**, *5*, 1945–1955. [CrossRef]
21. Parsons, M.E.M.; Szklanna, P.B.; Guererro, J.A.; Wynne, K.; Dervin, F.; O'Connell, K.; Allen, S.; Egan, K.; Bennett, C.; McGuigan, C.; et al. Platelet releasate proteome profiling reveals a core set of proteins with low variance between healthy adults. *Proteomics* **2018**, *18*, e1800219. [CrossRef] [PubMed]
22. Pienimaeki-Roemer, A.; Konovalova, T.; Musri, M.M.; Sigruener, A.; Boettcher, A.; Meister, G.; Schmitz, G. Transcriptomic profiling of platelet senescence and platelet extracellular vesicles. *Transfusion* **2017**, *57*, 144–156. [CrossRef] [PubMed]
23. Ambrose, A.R.; Alsahli, M.A.; Kurmani, S.A.; Goodall, A.H. Comparison of the release of micrornas and extracellular vesicles from platelets in response to different agonists. *Platelets* **2018**, *29*, 446–454. [CrossRef]
24. Liang, H.; Yan, X.; Pan, Y.; Wang, Y.; Wang, N.; Li, L.; Liu, Y.; Chen, X.; Zhang, C.Y.; Gu, H.; et al. Microrna-223 delivered by platelet-derived microvesicles promotes lung cancer cell invasion via targeting tumor suppressor epb41l3. *Mol. Cancer* **2015**, *14*, 58. [CrossRef] [PubMed]
25. Tang, M.; Jiang, L.; Lin, Y.; Wu, X.; Wang, K.; He, Q.; Wang, X.; Li, W. Platelet microparticle-mediated transfer of mir-939 to epithelial ovarian cancer cells promotes epithelial to mesenchymal transition. *Oncotarget* **2017**, *8*, 97464–97475. [CrossRef] [PubMed]
26. Michael, J.V.; Wurtzel, J.G.T.; Mao, G.F.; Rao, A.K.; Kolpakov, M.A.; Sabri, A.; Hoffman, N.E.; Rajan, S.; Tomar, D.; Madesh, M.; et al. Platelet microparticles infiltrating solid tumors transfer mirnas that suppress tumor growth. *Blood* **2017**, *130*, 567–580. [CrossRef]

27. Cassier, P.A.; Castets, M.; Belhabri, A.; Vey, N. Targeting apoptosis in acute myeloid leukaemia. *Br. J. Cancer* **2017**, *117*, 1089–1098. [CrossRef]

28. Ashkenazi, A.; Fairbrother, W.J.; Leverson, J.D.; Souers, A.J. From basic apoptosis discoveries to advanced selective bcl-2 family inhibitors. *Nat. Rev. Drug Discov.* **2017**, *16*, 273–284. [CrossRef]

29. Velez, J.; Enciso, L.J.; Suarez, M.; Fiegl, M.; Grismaldo, A.; López, C.; Barreto, A.; Cardozo, C.; Palacios, P.; Morales, L.; et al. Platelets promote mitochondrial uncoupling and resistance to apoptosis in leukemia cells: A novel paradigm for the bone marrow microenvironment. *Cancer Microenviron.* **2014**, *7*, 79–90. [CrossRef]

30. Bai, H.; Zhou, L.; Wang, C.; Xu, X.; Jiang, J.; Qin, Y.; Wang, X.; Zhao, C.; Shao, S. Involvement of mir-125a in resistance to daunorubicin by inhibiting apoptosis in leukemia cell lines. *Tumour Biol.* **2017**, *39*, 1010428317695964. [CrossRef]

31. Zhou, L.; Bai, H.; Wang, C.; Wei, D.; Qin, Y.; Xu, X. Microrna125b promotes leukemia cell resistance to daunorubicin by inhibiting apoptosis. *Mol. Med. Rep.* **2014**, *9*, 1909–1916. [CrossRef] [PubMed]

32. Gabra, M.M.; Salmena, L. Micrornas and acute myeloid leukemia chemoresistance: A mechanistic overview. *Front. Oncol.* **2017**, *7*, 255. [CrossRef] [PubMed]

33. Garzon, R.; Volinia, S.; Liu, C.G.; Fernandez-Cymering, C.; Palumbo, T.; Pichiorri, F.; Fabbri, M.; Coombes, K.; Alder, H.; Nakamura, T.; et al. Microrna signatures associated with cytogenetics and prognosis in acute myeloid leukemia. *Blood* **2008**, *111*, 3183–3189. [CrossRef]

34. Lim, E.L.; Trinh, D.L.; Ries, R.E.; Wang, J.; Gerbing, R.B.; Ma, Y.; Topham, J.; Hughes, M.; Pleasance, E.; Mungall, A.J.; et al. Microrna expression-based model indicates event-free survival in pediatric acute myeloid leukemia. *J. Clin. Oncol.* **2017**, *35*, 3964–3977. [CrossRef] [PubMed]

35. Bruserud, O.; Gjertsen, B.T.; Foss, B.; Huang, T.S. New strategies in the treatment of acute myelogenous leukemia (aml): In vitro culture of aml cells—The present use in experimental studies and the possible importance for future therapeutic approaches. *Stem Cells (Dayt. Ohio)* **2001**, *19*, 1–11. [CrossRef]

36. Chen, Y.; Wang, X. Mirdb: An online database for prediction of functional microrna targets. *Nucleic Acids Res.* **2020**, *48*, D127–D131. [CrossRef] [PubMed]

37. Agarwal, V.; Bell, G.W.; Nam, J.W.; Bartel, D.P. Predicting effective microrna target sites in mammalian mrnas. *Elife* **2015**, *4*, e05005. [CrossRef] [PubMed]

38. Youle, R.J.; Strasser, A. The bcl-2 protein family: Opposing activities that mediate cell death. *Nat. Rev. Mol. Cell Biol.* **2008**, *9*, 47–59. [CrossRef]

39. Trotter, E.W.; Hagan, I.M. Release from cell cycle arrest with cdk4/6 inhibitors generates highly synchronized cell cycle progression in human cell culture. *Open Biol.* **2020**, *10*, 200200. [CrossRef]

40. Skowron, M.A.; Vermeulen, M.; Winkelhausen, A.; Becker, T.K.; Bremmer, F.; Petzsch, P.; Schönberger, S.; Calaminus, G.; Köhrer, K.; Albers, P.; et al. Cdk4/6 inhibition presents as a therapeutic option for paediatric and adult germ cell tumours and induces cell cycle arrest and apoptosis via canonical and non-canonical mechanisms. *Br. J. Cancer* **2020**, *123*, 378–391. [CrossRef]

41. Daigneault, M.; Preston, J.A.; Marriott, H.M.; Whyte, M.K.; Dockrell, D.H. The identification of markers of macrophage differentiation in pma-stimulated thp-1 cells and monocyte-derived macrophages. *PLoS ONE* **2010**, *5*, e8668. [CrossRef]

42. Bogason, A.; Quartino, A.L.; Lafolie, P.; Masquelier, M.; Karlsson, M.O.; Paul, C.; Gruber, A.; Vitols, S. Inverse relationship between leukaemic cell burden and plasma concentrations of daunorubicin in patients with acute myeloid leukaemia. *Br. J. Clin. Pharmacol.* **2011**, *71*, 514–521. [CrossRef]

43. Burnett, A.K.; Russell, N.H.; Hills, R.K.; Kell, J.; Cavenagh, J.; Kjeldsen, L.; McMullin, M.F.; Cahalin, P.; Dennis, M.; Friis, L.; et al. A randomized comparison of daunorubicin 90 mg/m2 vs. 60 mg/m^2 in aml induction: Results from the uk ncri aml17 trial in 1206 patients. *Blood* **2015**, *125*, 3878–3885. [CrossRef] [PubMed]

44. O'Neill, K.L.; Huang, K.; Zhang, J.; Chen, Y.; Luo, X. Inactivation of prosurvival bcl-2 proteins activates bax/bak through the outer mitochondrial membrane. *Genes Dev.* **2016**, *30*, 973–988. [CrossRef] [PubMed]

45. Yan, L.; Yu, M.C.; Gao, G.L.; Liang, H.W.; Zhou, X.Y.; Zhu, Z.T.; Zhang, C.Y.; Wang, Y.B.; Chen, X. Mir-125a-5p functions as a tumour suppressor in breast cancer by downregulating bap1. *J. Cell. Biochem.* **2018**, *119*, 8773–8783. [CrossRef]

46. Tong, Z.; Liu, N.; Lin, L.; Guo, X.; Yang, D.; Zhang, Q. Mir-125a-5p inhibits cell proliferation and induces apoptosis in colon cancer via targeting bcl2, bcl2l12 and mcl1. *Biomed. Pharmacother.* **2015**, *75*, 129–136. [CrossRef] [PubMed]

47. Marinello, J.; Delcuratolo, M.; Capranico, G. Anthracyclines as topoisomerase ii poisons: From early studies to new perspectives. *Int. J. Mol. Sci.* **2018**, *19*, 3480. [CrossRef]

48. Al-Aamri, H.M.; Ku, H.; Irving, H.R.; Tucci, J.; Meehan-Andrews, T.; Bradley, C. Time dependent response of daunorubicin on cytotoxicity, cell cycle and DNA repair in acute lymphoblastic leukaemia. *Bmc Cancer* **2019**, *19*, 179. [CrossRef] [PubMed]

49. Gažová, I.; Lefevre, L.; Bush, S.J.; Clohisey, S.; Arner, E.; de Hoon, M.; Severin, J.; van Duin, L.; Andersson, R.; Lengeling, A.; et al. The transcriptional network that controls growth arrest and macrophage differentiation in the human myeloid leukemia cell line thp-1. *Front. Cell Dev. Biol.* **2020**, *8*, 498. [CrossRef]

50. Bernardi, M.; Agostini, F.; Chieregato, K.; Amati, E.; Durante, C.; Rassu, M.; Ruggeri, M.; Sella, S.; Lombardi, E.; Mazzucato, M.; et al. The production method affects the efficacy of platelet derivatives to expand mesenchymal stromal cells in vitro. *J. Transl. Med.* **2017**, *15*, 90. [CrossRef]

51. Agostini, F.; Polesel, J.; Battiston, M.; Lombardi, E.; Zanolin, S.; Da Ponte, A.; Astori, G.; Durante, C.; Mazzucato, M. Standardization of platelet releasate products for clinical applications in cell therapy: A mathematical approach. *J. Transl. Med.* **2017**, *15*, 107. [CrossRef]
52. Plantureux, L.; Mège, D.; Crescence, L.; Carminita, E.; Robert, S.; Cointe, S.; Brouilly, N.; Ezzedine, W.; Dignat-George, F.; Dubois, C.; et al. The interaction of platelets with colorectal cancer cells inhibits tumor growth but promotes metastasis. *Cancer Res.* **2020**, *80*, 291–303. [CrossRef]
53. Żmigrodzka, M.; Witkowska-Piłaszewicz, O.; Winnicka, A. Platelets extracellular vesicles as regulators of cancer progression-an updated perspective. *Int. J. Mol. Sci.* **2020**, *21*, 5195. [CrossRef]
54. Giacomazzi, A.; Degan, M.; Calabria, S.; Meneguzzi, A.; Minuz, P. Antiplatelet agents inhibit the generation of platelet-derived microparticles. *Front. Pharmacol.* **2016**, *7*, 314. [CrossRef] [PubMed]
55. Chan, A.T.; Ogino, S.; Fuchs, C.S. Aspirin use and survival after diagnosis of colorectal cancer. *JAMA* **2009**, *302*, 649–658. [CrossRef]
56. Bains, S.J.; Mahic, M.; Myklebust, T.A.; Smastuen, M.C.; Yaqub, S.; Dorum, L.M.; Bjornbeth, B.A.; Moller, B.; Brudvik, K.W.; Tasken, K. Aspirin as secondary prevention in patients with colorectal cancer: An unselected population-based study. *J. Clin. Oncol.* **2016**, *34*, 2501–2508. [CrossRef] [PubMed]
57. Cao, Y.; Nishihara, R.; Wu, K.; Wang, M.; Ogino, S.; Willett, W.C.; Spiegelman, D.; Fuchs, C.S.; Giovannucci, E.L.; Chan, A.T. Population-wide impact of long-term use of aspirin and the risk for cancer. *JAMA Oncol.* **2016**, *2*, 762–769. [CrossRef] [PubMed]
58. Qiao, Y.; Yang, T.; Gan, Y.; Li, W.; Wang, C.; Gong, Y.; Lu, Z. Associations between aspirin use and the risk of cancers: A meta-analysis of observational studies. *Bmc Cancer* **2018**, *18*, 288. [CrossRef]
59. Poncelet, P.; Robert, S.; Bouriche, T.; Bez, J.; Lacroix, R.; Dignat-George, F. Standardized counting of circulating platelet microparticles using currently available flow cytometers and scatter-based triggering: Forward or side scatter? *Cytom. Part A* **2016**, *89*, 148–158. [CrossRef]
60. Ludwig, L.M.; Maxcy, K.L.; LaBelle, J.L. Flow cytometry-based detection and analysis of bcl-2 family proteins and mitochondrial outer membrane permeabilization (momp). *Methods Mol. Biol. (Cliftonn. J.)* **2019**, *1877*, 77–91.
61. Schneider, C.A.; Rasband, W.S.; Eliceiri, K.W. Nih image to imagej: 25 years of image analysis. *Nat. Methods* **2012**, *9*, 671–675. [CrossRef] [PubMed]

Review

Surface-Enhanced Raman Scattering (SERS) Spectroscopy for Sensing and Characterization of Exosomes in Cancer Diagnosis

Luca Guerrini [1,*], **Eduardo Garcia-Rico** [2,3], **Ana O'Loghlen** [4], **Vincenzo Giannini** [5,6] **and Ramon A. Alvarez-Puebla** [1,7,*]

[1] Department of Physical and Inorganic Chemistry, Universitat Rovira i Virgili, Carrer de Marcel·li Domingo s/n, 43007 Tarragona, Spain
[2] Fundación de Investigación HM Hospitales, San Bernardo 101, 28015 Madrid, Spain; egarcia@hmhospitales.com
[3] School of Medicine, San Pablo CEU, Calle Julian Romea, 18, 28003 Madrid, Spain
[4] Epigenetics & Cellular Senescence Group, Blizard Institute, Barts and The London School of Medicine and Dentistry, Queen Mary University of London, London E1 2AT, UK; a.ologhlen@qmul.ac.uk
[5] Instituto de Estructura de la Materia (IEM-CSIC), Consejo Superior de Investigaciones Científicas, 28006 Madrid, Spain; v.giannini@csic.es
[6] Technology Innovation Institute, Masdar City, Abu Dhabi 9639, United Arab Emirates
[7] ICREA, Passeig Lluis Companys 23, 08010 Barcelona, Spain
* Correspondence: luca.guerrini@urv.cat (L.G.); ramon.alvarez@urv.cat (R.A.A.-P.)

Citation: Guerrini, L.; Garcia-Rico, E.; O'Loghlen, A.; Giannini, V.; Alvarez-Puebla, R.A. Surface-Enhanced Raman Scattering (SERS) Spectroscopy for Sensing and Characterization of Exosomes in Cancer Diagnosis. *Cancers* **2021**, *13*, 2179. https://doi.org/10.3390/cancers13092179

Academic Editor: Farrukh Aqil

Received: 24 March 2021
Accepted: 26 April 2021
Published: 30 April 2021

Publisher's Note: MDPI stays neutral with regard to jurisdictional claims in published maps and institutional affiliations.

Simple Summary: The distinct molecular and biological properties of exosomes, together with their abundance and stability, make them an ideal target in liquid biopsies for early diagnosis and disease monitoring. On the other hand, in recent years, nanomaterial-based optical biosensors have been extensively investigated as novel, rapid and sensitive tools for exosome detection and discrimination. The scope of this review is to summarize and coherently discussed the diverse applications, challenges and limitations of nanosensors based on surface-enhanced Raman spectroscopy (SERS) as the optosensing technique.

Abstract: Exosomes are emerging as one of the most intriguing cancer biomarkers in modern oncology for early cancer diagnosis, prognosis and treatment monitoring. Concurrently, several nanoplasmonic methods have been applied and developed to tackle the challenging task of enabling the rapid, sensitive, affordable analysis of exosomes. In this review, we specifically focus our attention on the application of plasmonic devices exploiting surface-enhanced Raman spectroscopy (SERS) as the optosensing technique for the structural interrogation and characterization of the heterogeneous nature of exosomes. We summarized the current state-of-art of this field while illustrating the main strategic approaches and discuss their advantages and limitations.

Keywords: exosomes; cancer diagnosis; sensing; early detection; plasmonics; nanoparticles; surface-enhanced Raman spectroscopy

1. Introduction

In the impending decades, cancer is set to become a major cause of morbidity and mortality across all regions of the globe [1], with an estimated 13.2 million related deaths by 2030 [1,2]. Thus, the development of more effective treatments and, fundamentally, new forms of prevention and early diagnosis are both necessary strategies to achieve a cure [3]. In fact, diagnosis at the very earliest stages improves cancer outcomes by prompting treatments aimed at preventing the disease development to incurable stages.

The prevailing theory about the origin of cancer indicates that a primary tumor develops for a long time, from a subclinical or microscopic level, before it spreads at distance (metastasis) [4]. To be clinically detectable, a tumor must reach a size of ca. 1 cm^3, which approximatively contain 10^9 cells [5]. Therefore, at the time of diagnosis, there is a high probability of prior dissemination. As a result, there is an urgent need for new technologies capable of detecting the presence of tumor cells before the disease emerges as clinically visible. In this regard, exosome-based liquid biopsy in peripheral blood and other body fluids is among the most promising techniques for pre-metastatic cancer diagnosis [6]. The validity of such an approach builds upon the current concept and understanding of metastasis [7–11]. Indeed, it has been recognized that before they spread to distant sites, the original or primary tumors "communicate" with cells and tissues of other organs, as well as their surrounding environment, to prepare what will be eventually a metastatic niche. This horizontal intercellular communication takes place through exosomes. In bone-marrow [7], for example, hematopoietic progenitor cells that express VEGFR1 are located in tumor premetastatic sites and form cellular clusters induced by exosomes originated in primary tumor cells. In these niches, such cells express VLA4 and certain integrins that facilitate the arrival of tumor cells, a process that is also mediated by exosomes [12]. On the other hand, by this communication, tumor stem cell deference can be induced from normal cells.

Exosomes were first described in the 1960s as vesicles related to coagulation processes derived from platelets and, two decades later, they were associated with enzymatic functions [13]. Subsequent observations showed that these vesicles were generated as cell desquamation in the reticulocyte maturation process [14]. Exosomes are small, single-membrane vesicles approximately between 30 and 150 nm diameter, secreted by practically all cells into the extracellular environment through the fusion of specific endosomes (multivesicular bodies, MVBs) with the plasma membrane. MVBs are formed by primary endosomes which are incorporated as "intraluminal vesicles" (ILV) via inward budding of the multivesicular body membrane [12]. They can follow this secretory pathway towards the extracellular environment or a degradative pathway through their fusion with lysosomes. Exosomes have been shown to intervene in multiple functions (e.g., immune response, healing, viral synthesis, antigenic presentation, etc.) [15,16]. In cancer, multiple functions have been attributed to exosome-mediated communication such as reprogramming of stromal cells, initiation of metastasis, preparation of metastatic niches, modelling of the immune response and extracellular matrix, drug resistance, antigen presentation, etc. [17]. Notably, such intercellular exosomal communication takes place in both directions: from tumor cell to normal cell and vice versa. Thus, tumor cells can gain capacities such as "invasiveness" or enhance their proliferative efficiency. An example of this reverse communication process has been observed for normal adipocytes, which secrete exosomes carrying proteins involved in the oxidation of fatty acids that are eventually incorporated into melanoma cells. Such process culminates in an increase of this function in malignant cells, which enhances their migration and invasion capabilities [18].

Exosome lumen and membrane carry biological and genetic information related to their parental cell types as they are selectively enriched of specific nucleic acids (e.g., mRNA, miRNA, tRNA, etc.), proteins (e.g., integrins, immunoglobulins, growth factors, cytoskeletal protein actin and tubulin, endosomal sorting complex required for transport ESCRT-related proteins, hsp90, hsp70, tetraspanin, major histocompatibility complex), lipids, metabolites and glycoconjugates (Figure 1) [19,20]. It is also worth stressing that the exosome biogenesis itself can significantly impact their composition and functionality [3]. In this sense, the most significant aspect is that exosome molecular composition is not a mere random replica of the original cell but is selected and specific. The overall mechanisms of such selection (sorting) are very complex and have been only recently being unveiled. In general terms, the integration of the molecular components into endosomes takes place selectively via recognition by specific sequences of nucleotides or peptides. This mechanism is similar to ubiquitination, a process that marks and selects proteins destined for endosomal degradation [21]. In this way, RNA molecules containing a specific sequence (EXOmotif) are recognized by certain proteins (hnRNPA2B1) that facilitate their entry into exosomes. These proteins also undergo an activation process through reactions known as sumoylation, which helps integration, or isegilation (incorporation of ISG15 into TSG101), which inhibits the generation of exosomes (or facilitates their elimination by fusion to lysosomes) [22,23]. As a result, characteristic miRNAs in exosomes nucleic acid cargoes, while not expressed in the corresponding healthy tissues, have been detected, for instance, in breast cancer [24] and lung adenocarcinoma [25], demonstrating their validity as unique disease markers. Similarly, exosome membrane protein composition has also shown to be correlated with the nature of the originating cell and the transformation events that have undergone [26,27], which also make them promising diagnostic biomarkers and therapeutic targets [26–28]. This is consistent with the central role played by surface membrane proteins of exosomes in malignant processes such as metastasis [29,30]. Moreover, glycans bound to surface proteins and outer lipids are also found onto the exosomal surfaces [19], and they have been reported to play an important biological role, among others, in the exosome uptake [31]. Overall, the unique distinct molecular and biological properties of exosomes, together with their abundance and stability, make them an ideal target in liquid biopsies not only for early diagnosis [6] but also for disease monitoring and, finally, an opportunity for cancer cure [32].

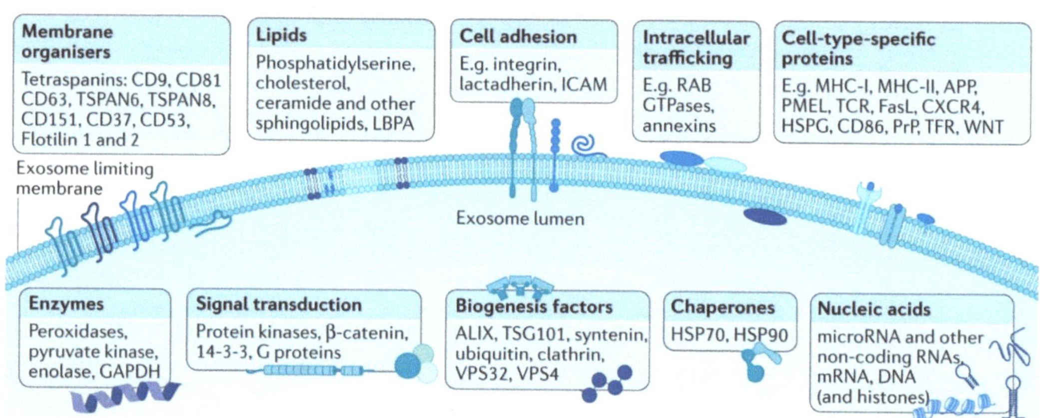

Figure 1. General outlook of the exosome membrane composition and different molecular cargoes in the lumen which can markedly vary based on the parental cell and vesicle biogenesis. Adapted with permission from [33]. Copyright 2018, Nature Publishing.

2. Isolation and Characterization of Exosomes

The pronounced molecular and size heterogeneity of exosomes, even for vesicles originating from the same parental cells, confers a central role to isolation methods in (i) separating exosomes from potentially interfering protein aggregates, lipoparticles, viruses and cell debris in cell culture supernatants or bodily fluids, and (ii) discriminating different exosome subpopulations that could be related with different pathological states and stages of disease progression [33]. The gold standard for exosome isolation is differential centrifugation which, through several centrifugation rounds (such as ultra-high-speed centrifugation or ultracentrifugation), selectively precipitates the vesicles of interest with high purity [34]. Ultracentrifugation, however, is a slow separation method with low recovery efficiency (<25%) that requires costly and bulky instrumentations and, thus, is not suitable for the point-of-care diagnosis [35]. Additional separation strategies, exploiting diverse physiochemical properties of exosomes, include size exclusion chromatography, ultrafiltration, immunoaffinity capturing, charge neutralization-based polymer precipitation, and microfluidic techniques, each of them with a characteristic set of advantages and disadvantages [35]. Physical characterization of the isolated vesicles (i.e., enumeration, size distribution and morphology) are commonly determined via nanoparticle tracking analysis (NTA), flow cytometry, microscopy methods (e.g., transmission electron microscopy, TEM; scanning electron microscopy, SEM) and dynamic light scattering [36–38]. On the other hand, exosome protein quantification is conventionally carried out via Western blotting and enzyme-linked immunosorbent assay (ELISA) [38]. However, western blotting typically requires complex and time-consuming procedures as well as relatively large volumes of biosamples; whereas ELISA fails to execute multiplexed analysis. Differently, nucleic acid cargoes, mainly RNAs, are commonly analyzed upon extraction via amplification and sequencing techniques (e.g., PCR, next-generation sequencing) [38].

In recent years, nanomaterial-based optical biosensors have been extensively investigated as novel, rapid and sensitive tools for exosome detection and discrimination [39–41]. Within the field of nanoplasmonic, surface-enhanced Raman spectroscopy (SERS) has emerged as a powerful optical technique for a very broad range of applications [42–45], with the most intriguing one being in biosensing and clinical diagnostic [46–50]. SERS is an analytical technique that relies on the excitation of strong electromagnetic fields (i.e., localized surface plasmon resonances, LSPRs) at the surface of plasmonic materials (mainly, silver and gold nanostructures) (Figure 2A). As a result of the excitation of the molecular species with the LSPR rather than with the illuminating light, the Raman scattering of molecules located in close contact or directly attached to the plasmonic substrate undergoes a notable amplification, up to a factor of ca. 10^{10}–10^{11} [51]. Thus, SERS simultaneously affords an ultra-sensitive optical response based on the plasmonic associated intensification and the intrinsically rich structural information contained in the Raman spectra. In Figure 2 we provide few illustrative examples with the aim of intuitively emphasize some of the key concepts of SERS spectroscopy which have major implications in the application to exosome analysis, as discussed later in the review. We refer the readers to references [46,51,52] for detailed insights on the theoretical and experimental aspects of SERS and related plasmonic substrates. Firstly, Figure 2B depicts the calculated electromagnetic field around a silver nanosphere of 45 nm diameter, hinting the distance-dependent nature of the SERS phenomenon. In fact, the local field enhancements swiftly drop at an increasing distance, d, from the metallic surface (the decay is $\sim 1/(a + d)^{12}$ for a nanosphere of radius a) [51]. In an explicative study, Kumari et al. [53] synthesized spherical silver colloids of increasing diameter and coat them with silica shells of progressively larger thicknesses (silica prevents the direct contact between the analyte and the nanoparticle). Results show that the SERS intensity decays exponentially for all nanoparticle size as silica shell thickness is increased. On the other hand, the enhancing properties of silver colloids improve with the nanoparticle size up to ca. 100 nm diameter before dropping due radiation effects that reduce the quality of the LSPRs (i.e., plasmon damping) for larger particles. Accordingly, the distance from the metallic surface up to which the SERS signal of the analyte can be observed (i.e., accessible distance) increases with the nanoparticle size up to a maximum of ca. 5 nm distance for ca. 90 nm size colloids. While nanosphere size plays a role in determining the final enhancing properties [54], the largest optical intensifications are, nonetheless, achieved at the tips of sharp protruding features [55,56] and, even more so, at nanometer-sized gaps between metal nanoparticles (i.e., hot-spots) due to interparticle plasmon coupling [52,57]. This latter effect is plainly visualized in Figure 2B–D, which compares the calculated electromagnetic fields in dimers and larger aggregates with that of their parental isolated nanosphere [58]. Notably, local enhancements at the gaps rapidly rise with the shortening of the interparticle distance (see the case of a silver nanoparticle dimer in Figure 2F). However, this simultaneously occurs at an increasing degree of spatial localization (i.e., higher enhancements extend over a smaller volume around the hotspot for smaller gaps) [51]. These aspects highlight the importance of an appropriate structural design of the plasmonic substrate as well as the successful localization of the target molecule within the volumes where the largest enhancements take place. It is also worth noting that, besides the dominant plasmonic-mediated amplification via an electromagnetic mechanism, additional enhancements can result from electron charge transfers between the surface and the analyte (i.e., chemical mechanisms).

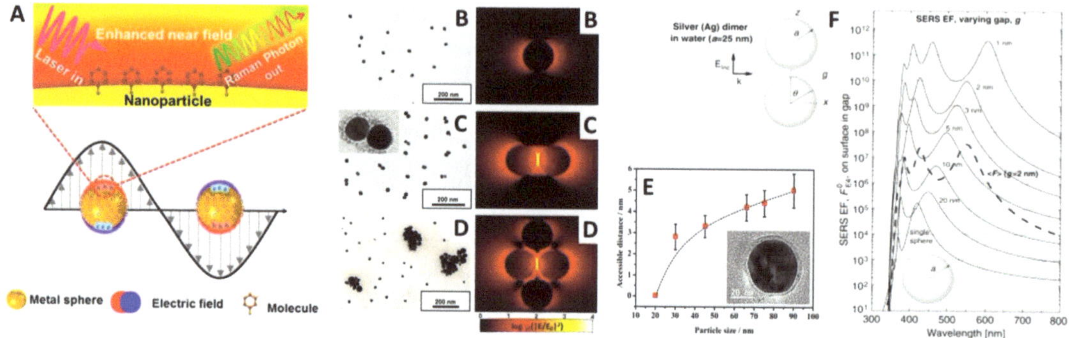

Figure 2. (**A**) Outline of the surface-enhanced Raman scattering effect due to the excitation of localized surface plasmon resonances at a gold nanosphere/air interface. Reprinted with permission from [59]. Copyright 2017, American Chemical Society. (**B–D**) TEM images and calculated electrical fields for an isolated silver nanoparticle of 45 nm diameter and its corresponding dimer and tetramer (as a representative example of a largercluster), respectively, under a 514 nm excitation laser (interparticle gap, g = 1.31 nm). Reprinted with permission from [58]. Copyright 2015, American Chemical Society. (**E**) TEM image of a silica-coated silver nanoparticle and plot of the maximum silica shell thickness at which the SERS signal of rhodamine 6G is still detectable (i.e., accessible distance) as a function of nanoparticle diameter. Adapted with permission from [53]. Copyright 2015, American Chemical Society. (**F**) Dimer composed of two identical silver nanospheres (radii a = 25 nm) separated by a gap g (the incoming wave is polarized along the axis of the dimer) and their theoretical SERS enhancement factors calculated at the point on the surface in the gap (i.e., hot-spot) as a function of the excitation wavelength for different gaps. The thick dashed line is the average SERS enhancement factor in the case of a 2 nm gap. Adapted with permission from [51]. Copyright 2009, Elsevier.

Most studies aimed at correlating the intercellular signalling and pathological responses of exosomes with their composition focused on analyzing their respective nucleic acid cargoes which often mirror the phenotypes of their parental cells [19,28,60,61]. SERS-based detection of RNA cargoes extracted from exosomes contained in blood samples of patients have been reported, for instance, for early detection of pancreatic cancer [62] and lung cancer [63]. In these studies, exosomes were separated from plasma and, subsequently, microRNA cargoes were isolated using available kits yielding miRNA elutes to be analyzed. Thus, the scientific challenges and sensing strategies of these approaches are by and large independent of the biomolecule source and fall within the field of nucleic acids SERS detection [64–67]. On the other hand, in this review, we will focus on the burgeoning body of work on SERS analysis of whole exosomes, a field of research that has been growing at an extremely fast rate in very recent years. For reasons that will be explained shortly, whole exosome SERS analysis mostly focuses and build upon the diversity in their membrane composition.

Broadly speaking, SERS detection approaches can be classified into two main configurations: direct and indirect schemes. In direct SERS, the signal read-out is provided by the acquisition of the intrinsic SERS spectrum of the analyte, which contains a wealth of structural information representative of the molecular structure and composition. This label-free analysis can be carried out with simple and inexpensive plasmonic materials. However, it is usually restricted to the interrogation of relatively pure samples to prevent competing co-adsorptions of other molecular species that could undermine the correct interpretation and reliability of the final SERS spectrum. Moreover, direct SERS characterization of large biomolecules or supramolecular structures such as exosome poses important challenges in terms of understanding and interpretation of complex and often highly similar vibrational patterns. On the other hand, indirect approaches are designed to monitor the extrinsic SERS signal of molecular labels for detection and quantification of the target species. The most common indirect strategy relies on the use of SERS-encoded nanoparticles (or SERS tags) combined with surface ligands for selective recognition (e.g., antibody, aptamers), as optical probes performing similar functions as fluorescent labels [68]. Although labelled methods typically demand elaborate and extensive fabrications of relatively expensive SERS substrates, they also feature ultrasensitivity, high-throughput screening, multiplexing abilities, robust quantitative response in complex media (e.g., biofluids) and suitability to be integrated into miniaturized devices for automated testing, especially at the point-of-care. For these reasons, they have been largely preferred for biosensing applications over direct approaches. Nonetheless, the current literature survey on exosome SERS analysis shows a larger number of reports based on direct approaches [69–91] than indirect ones [92–101]. This apparent anomaly may be explained by taking into considerations the intrinsic exosomal heterogeneity that currently burdens the identification of specific disease-related biomarkers for selective separation and labelling of clinically relevant exosome subpopulations [90,99,102]. Thus, the more holistic approach of acquiring the vibrational fingerprint of the whole ensemble of molecular constituents, including known and unknown biomarkers, remains a very valuable and effective sensing strategy as compared to indirect analytical methods that selectively inform about one or few structural features.

3. Direct Label-Free SERS Analysis of Exosomes

The acquisition of intense, well-defined and reproducible vibrational spectra is key to use direct SERS for sensing purposes. Overall, several factors determine the final intensification and ultimate spectral profile (i.e., band centers, relative intensities, bandwidths) of the vibrational fingerprint. Among those, we can identify inherent features of each element of the SERS analysis, such as the optical properties of the plasmonic material, the Raman cross-section of the target analyte and the experimental set-up (e.g., laser excitation wavelength) [51]. On the other hand, more intertwined variables play also a central role, such as the extent of analyte surface coverage and the relative spatial localization of the analyte with respect to the metallic surface. As previously discussed, the plasmon-mediated electromagnetic enhancement dramatically declines with the distance from the plasmonic surface. This phenomenon accounts for the observation that SERS spectra are typically dominated by the contributions of the first layer of molecules directly exposed to the metal surface. Furthermore, the adsorption of the molecular entity onto the metallic surface may induce both specific orientations and potential alterations of its Raman polarizability which can severely impact the spectral profile of the SERS signal [51]. Thus, the acquisition of reliable SERS spectra is mostly constrained by the careful control and knowledge of all these parameters.

In the liquid phase, the close contact between the analyte and the plasmonic surface is commonly achieved by exploiting the intrinsic chemical affinity of the molecule for gold or silver surfaces, mainly via the formation of metal-O, metal-N, and metal-S bonds (in the typical order of relative increasing strength) or via electrostatic interactions. Alternatively, molecular adhesion can be forced via physical evaporation of the sample solution onto the plasmonic substrate. In this regard, it is worth stressing that for large biomolecules, such as the exosome components (proteins, nucleic acids, etc.) and even further to micro-entities such as cells, a transition from a hydrated to a dried state often results in major structural alterations that usually increase the intra-sample spectral variability [103,104]. All these considerations also justify the need for pre-isolation steps to extract the target biomolecules from complex biological environments containing a multitude of other molecular species which would otherwise compete for the adsorption onto the metallic surface and, eventually, yield unintelligible SERS spectra.

In the specific case of exosomes analysis, several additional features further increase the complexity of common direct SERS analysis. Firstly, the complex composition of exosomes, which mostly includes a large fraction of biomolecules with typically weak spontaneous Raman scattering (e.g., lipids, proteins), intrinsically generates an intricate vibrational spectral pattern comprising overlapping and often broad features. As a representative example, in Figure 3 we report a normal Raman spectrum of exosomes isolated from rat hepatocytes together with a table illustrating the main vibrational features and their assignment to dominant molecular contributions. As exosomes differentiation occurs via recognition of subtle spectral differences, the use of multivariate mathematic and statistic methods is commonly required for a more accurate spectral analysis. These mathematical methods (e.g., principal component analysis, PCA; partial least square discriminant analysis, PLS-DA, etc.) reduce the high multidimensionality of the large set of vibrational data by identifying a dominant, smaller group of variables that still retains most of the key information of the initial large data set. It is worth noting that the spectral complexity may be also exacerbated by the inherited heterogeneity of the exosome particles, which further stresses the central role of efficient and reliable isolation methods to yield relative pure fractions of exosomes for direct SERS interrogation.

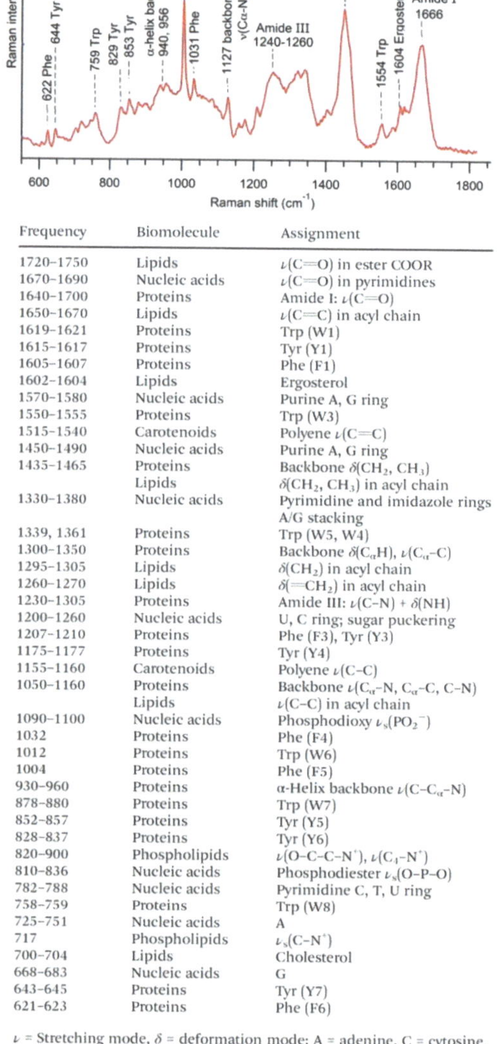

Frequency	Biomolecule	Assignment
1720–1750	Lipids	ν(C=O) in ester COOR
1670–1690	Nucleic acids	ν(C=O) in pyrimidines
1640–1700	Proteins	Amide I: ν(C=O)
1650–1670	Lipids	ν(C=C) in acyl chain
1619–1621	Proteins	Trp (W1)
1615–1617	Proteins	Tyr (Y1)
1605–1607	Proteins	Phe (F1)
1602–1604	Lipids	Ergosterol
1570–1580	Nucleic acids	Purine A, G ring
1550–1555	Proteins	Trp (W3)
1515–1540	Carotenoids	Polyene ν(C=C)
1450–1490	Nucleic acids	Purine A, G ring
1435–1465	Proteins	Backbone δ(CH$_2$, CH$_3$)
	Lipids	δ(CH$_2$, CH$_3$) in acyl chain
1330–1380	Nucleic acids	Pyrimidine and imidazole rings A/G stacking
1339, 1361	Proteins	Trp (W5, W4)
1300–1350	Proteins	Backbone δ(C$_\alpha$H), ν(C$_\alpha$–C)
1295–1305	Lipids	δ(CH$_2$) in acyl chain
1260–1270	Lipids	δ(=CH$_2$) in acyl chain
1230–1305	Proteins	Amide III: ν(C–N) + δ(NH)
1200–1260	Nucleic acids	U, C ring; sugar puckering
1207–1210	Proteins	Phe (F3), Tyr (Y3)
1175–1177	Proteins	Tyr (Y4)
1155–1160	Carotenoids	Polyene ν(C–C)
1050–1160	Proteins	Backbone ν(C$_\alpha$–N, C$_\alpha$–C, C–N)
	Lipids	ν(C–C) in acyl chain
1090–1100	Nucleic acids	Phosphodioxy ν_s(PO$_2^-$)
1032	Proteins	Phe (F4)
1012	Proteins	Trp (W6)
1004	Proteins	Phe (F5)
930–960	Proteins	α-Helix backbone ν(C–C$_\alpha$–N)
878–880	Proteins	Trp (W7)
852–857	Proteins	Tyr (Y5)
828–837	Proteins	Tyr (Y6)
820–900	Phospholipids	ν(O–C–C–N⁺), ν(C$_1$–N⁺)
810–836	Nucleic acids	Phosphodiester ν_s(O–P–O)
782–788	Nucleic acids	Pyrimidine C, T, U ring
758–759	Proteins	Trp (W8)
725–751	Nucleic acids	A
717	Phospholipids	ν_s(C–N⁺)
700–704	Lipids	Cholesterol
668–683	Nucleic acids	G
643–645	Proteins	Tyr (Y7)
621–623	Proteins	Phe (F6)

ν = Stretching mode, δ = deformation mode; A = adenine, C = cytosine, G = guanine, T = thymine, U = uracil; Trp = tryptophan, Tyr = tyrosine, Phe = phenylalanine.

Figure 3. Normal Raman spectrum of exosomes from rat hepatocytes (water buffer contribution was removed via subtraction) and vibrational assignment of the main bands. Adapted with permission from [105]. Copyright 2019, Royal Society of Chemistry.

Secondly, the size range itself of exosomes (ca. 30–150 nm) poses additional challenges. In fact, exosomes are large enough to prevent their optimum trapping into nanometric plasmonic gaps (hot spots) capable of concentrating extremely high intense EM fields in the whole analyte volume. Thus, the design and choice of the plasmonic substrate and experimental set-up analysis face two contrasting needs: (i) the necessity to expose exosomes to high electromagnetic enhancements to improve the amplification of their relatively weak Raman scattering; and (ii) the importance to immerse the vesicle in a relatively uniform electromagnetic field so that to minimize heterogeneous enhancements of random portions of exosomes due to different spatial arrangements onto the metallic surface. Indeed, such uneven exposures may lead to high spectral variability even within the same population of exosomes, an outcome that can be further aggravated by the heterogeneous distribution of the diverse molecular components on the exosome surface. Clearly, the reproducibility issue becomes particularly relevant when the SERS spectra are acquired from single/few exosomes rather than large ensembles of particles (i.e., single vs bulk analysis). For instance, Russo et al. [69] observed a significant loss in intra-sample spectral reproducibility, as compared to normal Raman spectroscopy, for drop-cast exosomes on non-uniform plasmonic substrates comprising randomly distributed gold nanostructures. The existence of such a significant number of variables arising from different sources (e.g., origin of the vesicle, isolation protocols, physiochemical characteristics of the plasmonic substrate, sample preparation, experimental set-up, etc.) is most likely the reason why we can observe, from study to study, marked fluctuations of the exosome spectral profiles that appear to go beyond the intrinsic biochemical nature of the interrogated vesicles.

Exosome sizes are, on the other hand, typically too small to facilitate single-particle Raman analysis. This explains why, for instance, single-cell Raman spectroscopy is a well-established and relatively straightforward tool for in vitro and in vivo interrogation of individual living cells while Raman characterization of individual/few exosomes is very limited and requires complex technologies such as exosome trapping via optical tweezers [105–107].

Finally, as the electromagnetic enhancement commonly declines very rapidly within few nanometers from the metallic surface, direct SERS analysis of whole exosomes yield spectra that are mostly dominated by the vibrational features of the molecular components of the outer membrane (mainly sugars and proteins). Thus, SERS spectra substantially disregard the lumen content as compared to normal Raman scattering of whole EVs, where nucleic acids contributions are distinguishable in the vibrational pattern [107]. While this aspect prevents the application of SERS as a technique for characterizing the global biomolecular composition of exosomes, it appears not to hamper its viability in diagnostic applications. In this regard, both normal Raman and SERS studies showed that trypsinization of exosomes drastically reduces the capability of differentiating subpopulations of exosomes, including vesicles from diverse cellular sources [73,107]. The enzymatic treatment of exosomes with trypsin promotes the cleavage of most surface membrane proteins and surface glycans, thereby exposing the intraluminal content to plasmonic-mediated signal enhancement. This result highlights the central role of the extraluminal domain for exosome differentiation and, in turn, the validity of the direct SERS approach for whole exosome classification.

Spherical-like gold and silver colloids synthesized via chemical reduction in solution are easily and reproducibly prepared in large batches at very low cost, yielding very amenable plasmonic materials for SERS analysis of exosomes. Most likely, the simplest and cheapest approach to generate SERS substrates rich in electromagnetic hot-spots is by the direct casting of colloids onto glass slides [79,80]. For instance, Choi and co-workers [80] dried 80 nm gold nanoparticles on a cover glass previously functionalized with 3-aminopropyltriethoxysilane (APTES) to yield a positive surface charge that would promote the adhesion of the negatively charged colloids via electrostatic binding (Figure 4A). Similarly, dried nanoparticle surfaces were further modified with cysteamine to promote the subsequent adsorption of the negatively charged exosomes (Figure 4B). SERS spectra of the vesicles were acquired at the edges of the dried spot (Figure 4C) where very dense nanoparticle clusters accumulate due to the coffee-stain effect (Figure 4D). Figure 4E shows representative SERS spectra of exosomes obtained via size-exclusion column chromatography from HPAEC (normal) and H1299, PC9 (lung cancer) cell lines (phosphate-buffered saline is used as a control). PCA score plot of the SERS data clearly shows the efficient discrimination between normal vs cancer cells-derived exosomes (Figure 4F). Besides a mere differentiation of distinct spectral patterns from normal and cancerous exosomes, the recognition of the molecular origin of the Raman markers at the core of such discrimination would provide a deeper understanding of their biochemical nature and, also, increase the diagnostic and prognostic value of direct SERS analysis. To this end, the authors performed a ratiometric analysis by acquiring the averaged SERS spectra of mixtures of normal and cancerous exosomes at different ratios (the total amount of exosomes was fixed at 10^8 particles/mL). Subsequently, they identified 13 bands that correlated well with the relative exosome content and which were used as Raman markers for non-small-cell lung carcinoma (NSCLC) derived exosomes (Figure 4G,H). These features were then compared to the vibrational profiles of clinically relevant exosomal protein markers (CD9, CD81, EpCAM, and EGFR). While all these individual protein markers display similar peak compositions, they diverge in relative band intensities thereby generating a unique spectral pattern. Notably, the ensemble of the Raman markers for NSCLC exosomes displayed low similarity for CD9, CD81 and EpCAM spectral fingerprints but high similarity for EGFR, indicating that EGFR expression is a primary variable of NSCLC exosome differentiation (Figure 4I), as further confirmed by immunoblotting analysis.

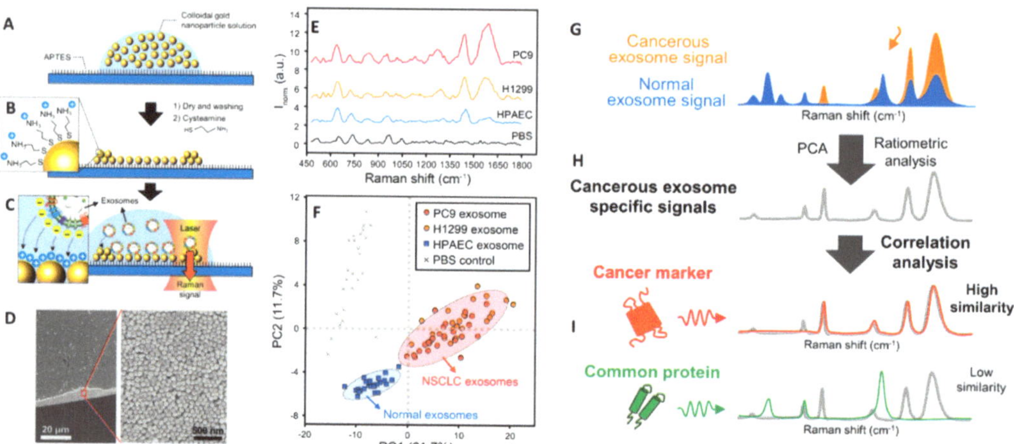

Figure 4. (**A–C**) Outline of the substrate fabrication and experimental set-up. (**D**) Scanning electron microscopy (SEM) image at an edge of the substrate. (**E**) SERS spectra of exosomes derived from HPAEC (normal) and H1299, PC9 (lung cancer) cell lines. Phosphate-buffered saline (PBS) was chosen as the experimental control. (**F**) PCA score plot of the SERS data and 90% confidence ellipses. (**G–I**) Schematic representation of the experimental process for the identification of unique SERS profile of lung cancer cell-derived exosomes followed by comparison to the profiles of their potential surface protein markers to determine their respective similarity. Adapted with permission from [80]. Copyright 2018, American Chemical Society.

Besides the direct casting of preformed colloids onto glass-slide surfaces, inexpensive SERS substrate can be generated by in-situ synthesis of plasmonic nanoparticles anchored onto the solid support. In this regard, Ferreira et al. [81] reported the simple fabrication of a hybrid SERS material via in situ silver nanoparticles growth into bacterial cellulose (BC), a low-cost and abundant support obtained from commercial nata de coco. The viability of the substrate for direct SERS analysis of exosomes was demonstrated in the efficient discrimination of exosome samples isolated from MCF-10A (nontumorigenic breast epithelium) and MDA-MB-231 (breast cancer) cell cultures.

A practical and straightforward alternative to promote exosome-nanoparticles interactions is by combining vesicles and plasmonic colloids in suspension before their deposition onto a support slide for SERS interrogation [71,79,87,88]. A common drawback of this method is the relatively low affinity of common negatively-charged gold and silver colloids (typically, citrate-stabilized) for similarly negatively-charged exosome membranes that reduces the extent of nanoparticle loading onto the vesicle surface [71]. To tackle this issue, Fraire et al. [88] modified gold nanoparticles with 4-(dimethylamino)pyridine (DMAP) to impart positive charge (DMAP-AuNPs) and, consequently, favor the electrostatic adhesion onto exosomes vesicles derived from B16F10 melanoma cells. In this regard, it is also worth noting that the largest enhancements of the exosome SERS signals have been observed for nanoparticle/exosome ratios yielding approximately 40% coverage, as higher nanoparticle coatings suffer from radiation damping. Regardless, DMAP yields intense bands that markedly overlap with the SERS signal from the vesicle (Figure 5). Such an issue has been circumvented by in situ overgrowing of a sufficiently thick Ag layer on Au nanoparticles (Au@AgNPs) previously attached to the exosomes. The outer metallic coating quenches the DMAP spectral contributions while further boosting the exosome signal by a factor of ca. 5. The acquisition of a "clean" exosome spectrum by removing the interfering DMAP features enabled a more reliable statistical classification of individual exosomes isolated from B16F10 melanoma cells and red blood cells.

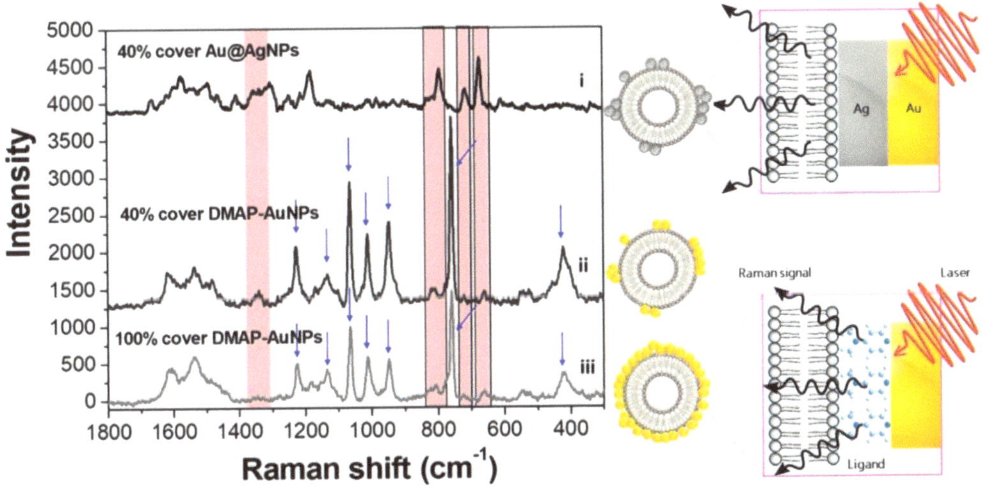

Figure 5. SERS spectra of B16F10 melanoma derived exosomes for 40% coverage with Au@AgNPs or DMAP–AuNPs, and 100% coverage with DMAP–AuNPs. Intense DMAP bands are indicated by blue arrows (these features disappear upon silver coating). An illustrative description of DMAP–AuNPs or Au@AgNPs attached on the exosomes surfaces is also included. Adapted with permission from [88]. Copyright 2019, American Chemical Society.

Direct interaction of exosomes with traditional gold and silver nanoparticles, either physically forced via evaporation onto a solid support or chemically-mediated in suspension, offers a very simple, inexpensive and straightforward strategy for direct SERS analysis. However, it inherently poses important challenges for obtaining reproducible and uniform SERS responses due to the irregular arrangement of the nanoparticles onto the exosome outer layer. To address these limitations, multiple examples of precisely tailored SERS substrates have been generated profiting from the continuous advances in very diverse areas of nanofabrication technologies [108–110]. While each methodology displays a characteristic set of drawbacks and advantages, the fine-tuning of the morphological features of plasmonic materials for maximizing the homogeneity and efficiency of the SERS performances typically takes place, as a rule of thumb, at an increasing price and technical complexity.

In this regard, Xie and co-workers [78] fabricated a substrate comprising a single-layer graphene overlaid on a periodic Au-pyramid nanostructure (Figure 6A). The graphene layer imparts a biocompatible and chemically stable surface while further boosting the amplification of the Raman signal via a chemical mechanism up to ca. 2 orders of magnitude [51,111]. In the same work, the authors highlighted the necessity of an efficient isolation procedure to enable a reliable exosome SERS fingerprinting analysis [78]. Exosomes from fetal bovine serum were isolated either via ultracentrifugation/filtration method or salting-out procedure using a commercial ExoQuick kit (System Biosciences LLC, Palo Alto, CA, USA). The former approach has the advantage of yielding purer samples while the second method is faster and capable of collect almost 1000 times more biomaterial but at the expenses of a lower purity. Particle size analysis of the two processed samples showed a similar mean diameter (ca. 135–143 nm range) but a narrower distribution for particles recovered by ultracentrifugation/filtration. Conversely, the outcome of the SERS analysis revealed many striking differences. Two μL of the exosome solutions were applied onto a hybrid plasmonic platform surface and allowed to air-dry before the measurement. A hundred of SERS spectra were collected on different spots of the platform, yielding reproducible fingerprint signatures for exosomes separated via ultracentrifugation/filtration (Figure 6B, see band assignment in Figure 6C) while ExoQuick-derived materials produced an ensemble of highly heterogenous vibrational profiles, preventing the acquisition of a unique and recognizable SERS spectrum (Figure 6D). The validity of the SERS platform for discriminating different populations of exosomes was demonstrated in combination with principal component analysis (PCA), using vesicles from different sources (fetal bovine serum vs human serum; and lung cancer cell lines HCC827 vs H1975). Interestingly, the authors also performed a dilution study to assess the possibility of performing single exosome analysis. SEM imaging was performed to visualize and count the exosomes localized over a specific area (Figure 6E). The so-estimated exosome density was correlated with the SERS mapping carried out on the same area (overlapping of adjacent laser spots for each SERS measurement was avoided). The results show a linear response of the overall SERS intensity with the change of the sample concentration, indirectly suggesting that individual SERS measurements possibly arise from the interrogation of single exosomes. In a separate work, Pramanik et al. [85] focused on maximizing the SERS response of a hybrid graphene-plasmonic substrate by embedding gold nanostars, one of most SERS efficient individual nanoparticles [55], into 2D graphene oxide structures. This hybrid substrate was successfully employed in the fingerprint identification and discrimination of exosomes derived from triple-negative breast cancer and HER2(+) breast cancer down to ca. 4×10^2 exosomes/mL.

In addition to the intrinsic qualities of the plasmonic substrate, practical issues associated with the sample preparation can significantly impact the overall sensitivity of the method, such as the capability of concentrating vesicles in highly localized and electromagnetically active spots of the substrate. For instance, drop-casting of exosome dispersion onto a surface is typically affected by the coffee-ring effect, leading to an uneven distribution of the vesicles over a relatively broad area. Technically, SERS mapping of large areas (e.g., in the upper micrometric ranges) with high spectral resolutions to maximize the collection of intense signals is feasible but is typically a rather time-consuming process unless state-of-the-art techniques (e.g., SERS holography) are used [112]. A convenient way to concentrate diluted solutions of biological samples onto a small area is integrating plasmonic features on micro- and nano-patterned surfaces with superhydrophobic properties [113]. Superhydrophobic substrates typically comprises micro- and nano-textured surface imparting superior non-adhesive properties via entrapment of air pockets underneath a liquid droplet deposited on top of it. Thus, a droplet retains a quasi-spherical shape during evaporation rather than spread all over the surface, which progressively minimizes the contact area and, in turn, concentrates the analytes [114] over a small spot (less than few microns) [77].

Di Fabrizio and co-workers pioneered such an approach for the direct SERS analysis of exosomes [77]. In their work, a superhydrophobic array of silicon micropillars decorated with silver nanostructures (Figure 7A,B) was designed to discriminate exosomes isolated from either healthy (CCD841-CoN) or tumor (HCT116) colon cells using a commercial ExoQuick kit. Small drops of exosome dispersions (~0.2 ng/mL) were deposited on the substrates (Figure 7C) and, through evaporation, the vesicles were conveyed into small plasmonic-active regions of the substrate (Figure 7D) for the acquisition of averaged SERS spectra (50 acquisitions for each sample). More recently, Suarasan et al. [74] reported a simple, cheaper superhydrophobic plasmonic platform for SERS interrogation of exosomes in small sample volume (as low as 0.5 µL). A PDMS substrate consisting of nano- and micro-bowl structures exhibiting superhydrophobic properties was fabricated via soft lithography. Silver nanoparticles were then grown in situ to impart SERS enhancing properties.

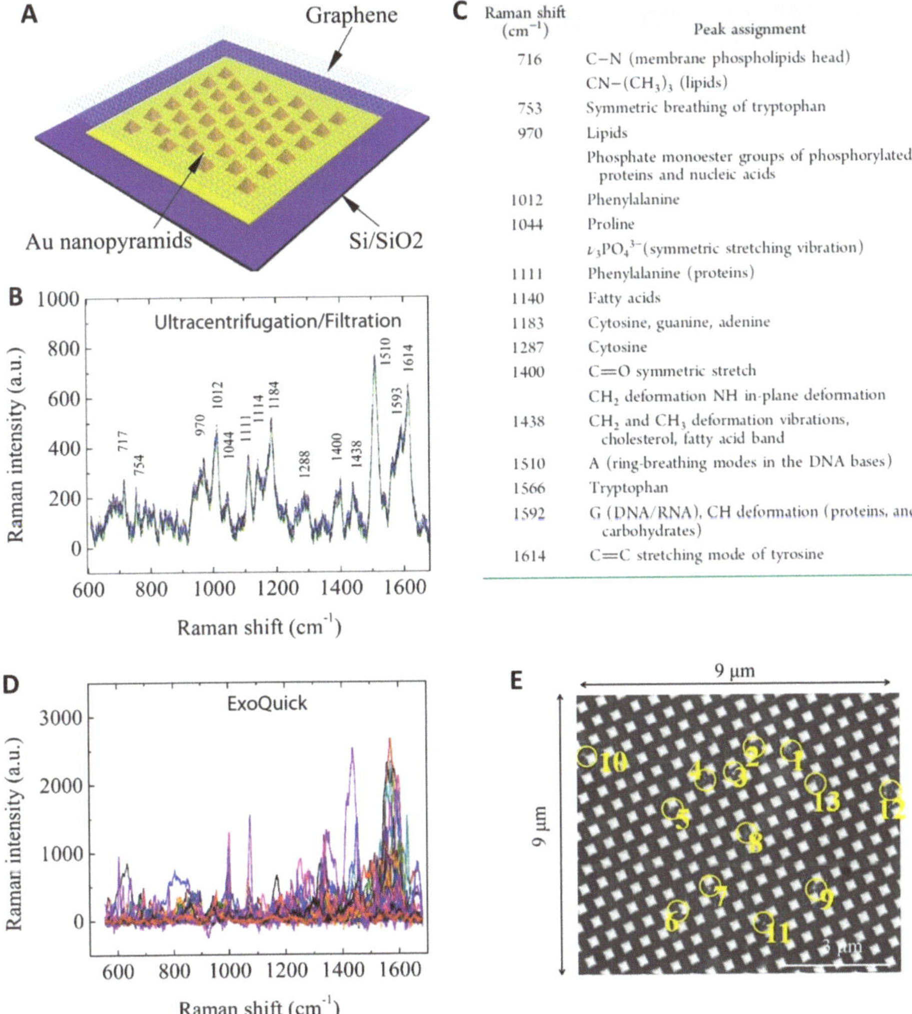

Figure 6. (**A**) Outline of the hybrid Au/graphene platform. (**B,D**) SERS spectra of exosomes isolated from fetal bovine serum using ultracentrifugation/filtration or the ExoQuick kit, respectively. (**C**) Assignment of the main SERS bands in (**B**). (**E**) A representative SEM micrograph of exosomes (circled in yellow) deposited onto the graphene-covered surface. Adapted with permission from [78]. Copyright 2019, American Chemical Society.

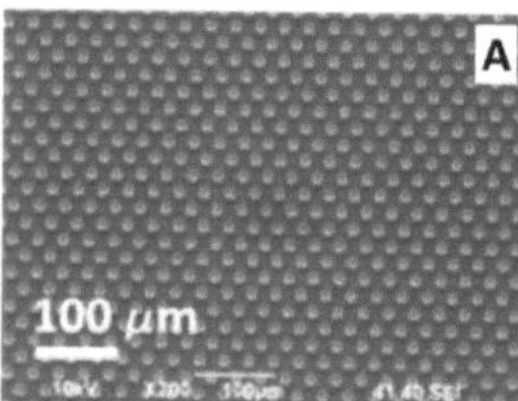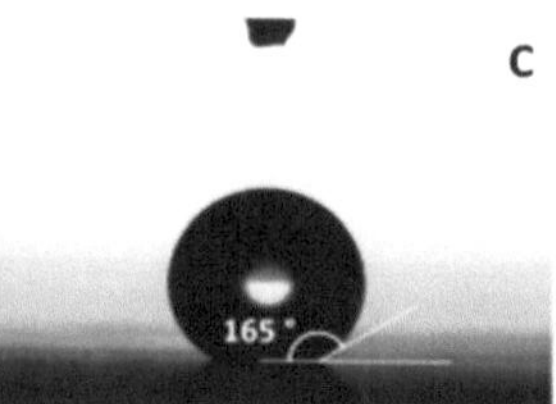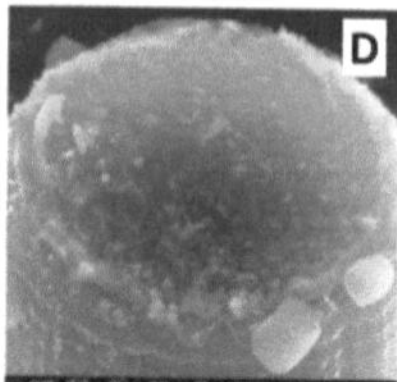

Figure 7. (**A**) A superhydrophobic surface consisting of periodic hexagonal patterns of cylindrical pillars. (**B**) A silicon micropillar with a randomly distributed silver nanograins. (**C**) A drop on top of the superhydrophobic surface displaying a contact angle as large as 165°. (**D**) Top view SEM image of exosomes on pillars. Adapted with permission from ref. [77]. Copyright 2012, Elsevier.

Alternatively, local enrichment of exosomes can be achieved via intracavity trapping. For instance, Xiao and co-workers [70] engineered a multifunctional 3D gold-coated TiO_2 macroporous inverse opal structure (Figure 8A) providing (i) an interconnected beehive-like pore networks for trapping exosomes to improve their separation from the medium; and (ii) enhanced signal amplification within the cavity volumes as compared to flat or non-cavity structures. This latter effect results from the superimposition of the field enhancements from both the dipole resonance of the spherical cavity, which amplifies the intensity of the normal Raman signal of exosomes, and the strong plasmonic resonances at the gold film surface, enabling the corresponding SERS magnification under a 633 nm excitation. As a result, these hybrid structures appear particularly suitable for the interrogation of molecular objects in the exosome size-range. The authors exploited these materials for discriminating exosomes from healthy donors and patients diagnosed with lung, liver and colon cancer using the intensity of the 1087 cm^{-1} band as the spectral biomarker. This feature has been ascribed to the vibration of the P-O bond in the phosphate groups of phosphoproteins, which have been described as a protein biomarker of breast cancer-derived exosomes [115]. Exosomes were isolated from peripheral blood samples of cancer patients, using the Total Exosome Isolation Reagent (Invitrogen, Carlsbad, CA, USA) and dispersed in deionized water. 50 μL were then dropped onto the substrate (25 mm × 25 mm) substrate and dried naturally. SERS mapping measurements were finally performed over an area of 16.5 μm × 11.5 μm at 0.4 μm intervals to yield the resulting average spectra (Figure 8B). The intensity of the 1087 cm^{-1} SERS peak from the exosomes secreted by most of these lung, liver, and colon cancer patients was at least two times of that from healthy individuals (Figure 8C) while displaying much larger intensity fluctuations. The validity of the 1087 cm^{-1} band intensity as a spectral biomarker was further corroborated by the analysis of exosomes from the prostate, lung, liver, colon cancer cell lines.

As previously exploited for promoting the adhesion of plasmonic colloids in suspension onto exosomes, electrostatic interactions of the vesicles onto solid supports can also be employed to promote their local accumulation. For instance, Carney and co-workers [73] described the fabrication of a simple, low-cost plasmonic material comprising a microscale biosilicate material decorated with silver nanoparticles for SERS analysis of ovarian and endometrial cancer exosomes. Metallic surfaces were functionalized with cysteamine to impart positive charge and, therefore, favor the accumulation of exosomes in suspension via electrostatic binding with their negatively charged outer shell. Exosomes were initially isolated via differential ultracentrifugation from serum of 8 patients (6 of them with different cancer subtypes) and resuspended in up to 100 μL of ultrapure water. The samples were diluted 1:100 in pH 6.4 buffer and 30 μL drops were pipetted onto 2 mm × 5 mm substrate elements. Pretreatment of the substrates with the slightly acidic buffer was also performed to maximize the protonation of the cysteamine amino groups onto the silver surface. Upon incubation, SERS spectra were acquired in liquid condition on 5–10 different random spatial locations of the biosilicate plasmonic platform. Multivariate data analysis was successfully applied to distinguish tumor samples from healthy ones in patients suspected of gynecologic malignancy. A limit of detection (LOD) of less than 600 vesicles/mL has been reported.

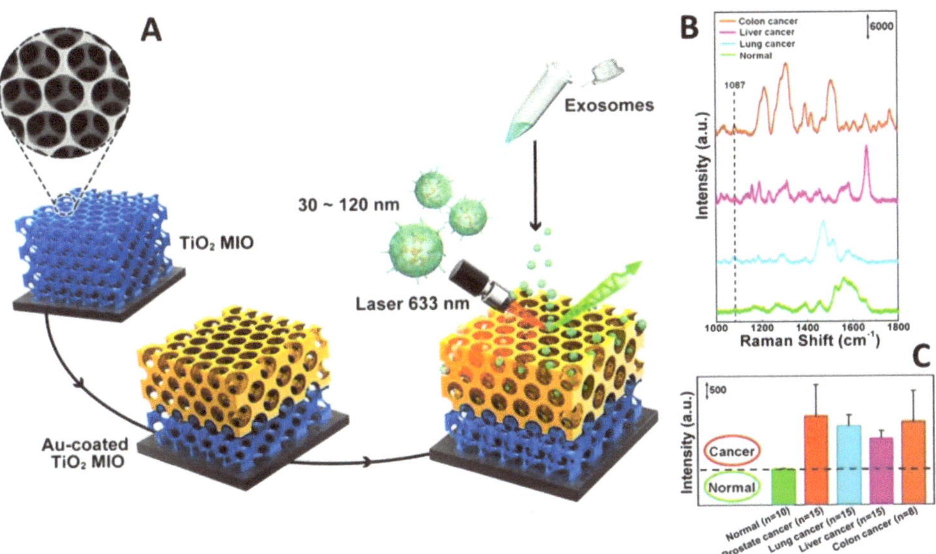

Figure 8. (**A**) Outline of the 3D gold-coated TiO$_2$ macroporous inverse opal structure. (**B**) Typical SERS spectra of exosomes separated from plasma of normal individual and lung, liver, and colon cancer patients. (**C**) Averaged SERS intensity at 1087 cm^{-1} from exosomes separated from normal individuals and 15 lung cancer, 15 liver cancer patients and 8 colon cancer patients. The black dashed line shows the intensity boundary of the 1087 cm^{-1} peak between normal individuals and cancer patients. Adapted with permission from ref. [70]. Copyright 2020, American Chemical Society.

Overall, direct SERS analysis in combination with multivariate statistical methods has fully demonstrated the consistent ability to discriminate exosomes isolated from different cell types. However, the complexity of exosome populations secreted from heterogeneous sources, such as those isolated from human blood, has significantly limited the viability of this approach as a diagnostic tool. As a striking example, Shin et al. [90] acquired the SERS spectra of cell-derived exosomes from healthy and cancer lung cell lines as well as human plasma exosomes from healthy controls and patients with different stages of lung cancer (Figure 9A). The average size of the examined exosomes fractions was similar (specifically, cell-derived exosomes = 139.6 ± 14.4 nm, human plasma-derived exosomes = 136.3 ± 3.2 nm). SERS measurements were performed by dry-cast exosome solutions onto gold nanoparticle decorated coverslips. Figure 9B illustrates the averaged SERS signals for normal cells and lung cancer cell exosomes which can be efficiently discriminated even by visual analysis. On the contrary, spectral differences between exosomes from plasma of healthy controls and lung cancer patients are negligible (Figure 9C), most likely due to the presence of a large number of exosomes from various organs that conceals the specific vibrational patterns of lung-derived exosomes. To overcome this limitation, the authors employed deep learning algorithms to analyze the spectroscopic signals of exosomes. Deep learning (DL) is a machine learning method based on artificial neural networks that effectively process big sensing data for complex matrices or samples, allowing classification, identification, and pattern recognition [116]. Notably, DL algorithms have shown to be extremely beneficial for analyzing spectroscopic data in biosensing applications [116]. In this abovementioned work [90], besides a mere classification of the SERS data from healthy controls and cancer patients, deep learning was used to establish a correlation between the exosome data from individual lung cells with the overall patient's histological characteristics. Specifically, the spectral data set of cell-derived exosomes were first used to train the DL models for binary classification of cell types which, subsequently, efficiently separated exosomes from human plasma (healthy vs cancer) into two clusters with an accuracy of 95% (Figure 9D). Finally, using PCA scores at the terminal fully connected layer, the Mahalanobis distance between plasma and cell exosome clusters is determined to quantitatively evaluate the resemblance of the data from plasma and cancer cell exosomes. The DL model predicted that 90.7% of plasma exosomes from 43 patients, including stage I and II cancer patients, had higher similarity to exosomes derived from lung cancer cell lines than the average of the healthy controls (Figure 9E). Remarkably, the degree of such similarity correlates to the progression of cancer. It is worth stressing that, besides reducing the need of acquiring a sufficiently large number of patient samples to generate a robust and reliable data set for discrimination, this approach provides the biological basis for classification. This reduces the impact of undetected potential experimental errors as the source of spectral differences.

Integration of SERS sensing and Raman components into multifunctional platforms with additional features (e.g., microfluidics and magnetic separation for sample handling, fluorescence spectroscopy for multimodal optical analysis, etc.) [62,117,118] has been intensely pursued in recent years to overcome intrinsic limitations of SERS as a stand-alone technique and, thus, pave the way for the fabrication of miniaturized biosensor for point-of-care testing. In particular, Raman spectroscopy can be easily combined with microfluidics, a technology that facilitates high throughput and automated analysis with very low sample consumption [62,117]. In exosome analysis, Hao et al. [89] combined acoustics and microfluidics technologies with dual fluorescence and SERS optical detection into a single analytical device. The main features of the fabricated acoustofluidic platform are outlined in Figure 10. The device generates surface acoustic waves (SAWs) that propagate toward the glass capillary microchannel where the exosome suspension is confined. This results in pressure fluctuations within the liquid that force the suspended particles to concentrate at the center of the fluid chamber (for immunofluorescent detection) or the edge near a plasmonic Ag nanoparticle-deposited ZnO nanorod arrays (for SERS analysis). To this end, CD63 aptamer-conjugated 400 nm silica nanoparticles were used to capture exosomes and enable their acoustic-assisted enrichment at the plasmonic substrates for highly sensitive label-free SERS detection down to ca. 20 exosomes per μL (from human plasma-derived exosome samples).

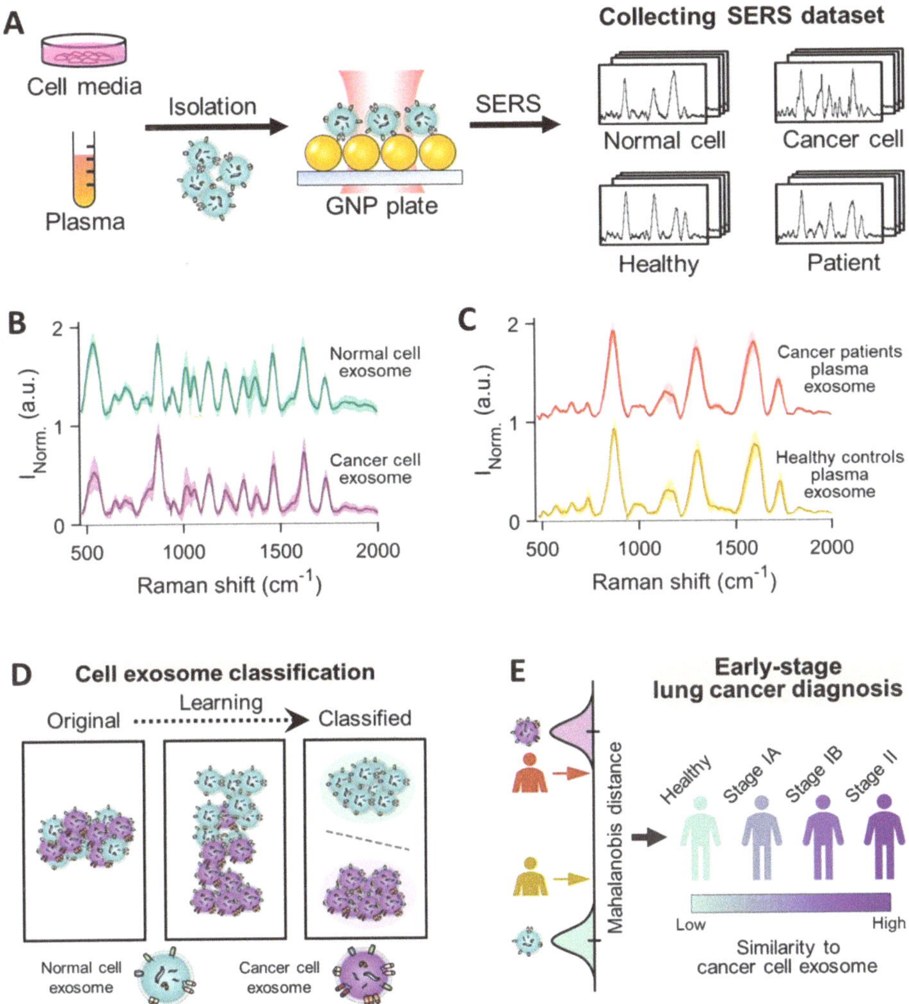

Figure 9. (**A**) Schematic of the collection of SERS spectra for exosomes isolated from different cell media and human plasma and dry-cast onto a gold-nanoparticle (100 nm) coated cover-slip. Specifically, cell-derived exosomes were isolated from lung-related cells: normal cell exosomes from human pulmonary alveolar epithelial cells (HPAEpiC) and cancer cell exosomes from A549, H460, H1299, H1763, and PC9 cells. Human plasma samples were collected from 20 healthy controls and 43 lung adenocarcinoma patients (22 patients in stage IA, 16 in stage IB, and 5 in stage IIB). (**B**,**C**) Average SERS signals of cell media supernatant-derived and human plasma-derived exosomes, respectively. (**D**,**E**) Overview of deep learning-based cell exosome classification and lung cancer diagnosis, respectively, using exosomal SERS signal patterns. Adapted with permission from ref. [90]. Copyright 2020, American Chemical Society.

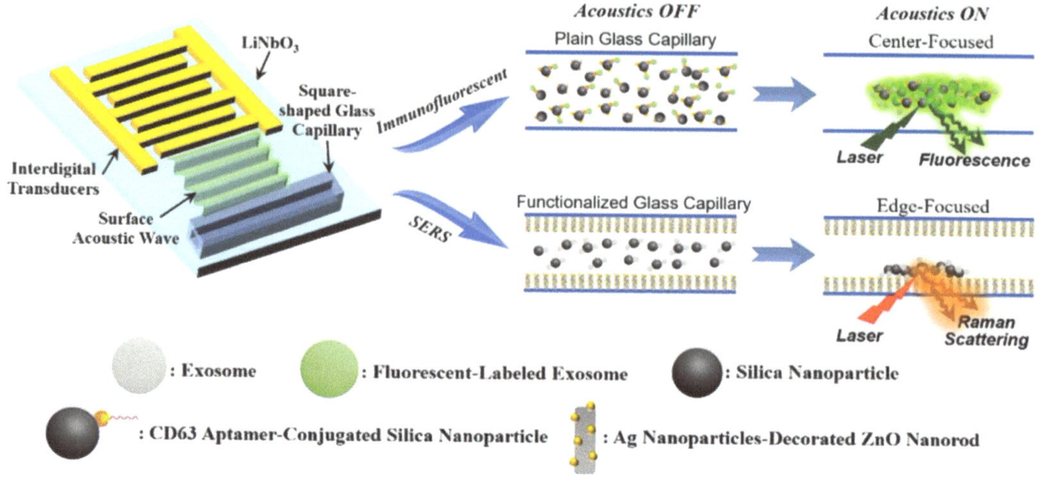

Figure 10. Outline of the acoustofluidic biosensor features and mechanisms of optical detections (immunofluorescence and SERS sensing). The device comprises a transparent piezoelectric lithium niobate (LiNbO₃) substrate with patterned interdigital transducers (IDTs) and a square-shaped glass capillary bonded to the substrate. Surface acoustic waves concentrate particles at either the center or the perimeter of a glass capillary. Adapted with permission from [89]. Copyright 2020, Wiley-VCH.

4. Indirect SERS Analysis of Exosomes Using SERS-Encoded Nanoparticles (SERS tags)

A large variety of SERS-encoded particles (SEPs), or also referred to as SERS tags, with different structural and chemical features has been reported in the literature [68,119–121]. Despite a broad range of diversity, it is possible to recognize the following key building units (Figure 11): (i) a nanoparticle-based core (typically, silver or gold) as the plasmonic enhancer, (ii) a dense collection of molecules with large Raman cross-sections (referred to as codes or labels or reporters) attached to the metallic surface to provide an intense and well-defined vibrational fingerprint, and (iii) a variety of surface molecular ligands (e.g., antibodies, aptamers, peptides) to impart selectivity toward a target analyte. These recognition elements are often conjugated onto the surface of a protective inert layer (e.g., silica) coating the SERS labelled plasmonic core, which is integrated into the nanomaterial to afford high stability in complex media and avoid leaking of the codes [119]. Indirect sensing with SEPs is entailed with multiplexing capabilities with single laser excitation, thanks to the unique vibrational fingerprints of each code, and quantitative response, as the SEP structure can be engineered to provide a SERS intensity that would scales linearly with the SEP content [119].

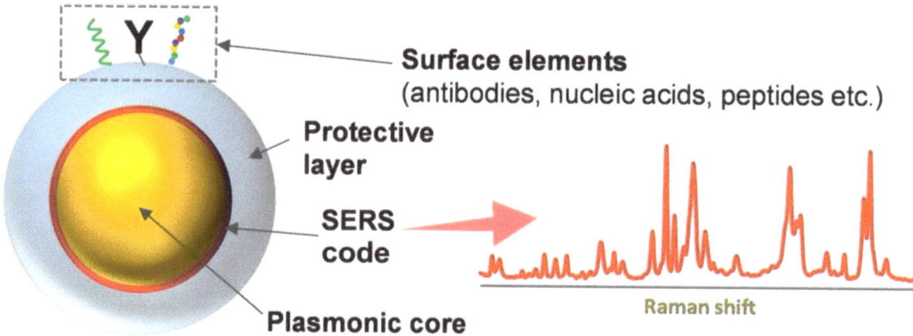

Figure 11. Depiction of a representative example of SERS-encoded particle (SEP) or SERS tag. Adapted with permission from [119]. Copyright 2017, Springer Nature.

In the indirect SERS analysis of exosomes, SEPs are conjugated with recognition elements that promote their selective accumulation at the surfaces of the vesicles. To facilitate the SERS interrogation, capturing substrates are also integrated into the sensing system to enable the separation and accumulation of the SEPs-decorated exosomes into a small area for ultrasensitive detection. Typically, capturing substrates consist of magnetic beads [92–94,96,97,122] or flat supports [98–100] functionalized with further recognition molecules for specific exosome binding. Notably, such an approach also removes the need for time-consuming, costly and complex exosome isolation procedures (e.g., ultracentrifugation) as SEPs and capturing substrates can be directly applied to biological media (e.g., conditioned medium, serum, blood, urine, saliva) for exosome binding and separation. The multiplexing capability of indirect SEP-based sensing makes this method particularly suited for phenotypic profiling of transmembrane proteins of cancer-derived exosomes [93,97,99]. Simultaneous evaluation of the expression levels of multiple surface proteins (phenotype) provides a much more reliable molecular description of the heterogeneous nature of tumour-derived exosomes as compared to a single marker characterization [97]. Notably, multidimensional phenotyping has shown to play a central role in improving diagnostic, drug treatment, disease monitoring and prognosis [47,123]. In a proof of concept study, Wang and co-workers [97] demonstrated the viability of such an approach by profiling three surface biomarkers of pancreatic cancer (Glypican-1, epithelial cell adhesion molecules—EpCAMs, and CD44 variant isoform 6—CD44V6) on exosomes secreted by a human pancreatic cancer cell line (Panc-1). Exosomes of ca. 130 nm size suspended in a conditioned medium were obtained upon removing cells from the culturing media. Aliquots of conditioned exosomes were either diluted in either PBS or plasma from healthy subjects and, subsequently, directly combined with an equimolar mixture of three classes of SEPs (Figure 12A). The different batches of SEPs (Au@MBA-EpCAM; Au@TFMBA-CD44V6 and Au@DTNB-MIL38) comprise 55 nm gold nanoparticles labelled with a unique molecular code (DTNB, MBA or TFMBA) and conjugated with one CD44V6, EpCAM and MIL38 monoclonal antibodies (MIL38 is specific to Glypican-1). The mixture was stirred for 1 h before adding exosome-specific CD63-modified magnetic beads to enable, after an additional hour of incubation, the magnetically-assisted separation of the sandwich-like immunocomplex (Figure 12A). An enriched immunocomplex suspension in PBS was then interrogated by SERS, providing highly averaged spectra (Figure 12B) where the intensities of marker bands of the three codes (1077, 1340 and 1380 cm^{-1} for MBA, DTNB and TFMBA, respectively) indirectly inform about the relative expression levels of surface protein biomarkers. The phenotype signature was expressed by normalizing the corresponding Raman signals (I_{EVs}) and SEPs ($I_{nanotags}$) using the same concentration of conditioned exosomes. The study was also repeated with C3 (bladder cancer) and SW480 (colorectal cancer)-derived exosomes. Figure 12C illustrates the corresponding phenotypic signatures in either PBS or exosome-spiked plasma from healthy people. The results display a different degree of both absolute and relative surface proteins levels, with EpCAM the more expressed one. This is consistent with the current knowledge of EpCAM as the most highly expressed cancer biomarker. Molecular profiles show high similarity in both PBS and plasma suspended exosomes, but with a consistently lower absolute intensity for exosome dispersed in the plasma medium. This has been tentatively ascribed to a dilution of the cancer-specific signals due to the presence of additional exosomes equipped with surface CD63 antigens that are also magnetically separated by the CD63-conjugated magnetic beads. The sensitivity of the assay was determined to be 2.3×10^6 particles/mL in PBS, which is below the average exosome concentration in most body fluids (ca. 10^8 EVs/mL and above), thereby meeting the clinical requirements.

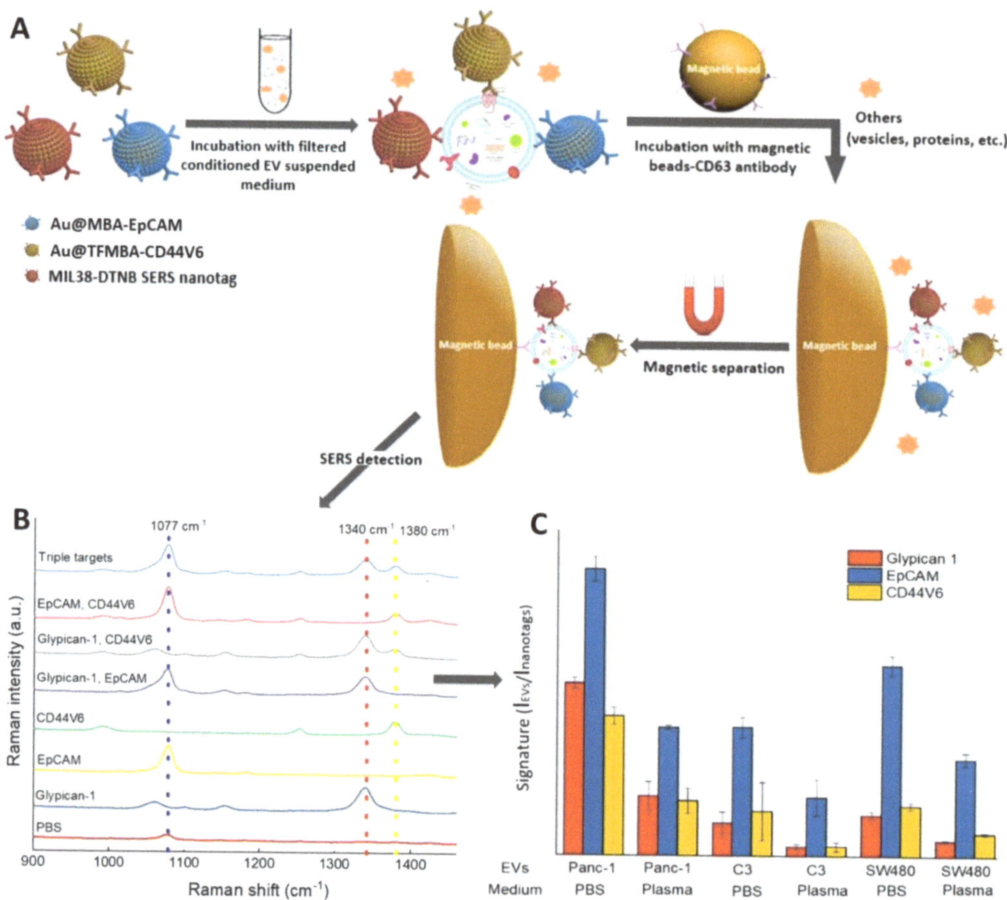

Figure 12. (**A**) Schematic illustration of molecular phenotype profiling of CD63-positive exosomes using CD63 antibody-functionalized magnetic beads and three classes of SERS-encoded nanoparticles separately conjugated with antibodies targeting Glypican-1, EpCAM, and CD44V6 surface biomarkers. SERS codes: 5,5′-dithiobis(2-nitrobenzoic acid) (DTNB); 4-mercaptobenzoic acid (MBA), and 2,3,5,6-Tetrafluoro-4-mercaptobenzonic acid (TFMBA). (**B**) SERS spectra for the simultaneous detection of three biomarkers on Panc-1-derived exosomes in PBS. Peaks at ca. 1077, 1340, and 1380 cm^{-1} are correlated with the presence of EpCAM, Glypican-1, and CD44V6, respectively. (**C**) Phenotypic signature of Panc-1-, C3-, and SW480-derived exosomes in PBS and plasma (n = 3). Adapted with permission from [97]. Copyright 2020, American Chemical Society.

The use of immunoaffinity magnetic beads for separation/enrichments of exosomes from biofluids is, however, often affected by issues related to limited reproducibility, long incubation time and low exosome yields (<50%) [95]. On the other hand, the use of capturing substrates with one, highly specific recognition element such as an antibody, can introduce biases in the exosome isolation, leading to the enrichment of distinct exosome subpopulations to the detriment of others which, however, may be critical to diagnosis [124]. To this end, Pang et al. [95] replaced immunoaffinity beads with TiO_2-coated magnetic particles, which allowed the indiscriminate, rapid and highly efficient removal of the exosomes from serum by exploiting the affinity of TiO_2 for binding the hydrophilic phosphate head of the exosomal phospholipids. In their study, the programmed death ligand 1 (PD-L1) protein biomarker on the exosomal membrane was subsequently targeted with SEPs modified with an anti-PD-L1 antibody (Figure 13A). Exosomes derived from adenocarcinoma human alveolar basal epithelial cells (A549) were initially used as model samples because of their PD-L1 expression level closely correlated with lung cancer stage. Figure 13B,C show SEM images that visualize the capturing of the exosomes onto the $Fe_3O_4@TiO_2$ particles and the subsequent binding of SEPs onto the exosomal surfaces. Laser interrogation of the aggregates yields SERS intensities that linearly scales with the exosome concentration in the 5×10^3 to 2×10^5 particles/mL range, with a detection limit of 1 PD-L1 + exosome/µL and an exosomal capture efficiency of 96.5%. The assay was finally tested with human serum samples from healthy donors (12) and NSCLC patients of early (7) and advanced stages (10). Figure 13D shows the scatter plots of the log SERS intensity for each group of samples. Clear separation can be observed for the healthy persons and the diagnosed patients, whereas discrimination has not been successfully achieved for this cohort of stage I-II and stage III-IV patients. Notably, this strategy allows the ultracentrifugation-free quantification of exosomal PD-L1 by using only 4 µL clinic serum sample and in less than 40 min in total (much lower than the 2–5 h time reported by other exosomes detection methods) [95].

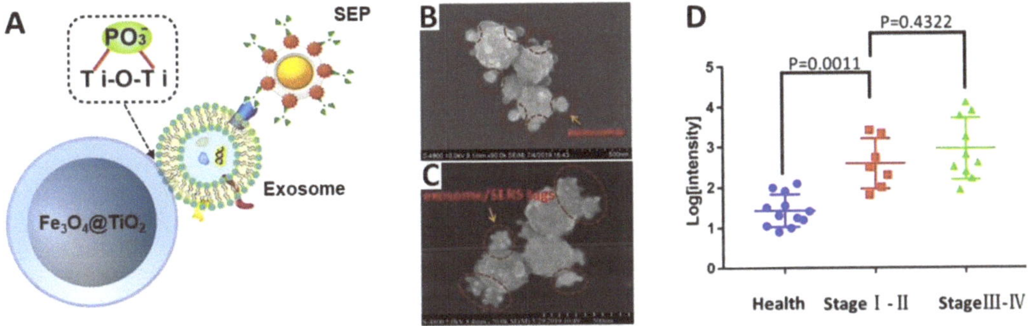

Figure 13. (**A**) Outline of the sandwich complex between TiO_2-coated Fe_3O_4 beads ($Fe_3O_4@TiO_2$), exosomes and SERS-encoded nanoparticles (SEPs) comprising a gold-silver core-shell nucleus functionalized with 4-mercaptobenzoic acid (MBA) as the SERS code and further conjugated with an anti-PD-L1 antibody. (**B,C**) SEM images of $Fe_3O_4@TiO_2$ + A549 exosome, and $Fe_3O_4@TiO_2$ + A549 exosome + SEPs, respectively.(**D**) Scatter plots of the log SERS intensity (MBA band at ca. 1074 cm^{-1}) in the serum samples from the controls and the early-stage (stage I/II) and advanced (stage III/IV) patients diagnosed with non-small-cell lung carcinoma (NSCLC). Adapted with permission from [95]. Copyright 2020, Elsevier.

Alternatively, Trau and co-workers [99] tackled the limitations of immune-affinity separation and slow binding kinetics by integrating a nanomixing strategy that improves exosome capture efficiency while reducing non-specific adsorption and incubation time. A chip implementing nanomixing forces was designed for the streamlined plasma exosome phenotype analysis in less than 40 min, as outlined in Figure 14A–C. Exosomes derived from melanoma cell lines of patients treated with the BRAF inhibitor were selected to evaluate responses to the treatment. BRAF inhibitor targets BRAF V600, a mutation found in ca. 40% of melanoma patients that promotes cell cycle progression and tumor growth. The cell culture medium or diluted patient plasma containing the exosomes are directly fed into the capturing chip without any previous purification and enrichment steps (Figure 14A). The capturing area was modified with an anti-CD63 antibody targeting a generic, non-cancer specific exosome biomarker to maximize the vesicle accumulation at the interrogation spot. Exosomes are then simultaneously targeted by a pool of four classes of SEPs (Figure 14B), which comprise gold nanoparticles labelled with unique SERS codes and tumour-specific antibodies targeting four biomarkers that have been previously shown to undergo changes in expression levels with treatment and melanoma progression (i.e., melanoma chondroitin sulfate proteoglycan—MCSP, melanoma cell adhesion molecule—MCAM, low-affinity nerve growth factor receptor—LNGFR, and receptor tyrosine protein kinase—ErbB3). SERS mapping of the surface capturing area (Figure 14C) collects spectral intensities that are proportional to the numbers of exosomes and their expressing biomarker levels. Thus, as previously discussed, the exosome phenotype can be extracted by determining the relative SERS intensities of the code marker bands. Characterization of the phenotypic changes during treatment was first demonstrated on exosomes from patient-derived melanoma cell lines harbouring either a BRAF mutation (e.g., LM-MEL-64) or an NRAS mutation in a BRAF wild type (experimental control). For instance, exosomes collected from LM-MEL-64 cells without drug treatment did not show any significant changes across four selected biomarkers (Figure 14D). Upon drug exposure, however, it is visible a radical reshaping of the protein expression levels followed by a general up-regulation of the MCSP and MCAM levels once the BRAF inhibitor treatment was interrupted (Figure 14E). It is worth noting that, when anti-CD63 antibody at the capturing area was replaced with anti-MCSP, exosomes cell-derived phenotypes (specifically, from SK-MEL-28 cell lines) were different, suggesting heterogeneity of secreted vesicle subpopulations. Such heterogeneity further reflects a potential genetic or epigenetic variability within the cell population. The method was finally validated by monitoring the evolution of cancer-specific exosomes phenotypes from the plasma of melanoma patients receiving targeted therapy, which reflects the potential of exosome phenotyping for monitoring treatment responses.

Increasing the density of biologically functioning antibodies at the capturing surface is also a central factor for improving the efficiency and sensitivity of the immunoassay. In this regard, Li et al. [100] reported the use of polydopamine (PDA) self-polymerizing on glass slides to generated a 50–100 nm thick, rough layer suitable for enhanced antibody anchoring. Similarly, SEPs were fabricated with a thin PDA shell for antibody conjugation. Overall, the PDA encapsulation yields a more uniform, mild and biocompatible surface functionalization which entails high antibody capture efficiency and high sensitivity for detecting cancer-derived exosomes. PDA technology was integrated into a miniaturized device for the monoplex SERS analysis of 2 μL samples from clinical serum collected from healthy donors and pancreatic cancer patients. The assay efficiently discriminated between healthy donors and patients as well as between patients with different stage tumors (discriminatory sensitivity = 95.7%). Also, the assay displayed high sensitivity with a detection limit estimated to be just one single exosome per 2 μL for cell line-derived samples.

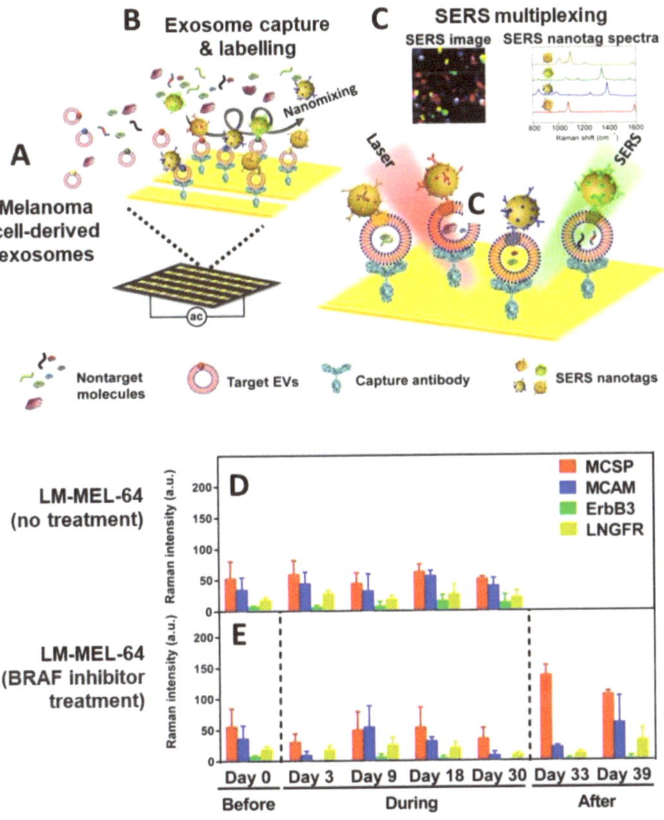

Figure 14. (**A–C**) Schematic of the exosome phenotyping using SERS-encoded nanoparticles: (**A**) Exosomes are secreted by melanoma cells with a BRAF V600E mutation in the culture medium or into circulation; (**B**) The exosome containing sample is injected, together with SERS tags, into the nanomixing chip equipped with capture antibodies; (**C**) Upon removal of non-target molecules (e.g., protein aggregates and apoptotic bodies) and unbound SERS tags, SERS mapping is performed to provide the SERS phenotyping of the captured exosomes. The false-color SERS image is generated from the characteristic peak intensities of each SERS tags (MCSP-MBA, red; MCAM-TFMBA, blue; ErbB3-DTNB, green; LNGFR-MPY, yellow). (**D,E**) Phenotypic alterations of exosomes derived from melanoma patient-derived LM-MEL-64 cell line in response to BRAF inhibitor treatment at different times (before, during and after treatment). Anti-CD63 antibodies were used in the capturing area. Adapted with permission from [99]. © The Authors, some rights reserved; exclusive licensee AAAS. Distributed under a Creative Commons Attribution NonCommercial License 4.0 (CC BY-NC). Available online: http://creativecommons.org/licenses/by-nc/4.0/ (accessed on 30 April 2021).

5. Future Challenges

In this review, we summarized and coherently discussed the diverse applications of SERS in the analysis of exosomes, with a special focus on the more recent and promising advances. We have also progressively highlighted current key challenges and limitations, which can be broadly associated with either the general application of SERS in biosensing and clinical diagnostic or the specific nature of exosomes as the biological target. In the first case, the translation of SERS-based analytical tools into competitive, commercial devices still faces important practical obstacles such as the production of cost-effective, robust and efficient plasmonic substrates at a large scale. Similarly, the fabrication of affordable and portable Raman spectrometers for fast data acquisition is critical for lowering the cost while providing manageable equipment for routine analysis in the clinical setting. In this regard, the integration of SERS substrates and Raman components into multifunctional platforms (e.g., microfluidics) is also pivotal for automatization and efficient standardization of the measuring procedures. Furthermore, as also stressed in the review, the efficient implementation of the most advanced chemometric tools appears to be the way to fully access the multidimensional information contained in large SERS data set. On the other hand, the intrinsic nature of these vesicles makes exosome-based diagnostics a difficult task, mainly due to their pronounced molecular heterogeneity and the requirement of determining the presence and relative distribution of different sub-populations, especially in complex biofluids. As pointed out, both improvement and standardization of the isolation procedures are needed to reproducibly supply exosomes with high purity in good yields, while identification of a much broader set of potential biomarkers is mandatory for enabling clinical applications.

Author Contributions: Conceptualization—L.G.; E.G.-R. and R.A.A.-P.; Writing—Review and editing, L.G., E.G.-R., A.O., V.G., R.A.A.-P. All authors have read and agreed to the published version of the manuscript.

Funding: This research was supported by the Spanish Ministerio de Economia y Competitividad (CTQ2017-88648R and RYC-2016-20331), the Generalitat de Cataluña (2017SGR883), and the Universitat Rovira I Virgili (FR 2019-B2).

Data Availability Statement: Not applicable.

Conflicts of Interest: The authors declare no conflict of interest.

References

1. Bray, F.; Jemal, A.; Grey, N.; Ferlay, J.; Forman, D. Global cancer transitions according to the Human Development Index (2008–2030): A population-based study. *Lancet Oncol.* **2012**, *13*, 790–801. [CrossRef]
2. Adjiri, A. Tracing the path of cancer initiation: The AA protein-based model for cancer genesis. *BMC Cancer* **2018**, *18*, 831. [CrossRef] [PubMed]
3. Jafari, M.; Hasanzadeh, M. Cell-Specific frequency as a new hallmark to early detection of cancer and efficient therapy: Recording of cancer voice as a new horizon. *Biomed. Pharmacother.* **2020**, *122*, 7. [CrossRef]
4. Klein, C.A. Parallel progression of primary tumours and metastases. *Nat. Rev. Cancer* **2009**, *9*, 302–312. [CrossRef] [PubMed]
5. Del Monte, U. Does the cell number 10^9 still really fit one gram of tumor tissue? *Cell Cycle* **2009**, *8*, 505–506. [CrossRef] [PubMed]
6. Zhou, B.; Xu, K.; Zheng, X.; Chen, T.; Wang, J.; Song, Y.; Shao, Y.; Zheng, S. Application of exosomes as liquid biopsy in clinical diagnosis. *Signal Transduct. Target. Ther.* **2020**, *5*, 144. [CrossRef] [PubMed]
7. Kaplan, R.N.; Riba, R.D.; Zacharoulis, S.; Bramley, A.H.; Vincent, L.; Costa, C.; MacDonald, D.D.; Jin, D.K.; Shido, K.; Kerns, S.A.; et al. VEGFR1-Positive haematopoietic bone marrow progenitors initiate the pre-metastatic niche. *Nature* **2005**, *438*, 820–827. [CrossRef] [PubMed]
8. Klein, C.A. Tumour cell dissemination and growth of metastasis. *Nat. Rev. Cancer* **2010**, *10*. [CrossRef]
9. Langley, R.R.; Fidler, I.J. The seed and soil hypothesis revisited-The role of tumor-stroma interactions in metastasis to different organs. *Int. J. Cancer* **2011**, *128*, 2527–2535. [CrossRef]
10. Psaila, B.; Lyden, D. The metastatic niche: Adapting the foreign soil. *Nat. Rev. Cancer* **2009**, *9*, 285–293. [CrossRef]
11. Fidler, I.J.; Kripke, M.L. Metastasis Results from Preexisting Variant Cells within a Malignant-Tumor. *Science* **1977**, *197*, 893–895. [CrossRef]
12. Tai, Y.L.; Chen, K.C.; Hsieh, J.T.; Shen, T.L. Exosomes in cancer development and clinical applications. *Cancer Sci.* **2018**, *109*, 2364–2374. [CrossRef]
13. Trams, E.G.; Lauter, C.J.; Salem, N.; Heine, U. Exfoliation of Membrane Ecto-Enzymes in the Form of Micro-Vesicles. *Biochim. Biophys. Acta* **1981**, *645*, 63–70. [CrossRef]
14. Johnstone, R.M.; Adam, M.; Hammond, J.R.; Orr, L.; Turbide, C. Vesicle Formation During Reticulocyte Maturation—Association of Plasma-Membrane Activities with Released Vesicles (Exosomes). *J. Biol. Chem.* **1987**, *262*, 9412–9420. [CrossRef]
15. Farooqi, A.A.; Desai, N.N.; Qureshi, M.Z.; Librelotto, D.R.N.; Gasparri, M.L.; Bishayee, A.; Nabavi, S.M.; Curti, V.; Daglia, M. Exosome biogenesis, bioactivities and functions as new delivery systems of natural compounds. *Biotechnol. Adv.* **2018**, *36*, 328–334. [CrossRef]
16. Théry, C.; Ostrowski, M.; Segura, E. Membrane vesicles as conveyors of immune responses. *Nat. Rev. Immunol.* **2009**, *9*, 581–593. [CrossRef]
17. Huang, T.; Deng, C.X. Current Progresses of Exosomes as Cancer Diagnostic and Prognostic Biomarkers. *Int. J. Biol. Sci.* **2019**, *15*, 1–11. [CrossRef]

18. Clement, E.; Lazar, I.; Attané, C.; Carrié, L.; Dauvillier, S.; Ducoux-Petit, M.; Esteve, D.; Menneteau, T.; Moutahir, M.; Le Gonidec, S.; et al. Adipocyte extracellular vesicles carry enzymes and fatty acids that stimulate mitochondrial metabolism and remodeling in tumor cells. *EMBO J.* **2020**, *39*, e102525. [CrossRef]
19. Pegtel, D.M.; Gould, S.J. Exosomes. *Annu. Rev. Biochem.* **2019**, *88*, 487–514. [CrossRef]
20. Nolte-'t Hoen, E.N.M.; Buermans, H.P.J.; Waasdorp, M.; Stoorvogel, W.; Wauben, M.H.M.; 't Hoen, P.A.C. Deep sequencing of RNA from immune cell-derived vesicles uncovers the selective incorporation of small non-coding RNA biotypes with potential regulatory functions. *Nucleic Acids Res.* **2012**, *40*, 9272–9285. [CrossRef]
21. Mehrtash, A.B.; Hochstrasser, M. Ubiquitin-Dependent protein degradation at the endoplasmic reticulum and nuclear envelope. *Semin. Cell Dev. Biol.* **2019**, 111–124. [CrossRef] [PubMed]
22. Wang, M.; Yu, F.; Li, P.F.; Wang, K. Emerging Function and Clinical Significance of Exosomal circRNAs in Cancer. *Mol. Ther. Nucleic Acids* **2020**, *21*, 367–383. [CrossRef] [PubMed]
23. Wei, H.; Chen, Q.; Lin, L.; Sha, C.; Li, T.; Liu, Y.; Yin, X.; Xu, Y.; Chen, L.; Gao, W.; et al. Regulation of exosome production and cargo sorting. *Int. J. Biol. Sci.* **2021**, *17*, 163–177. [CrossRef] [PubMed]
24. Wu, C.; Du, S.; Zhang, J.; Liang, A.; Liu, Y. Exosomes and breast cancer: A comprehensive review of novel therapeutic strategies from diagnosis to treatment. *Cancer Gene Ther.* **2017**, *24*, 6–12. [CrossRef]
25. Chen, F.; Huang, C.; Wu, Q.; Jiang, L.; Chen, S.; Chen, L. Circular RNAs expression profiles in plasma exosomes from early-stage lung adenocarcinoma and the potential biomarkers. *J. Cell. Biochem.* **2020**, *121*, 2525–2533. [CrossRef]
26. Jakobsen, K.R.; Paulsen, B.S.; Bæk, R.; Varming, K.; Sorensen, B.S.; Jørgensen, M.M. Exosomal proteins as potential diagnostic markers in advanced non-small cell lung carcinoma. *J. Extracell. Vesicles* **2015**, *4*, 26659. [CrossRef]
27. Bilen, M.A.; Pan, T.; Lee, Y.-C.; Lin, S.-C.; Yu, G.; Pan, J.; Hawke, D.; Pan, B.-F.; Vykoukal, J.; Gray, K.; et al. Proteomics Profiling of Exosomes from Primary Mouse Osteoblasts under Proliferation versus Mineralization Conditions and Characterization of Their Uptake into Prostate Cancer Cells. *J. Proteome Res.* **2017**, *16*, 2709–2728. [CrossRef]
28. Ji, H.; Greening, D.W.; Barnes, T.W.; Lim, J.W.; Tauro, B.J.; Rai, A.; Xu, R.; Adda, C.; Mathivanan, S.; Zhao, W.; et al. Proteome profiling of exosomes derived from human primary and metastatic colorectal cancer cells reveal differential expression of key metastatic factors and signal transduction components. *Proteomics* **2013**, *13*, 1672–1686. [CrossRef]
29. Tang, M.K.S.; Yue, P.Y.K.; Ip, P.P.; Huang, R.-L.; Lai, H.-C.; Cheung, A.N.Y.; Tse, K.Y.; Ngan, H.Y.S.; Wong, A.S.T. Soluble E-cadherin promotes tumor angiogenesis and localizes to exosome surface. *Nat. Commun.* **2018**, *9*, 2270. [CrossRef]
30. Hoshino, A.; Costa-Silva, B.; Shen, T.L.; Rodrigues, G.; Hashimoto, A.; Tesic Mark, M.; Molina, H.; Kohsaka, S.; Di Giannatale, A.; Ceder, S.; et al. Tumour exosome integrins determine organotropic metastasis. *Nature* **2015**, *527*, 329–335. [CrossRef]
31. Williams, C.; Pazos, R.; Royo, F.; González, E.; Roura-Ferrer, M.; Martinez, A.; Gamiz, J.; Reichardt, N.-C.; Falcón-Pérez, J.M. Assessing the role of surface glycans of extracellular vesicles on cellular uptake. *Sci. Rep.* **2019**, *9*, 11920. [CrossRef]
32. Makler, A.; Asghar, W. Exosomal biomarkers for cancer diagnosis and patient monitoring. *Expert Rev. Mol. Diagn.* **2020**, *20*, 387–400. [CrossRef]
33. Van Niel, G.; D'Angelo, G.; Raposo, G. Shedding light on the cell biology of extracellular vesicles. *Nat. Rev. Mol. Cell Biol.* **2018**, *19*, 213–228. [CrossRef]
34. Livshits, M.A.; Khomyakova, E.; Evtushenko, E.G.; Lazarev, V.N.; Kulemin, N.A.; Semina, S.E.; Generozov, E.V.; Govorun, V.M. Isolation of exosomes by differential centrifugation: Theoretical analysis of a commonly used protocol. *Sci. Rep.* **2015**, *5*, 17319. [CrossRef]
35. Yang, D.; Zhang, W.; Zhang, H.; Zhang, F.; Chen, L.; Ma, L.; Larcher, L.M.; Chen, S.; Liu, N.; Zhao, Q.; et al. Progress, opportunity, and perspective on exosome isolation—Efforts for efficient exosome-based theranostics. *Theranostics* **2020**, *10*, 3684–3707. [CrossRef]
36. Mateescu, B.; Kowal, E.J.K.; van Balkom, B.W.M.; Bartel, S.; Bhattacharyya, S.N.; Buzás, E.I.; Buck, A.H.; de Candia, P.; Chow, F.W.N.; Das, S.; et al. Obstacles and opportunities in the functional analysis of extracellular vesicle RNA—An ISEV position paper. *J. Extracell. Vesicles* **2017**, *6*, 1286095. [CrossRef]
37. Coumans, F.A.W.; Brisson, A.R.; Buzas, E.I.; Dignat-George, F.; Drees, E.E.E.; El-Andaloussi, S.; Emanueli, C.; Gasecka, A.; Hendrix, A.; Hill, A.F.; et al. Methodological guidelines to study extracellular vesicles. *Circ. Res.* **2017**, *120*, 1632–1648. [CrossRef]
38. Shao, H.; Im, H.; Castro, C.M.; Breakefield, X.; Weissleder, R.; Lee, H. New Technologies for Analysis of Extracellular Vesicles. *Chem. Rev.* **2018**, *118*, 1917–1950. [CrossRef]
39. Ferhan, A.R.; Jackman, J.A.; Park, J.H.; Cho, N.J. Nanoplasmonic sensors for detecting circulating cancer biomarkers. *Adv. Drug Deliv. Rev.* **2018**, *125*, 48–77. [CrossRef]
40. Chin, L.K.; Son, T.; Hong, J.S.; Liu, A.Q.; Skog, J.; Castro, C.M.; Weissleder, R.; Lee, H.; Im, H. Plasmonic sensors for extracellular vesicle analysis: From scientific development to translational research. *ACS Nano* **2020**, *14*, 14528–14548. [CrossRef]
41. Shao, B.; Xiao, Z. Recent achievements in exosomal biomarkers detection by nanomaterials-based optical biosensors—A review. *Anal. Chim. Acta* **2020**, *1114*, 74–84. [CrossRef]
42. Ong, T.T.X.; Blanch, E.W.; Jones, O.A.H. Surface Enhanced Raman Spectroscopy in environmental analysis, monitoring and assessment. *Sci. Total Environ.* **2020**, *720*, 12. [CrossRef]
43. Guerrini, L.; Alvarez-Puebla, R.A. Surface-Enhanced Raman Scattering Sensing of Transition Metal Ions in Waters. *ACS Omega* **2021**, *6*, 1054–1063. [CrossRef]

44. Zhang, H.; Duan, S.; Radjenovic, P.M.; Tian, Z.Q.; Li, J.F. Core-Shell Nanostructure-Enhanced Raman Spectroscopy for Surface Catalysis. *Acc. Chem. Res.* **2020**, *53*, 729–739. [CrossRef]
45. Phan-Quang, G.C.; Han, X.M.; Koh, C.S.L.; Sim, H.Y.F.; Lay, C.L.; Leong, S.X.; Lee, Y.H.; Pazos-Perez, N.; Alvarez-Puebla, R.A.; Ling, X.Y. Three-Dimensional Surface-Enhanced Raman Scattering Platforms: Large-Scale Plasmonic Hotspots for New Applications in Sensing, Microreaction, and Data Storage. *Acc. Chem. Res.* **2019**, *52*, 1844–1854. [CrossRef]
46. Langer, J.; de Aberasturi, D.J.; Aizpurua, J.; Alvarez-Puebla, R.A.; Auguié, B.; Baumberg, J.J.; Bazan, G.C.; Bell, S.E.J.; Boisen, A.; Brolo, A.G.; et al. Present and future of surface-enhanced Raman scattering. *ACS Nano* **2020**, *14*, 28–117. [CrossRef]
47. Guerrini, L.; Alvarez-Puebla, R.A. Surface-Enhanced Raman Spectroscopy in Cancer Diagnosis, Prognosis and Monitoring. *Cancers* **2019**, *11*, 748. [CrossRef]
48. Wang, J.; Koo, K.M.; Wang, Y.L.; Trau, M. Engineering State-of-the-Art Plasmonic Nanomaterials for SERS-Based Clinical Liquid Biopsy Applications. *Adv. Sci.* **2019**, *6*, 35. [CrossRef]
49. Pazos, E.; Garcia-Algar, M.; Penas, C.; Nazarenus, M.; Torruella, A.; Pazos-Perez, N.; Guerrini, L.; Vázquez, M.E.; Garcia-Rico, E.; Mascareñas, J.L.; et al. Surface-Enhanced Raman Scattering Surface Selection Rules for the Proteomic Liquid Biopsy in Real Samples: Efficient Detection of the Oncoprotein c-MYC. *J. Am. Chem. Soc.* **2016**, *138*, 14206–14209. [CrossRef]
50. Catala, C.; Mir-Simon, B.; Feng, X.; Cardozo, C.; Pazos-Perez, N.; Pazos, E.; Gómez-de Pedro, S.; Guerrini, L.; Soriano, A.; Vila, J.; et al. Online SERS Quantification of *Staphylococcus aureus* and the Application to Diagnostics in Human Fluids. *Adv. Mater. Technol.* **2016**, *1*, 1600163. [CrossRef]
51. Le Ru, E.C.; Etchegoin, P.G. *Principles of Surface-Enhanced Raman Spectroscopy*; Elsevier: Amsterdam, The Netherlands, 2009. [CrossRef]
52. Schlücker, S. Surface-Enhanced Raman Spectroscopy: Concepts and Chemical Applications. *Angew. Chem. Int. Ed.* **2014**, *53*, 4756–4795. [CrossRef] [PubMed]
53. Kumari, G.; Kandula, J.; Narayana, C. How Far Can We Probe by SERS? *J. Phys. Chem. C* **2015**, *119*, 20057–20064. [CrossRef]
54. Mir-Simon, B.; Morla-Folch, J.; Gisbert-Quilis, P.; Pazos-Perez, N.; Xie, H.-N.; Bastús, N.G.; Puntes, V.; Alvarez-Puebla, R.A.; Guerrini, L. SERS efficiencies of micrometric polystyrene beads coated with gold and silver nanoparticles: The effect of nanoparticle size. *J. Opt.* **2015**, *17*, 114012. [CrossRef]
55. Morla-Folch, J.; Guerrini, L.; Pazos-Perez, N.; Arenal, R.; Alvarez-Puebla, R.A. Synthesis and Optical Properties of Homogeneous Nanoshurikens. *ACS Photonics* **2014**, *1*, 1237–1244. [CrossRef]
56. Liebtrau, M.; Sivis, M.; Feist, A.; Lourenço-Martins, H.; Pazos-Pérez, N.; Alvarez-Puebla, R.A.; de Abajo, F.J.G.; Polman, A.; Ropers, C. Spontaneous and stimulated electron–photon interactions in nanoscale plasmonic near fields. *Light Sci. Appl.* **2021**, *10*, 82. [CrossRef]
57. Guerrini, L.; Graham, D. Molecularly-mediated assemblies of plasmonic nanoparticles for Surface-Enhanced Raman Spectroscopy applications. *Chem. Soc. Rev.* **2012**, *41*, 7085–7107. [CrossRef]
58. Correa-Duarte, M.A.; Pazos Perez, N.; Guerrini, L.; Giannini, V.; Alvarez-Puebla, R.A. Boosting the Quantitative Inorganic Surface-Enhanced Raman Scattering Sensing to the Limit: The Case of Nitrite/Nitrate Detection. *J. Phys. Chem. Lett.* **2015**, *6*, 868–874. [CrossRef]
59. Li, J.-F.; Zhang, Y.-J.; Ding, S.-Y.; Panneerselvam, R.; Tian, Z.-Q. Core–Shell Nanoparticle-Enhanced Raman Spectroscopy. *Chem. Rev.* **2017**, *117*, 5002–5069. [CrossRef]
60. Siravegna, G.; Marsoni, S.; Siena, S.; Bardelli, A. Integrating liquid biopsies into the management of cancer. *Nat. Rev. Clin. Oncol.* **2017**, *14*, 531–548. [CrossRef]
61. Hu, Q.; Su, H.; Li, J.; Lyon, C.; Tang, W.; Wan, M.; Hu, T.Y. Clinical applications of exosome membrane proteins. *Precis. Clin. Med.* **2020**, *3*, 54–66. [CrossRef]
62. Pang, Y.F.; Wang, C.G.; Lu, L.C.; Wang, C.W.; Sun, Z.W.; Xiao, R. Dual-SERS biosensor for one-step detection of microRNAs in exosome and residual plasma of blood samples for diagnosing pancreatic cancer. *Biosens. Bioelectron.* **2019**, *130*, 204–213. [CrossRef]
63. Ma, D.; Huang, C.; Zheng, J.; Tang, J.; Li, J.; Yang, J.; Yang, R. Quantitative detection of exosomal microRNA extracted from human blood based on surface-enhanced Raman scattering. *Biosens. Bioelectron.* **2018**, *101*, 167–173. [CrossRef]
64. Cheng, Y.; Dong, L.; Zhang, J.; Zhao, Y.; Li, Z. Recent advances in microRNA detection. *Analyst* **2018**, *143*, 1758–1774. [CrossRef]
65. Morla-Folch, J.; Xie, H.-N.; Alvarez-Puebla, R.A.; Guerrini, L. Fast Optical Chemical and Structural Classification of RNA. *ACS Nano* **2016**, *10*, 2834–2842. [CrossRef]
66. Sun, Y.; Shi, L.; Mi, L.; Guo, R.; Li, T. Recent progress of SERS optical nanosensors for miRNA analysis. *J. Mater. Chem. B* **2020**, *8*, 5178–5183. [CrossRef] [PubMed]
67. Zong, C.; Xu, M.X.; Xu, L.J.; Wei, T.; Ma, X.; Zheng, X.S.; Hu, R.; Ren, B. Surface-Enhanced Raman Spectroscopy for Bioanalysis: Reliability and Challenges. *Chem. Rev.* **2018**, *118*, 4946–4980. [CrossRef]
68. Fabris, L. SERS Tags: The Next Promising Tool for Personalized Cancer Detection? *ChemNanoMat* **2016**, *2*, 249–258. [CrossRef]
69. Russo, M.; Tirinato, L.; Scionti, F.; Coluccio, M.L.; Perozziello, G.; Riillo, C.; Mollace, V.; Gratteri, S.; Malara, N.; Di Martino, M.T.; et al. Raman Spectroscopic Stratification of Multiple Myeloma Patients Based on Exosome Profiling. *ACS Omega* **2020**, *5*, 30436–30443. [CrossRef]

70. Dong, S.; Wang, Y.; Liu, Z.; Zhang, W.; Yi, K.; Zhang, X.; Zhang, X.; Jiang, C.; Yang, S.; Wang, F.; et al. Beehive-Inspired Macroporous SERS Probe for Cancer Detection through Capturing and Analyzing Exosomes in Plasma. *ACS Appl. Mater. Interfaces* **2020**, *12*, 5136–5146. [CrossRef]
71. Chalapathi, D.; Padmanabhan, S.; Manjithaya, R.; Narayana, C. Surface-Enhanced Raman Spectroscopy as a Tool for Distinguishing Extracellular Vesicles under Autophagic Conditions: A Marker for Disease Diagnostics. *J. Phys. Chem. B* **2020**, *124*, 10952–10960. [CrossRef]
72. Krafft, C.; Osei, E.B.; Popp, J.; Nazarenko, I. Raman and SERS Spectroscopy for Characterization of Extracellular Vesicles from Control and Prostate Carcinoma Patients. *SPIE BiOS* **2020**, 11236.
73. Rojalin, T.; Koster, H.J.; Liu, J.; Mizenko, R.R.; Tran, D.; Wachsmann-Hogiu, S.; Carney, R.P. Hybrid Nanoplasmonic Porous Biomaterial Scaffold for Liquid Biopsy Diagnostics Using Extracellular Vesicles. *ACS Sens.* **2020**, *5*, 2820–2833. [CrossRef] [PubMed]
74. Suarasan, S.; Liu, J.; Imanbekova, M.; Rojalin, T.; Hilt, S.; Voss, J.C.; Wachsmann-Hogiu, S. Superhydrophobic bowl-like SERS substrates patterned from CMOS sensors for extracellular vesicle characterization. *J. Mater. Chem. B* **2020**, *8*, 8845–8852. [CrossRef] [PubMed]
75. Sivashanmugan, K.; Huang, W.L.; Lin, C.H.; Liao, J.D.; Lin, C.C.; Su, W.C.; Wen, T.C. Bimetallic nanoplasmonic gap-mode SERS substrate for lung normal and cancer-derived exosomes detection. *J. Taiwan Inst. Chem. Eng.* **2017**, *80*, 149–155. [CrossRef]
76. Lee, C.; Carney, R.P.; Hazari, S.; Smith, Z.J.; Knudson, A.; Robertson, C.S.; Lam, K.S.; Wachsmann-Hogiu, S. 3D plasmonic nanobowl platform for the study of exosomes in solution. *Nanoscale* **2015**, *7*, 9290–9297. [CrossRef]
77. Tirinato, L.; Gentile, F.; Di Mascolo, D.; Coluccio, M.L.; Das, G.; Liberale, C.; Pullano, S.A.; Perozziello, G.; Francardi, M.; Accardo, A.; et al. SERS analysis on exosomes using super-hydrophobic surfaces. *Microelectron. Eng.* **2012**, *97*, 337–340. [CrossRef]
78. Yan, Z.B.; Dutta, S.; Liu, Z.R.; Yu, X.K.; Mesgarzadeh, N.; Ji, F.; Bitan, G.; Xie, Y.H. A Label-Free Platform for Identification of Exosomes from Different Sources. *ACS Sens.* **2019**, *4*, 488–497. [CrossRef]
79. Park, J.; Hwang, M.; Choi, B.; Jeong, H.; Jung, J.H.; Kim, H.K.; Hong, S.; Park, J.H.; Choi, Y. Exosome Classification by Pattern Analysis of Surface-Enhanced Raman Spectroscopy Data for Lung Cancer Diagnosis. *Anal. Chem.* **2017**, *89*, 6695–6701. [CrossRef]
80. Shin, H.; Jeong, H.; Park, J.; Hong, S.; Choi, Y. Correlation between Cancerous Exosomes and Protein Markers Based on Surface-Enhanced Raman Spectroscopy (SERS) and Principal Component Analysis (PCA). *ACS Sens.* **2018**, *3*, 2637–2643. [CrossRef]
81. Ferreira, N.; Marques, A.; Águas, H.; Bandarenka, H.; Martins, R.; Bodo, C.; Costa-Silva, B.; Fortunato, E. Label-Free Nanosensing Platform for Breast Cancer Exosome Profiling. *ACS Sens.* **2019**, *4*, 2073–2083. [CrossRef]
82. Carmicheal, J.; Hayashi, C.; Huang, X.; Liu, L.; Lu, Y.; Krasnoslobodtsev, A.; Lushnikov, A.; Kshirsagar, P.G.; Patel, A.; Jain, M.; et al. Label-Free characterization of exosome via surface enhanced Raman spectroscopy for the early detection of pancreatic cancer. *Nanomed. Nanotechnol. Biol. Med.* **2019**, *16*, 88–96. [CrossRef]
83. Avella-Oliver, M.; Puchades, R.; Wachsmann-Hogiu, S.; Maquieira, A. Label-Free SERS analysis of proteins and exosomes with large-scale substrates from recordable compact disks. *Sens. Actuators B Chem.* **2017**, *252*, 657–662. [CrossRef]
84. Kaufman, L.; Cooper, T.; Wallace, G.; Hawke, D.; Betts, D.; Hess, D.; Lagugné-Labarthet, F. Trapping and SERS Identification of Extracellular Vesicles Using Nanohole Arrays. *SPIE BiOS* **2020**, 10894. [CrossRef]
85. Pramanik, A.; Mayer, J.; Patibandla, S.; Gates, K.; Gao, Y.; Davis, D.; Seshadri, R.; Ray, P.C. Mixed-Dimensional Heterostructure Material-Based SERS for Trace Level Identification of Breast Cancer-Derived Exosomes. *ACS Omega* **2020**, *5*, 16602–16611. [CrossRef]
86. Zhang, P.; Wang, L.; Fang, Y.; Zheng, D.; Lin, T.; Wang, H. Label-Free exosomal detection and classification in rapid discriminating different cancer types based on specific Raman phenotypes and multivariate statistical analysis. *Molecules* **2019**, *24*, 2947. [CrossRef]
87. Stremersch, S.; Marro, M.; Pinchasik, B.E.; Baatsen, P.; Hendrix, A.; De Smedt, S.C.; Loza-Alvarez, P.; Skirtach, A.G.; Raemdonck, K.; Braeckmans, K. Identification of individual exosome-like vesicles by surface enhanced raman spectroscopy. *Small* **2016**, *12*, 3292–3301. [CrossRef]
88. Fraire, J.C.; Stremersch, S.; Bouckaert, D.; Monteyne, T.; De Beer, T.; Wuytens, P.; De Rycke, R.; Skirtach, A.G.; Raemdonck, K.; De Smedt, S.; et al. Improved Label-Free Identification of Individual Exosome-like Vesicles with Au@Ag Nanoparticles as SERS Substrate. *ACS Appl. Mater. Interfaces* **2019**, *11*, 39424–39435. [CrossRef]
89. Hao, N.; Pei, Z.; Liu, P.; Bachman, H.; Naquin, T.D.; Zhang, P.; Zhang, J.; Shen, L.; Yang, S.; Yang, K.; et al. Acoustofluidics-Assisted Fluorescence-SERS Bimodal Biosensors. *Small* **2020**, *16*. [CrossRef]
90. Shin, H.; Oh, S.; Hong, S.; Kang, M.; Kang, D.; Ji, Y.G.; Choi, B.H.; Kang, K.W.; Jeong, H.; Park, Y.; et al. Early-Stage Lung Cancer Diagnosis by Deep Learning-Based Spectroscopic Analysis of Circulating Exosomes. *ACS Nano* **2020**, *14*, 5435–5444. [CrossRef]
91. Koponen, A.; Kerkelä, E.; Rojalin, T.; Lázaro-Ibáñez, E.; Suutari, T.; Saari, H.O.; Siljander, P.; Yliperttula, M.; Laitinen, S.; Viitala, T. Label-Free characterization and real-time monitoring of cell uptake of extracellular vesicles. *Biosens. Bioelectron.* **2020**, *168*. [CrossRef]
92. Tian, Y.F.; Ning, C.F.; He, F.; Yin, B.C.; Ye, B.C. Highly sensitive detection of exosomes by SERS using gold nanostar@Raman reporter@nanoshell structures modified with a bivalent cholesterol-labeled DNA anchor. *Analyst* **2018**, *143*, 4915–4922. [CrossRef]
93. Weng, Z.; Zong, S.; Wang, Y.; Li, N.; Li, L.; Lu, J.; Wang, Z.; Chen, B.; Cui, Y. Screening and multiple detection of cancer exosomes using an SERS-based method. *Nanoscale* **2018**, *10*, 9053–9062. [CrossRef]

94. Zhang, X.; Liu, C.; Pei, Y.; Song, W.; Zhang, S. Preparation of a Novel Raman Probe and Its Application in the Detection of Circulating Tumor Cells and Exosomes. *ACS Appl. Mater. Interfaces* **2019**, *11*, 28671–28680. [CrossRef]

95. Pang, Y.; Shi, J.; Yang, X.; Wang, C.; Sun, Z.; Xiao, R. Personalized detection of circling exosomal PD-L1 based on Fe_3O_4@TiO_2 isolation and SERS immunoassay. *Biosens. Bioelectron.* **2020**, *148*, 111800. [CrossRef]

96. Zong, S.; Wang, L.; Chen, C.; Lu, J.; Zhu, D.; Zhang, Y.; Wang, Z.; Cui, Y. Facile detection of tumor-derived exosomes using magnetic nanobeads and SERS nanoprobes. *Anal. Methods* **2016**, *8*, 5001–5008. [CrossRef]

97. Zhang, W.; Jiang, L.; Diefenbach, R.J.; Campbell, D.H.; Walsh, B.J.; Packer, N.H.; Wang, Y. Enabling Sensitive Phenotypic Profiling of Cancer-Derived Small Extracellular Vesicles Using Surface-Enhanced Raman Spectroscopy Nanotags. *ACS Sens.* **2020**, *5*, 764–771. [CrossRef]

98. Kwizera, E.A.; O'Connor, R.; Vinduska, V.; Williams, M.; Butch, E.R.; Snyder, S.E.; Chen, X.; Huang, X. Molecular detection and analysis of exosomes using surface-enhanced Raman scattering gold nanorods and a miniaturized device. *Theranostics* **2018**, *8*, 2722–2738. [CrossRef]

99. Wang, J.; Wuethrich, A.; Sina, A.A.I.; Lane, R.E.; Lin, L.L.; Wang, Y.; Cebon, J.; Behren, A.; Trau, M. Tracking extracellular vesicle phenotypic changes enables treatment monitoring in melanoma. *Sci. Adv.* **2020**, *6*, eaax3223. [CrossRef]

100. Li, T.-D.; Zhang, R.; Chen, H.; Huang, Z.-P.; Ye, X.; Wang, H.; Deng, A.-M.; Kong, J.-L. An ultrasensitive polydopamine bi-functionalized SERS immunoassay for exosome-based diagnosis and classification of pancreatic cancer. *Chem. Sci.* **2018**, *9*, 5372–5382. [CrossRef]

101. Hou, M.; He, D.; Bu, H.; Wang, H.; Huang, J.; Gu, J.; Wu, R.; Li, H.W.; He, X.; Wang, K. A sandwich-type surface-enhanced Raman scattering sensor using dual aptamers and gold nanoparticles for the detection of tumor extracellular vesicles. *Analyst* **2020**, *145*, 6232–6236. [CrossRef]

102. Shin, H.; Seo, D.; Choi, Y. Extracellular Vesicle Identification Using Label-Free Surface-Enhanced Raman Spectroscopy: Detection and Signal Analysis Strategies. *Molecules* **2020**, *25*, 5209. [CrossRef] [PubMed]

103. Moorthy, B.S.; Iyer, L.K.; Topp, E.M. Characterizing protein structure, dynamics and conformation in lyophilized solids. *Curr. Pharm. Des.* **2015**, *21*, 5845–5853. [CrossRef] [PubMed]

104. Wood, B.R. The importance of hydration and DNA conformation in interpreting infrared spectra of cells and tissues. *Chem. Soc. Rev.* **2016**, *45*, 1980–1998. [CrossRef] [PubMed]

105. Kruglik, S.G.; Royo, F.; Guigner, J.-M.; Palomo, L.; Seksek, O.; Turpin, P.-Y.; Tatischeff, I.; Falcón-Pérez, J.M. Raman tweezers microspectroscopy of circa 100 nm extracellular vesicles. *Nanoscale* **2019**, *11*, 1661–1679. [CrossRef] [PubMed]

106. Lee, W.; Nanou, A.; Rikkert, L.; Coumans, F.A.W.; Otto, C.; Terstappen, L.W.M.M.; Offerhaus, H.L. Label-Free Prostate Cancer Detection by Characterization of Extracellular Vesicles Using Raman Spectroscopy. *Anal. Chem.* **2018**, *90*, 11290–11296. [CrossRef] [PubMed]

107. Smith, Z.J.; Lee, C.; Rojalin, T.; Carney, R.P.; Hazari, S.; Knudson, A.; Lam, K.; Saari, H.; Ibañez, E.L.; Viitala, T.; et al. Single exosome study reveals subpopulations distributed among cell lines with variability related to membrane content. *J. Extracell. Vesicles* **2015**, *4*, 28533. [CrossRef] [PubMed]

108. Lee, H.K.; Lee, Y.H.; Koh, C.S.L.; Gia, C.P.Q.; Han, X.M.; Lay, C.L.; Sim, H.Y.F.; Kao, Y.C.; An, Q.; Ling, X.Y. Designing surface-enhanced Raman scattering (SERS) platforms beyond hotspot engineering: Emerging opportunities in analyte manipulations and hybrid materials. *Chem. Soc. Rev.* **2019**, *48*, 731–756. [CrossRef]

109. Cardinal, M.F.; Vander Ende, E.; Hackler, R.A.; McAnally, M.O.; Stair, P.C.; Schatz, G.C.; Van Duyne, R.P. Expanding applications of SERS through versatile nanomaterials engineering. *Chem. Soc. Rev.* **2017**, *46*, 3886–3903. [CrossRef]

110. Blanco-Formoso, M.; Pazos-Perez, N.; Alvarez-Puebla, R.A. Fabrication and SERS properties of complex and organized nanoparticle plasmonic clusters stable in solution. *Nanoscale* **2020**, *12*, 14948–14956. [CrossRef]

111. Huang, S.; Ling, X.; Liang, L.; Song, Y.; Fang, W.; Zhang, J.; Kong, J.; Meunier, V.; Dresselhaus, M.S. Molecular Selectivity of Graphene-Enhanced Raman Scattering. *Nano Lett.* **2015**, *15*, 2892–2901. [CrossRef]

112. Liebel, M.; Pazos-Perez, N.; van Hulst, N.F.; Alvarez-Puebla, R.A. Surface-Enhanced Raman scattering holography. *Nat. Nanotechnol.* **2020**, *15*, 1005–1011. [CrossRef]

113. Gentile, F.; Coluccio, M.L.; Accardo, A.; Asande, M.; Cojoc, G.; Mecarini, F.; Das, G.; Liberale, C.; De Angelis, F.; Candeloro, P.; et al. Nanoporous- micropatterned- superhydrophobic surfaces as harvesting agents for few low molecular weight molecules. *Microelectron. Eng.* **2011**, *88*, 1749–1752. [CrossRef]

114. Accardo, A.; Tirinato, L.; Altamura, D.; Sibillano, T.; Giannini, C.; Riekel, C.; Di Fabrizio, E. Superhydrophobic surfaces allow probing of exosome self organization using X-ray scattering. *Nanoscale* **2013**, *5*, 2295–2299. [CrossRef]

115. Chen, I.-H.; Xue, L.; Hsu, C.-C.; Paez, J.S.P.; Pan, L.; Andaluz, H.; Wendt, M.K.; Iliuk, A.B.; Zhu, J.-K.; Tao, W.A. Phosphoproteins in extracellular vesicles as candidate markers for breast cancer. *Proc. Natl. Acad. Sci. USA* **2017**, *114*, 3175–3180. [CrossRef]

116. Cui, F.; Yue, Y.; Zhang, Y.; Zhang, Z.; Zhou, H.S. Advancing Biosensors with Machine Learning. *ACS Sens.* **2020**, *5*, 3346–3364. [CrossRef]

117. Li, G.; Tang, W.; Yang, F. Cancer Liquid Biopsy Using Integrated Microfluidic Exosome Analysis Platforms. *Biotechnol. J.* **2020**, *15*. [CrossRef]

118. Alvarez-Puebla, R.A.; Pazos-Perez, N.; Guerrini, L. SERS-Fluorescent encoded particles as dual-mode optical probes. *Appl. Mater. Today* **2018**, *13*, 1–14. [CrossRef]

119. Guerrini, L.; Pazos-Perez, N.; Garcia-Rico, E.; Alvarez-Puebla, R. Cancer characterization and diagnosis with SERS-encoded particles. *Cancer Nanotechnol.* **2017**, *8*, 5. [CrossRef]
120. Mir-Simon, B.; Reche-Perez, I.; Guerrini, L.; Pazos-Perez, N.; Alvarez-Puebla, R.A. Universal One-Pot and Scalable Synthesis of SERS Encoded Nanoparticles. *Chem. Mater.* **2015**, *27*, 950–958. [CrossRef]
121. Wang, Y.Q.; Yan, B.; Chen, L.X. SERS Tags: Novel Optical Nanoprobes for Bioanalysis. *Chem. Rev.* **2013**, *113*, 1391–1428. [CrossRef]
122. Wang, Y.; Li, Q.; Shi, H.; Tang, K.; Qiao, L.; Yu, G.; Ding, C.; Yu, S. Microfluidic Raman biochip detection of exosomes: A promising tool for prostate cancer diagnosis. *Lab Chip* **2020**, *20*, 4632–4637. [CrossRef] [PubMed]
123. Dai, J.; Su, Y.; Zhong, S.; Cong, L.; Liu, B.; Yang, J.; Tao, Y.; He, Z.; Chen, C.; Jiang, Y. Exosomes: Key players in cancer and potential therapeutic strategy. *Signal Transduct. Target. Ther.* **2020**, *5*, 145. [CrossRef] [PubMed]
124. Lewis, J.M.; Vyas, A.D.; Qiu, Y.; Messer, K.S.; White, R.; Heller, M.J. Integrated Analysis of Exosomal Protein Biomarkers on Alternating Current Electrokinetic Chips Enables Rapid Detection of Pancreatic Cancer in Patient Blood. *ACS Nano* **2018**, *12*, 3311–3320. [CrossRef] [PubMed]

MDPI AG
Grosspeteranlage 5
4052 Basel
Switzerland
Tel.: +41 61 683 77 34

Cancers Editorial Office
E-mail: cancers@mdpi.com
www.mdpi.com/journal/cancers